STUDY GUIDE

AND

SOLUTIONS MANUAL

iGenetics

Bruce A. Chase
University of Nebraska at Omaha

PEARSON

Benjamin
Cummings

San Francisco • Boston • New York
Cape Town • Hong Kong • London • Madrid • Mexico City
Montreal • Munich • Paris • Singapore • Sydney • Tokyo • Toronto

Sponsoring Editor: Susan Winslow
Project Editor: Susan Minarcin
Associate Editor: Alissa Anderson
Editorial Assistant: Haig MacGregor
Senior Production Supervisor: Corinne Benson
Production Coordinator: Mary O'Connell
Compositor: Progressive Information Technologies
Production Services: Progressive Publishing Alternatives
Manufacturing Supervisor: Evelyn Beaton
Senior Marketing Manager: Scott Dustan

ISBN 0-8053-4675-9

5 6 7 8 9 10 -BB- 08

www.aw-bc.com

CONTENTS

PREFACE

Genetics is not only a fascinating subject, its fundamental principles and applications are now used throughout the life sciences and in many areas that affect our everyday lives. However, gaining a solid understanding of genetic principles and applications presents challenges. I have written this guide to help students overcome these challenges and to provide multiple, productive strategies for gaining a solid understanding of genetics.

Each chapter of the guide contains a set of features that will help students acquire a well-rounded, thorough understanding of genetic principles and their applications. First, I provide an outline of the text for an organizational overview. Second, I group key terms contextually to challenge the reader to use and distinguish between related terms by developing concept maps. Third, I present suggestions for analytical approaches and strategies for problem solving. Fourth, I offer a set of multiple-choice questions—some designed to stimulate recall of text material, others designed to stimulate clearer thinking and to probe for a deeper understanding. Fifth, I offer a set of thought questions for the text and accompanying media that ask the reader to garner evidence, argue for a viewpoint, and generate hypotheses. Sixth, I provide complete solutions to all of the text problems. I have written these solutions to anticipate student questions and present the logic behind multiple approaches. Students arriving at a correct answer, but without clear explanations for their answer, are encouraged to consult these solutions, as it is not just the answer but also the logic leading to the answer that adds value to their understanding. Moreover, it is imperative that students realize that just reading the solutions without independently attempting to solve the problems—using the guide primarily as an answer key—does not provide a shortcut to understanding. Students who want to gain a solid understanding of genetics will find no substitute for independent and systematic problem solving efforts. Indeed, I intend this guide as a resource to facilitate and enhance these efforts.

I am indebted to Peter Russell for providing exceptionally clear, well organized, and thoughtful material with which to work; to users of previous editions of this guide for their comments; to Susan Minarcin, Jean Sims Fornango, and Alissa Anderson at Benjamin Cummings, and to Crystal Clifton at Progressive Publishing Alternatives for their comments, diligence, care, and support during the preparation of this guide. Finally, I welcome comments and corrections from the users of the guide.

Bruce Chase
bchase@mail.unomaha.edu

1

GENETICS: AN INTRODUCTION

CHAPTER OUTLINE

BUILDING CONCEPT MAPS USING GENETICS TERMS

Every chapter of the text introduces new terms used in the "language of genetics." Like new words of a foreign language, each term has a precise meaning and contextual use. To help develop your ability to "speak" the language of genetics, each chapter of this guide will provide lists of key terms, symbols, and concepts. After reading a chapter in the text, review the lists. Without consulting the text and in your own words, write a brief, precise definition of each term. Check your definition against that in the text. Then visualize the relationships between the terms by constructing a concept map using the terms in each list.

What is a concept map? A concept map is a visual image that depicts the relationships between ideas. Terms are arranged on a page, and related terms are connected with lines. A word or phrase that explains the relationship of the concepts is placed next to the lines. Different concept maps can be constructed from the same list of terms, depending on how the relationships between the concepts are viewed. Here is a sample concept map illustrating relationships between ancestral canines and modern dogs.

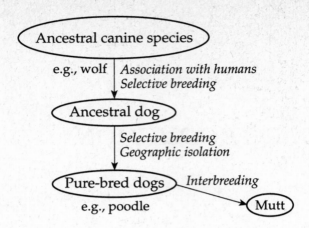

Here are some hints* to help you build concept maps using the provided lists of terms:

1. Write each term on a small piece of paper. In some lists, specific examples of a term will be given. For example, in this chapter, one list contains "genetic material," "DNA," and "RNA." DNA and RNA could be considered examples of genetic material. Keep the examples separate, perhaps even using different colored paper for them.

2. Arrange the terms on a larger sheet of paper, with the broadest or most abstract ideas at the top and the most specific ideas at the bottom.

3. Arrange the concepts so that two related ideas are placed near (e.g., under) each other.

4. Draw lines between related concepts. Rearrange the pieces of paper as you see fit.

5. On the connecting lines, write words or phrases that explain the relationship between the concepts. Be as precise and succinct as you can. This is challenging, so be patient.

6. Put examples under the concepts they belong with. Connect examples to a concept using an explanatory phrase—"e.g." or "specifically."

7. Copy the results onto a single sheet of paper. Draw circles around the concepts, but not the examples.

While concept maps can be generated with pencil and paper, computer tools for concept mapping also exist. The site http://users.edte.utwente.nl/lanzing/cm_home.htm provides a useful list of links.

REVIEW OF KEY TERMS, SYMBOLS, AND CONCEPTS

In your own words, write a brief, precise definition of each term in the groups below. Check your definitions using the text. Then develop a concept map using the terms in each list.

1	2	3
genome	gene	genetics
eukaryote	gene expression	genome project
nucleus	enzyme	genetic map
prokaryote	polypeptide	map unit
bacteria	amino acid	gene locus

* Adapted from *http://www.gpc.edu/~shale/humanities/composition/handouts/concept.html.*

1	2	3
eubacteria	protein	mutation
archaebacteria	one gene–one enzyme	recombination
chromosome	hypothesis	selection
gene	one gene–one polypeptide	hypothetico-deductive
allele	hypothesis	investigation
genotype	operon	transmission genetics
phenotype	messenger RNA	molecular genetics
homozygous	ribosomal RNA	population genetics
heterozygous	transfer RNA	quantitative genetics
dominant	small nuclear RNA	recombinant DNA technology
recessive	ribosome	PCR
genetic material	nucleolus	basic research
DNA	transcription	applied research
RNA	translation	model organism
nucleotide	genetic code	clone
ribose	codon	DNA typing
deoxyribose	amino acid	genetic database
adenine (A)	RNA polymerase	Entrez
guanine (G)	centriole	GenBank
cytosine (C)	endoplasmic reticulum	OMIM
thymine (T)	mitochondria	NCBI
uracil (U)	chloroplast	BLAST
		PubMed

QUESTIONS FOR PRACTICE

Multiple Choice Questions

1. Transmission genetics is primarily concerned with
   a. the distribution and behavior of genes in populations
   b. the passing of genes from generation to generation and their recombination
   c. the structure and function of genes at the molecular level
   d. the means by which mutations are retained in nature

2. Which one of the following is not a criterion to choose an organism for genetic experimentation?
   a. The organism should be able to be used in applied research.
   b. The organism should be easy to handle.
   c. The organism should exhibit genetic variation.
   d. The organism should have a relatively short life cycle.

3. Which one of the following eukaryotic cell structures does not contain DNA?
   a. a nucleus
   b. a mitochondrion
   c. the endoplasmic reticulum
   d. a chloroplast

4. Which one of the following is not an accurate description of a chromosome?
   a. It is a colored body localized in the nucleus.
   b. It is a protein and nucleic acid complex.

   c. It is the cellular structure that contains the genetic material.

   d. In eukaryotes, it is composed of many DNA molecules attached end to end.

5. Which one of the following is not true?

   a. Genes encode proteins.

   b. Genes are proteins.

   c. Genes are transcribed into RNA molecules.

   d. mRNA molecules are translated into proteins on ribosomes.

6. A eukaryotic organism is one that

   a. is multicellular

   b. has a plasma membrane

   c. has genetic material encapsulated in a nuclear membrane

   d. all of the above

7. A centriole is an organelle that is

   a. present in the center of a cell's cytoplasm

   b. composed of microtubules and important for organizing the spindle fibers

   c. surrounded by a membrane

   d. part of a chromosome

8. The rough endoplasmic reticulum is

   a. an intracellular double-membrane system to which ribosomes are attached

   b. an intracellular membrane that is studded with microtubular structures

   c. a membranous structure found within mitochondria

   d. only found in prokaryotic cells

9. Which one of the following is available at the NCBI website and is a search tool that is useful to compare a nucleotide sequence with other nucleotide sequences in a database?

   a. PubMed

   b. BLAST

   c. OMIM

   d. GenBank

   e. PCR

10. Which one of the following can be translated?

   a. rRNA

   b. tRNA

   c. snRNA

   d. mRNA

   e. DNA

Answers: 1b, 2a, 3c, 4d, 5b, 6c, 7b, 8a, 9b, 10d

Thought Questions

1. Why is it difficult to draw a sharp boundary between molecular genetics, transmission genetics, population genetics, and quantitative genetics?

2. Geneticists use the hypothetico-deductive method of investigation. How does this rational approach still allow research projects to go in exciting, unpredictable directions?

3. How are basic and applied research interrelated in the following research areas?

   a. studies of mutations that affect the production of alcohol (ethanol) in yeast

   b. studies of mutations that result in pesticide resistance in *Drosophila melanogaster*

    c. studies of mutations affecting the timing of seed ripening in *Arabidopsis*

    d. studies of transposons in corn

4. What features make an organism well suited for genetic experimentation?

5. What is the relationship between a genotype and a phenotype?

6. What is a gene?

7. What different types of analyses can be made using genetic databases?

8. How are genetic maps used?

9. What hopes and concerns are associated with the field of genomics?

10. How has recombinant DNA technology led to advances in both basic and applied research?

Thought Questions for Media

After reviewing the media on the *iGenetics* CD-ROM, try to answer the following questions.

1. Geneticists investigate traits in families as well as in populations. In what different ways does studying traits in these different settings contribute to our understanding of genetic traits?

2. Which types of studies allow geneticists to determine that a trait has a genetic basis: studies of families, studies of populations, or both? Explain your answer.

3. Which types of studies allow geneticists to determine that a genetic trait is dominant or recessive: studies of families, studies of populations, or both? Explain your answer.

4. How are pedigrees used to demonstrate that a trait has a genetic basis?

5. What processes contribute to the differences in the frequency of a genetic trait in geographically different populations?

2

MENDELIAN GENETICS

CHAPTER OUTLINE

REVIEW OF KEY TERMS, SYMBOLS, AND CONCEPTS

In your own words, write a brief, precise definition of each term in the groups on the following page. Check your definitions using the text. Then develop a concept map using the terms in each list.

THINKING ANALYTICALLY

There are different levels of understanding of Mendelian genetics. You can certainly obtain some understanding of this material from a careful reading and outlining of the chapter, especially if you pay attention to detail and the logic of Mendel's experiments. Indeed, understanding why Mendel's approach gave insights into the principles of inheritance, when

1	2
hereditary trait	probability
gene, particulate factor	product rule
locus, loci	sum rule
allele, allelomorph	principle of segregation
genotype	principle of independent assortment
phenotype	branch diagram, Punnett square
dominant	chi-square (χ^2)test
recessive	goodness-of-fit test
heterozygote, heterozygous	null hypothesis
homozygote, homozygous	expected values
self-fertilization, selfing	P
cross-fertilization, cross	significance
true-breeding, pure-breeding	degrees of freedom
P, F_1, F_2, F_3 generation, cross	pedigree analysis
monohybrid, dihybrid, reciprocal cross	proband, propositus, proposita
gamete, zygote	$1:1, 3:1$, all : none ratios
hybrid, dihybrid, trihybrid individual, cross	wild-type allele
testcross	rare homozygote
haploid, diploid	$AA, A-, Aa, aa$
gene segregation	loss-of-function, gain-of-function mutation
principle of segregation	recessive, dominant trait
principle of independent assortment	recessive, dominant pedigree

those for thousands of years before him failed, is a substantial accomplishment. However, it is through solving problems that you will develop a practical, useful understanding of this material. Solving the problems in this and every chapter will not only reinforce what you have read, but will help you develop an understanding of the relationships between terms, ideas, and facts.

With some effort, you can discover one of the multiple "right" approaches to solving a Mendelian genetics problem. Pursuing any number of "right" approaches will reinforce your understanding of Mendelian genetic principles and help you gain a deeper understanding of relationships you read about in the chapter. Be warned, however, that there is also a "wrong" approach. It is easy to describe. In the wrong approach, you read through the problem and then see if you can understand an answer presented in this guide or another key. This is about as beneficial as reading today's weather forecast after you have already stepped out into the rain. You have learned that it's raining, but you're already wet. What use is your new knowledge? What will you do tomorrow? When you read a solution to a problem, you have not yourself devised the solution, so you have not developed the skills to analyze the situation, apply your knowledge of Mendelian principles, and identify a path that leads to a solution. You may understand the answer, but you won't have developed any of the skills to solve a similar problem. On an exam you will be stuck, and the purpose of doing the problems—to gain a deeper understanding of Mendelian principles—will have been thwarted.

What, then, is a "right" approach? It's not the same thing as finding *the* correct answer, usually because there is more than one approach to solving the problem. Here are some pointers:

1. Read the problem straight through without pausing. Get the sense of what it is about.
2. Then read it again, slowly and critically. This time,
   a. Jot down pertinent information.
   b. Assign descriptive gene symbols. Use a dash (e.g., − − or *A* −) when unsure of a genotype.
   c. Scan the problem from start to finish and then from the end to the beginning, and carefully analyze the *terms* used in the problem. Often, clues to the answer can be found in the way the problem is stated. (Geneticists use very precise language!)
3. If you are unsure how to continue, ask yourself, "What are the options here?"
   a. Write down all the options you perceive.
   b. Carefully analyze whether a particular option will provide a solution.
4. When you think you have a solution, read through the problem once more and make sure that your analysis is consistent with *all* the data in the problem. Try to see if there is a more general principle behind the solution or, for that matter, a clearer or more straightforward solution.
5. If you do get stuck (and everyone does!), DO NOT GO IMMEDIATELY TO THE ANSWER KEY. This will only give you the *illusion* of having solved the problem. On an exam, or in real life, there won't be an answer key available. Learn to control the temptation "to just know the answer" and go on to another problem. Come back to this one another time, perhaps the next day. Your mind may be able to use the time to sort through loose ends.
6. If you cannot solve the problem after coming back to it a second time, read through the answer key. Then close the answer key and try the problem once more. The next day, try it a third time without the answer key.

APPLYING RULES OF PROBABILITY

To solve some problems in this and subsequent chapters, you will need to apply rules of probability. Two rules are commonly employed in genetics:

Product rule: The probability of two independent events occurring simultaneously is the product of their individual probabilities.

Sum rule: The probability of either one of two independent, mutually exclusive events occurring is the sum of their individual probabilities.

To apply these rules, first consider the possible outcomes of a situation. Then identify the mutually exclusive independent events leading to these outcomes. Apply the rules after rephrasing the question. Suppose there are two mutually exclusive independent events, A and B. If you can rephrase the question to ask for the chance that *both A and B* occur (together), multiply the individual probabilities: $P(A \text{ and } B) = P(A) \times P(B)$. If you can rephrase the question or statement to ask for the chance that *either A or B* occurs (separately), add the individual probabilities: $P(A \text{ or } B) = P(A) + P(B)$.

QUESTIONS FOR PRACTICE

Multiple Choice Questions

1. A dominant gene is one that
   a. suppresses the expression of genes at all loci

    b. masks the expression of neighboring gene loci
    c. masks the expression of a recessive allele
    d. masks the expression of all the foregoing

2. A dihybrid is an individual that
    a. is heterozygous for every gene
    b. is heterozygous for two genes under study
    c. is the result of a testcross
    d. is used for a testcross

3. The cross of an uncertain genotype with a homozygous recessive genotype at the same locus is a
    a. pure-breeding cross
    b. monohybrid cross
    c. testcross
    d. dihybrid cross

4. The genotypic ratio of the progeny of a monohybrid cross is typically
    a. 1:2:1
    b. 9:3:3:1
    c. 27:9:9:9:3:3:3:1
    d. 3:1

5. A typical phenotypic ratio of a dihybrid cross with dominant and recessive alleles is
    a. 9:1
    b. 1:2:1
    c. 3:1
    d. 9:3:3:1

6. Pedigrees showing rare recessive traits
    a. have about half of the progeny affected when one parent is affected
    b. have heterozygotes that are phenotypically affected
    c. have about ¾ of the progeny affected when both parents are affected
    d. often skip a generation

7. In a chi-square test, a P value equal to 0.04 tells one that
    a. there is a 4 percent chance the hypothesis is correct
    b. there is a 4 percent chance the hypothesis is incorrect
    c. if the experiment were repeated, chance deviations from the expected values as large as those observed would be seen only 4 percent of the time
    d. if the experiment were repeated, chance deviations from the expected values as large as those observed would be seen at least 96 percent of the time

8. In a chi-square test, a P value equal to 0.04 indicates that
    a. the hypothesis is unlikely to be true
    b. the hypothesis is false and must be rejected
    c. the hypothesis is true
    d. the hypothesis is likely to be true

9. Gain-of-function mutations are associated with all of the following except
    a. dominant traits
    b. new properties in a gene
    c. phenotypes in heterozygous individuals
    d. a decrease in the normal activity of a gene

10. Two parents affected with a genetic disease have five children, all of whom are affected. Which one statement must be true?
    a. Both parents must have at least one mutant allele; the trait might be dominant or recessive.
    b. Both parents must be homozygous for a recessive trait.
    c. At least one parent must be homozygous.
    d. At least one parent must be homozygous for a dominant trait.

11. How many different genotypes are obtainable from the cross *Aa Bb* × *Aa Bb*?
    a. 3
    b. 4
    c. 9
    d. 16

12. What is the chance of obtaining a heterozygous individual from a testcross?
    a. 0%
    b. 50%
    c. at least 50%
    d. 100%

Answers: 1c, 2b, 3c, 4a, 5d, 6d, 7c, 8a, 9d, 10a, 11c, 12c

Thought Questions

1. Over a period of many years, numerous attempts were made to understand how physical traits are passed from one generation to the next. Gregor Mendel was first to make a breakthrough. How do you account for his success, in light of years of failure before him?

2. What led to the rediscovery of Mendel's work?

3. a. Clearly distinguish between an allele, a gene, and a locus.
   b. Can you see an analogy between a gene and an allele and
      • a digit and an index finger?
      • a canine and a German shepherd?
      • a dime and a mint 1954 dime?
      Where do these analogies break down?

4. Why might the term *degrees of freedom* be so named, and why (for the problems in this chapter) is it equal to $n-1$?

5. Why are the frequencies of harmful recessive mutant alleles usually higher than the frequencies of harmful dominant mutations, even though individuals with the recessive trait are rare?

6. Construct a decision-making tree (a flowchart) that can be used to decide if a pedigree shows a dominant or recessive trait.

7. Why is it important to know if a trait is common or rare when analyzing a pedigree?

8. Must wild-type alleles always be dominant?

9. Why are loss-of-function mutations usually recessive and gain-of-function mutations usually dominant?

Thought Questions for Media

After reviewing the media on the *iGenetics* CD-ROM, try to answer these questions.

1. What is the difference between a locus, a gene, and an allele? Can these terms be used interchangeably?

2. What is the difference between a phenotypic ratio and a genotypic ratio? Can these ever be the same in the offspring of a cross?

3. When a true-breeding plant is self-fertilized, why aren't plants with different traits ever produced? If two true-breeding plants are crossed, why do all of the offspring have just one phenotype?

4. In a cross between two true-breeding plants differing in two characteristics, will the F_1 offspring always exhibit both traits of just one parent? Will they always be dihybrids?

5. What does it mean for "genes on different chromosomes to behave independently in the production of gametes"?

6. How many types of gametes are produced by a monohybrid? By a dihybrid?

7. How many different zygote genotypes can be produced in a monohybrid cross? In a dihybrid cross?

8. How many different zygote phenotypes can be produced in a monohybrid cross? In a dihybrid cross?

9. Why doesn't a recessive trait ever appear in a heterozygote? How do we know that the recessive allele is still present in a heterozygote?

1. Two tagged tribbles wander out of the iActivity tribble laboratory and show up in your bed. One is tagged "male *bb ss* (solid yellow)," and one is tagged "female *Bb Ss* (spotted brown)." You recall that spotted (*S*) is dominant to solid (*s*) and brown (*B*) is dominant to yellow (*b*) and that tribbles reproduce quickly. You seize this opportunity to make a (small) fortune selling spotted yellow tribbles. What crosses would you do to develop a true-breeding, spotted yellow strain? Since it is important to keep the population of non-spotted yellow tribbles to a minimum, you want to be sure that the animals chosen for the strain are true breeding. How can you do this, since they cannot be selfed?

2. As you read the morning paper at breakfast, you notice that a (tiny) reward has been posted by the iActivity tribble laboratory for three missing tribbles. All the ad says in describing them is that one is a heterozygote and two are homozygotes. Midway through breakfast, three untagged tribbles waddle across your plate. One is a solid brown male, one is a solid brown female, and the third is a spotted brown male. Assuming that these animals were not just stained with coffee, could they be the missing tribbles? If so, how would you determine, using crosses with just these animals, which animal is heterozygous, which animals are homozygous, and if the homozygotes are both homozygous dominant, both homozygous recessive, or one is homozygous dominant and one is homozygous recessive?

SOLUTIONS TO TEXT PROBLEMS

2.1 In tomatoes, red fruit color is dominant to yellow. Suppose a tomato plant homozygous for red is crossed with one homozygous for yellow. Determine the appearance of

a. the F_1 tomatoes

b. the F_2 tomatoes

c. the offspring of a cross of the F_1 tomatoes back to the red parent

d. the offspring of a cross of the F_1 tomatoes back to the yellow parent

Answer: Consider two ways to solve this problem.

1. First, notice that this is a situation akin to Mendel's crosses. In such crosses, dominant traits mask the appearance of recessive traits, and when a true-breeding dominant plant is crossed to a true-breeding recessive plant, all the F_1 progeny show the dominant trait. Here the dominant trait is red, the recessive trait is yellow, and the parents are true breeding, since they are homozygous.

   a. Just as in Mendel's crosses, the F_1 plants will be heterozygous and all show the dominant red trait.

   b. The F_2 plants that result from the cross of two F_1 heterozygotes will also show the same ratios seen by Mendel, that is, 3 dominant : 1 recessive, or 3 red : 1 yellow.

   c. When an F_1 plant is crossed back to the red parent, a heterozygous plant is crossed to a true-breeding dominant plant. Here, each offspring receives a dominant allele from the red parent, and so all the progeny must show the dominant red phenotype.

   d. When an F_1 plant is crossed back to the yellow parent, a heterozygous plant is crossed back to a true-breeding recessive plant. This is akin to one of Mendel's test-crosses, and so the progeny should show a 1 : 1 ratio of red : yellow plants.

2. Assign R as the allele symbol for dominant red color, and r as the allele symbol for recessive yellow color.

   a. Then the initial cross between two homozygous plants can be depicted as $RR \times rr$, and the F_1 progeny obtain an R allele from the red parent and a r allele from the yellow parent. Therefore, they are all heterozygotes and are Rr. As the R (red) allele is dominant to the r (yellow) allele, the Rr progeny are red.

   b. The F_2 are obtained from $Rr \times Rr$ and so will be composed of 1 RR : 2 Rr : 1 rr types of progeny. There will be 3 red (RR or Rr) : 1 yellow (rr).

   c. The F_1 crossed back to the red parent can be depicted as $Rr \times RR$. All the progeny will obtain an R allele from the red parent; the progeny can be written as $R-$ (either RR or Rr), and all will be red.

   d. The F_1 crossed back to the yellow parent can be depicted as $Rr \times rr$. There will be two equally frequent types of progeny, Rr and rr. Thus, half the progeny will be red and half yellow.

2.2 In maize, a dominant allele A is necessary for seed color, as opposed to colorless (a). Another gene has a recessive allele wx that results in waxy starch, as opposed to normal starch (Wx). The two genes segregate independently. An $Aa\,WxWx$ plant is testcrossed. What are the phenotypes and relative frequencies of offspring?

Answer: A plant that is genotypically $Aa\,WxWx$ has two types of gametes: $A\,Wx$ and $a\,Wx$ (notice that Wx is a symbol for one allele, not two). In a testcross, this plant is crossed to one homozygous for recessive alleles at both the color gene and the waxy gene, $aa\,wxwx$.

The gametes of this plant are all *a wx*. This cross can be illustrated in the following Punnett square:

		Gametes of *Aa WxWx*	
		A Wx	*a Wx*
Gametes of *aa wxwx*	*a wx*	*Aa Wxwx*	*aa Wxwx*

Progeny will be of two equally frequent genotypes: *Aa Wxwx* and *aa Wxwx*. Thus, half will be colored (*Aa*), half will be colorless (*aa*), and all will have normal starch (*Wxwx*).

2.3 F_2 plants segregate ¾ colored : ¼ colorless. If a colored plant is picked at random and selfed, what is the probability that both colored and colorless plants will be seen among a large number of its progeny?

Answer: In the F_2, there is a 3:1 colored:colorless phenotypic ratio. This reflects a 1 *CC* : 2 *Cc* : 1 *cc* genotypic ratio, where colored (*C*) is dominant to colorless (*c*). Thus, there are two types of colored plants (*CC, Cc*) that are present in a 1 *CC* : 2 *Cc* ratio. If one is picked at random, there is a ⅓ chance of picking a homozygous *CC* plant, and a ⅔ chance of picking a heterozygous *Cc* plant. If selfed, the *CC* plant will produce only *CC* (colored) plants, while the *Cc* plant will produce both colored and colorless plants (in a 3:1 ratio). Thus, to see more than one type of plant in the progeny, a *Cc* plant must be chosen initially. The chance of this is ⅔. (Notice that the question asks for the chance that a particular colored plant will have two types of progeny, and *not* for the ratio of progeny types when two types are seen.)

2.4   In guinea pigs, rough coat (*R*) is dominant over smooth coat (*r*). A rough-coated guinea pig is bred to a smooth one, giving eight rough and seven smooth progeny in the F_1 generation.
a. What are the genotypes of the parents and their offspring?
b. If one of the rough F_1 animals is mated to its rough parent, what progeny would you expect?

Answer:
a. Since rough (*R*) is dominant over smooth (*r*), a rough parent is *RR* or *Rr* (i.e., *R–*). A cross between a rough and smooth guinea pig is *R– × rr*. If the cross is *RR × rr*, then all the progeny will be *Rr* and be rough. If the cross is *Rr × rr*, then half the progeny will be *Rr* (rough) and half will be *rr* (smooth). The 8 rough : 7 smooth progeny ratio approximates a 1:1 *Rr : rr* ratio, and so the initial cross must have been *Rr × rr*. The rough progeny must be *Rr*, while the smooth progeny must be *rr*.
b. If one of the rough F_1 animals is mated back to its rough parent, the cross is *Rr × Rr*. This cross would produce both rough (*R–*) and smooth (*rr*) progeny in a 3:1 ratio.

2.5   In cattle, the polled (hornless) condition (*P*) is dominant over the horned (*p*) phenotype. A particular polled bull is bred to three cows. Cow A, which is horned, produces a horned calf; polled cow B produces a horned calf; and horned cow C produces a polled calf. What are the genotypes of the bull and the three cows, and what phenotypic ratios do you expect in the offspring of these three matings?

Answer: First, depict the phenotypes as genotypes: A polled animal exhibiting the dominant, hornless condition can be depicted as $P-$, while a horned animal exhibits the recessive condition and must be *pp*. Therefore, the three crosses and their progeny can be depicted as

Cow		×	Polled Bull	Progeny	
Phenotype	Genotype		Genotype	Phenotype	Genotype
A: horned	*pp*	×	*P–*	horned	*pp*
B: polled	*P–*	×	*P–*	horned	*pp*
C: horned	*pp*	×	*P–*	polled	*P–*

Now follow how each set of parents contributed alleles to their progeny. In the crosses with cows A and B, a horned, *pp* offspring can be obtained only if each parent contributes a recessive *p* allele to the progeny. Therefore both the polled bull and cow B must be heterozygous: Each is *Pp*. With this information, both the crosses with cows A and C appear to be testcrosses ($Pp \times pp$) and will produce 1:1 phenotypic ratios of polled (*Pp*) and horned (*pp*) progeny. The cross with cow B is a cross between two heterozygotes ($Pp \times Pp$) and will produce a 3:1 phenotypic ratio of polled to horned progeny.

2.6 In jimsonweed, purple flowers are dominant to white. Self-fertilization of a particular purple-flowered jimsonweed produces 28 purple-flowered and 10 white-flowered progeny. What proportion of the purple-flowered progeny will breed true?

Answer: The ratio of 28 purple to 10 white plants is close to 3:1. A 3:1 ratio means that the purple plant that was initially selfed must have been heterozygous, or *Pp*. A cross of $Pp \times Pp$ would yield progeny with a genotypic ratio of 1 *PP*:2 *Pp*:1 *pp*. Thus, among the purple progeny, there would be ⅓ *PP* homozygotes and ⅔ *Pp* heterozygotes. Since the homozygous *PP* plants are the only purple plants that will breed true, only ⅓ of the purple-flowered progeny will breed true.

2.7 Two black female mice are crossed with the same brown male. In a number of litters, female X produced 9 blacks and 7 browns, and female Y produced 14 blacks. What is the mechanism of inheritance of black and brown coat color in mice? What are the genotypes of the parents?

Answer: Notice that the two crosses give very different results. Use both sets of results to answer the question. In the cross of black female X with the brown male, the 9 black and 7 brown progeny approximate a 1:1 phenotypic ratio. This suggests that this cross is similar to a testcross, and might be depicted $Bb \times bb$. However, it is not clear from this cross alone which color trait is dominant. This question can be answered by considering the cross of black female Y with the brown male, where only black progeny are seen. This is like a cross between two true-breeding individuals where black is dominant to brown, $BB \times bb$. Thus, the brown male is homozygous recessive (*bb*), female X is heterozygous (*Bb*), and female Y is homozygous dominant (*BB*).

2.8 Bean plants may have different symptoms when infected with a virus. Some show local lesions that do not seriously harm the plant; others show general systemic infection. The following genetic analysis was made:

P   local lesions × systemic infection
F$_1$ all local lesions
F$_2$ 785 local lesions : 269 systemic infection

What is the likely genetic basis of this difference in beans? Assign gene symbols to all the genotypes occurring in the genetic analysis. Design a testcross to verify your assumptions.

Answer: In this question, note that the trait that is being examined is the *response* of plants to a virus. One envisions that crosses are performed between plants, and then, instead of looking at flower color or seed shape to score a trait, plants are individually tested for their response to viral infection. A cross of a plant showing only local lesions to one showing systemic lesions gives only plants that show local lesions, indicating that a local lesion response might be considered to be dominant to a systemic lesion response. If the allele for the local lesion response is depicted as *L*, and the allele for the systemic lesion response as *l*, the parental generation can be written as *LL* × *ll*. Then, the F$_1$ would be all *Ll* and the F$_2$ would be 1 *LL* : 2 *Ll* : 1 *ll*. This fits with the observed numbers of 785 local lesion and 269 systemic lesion (3 *L*− : 1 *ll*) plants.

To test the hypothesis that the local lesion response is dominant to the systemic lesion response, testcross the F$_2$ plants to true-breeding plants that show a systemic lesion response (i.e., are *ll*). The progeny of this testcross can then be assayed for their response to viral infection so that the genotype of the F$_2$ can be determined. One would expect ⅔ of those F$_2$ plants showing a local lesion response to be heterozygotes, i.e., *Ll*. When they are testcrossed, half of their progeny should show local responses and half systemic responses. The remaining ⅓ will be *LL* and give only progeny with local responses.

2.9   A normal *Drosophila* (fruit fly) has both brown and scarlet pigment granules in its eyes, which appear red as a result. Brown (*bw*) is a recessive allele on chromosome 2 that, when homozygous, results in brown eyes because of the absence of scarlet pigment granules. Scarlet (*st*) is a recessive allele on chromosome 3 that, when homozygous, results in scarlet eyes because of the absence of brown pigment granules. Any fly homozygous for recessive alleles at both genes produces no eye pigment and has white eyes. The following results were obtained from crosses:

P   brown-eyed fly × scarlet-eyed fly
F$_1$ red eyes (both brown and scarlet pigment present)
F$_2$ ⁹⁄₁₆ red : ³⁄₁₆ scarlet : ³⁄₁₆ brown : ¹⁄₁₆ white

a. Assign genotypes to the P and F$_1$ generations.

b. Design a testcross to verify the F$_1$ genotype, and predict the results.

Answer: Here, two independently assorting genes affecting eye pigment are being followed. The symbols suggested in the problem follow the convention that symbols are based on the phenotypes of mutants. Here, *bw* is used to represent the recessive, mutant allele causing brown eyes. *Bw* represents the dominant, normal allele at this locus. We infer that this gene controls the production of scarlet granules since they are lacking in the brown-eyed mutant. By the same reasoning, *st* is used to represent the recessive, mutant allele causing scarlet eyes, and *St* is the dominant, normal allele at this locus. We infer that the locus controls the production of brown granules. When writing the genotypes of the flies being crossed, be certain to write the alleles present at both genes.

a. Our first task is to assign genotypes to the P and F$_1$ generations. Brown-eyed flies lack scarlet granules, so they must be *bwbw* (no scarlet) and *St*−; these flies do make brown granules. Note that they can be either *bwbw StSt* or *bwbw Stst*, and that we cannot tell their exact genotype yet. Using similar reasoning, scarlet-eyed flies must be *Bw– stst*. We can determine the exact genotype of these P generation flies by considering what progeny they produce. Since each parent can be either of two genotypes, four different crosses are possible. These are illustrated in the following branch diagrams:

(i)   brown eyes × scarlet eyes
       bwbw StSt × BwBw stst

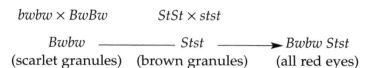

(ii)   brown eyes × scarlet eyes
       bwbw Stst × BwBw stst

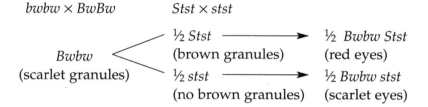

(iii) brown eyes × scarlet eyes
       bwbw StSt × Bwbw stst

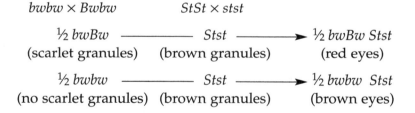

(iv)   brown eyes × scarlet eyes
       bwbw Stst × Bwbw stst

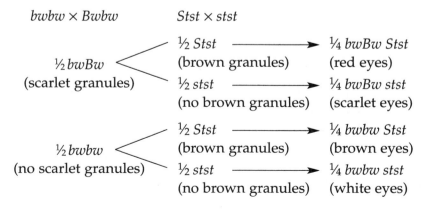

Notice that even though the parents in each of the above crosses look the same, their progeny are not. The only way all red-eyed progeny can be obtained is if the cross is that shown in (i). To see if the observed F_2 is consistent with this view, use a branch diagram:

red eyes × red eyes
Bwbw Stst × Bwbw Stst

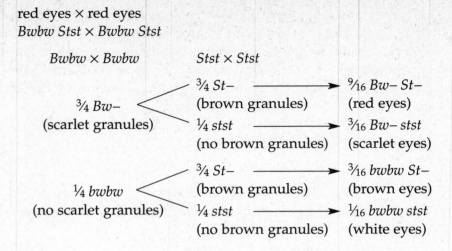

b. To verify the F$_1$ genotype, cross the red-eyed F$_1$ to doubly homozygous recessive white-eyed flies of the genotype *bwbw stst*. One expects to see a 1:1:1:1 ratio of red: brown:scarlet:white flies, as shown in the following branch diagram.

red eyes × white eyes
Bwbw Stst × bwbw stst

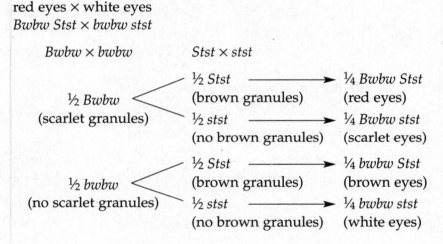

2.10   Grey seed color (*G*) in garden peas is dominant to white seed color (*g*). In the following crosses, the indicated parents with known phenotypes but unknown genotypes produced the progeny listed:

Parents	Progeny		Female Parent
Female × Male	Grey	White	Genotype
grey × white	81	82	?
grey × grey	118	39	?
grey × white	74	0	?
grey × grey	90	0	?

On the basis of the segregation data, give the possible genotypes of each female parent.

Answer: First notice that white seeds are always *gg* while grey seeds are either *GG* or *Gg*. Thus, one can assign partial genotypes (either *G–* or *gg*) based on the phenotypes of the parents. The problem is to determine whether a grey parent is *GG* or *Gg*. In monohybrid crosses, there are three kinds of phenotypic ratios that can be generated. The cross *Gg* × *Gg* gives a 3 grey ($G-$):1 white (gg) ratio, the cross $Gg \times gg$ gives a 1 grey (Gg):1 white (gg) ratio, and the crosses $GG \times gg$ and $GG \times G-$ give all grey ($G-$) progeny. As the first cross of grey × white gives a 1:1 ratio, it must be $Gg \times gg$. As the second cross of grey × grey gives a 3:1 ratio, it must be $Gg \times Gg$. As the third cross of grey × white gives all grey, it must be $GG \times gg$. In the fourth cross, only grey progeny are produced. The genotype of one of the parents must be GG, but the genotype of the other can only be specified as $G-$. One has:

Parents	Progeny		Female Parent
Female × Male	Grey	White	Genotype
grey × white	81	82	Gg
grey × grey	118	39	Gg
grey × white	74	0	GG
grey × grey	90	0	GG or $G-$

2.11 Fur color in the babbit, a furry little animal and popular pet, is determined by a pair of alleles, B and b. BB and Bb babbits are black, and bb babbits are white. A farmer wants to breed babbits for sale. True-breeding white (bb) female babbits breed poorly. The farmer purchases a pair of black babbits, and these mate and produce six black and two white offspring. The farmer immediately sells his white babbits and then consults you for a breeding strategy to produce more white babbits.

a. If he performed random crosses between pairs of F_1 black babbits, what proportion of the F_2 progeny would be white?

b. If he crossed an F_1 male to the parental female, what is the probability that this cross would produce white progeny?

c. What would be the farmer's best strategy to maximize the production of white babbits?

Answer: As the farmer starts with a pair of black babbits, he must be starting with animals that are either BB or Bb. Since his pair is not true breeding, and indeed gives a 3:1 ratio of black to white progeny, he must have two heterozygotes, or Bb babbits. Therefore, one expects to find a 1 BB:2 Bb genotypic ratio among the black F_1 progeny that were not sold.

a. Consider that for a white babbit to be obtained in a cross between a pair of F_1 babbits, both babbits must be Bb in genotype, and a bb offspring must be produced by these parents. Now ascertain the chance of these events occurring. If one picks randomly among the black F_1 progeny, there will be a ⅓ chance of picking a BB individual and a ⅔ chance of picking a Bb individual. The chance of two Bb parents giving a bb offspring is ¼. Using the product rule, one has

P(white offspring) = P(both F_1 babbits are Bb and a bb offspring is produced)

= P(both F_1 babbits are Bb) × P(bb offspring)

= (⅔ × ⅔) × (¼)

= ⅑

b. If he crosses an F_1 male (Bb or BB) to the parental female (Bb), two types of crosses are possible. The crosses and probabilities are

(i) Bb (F_1 male) × Bb (parental female) P = ⅔ × 1 = ⅔
(ii) Bb (F_1 male) × Bb (parental female) P = ⅓ × 1 = ⅓

Here, only the first cross can produce white progeny, ¼ of the time. Using the product rule, the chance that this strategy will yield white progeny is

p = ⅔ (chance of $Bb \times Bb$ cross) × ¼ (chance of bb offspring) = ⅙.

c. As neither of the strategies in (b) work very well, evaluate other potential strategies (use trial and error). One that works better in the long run, but is more work initially, is as follows. Re-mate the initial two black babbits (which we have determined to be *Bb*) to obtain a white male offspring [$P = \frac{1}{4}$ (white *bb*) $\times \frac{1}{2}$ (male) $= \frac{1}{8}$]. Retain this male (note that the fertility of white males is not affected) and breed it back to its mother. This cross would be *Bb* $\times$ *bb* and give $\frac{1}{2}$ white (*bb*) and $\frac{1}{2}$ black (*Bb*) offspring. The progeny of this cross could be used to develop a "breeding colony" consisting of black (*Bb*) females and white (*bb*) males. These would consistently produce half white and half black offspring.

2.12 In jimsonweed, purple flower (*P*) is dominant to white (*p*) and spiny pods (*S*) are dominant to smooth (*s*). A true-breeding plant with white flowers and spiny pods is crossed to a true-breeding plant with purple flowers and smooth pods. Determine the phenotype of

a. the F_1 generation
b. the F_2 generation
c. the progeny of a cross of the F_1 back to the white, spiny parent
d. the progeny of a cross of the F_1 back to the purple, smooth parent

Answer: Since you are told that each parent is homozygous, the parental genotypes can be determined from their phenotypes. One has a white (*pp*) spiny (*SS*) plant crossed to a purple (*PP*) smooth (*ss*) plant. This initial information can be used with branch diagrams to solve each part of the problem as follows:

a.   P: white, spiny $\times$ purple, smooth
                    pp SS $\times$ *PP ss*

    Gametes: *p S*   *P s*
            F_1: *Pp Ss* (purple, spiny)
b. $F_1 \times F_1$: *Pp Ss* $\times$ *Pp Ss*

        Pp $\times$ *Pp*   *Ss* $\times$ *Ss*

                                        $\frac{3}{4}$ *S–* $\longrightarrow$ $\frac{9}{16}$ *P– S–* (purple, spiny)
            $\frac{3}{4}$ *P–*
                                        $\frac{1}{4}$ *ss* $\longrightarrow$ $\frac{3}{16}$ *P– ss* (purple, smooth)

                                        $\frac{3}{4}$ *S–* $\longrightarrow$ $\frac{3}{16}$ *pp S–* (white, spiny)
            $\frac{1}{4}$ *pp*
                                        $\frac{1}{4}$ *ss* $\longrightarrow$ $\frac{1}{16}$ *pp ss* (white, smooth)

c. $F_1 \times$ white, spiny: *Pp Ss* $\times$ *pp SS*

        Pp $\times$ *pp*   *Ss* $\times$ *SS*

            $\frac{1}{2}$ *Pp* ——— *S–* $\longrightarrow$ $\frac{1}{2}$ *Pp S–* (purple, spiny)
            $\frac{1}{2}$ *pp* ——— *S–* $\longrightarrow$ $\frac{1}{2}$ *pp S–* (white, spiny)

d. $F_1 \times$ purple, smooth: *Pp Ss* $\times$ *PP ss*

        Pp $\times$ *PP*   *Ss* $\times$ *ss*

                                        $\frac{1}{2}$ *Ss* $\longrightarrow$ $\frac{1}{2}$ *P– Ss* (purple, spiny)
            P–
                                        $\frac{1}{2}$ *ss* $\longrightarrow$ $\frac{1}{2}$ *P– ss* (purple, smooth)

2.13   Use the information in Problem 2.12 to determine what progeny you would expect from the following jimsonweed crosses (you are encouraged to use the branch diagram approach):

a. *PP ss* × *pp SS*

b. *Pp SS* × *pp ss*

c. *Pp Ss* × *Pp SS*

d. *Pp Ss* × *Pp Ss*

e. *Pp Ss* × *Pp ss*

f. *Pp Ss* × *pp ss*

Answer: Using branch diagrams, one has:

a. *PP ss* × *pp SS*

   PP × *pp*   *ss* × *SS*

   Pp ——— *Ss* ——→ all *Pp Ss* (purple, spiny)

b. *Pp SS* × *pp ss*

   Pp × *pp*   *SS* × *ss*

   ½ *Pp* ——— *Ss* ——→ ½ *Pp Ss* (purple, spiny)

   ½ *pp* ——— *Ss* ——→ ½ *pp Ss* (white, spiny)

c. *Pp Ss* × *Pp SS*

   Pp × *Pp*   *Ss* × *SS*

   ¾ *P–* ——— *S–* ——→ ¾ *P– S–* (purple, spiny)

   ¼ *pp* ——— *S–* ——→ ¼ *pp S–* (white, spiny)

d. *Pp Ss* × *Pp Ss*

   Pp × *Pp*   *Ss* × *Ss*

   ¾ *P–* ⟨ ¾ *S–* ——→ $9/16$ *P– S–* (purple, spiny)

   ¼ *ss* ——→ $3/16$ *P– ss* (purple, smooth)

   ¼ *pp* ⟨ ¾ *S–* ——→ $3/16$ *pp S–* (white, spiny)

   ¼ *ss* ——→ $1/16$ *pp ss* (white, smooth)

e. *Pp Ss* × *Pp ss*

   Pp × *Pp*   *Ss* × *ss*

   ¾ *P–* ⟨ ½ *Ss* ——→ $3/8$ *P– Ss* (purple, spiny)

   ½ *ss* ——→ $3/8$ *P– ss* (purple, smooth)

   ¼ *pp* ⟨ ½ *Ss* ——→ $1/8$ *pp Ss* (white, spiny)

   ½ *ss* ——→ $1/8$ *pp ss* (white, smooth)

f. *Pp Ss × pp ss*

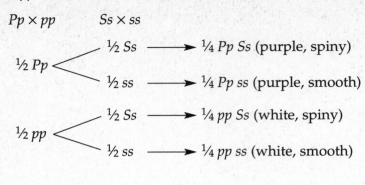

2.14 Cleopatra is normally a very refined cat. When she finds even a small amount of catnip, however, she purrs madly, rolls around in the catnip, becomes exceedingly playful, and appears intoxicated. Cleopatra and Antony, who walks past catnip with an air of indifference, have produced five kittens who respond to catnip just as Cleopatra does. When the kittens mature, two of them mate and produce four kittens that respond to catnip and one that does not. When another of Cleopatra's daughters mates with Augustus (a nonrelative), who behaves just like Antony, three catnip-sensitive and two catnip-insensitive kittens are produced. Propose a hypothesis for the inheritance of catnip sensitivity that explains these data.

Answer: Try fitting the data to a model in which catnip sensitivity/insensitivity is controlled by a pair of alleles at one gene. Hypothesize that since sensitivity is seen in all of the progeny of the initial mating between catnip-sensitive Cleopatra and catnip-insensitive Antony, sensitivity is dominant. Let *S* represent the sensitive allele, and *s* the insensitive allele. Then the initial cross would have been *S– × ss*, and the progeny are *Ss*. If two of the *Ss* kittens mate, one would expect 3 *Ss* (sensitive) : 1 *ss* (insensitive) kittens. In the mating with Augustus, the cross would be *Ss × ss*, and should give a 1 *Ss* (sensitive) : 1 *ss* (insensitive) progeny ratio. The observed progeny ratios are not far off from these expectations.

An alternative hypothesis is that sensitivity (*s*) is recessive, and insensitivity (*S*) is dominant. For Antony and Cleopatra to have sensitive (*ss*) offspring, they would need to be *Ss* and *ss*, respectively. When two of their *ss* progeny mate, only sensitive, *ss* offspring should be produced. Since this is not observed, this hypothesis does not explain the data.

2.15 In summer squash, white fruit (*W*) is dominant over yellow (*w*), and disk-shaped fruit (*D*) is dominant over sphere-shaped fruit (*d*). Determine the genotypes of the parents in each of the following crosses:
a. White, disk × yellow, sphere gives ½ white, disk and ½ white, sphere.
b. White, sphere × white, sphere gives ¾ white, sphere and ¼ yellow, sphere.
c. Yellow, disk × white, sphere gives all white, disk progeny.
d. White, disk × yellow, sphere gives ¼ white, disk; ¼ white, sphere; ¼ yellow, disk; and ¼ yellow, sphere.
e. White, disk × white, sphere gives ⅜ white, disk; ⅜ white, sphere; ⅛ yellow, disk; and ⅛ yellow, sphere.

Answer: First use the symbols for the alleles specifying each trait and the information about which allele is dominant or recessive to make initial assignments of possible genotypes. For example, if a plant is white, it must have a dominant *W* allele, but as it may be either *WW* or *Ww*, it would be initially noted as *W*–. If a plant is yellow, it has to be *ww*. Then, by considering just one pair of allelic traits at a time, and recalling the Mendelian progeny ratios you have seen [a 1:1 ratio follows from a testcross (*Aa* × *aa*), an all-to-none ratio follows if at least one parent is homozygous dominant (*AA* × *A*– or *aa*), and a 3:1 ratio follows from a monohybrid cross (*Aa* × *Aa*)], you can ascertain whether a parental *W*– plant is *Ww* or *WW*.

Cross	Parents	Progeny
a.	white, disk × yellow, sphere *WW Dd*　　　*ww dd*	½ white, disk; ½ white, sphere
b.	white, sphere × white, sphere *Ww dd*　　　*Ww dd*	¾ white, sphere; ¼ yellow, sphere
c.	yellow, disk × white, sphere *ww DD*　　　*WW dd*	all white, disk
d.	white, disk × yellow, sphere *Ww Dd*　　　*ww dd*	¼ white, disk; ¼ white, sphere; ¼ yellow, disk; ¼ yellow, sphere
e.	white, disk × white, sphere *Ww Dd*　　　*Ww dd*	⅜ white, disk; ⅜ white, sphere; ⅛ yellow, disk; ⅛ yellow, sphere

2.16　Genes *a*, *b*, and *c* assort independently and are recessive to their respective alleles *A*, *B*, and *C*. Two triply heterozygous (*Aa Bb Cc*) individuals are crossed.

　　a. What is the probability that a given offspring will be phenotypically *A B C*—that is, will exhibit all three dominant traits?

　　b. What is the probability that a given offspring will be homozygous for all three dominant alleles?

Answer: First consider just one pair of alleles, say *A* and *a*. Since the cross is *Aa* × *Aa*, the progeny will be 1 *AA* : 2 *Aa* : 1 *aa*, or 3 *A*– : 1 *aa*.

　　a. Since the three genes assort independently, one can use the product rule to determine the chance of obtaining an *A*– *B*– *C*– offspring.

　　　　P = (chance of *A*–)(chance of *B*–)(chance of *C*–)

　　　　P = (¾)(¾)(¾)

　　　　P = ($^{27}/_{64}$)

　　b. Since the chance of obtaining an *AA* offspring will be ¼, and the three genes assort independently, use the product rule to determine the chance of obtaining an *AA BB CC* offspring.

　　　　P = (chance of *AA*)(chance of *BB*)(chance of *CC*)

　　　　P = (¼)(¼)(¼)

　　　　P = ($^{1}/_{64}$)

You can also solve this problem by setting up a branch diagram, or much more laboriously, by setting up a Punnett square.

2.17 In garden peas, tall stem (T) is dominant over short stem (t), green pods (G) are dominant over yellow pods (g), and smooth seeds (S) are dominant over wrinkled seeds (s). Suppose a homozygous short, green, wrinkled pea plant is crossed with a homozygous tall, yellow, smooth one.

a. What will be the appearance of the F_1 generation?

b. If the F_1 plants are interbred, what will be the appearance of the F_2 generation?

c. What will be the appearance of the offspring of a cross of the F_1 back to their short, green, wrinkled parent?

d. What will be the appearance of the offspring of a cross of the F_1 back to their tall, yellow, smooth parent?

Answer: First note that the problem asks only for the *appearance* of the offspring of the various crosses, and not the genotypes. Then write out the genotype of the parental cross: *tt GG ss* × *TT gg SS*.

a. For each pair of traits, a homozygous dominant plant is crossed to a homozygous recessive plant. Thus, all the F_1 progeny will show the dominant trait, and be tall, green, and smooth (*Tt Gg Ss*).

b. The F_2 results from selfing the F_1, i.e., *Tt Gg Ss* × *Tt Gg Ss*. The appearance of the F_2 is most readily determined by employing a branch diagram.

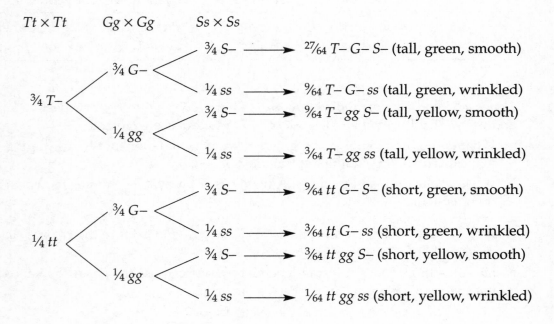

c. The branch diagram of the cross *Tt Gg Ss* × *tt GG ss* is:

d. The branch diagram of the cross *Tt Gg Ss* × *TT gg SS* is:

$Tt × TT$ $Gg × gg$ $Ss × SS$

all *T*– ⟨
  ½ *Gg* ——— all *S*– ⟶ ½ *T*– *Gg S*– (tall, green, smooth)
  ½ *gg* ——— all *S*– ⟶ ½ *T*– *gg S*– (tall, yellow, smooth)

2.18 *C* and *c*, *O* and *o*, and *I* and *i* are three independently segregating pairs of alleles in chickens. *C* and *O* are dominant alleles, both of which are necessary for pigmentation. *I* is a dominant inhibitor of pigmentation. Individuals of genotype *cc*, *oo*, *Ii*, or *II* are white, regardless of what other genes they possess. Assume that White Leghorns are *CC OO II*, White Wyandottes are *cc OO ii*, and White Silkies are *CC oo ii*. What types of offspring (white or pigmented) are possible, and what is the probability of each, from the following crosses?

 a. White Silkie × White Wyandotte
 b. White Leghorn × White Wyandotte
 c. (Wyandotte-Silkie F₁) × White Silkie

Answer: In order for an individual to be pigmented, they must have one of each of the dominant *C* and *O* alleles and not have any of the dominant *I* alleles. That is, they must be *C*– *O*– *ii*.

 a. White Silkie (*CC oo ii*) × White Wyandotte (*cc OO ii*). These are two true-breeding strains at each of three genes, and differ from each other at two loci, the *C/c* and *O/o* loci. The progeny will be dihybrids, are all *Cc Oo ii*, and are all pigmented.
 b. White Leghorn (*CC OO II*) × White Wyandotte (*cc OO ii*). These two are also true breeding at each of three genes, and differ from each other at two loci, the *C/c* and *I/i* loci. Their progeny will be dihybrids, are all *Cc OO Ii*, and are all white (because of *Ii*).
 c. Wyandotte-Silkie F₁ (*Cc Oo ii*) × White Silkie (*CC oo ii*). As both parents are *ii*, all the progeny will be as well, and so if they are *C*– *O*–, they will be pigmented. The proportions of white and pigmented progeny can be determined by using a branch diagram or considering each locus separately as follows: At the *C/c* locus, the cross is *Cc* × *CC*, so all the progeny will be *C*–. At the *O* locus, the cross is *Oo* × *oo*, a testcross, so half the progeny will be *Oo* and half will be *oo*. Thus, half of the progeny will be *C*– *Oo ii* and be pigmented, and half will be *C*– *oo ii* and be white.

2.19 Two homozygous strains of corn are hybridized. They are distinguished by six different pairs of genes, all of which assort independently and produce an independent phenotypic effect. The F₁ hybrid is selfed to give an F₂ generation.

 a. What is the number of possible genotypes in the F₂ plants?
 b. How many of these genotypes will be homozygous at all six gene loci?
 c. If all gene pairs act in a dominant-recessive fashion, what proportion of the F₂ plants will be homozygous for all dominants?
 d. What proportion of the F₂ will show all dominant phenotypes?

Answer: Although this problem can be solved using a branch diagram, the branch diagram analysis of six genes is tedious. Therefore, approach it by considering the general relationship that exists between the number of possible genotypes and phenotypes. At any one locus having alleles that act in a dominant-recessive fashion, there are two possible phenotypes (*A*– and *aa*) and three possible genotypes (*AA, Aa,* and *aa*). When several independently assorting loci are considered, the number of possibilities grows according to the number of combinations of alleles possible. For two loci, independent assortment of the alleles at each locus allows for $2 \times 2 = 4$ phenotypes (two possible phenotypes at each locus assorted with either of two possible phenotypes at a second locus) and 3×3 genotypes (three possible genotypes at one locus assorted with any of three possible genotypes at a second locus). For three loci, independent assortment of the alleles at each locus allows for $2 \times 2 \times 2 = 8$ phenotypes and $3 \times 3 \times 3 = 27$ genotypes. For *n* loci, the relationship becomes 2^n possible phenotypes and 3^n possible genotypes.

To determine the frequency of a particular phenotypic or genotypic class, consider that at one locus, an F_1 hybrid cross gives ¾ phenotypically dominant and ¼ phenotypically recessive progeny and ¼ homozygous dominant, ½ heterozygous and ¼ homozygous recessive progeny. The fraction of progeny of a particular genotypic or phenotypic class can be determined by considering the combinations that are possible. For example, for two independently assorting loci *A/a* and *B/b*, ¾ of the progeny of a cross of *Aa Bb* $\times$ *Aa Bb* will show the *A*– phenotype and ¾ of the progeny will show the *B*– phenotype. When independent assortment of these two loci is considered, $¾ \times ¾ = \frac{9}{16}$ of the progeny will show both the *A*– and the *B*– phenotypes. (Multiply the individual probabilities using the product rule.) Thus, when *n* loci are involved, $(¾)^n$ of the progeny will show all dominant phenotypes, $(¼)^n$ of the progeny will show all recessive phenotypes, $(¼)^n$ of the progeny will be homozygous dominant at all loci, $(¼)^n$ of the progeny will be homozygous recessive at all loci, and $(½)^n$ of the progeny will be heterozygous at all loci.

a. For six loci, $3^6 = 729$ possible genotypes will be seen.

b. Notice that the question asks for how many of the *genotypes* will be homozygous. For just one locus, two (*AA, aa*) of three possible (*AA, Aa, aa*) genotypes are homozygous. Thus, for six loci, $2^6 = 64$ genotypes will be homozygous for all loci. Note that included in these genotypes are those that are homozygous dominant at each locus, those that are homozygous recessive at each locus, and those that are homozygous dominant at one or more loci and homozygous recessive at the other loci.

c. In a monohybrid cross, ¼ of the progeny will be homozygous dominant. For the F_2 in this cross, $(¼)^6 = 1/4{,}096$ of the progeny will be homozygous dominant. Notice that this fraction of the progeny derives from a single genotype.

d. In a monohybrid cross, ¾ of the progeny show a dominant phenotype. For the F_2 in this cross, $(¾)^6 = 729/4{,}096 = 17\%$ of the progeny will show all six dominant phenotypes.

2.20   The coat color of mice is controlled by several genes. The agouti pattern, characterized by a yellow band of pigment near the tip of the hairs, is produced by the dominant allele *A*; homozygous *aa* mice do not have the band and are nonagouti. The dominant allele *B* determines black hairs, and the recessive allele *b* determines brown. Homozygous $c^h c^h$ individuals allow pigments to be deposited only at the extremities (e.g., feet, nose, and ears) in a pattern called Himalayan. The genotype *C*– allows pigment to be distributed over the entire body.

a. If a true-breeding black mouse is crossed with a true-breeding brown, agouti, Himalayan mouse, what will be the phenotypes of the F_1 and F_2 generation?

b. What proportion of the non-Himalayan black agouti F_2 animals will be *Aa BB Cc^h*?

c. What proportion of the Himalayan mice in the F_2 generation is expected to show brown pigment?

d. What proportion of all agoutis in the F_2 generation is expected to show black pigment?

Answer:

a. The cross is *aa BB CC* × *AA bb $c^h c^h$*. Since the parents are true breeding at each of three different loci, the F_1 must be a trihybrid, or *Aa Bb Cc^h*. The F_1 will be agouti with black hairs and pigmented over the entire body. In the F_2, a branch diagram will identify eight possible phenotypic classes: agouti, black, full pigmentation (*A– B– C–*); nonagouti, black, full (*aa B– C–*); agouti, brown, full (*A– bb C–*); nonagouti, brown, full (*aa bb C–*); agouti, black, Himalayan (*A– B– $c^h c^h$*); nonagouti, black, Himalayan (*aa B– $c^h c^h$*); agouti, brown, Himalayan (*A– bb $c^h c^h$*); and nonagouti, brown, Himalayan (*aa bb $c^h c^h$*).

b. F_2 animals that are non-Himalayan, black, and agouti have the genotype *A– B– C–*. Among the *A–* animals, $\frac{2}{3}$ are *Aa*. Among the *B–* animals, $\frac{1}{3}$ are *BB*. Among the *C–* animals, $\frac{2}{3}$ are Cc^h. Thus, among all of the non-Himalayan black agouti F_2, $\frac{2}{3} \times \frac{1}{3} \times \frac{2}{3} = \frac{4}{27}$ are *Aa BB Cc^h*.

c. From the cross *Aa Bb Cc^h* × *Aa Bb Cc^h*, $\frac{1}{4}$ of the progeny will be *bb* and show brown pigment. This will be the case regardless of whether the animals are pigmented over their entire body or are Himalayan. Thus, $\frac{1}{4}$ of the Himalayan mice will show brown pigment. (Be careful not to misread this question: It does *not* ask what proportion of the progeny are both brown *and* Himalayan.)

d. From the cross *Aa Bb Cc^h* × *Aa Bb Cc^h*, $\frac{3}{4}$ of the progeny will be *B–* and show black pigment. This will be the case regardless of whether the animals are agouti or nonagouti. Thus, $\frac{3}{4}$ of the agouti mice will show black pigment. (Be careful not to misread this question: It does *not* ask what proportion of the progeny are both black *and* agouti.)

2.21 In cocker spaniels, solid coat color is dominant over spotted coat. Suppose a true-breeding, solid-colored dog is crossed with a spotted dog, and the F_1 dogs are interbred.

a. What is the probability that the first puppy born will have a spotted coat?

b. What is the probability that, if four puppies are born, all of them will have solid coats?

Answer: First, assign symbols to the alleles, and write down the cross. Let solid color be *S* and spotted be *s*. Then, if a true-breeding, solid-colored dog is bred to a spotted dog, the cross is *SS* × *ss*, the F_1 all *Ss*, and interbreeding the F_1 will give $\frac{3}{4}$ *S–* (solid) and $\frac{1}{4}$ *ss* (spotted) progeny.

a. The chance that the first puppy born is spotted is just the chance of getting an *ss* offspring from an *Ss* × *Ss* cross, or $\frac{1}{4}$.

b. The chance of getting four puppies all having solid coats is the chance of getting an *S–* offspring the first time, *and* the second time, *and* the third time, *and* the fourth time. Apply the product rule to get $p = \frac{3}{4} \times \frac{3}{4} \times \frac{3}{4} \times \frac{3}{4} = \frac{81}{256}$.

2.22 In the F_2 of his cross of red-flowered × white-flowered *Pisum* (pea plant), Mendel obtained 705 plants with red flowers and 224 with white.

a. Is this result consistent with his hypothesis of factor segregation, which predicts a $3:1$ ratio?

b. In how many similar experiments would a deviation as great as or greater than this one be expected? (Calculate χ^2 and obtain the approximate value of *P* from Table 2.5.)

Answer:

a. It is important to *think through* this problem, and not just plug numbers into a chi-square formula. Here, we want to test the hypothesis of factor segregation. The hypothesis states that each allele in a monohybrid will segregate into gametes independently, so that zygotes have a half-chance of obtaining either allele from a heterozygous parent. Drawing a Punnett square shows that a 3:1 ratio of progeny phenotypes would be expected from an $F_1 \times F_1$ cross. Mendel observed 705 red and 224 white plants in the F_2, a ratio of 3.14:1. Thus, this ratio seems consistent with his hypothesis.

b. To test *how* significant this result is, use the χ^2 test to determine how frequently these types of numbers would be obtained in similar experiments.

Class	Observed	Expected	d	d^2	d^2/e
red	705	697	8	64	0.09
white	224	232	−8	64	0.28
Total	929	929	0	−	$\chi^2 = 0.37$

$\chi^2 = 0.37$; df = 1; $0.50 < P < 0.70$

According to the χ^2 test, then, Mendel's result is consistent with this hypothesis. More specifically, in approximately 60 percent of similar experiments, one would expect a deviation (i.e., a value of χ^2) as great as or greater than this one. One therefore fails to reject the hypothesis.

2.23   In tomatoes, cut leaf and potato leaf are alternative characters, with cut (*C*) dominant to potato (*c*). Purple stem and green stem are another pair of alternative characters, with purple (*P*) dominant to green (*p*). A true-breeding cut, green tomato plant is crossed with a true-breeding potato, purple plant, and the F_1 plants are allowed to interbreed. The 320 F_2 plants were phenotypically 189 cut, purple; 67 cut, green; 50 potato, purple; and 14 potato, green. Propose a hypothesis to explain the data, and use the χ^2 test to test the hypothesis.

Answer: First, use the symbols defined in the problem to write the genotypes of the parents and expected offspring if the two loci assort independently. One has

P: *CC pp* × *cc PP*

$F_1 \times F_1$:   *Cc Pp* × *Cc Pp*

F_2:   9 *C– P–*:3 *C– pp*:3 *cc P–*:1 *cc pp*

(9 cut, purple:3 cut, green:3 potato, purple:1 potato, green)

If the two loci assort independently then, one expects to see a 9:3:3:1 phenotypic ratio in the F_2. Therefore, test the hypothesis of independent assortment using the χ^2 test.

Class	Observed	Expected	d	d^2	d^2/e
cut, purple	189	180	9	81	0.45
cut, green	67	60	7	49	0.81
potato, purple	50	60	−10	100	1.66
potato, green	14	20	−6	36	1.80
Total	320	320	0	−	$\chi^2 = 4.72$

$\chi^2 = 4.72$; df = 3; $0.10 < P < 0.20$

Therefore, in experiments similar to this one, a deviation at least as great as that observed here would be seen about 20 percent of the time. Thus, the hypothesis of two independently assorting genes is accepted as being possible.

2.24   The simple case of just two mating types (male and female) is by no means the only sexual system known. The ciliated protozoan *Paramecium bursaria* has a system of four mating types, controlled by two genes (*A* and *B*). Each gene has a dominant and a recessive allele. The four mating types are expressed according to the following scheme:

Genotype	Mating Type
AA BB	*A*
Aa BB	*A*
AA Bb	*A*
Aa Bb	*A*
AA bb	*D*
Aa bb	*D*
aa BB	*B*
aa Bb	*B*
aa bb	*C*

It is clear, therefore, that some of the mating types result from more than one possible genotype. We have four strains of known mating type—"A," "B," "C," and "D"—but unknown genotype. The following crosses were made, with the indicated results:

Cross	Mating Type of Progeny			
	A	*B*	*C*	*D*
"*A*" × "*B*"	24	21	14	18
"*A*" × "*C*"	56	76	55	41
"*A*" × "*D*"	44	11	19	33
"*B*" × "*C*"	0	40	38	0
"*B*" × "*D*"	6	8	14	10
"*C*" × "*D*"	0	0	45	45

Assign genotypes to "A," "B," "C," and "D."

Answer: First summarize what you know about the mating types:

"A" = A– B– "B" = aa B– "C" = aa bb "D" = A– bb

Then rewrite the table in terms of what is known.

| Cross | Mating Type of Progeny | | | |
	A A– B–	B aa B–	C aa bb	D A– bb
"A" × "B" A– B– × aa B–	24	21	14	18
"A" × "C" A– B– × aa bb	56	76	55	41
"A" × "D" A– B– × A– bb	44	11	19	33
"B" × "C" aa B– × aa bb	0	40	38	0
"B" × "D" aa B– × A– bb	6	8	14	10
"C" × "D" aa bb × A– bb	0	0	45	45

Notice that a cross involving mating type C is similar to a testcross, as C is *aa bb*. The progeny ratios in crosses involving "C" will therefore reflect the gametes from the non-"C" parent. In the cross of "B" × "C," both the parents are *aa*, and the $1B:1C$ (1 *aa B–* : 1 *aa bb*) ratio in the progeny reflects the fact that "B" is *Bb*. Therefore, "B" is *aa Bb*. In the cross "C" × "D," both parents are *bb*, and the $1C:1D$ progeny ratio indicates that "D" is *Aa*. "D" is therefore *Aa bb*. To determine the genotype of "A," consider that the cross of "A" × "C" produces progeny of all four mating types. This would be expected only if "A" were *Aa Bb*.

2.25 In bees, males (drones) develop from unfertilized eggs and are haploid. Females (workers and queens) are diploid and come from fertilized eggs. *W* (black eyes) is dominant over *w* (white eyes). Workers of genotype *RR* or *Rr* use wax to seal crevices in the hive; *rr* workers use resin instead. A *Ww Rr* queen founds a colony after being fertilized by a black-eyed drone bearing the *r* allele.

a. What will be the appearance and behavior of workers in the new hive, and what are their relative frequencies?

b. Give the genotypes of male offspring, with relative frequencies.

c. Fertilization normally takes place in the air during a nuptial flight, and any bee unable to fly would effectively be rendered sterile. Suppose that a recessive mutation, *c*, occurs spontaneously in a sperm that fertilizes a normal egg, and suppose also that the effect of the mutant gene is to cripple the wings of any adult not bearing the normal allele *C*. The fertilized egg develops into a normal queen named Madonna. What is the probability that wingless males will be *found in a hive* founded two generations later by one of Madonna's granddaughters?

d. By one of Madonna's great-great-granddaughters?

Answer: The initial cross is *Ww Rr* × *W r*. Note that since males are haploid, they have only one set of chromosomes. As a consequence, their gametes are only of one kind (e.g., the *W r* male will have only *W r* gametes) and their phenotype can be used to directly infer their genotype.

a. The progeny females will be ½ *W– Rr* (black-eyed, wax sealers) and ½ *W– rr* (black-eyed, resin sealers).

b. As males arise from unfertilized eggs, they receive chromosomes only from their mother. The progeny males will be ¼ *W R* (black-eyed, wax sealers), ¼ *W r* (black-eyed, resin sealers), ¼ *w r* (white-eyed, resin sealers), and ¼ *w R* (white-eyed, wax sealers).

c. The egg fertilized by the mutation-bearing sperm results in a *Cc* female. Since fertilization occurs in flight, only males that are *C* contribute genes to the next generation. Hence the first generation arises from the cross *Cc* × *C*. There is a ½ chance of obtaining daughters that are *Cc*. Such a daughter can be fertilized only by a *C* male, so that the chance of her having a *Cc* daughter is also ½. The chance of having a *Cc* granddaughter is thus ½ × ½ = ¼. Such granddaughters will have wingless males if they are prolific. Thus, the probability that wingless males will be found in a hive founded two generations later by one of Madonna's granddaughters is the probability that the granddaughter is *Cc*, or ¼.

d. $(\frac{1}{2})^4 = \frac{1}{16}$.

2.26 Consider the following pedigree, in which the allele responsible for the trait (*a*) is recessive to the normal allele (*A*):

Generation

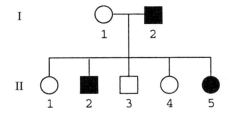

a. What is the genotype of the mother?
b. What is the genotype of the father?
c. What are the genotypes of the children?
d. Given the mechanism of inheritance involved, does the ratio of children with the trait to children without the trait match what would be expected?

Answer:

a. If the allele responsible for the trait is recessive to the normal allele, then to have affected offspring in generation II, the mother (individual I-1) must be heterozygous (i.e., be *Aa*). If she were homozygous for the normal allele (i.e., *AA*), all the offspring would be $A-$ and be normal. If she were heterozygous, the cross in the first generation would be $Aa \times aa$, and one would expect about 50 percent affected and 50 percent normal offspring. This is close to what is observed, and so the mother is Aa.

b. As the father is affected with a recessive trait, he must be aa.

c. Since the father is homozygous for the recessive allele, all the children must have inherited an abnormal allele from their father. If they inherited a normal, dominant allele from their mother, they will not be affected (Aa: II-1, II-3, II-4), whereas if they inherited the abnormal, recessive allele from their mother, they will be affected (aa: II-2, II-5).

d. Given that the parents are $Aa \times aa$, one expects a 1:1 ratio of affected to unaffected children. For the five offspring shown, one sees a ratio of 2:3, which, given the number of offspring, approximates a 1:1 ratio. Remember that each birth is independent of the others, and while, if many progeny are seen, one should see about a 1:1 ratio, for small numbers of progeny, the ratios may be significantly off from what is expected. Indeed, even if the couple were to have more unaffected children, the pedigree would still be consistent with the inheritance of a recessive trait. (Can you show that if the pedigree is taken on its own, *without any information about whether the trait is dominant or recessive*, the pedigree is also consistent with the inheritance of a dominant trait?)

2.27 For the following pedigrees A and B, indicate whether the trait involved in each case could be recessive or dominant, and explain your answers:

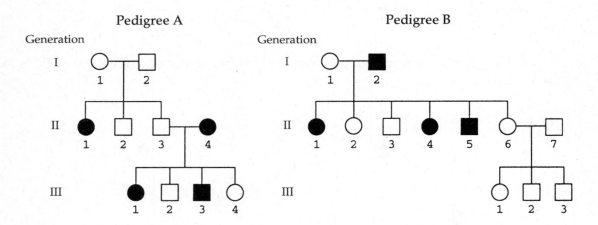

Answer: For pedigree A, parents not showing the trait in generation I have a daughter that exhibits the trait. This is typical of a cross of two heterozygotes giving a homozygous recessive, affected individual: $Aa \times Aa$ (two normal parents) give 3 $A-$ (normal) : 1 aa (affected). If this were the case, individuals II-2 and II-3 could be carriers (i.e., be Aa). If individual II-3 is a carrier and his spouse is homozygous recessive (aa), generation III would arise from the cross $Aa \times aa$, and the 1:1 ratio of affected to unaffected individuals seen would be expected. Thus, pedigree A is consistent with the trait being recessive.

In pedigree B, the trait is present in half of the individuals in generations I and II, but in none of those of generation III. However, the parents of generation III do not themselves exhibit the trait. This is consistent with the trait being dominant. With this view, individual I-1 would be aa and individual I-2 would be Aa. Half of their progeny would be expected to show the trait, as is observed. Furthermore, unaffected parents should have unaffected offspring, as is seen in generation III. Notice, however, that the pedigree is also consistent with a recessive trait that is not rare. In this case, individual I-1 would be heterozygous, and the cross in generation I would be *Aa* × *aa*, giving the 50 percent affected progeny seen in generation II. Although individual II-6 would have to be *Aa*, if her spouse was *AA*, all the children would be normal.

2.28 After a few years of marriage, a woman comes to believe that, among all of the reasonable relatives in her and her husband's families, her husband, her mother-in-law, and her father have so many similarities in their unreasonableness that they must share a mutation. A friend taking a course in genetics assures her that it is unlikely that this trait has a genetic basis and that, even if it did, all of her children would be reasonable. Diagram and analyze the relevant pedigree to evaluate whether the friend's advice is accurate.

Answer: First draw the pedigree. Then try to fit the "unreasonableness" trait to a dominant (*A*− = affected, *aa* = unaffected) or recessive character (*A*− = unaffected, *aa* = affected), as shown below:

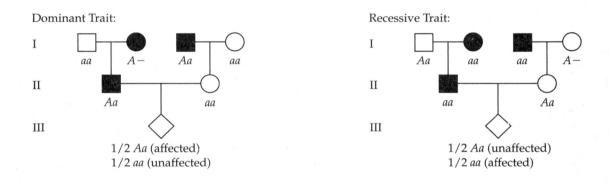

In the diagrams above, the woman is II-2, her husband is II-1, her father is I-3, and her mother-in-law is I-2. If the trait indeed has a genetic basis, it could be modeled as either a dominant or a recessive trait. Notice that one can infer whether an *A*− individual is *AA* or *Aa* based on the offspring that are produced. Independent of whether the trait is dominant or recessive, about half of the woman's children should be "unreasonable" in this particular situation.

2.29 Gaucher disease is caused by a chronic enzyme deficiency that is more common among Ashkenazi Jews than it is in the general population. A Jewish man has a sister afflicted with the disease. His parents, grandparents, and three siblings are not affected. Discussions with relatives in his wife's family reveal that the disease is not likely to be present in her family, although some relatives recall that the brother of his wife's paternal grandmother suffered from a very similar disease. Diagram and analyze the relevant pedigree to determine (a) the genetic basis for inheriting the trait for Gaucher disease and (b) the highest probability that, if this couple has a child, the child will be affected (i.e., what is the chance of the worst-case scenario occurring?).

Answer: An uncertainty exists about whether the brother of the man's wife's paternal grandmother had Gaucher's disease. If this distant relative had the disease, a disease allele might have been passed on to the man's wife. Therefore, in a worst-case scenario, consider this distant relative to have had the disease. Under this scenario, the pedigree is as shown below:

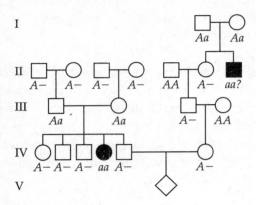

In this pedigree, the man is IV-5, his affected sister is IV-4, his wife is IV-6, and the brother of his wife's paternal grandmother is II-7. Since II-7 is affected but his parents are not, the disease must be a recessive trait and each of his parents must be heterozygous.

To calculate the chance that, if the couple has a child, the child will be affected, one needs to calculate the probability that V-1 can receive an a allele from each of its parents. For each parent, this is ½ the chance a parent is Aa.

Since we know the parents of IV-5 must be Aa, and IV-5 is $A-$, there is a ⅔ chance that IV-5 is Aa. Thus, there is a ½ × ⅔ = ⅓ chance that IV-5 will pass the a allele to V-1.

The probability that IV-6 is Aa is more complex, and is:

P(III-3 was Aa and III-3 passed the a allele on to IV-6)

= P[(II-6 was Aa and II-6 passed a to III-3) and (III-3 passed a to IV-6)]

= [(⅔ × ½) × ½] = ⅙.

Thus, there is a ⅙ × ½ = 1/12 chance that IV-6 will pass the a allele to V-1. In this worst-case scenario, the chance that both parents will pass on an a allele and have an affected child is 1/12 × ⅓ = 1/36.

If the brother of the wife's paternal grandmother did not have the disease, IV-6 would be AA. The child would be $A-$ and be phenotypically normal.

3

CHROMOSOMAL BASIS OF INHERITANCE

CHAPTER OUTLINE

CHROMOSOMES AND CELLULAR REPRODUCTION
Eukaryotic Chromosomes
Mitosis
Meiosis

CHROMOSOME THEORY OF INHERITANCE
Sex Chromosomes
Sex Linkage
Nondisjunction of X Chromosomes

SEX CHROMOSOMES AND SEX DETERMINATION
Genotypic Sex Determination
Genic Sex Determination

ANALYSIS OF SEX-LINKED TRAITS IN HUMANS
X-linked Recessive Inheritance
X-linked Dominant Inheritance
Y-linked Inheritance

REVIEW OF KEY TERMS, SYMBOLS, AND CONCEPTS

In your own words, write a brief, precise definition of each term in the groups on the following page. Check your definitions using the text. Then develop a concept map using the terms in each list.

THINKING ANALYTICALLY

The problems on Mendelian genetics in this chapter are more complicated than those in Chapter 2, and present more of a challenge to your analytical reasoning skills. Review the "Thinking Analytically" hints in Chapter 2 of this guide, and approach the problems analytically and systematically. After you have arrived at a solution, always go back and test it against the data presented in the problem. Try to avoid falling into a trap: Don't look at the answer presented here before you have attempted to solve the problem independently. Being able to understand

1	2	3
eukaryote	mitosis	meiosis I, II
diploid	karyokinesis	prophase I, II
haploid	cytokinesis	metaphase I, II
genome	cell cycle	anaphase I, II
gamete	M, G_1, G_0, S, G_2	telophase I, II
zygote	sister chromatid	leptonema
homolog	daughter chromosome	zygonema
homologous chromosome	prophase	pachynema
nonhomologous chromosome	prometaphase	diplonema
sex chromosome	metaphase	diakinesis
autosome	anaphase	pairing, synapsis
centromere	telophase	crossing-over
p, q arms	mitotic spindle	synaptonemal complex
metacentric chromosome	tubulin, microtubule	bivalent
submetacentric chromosome	centriole	tetrad
acrocentric chromosome	kinetochore	bouquet
telocentric chromosome	metaphase plate	chiasma, chiasmata
karyotype	scaffold	genetic recombination
band	disjunction	recombinant chromosome
G banding	cell plate	pseudoautosomal region (PAR)

4	5	6
meiosis	meiosis	chromosome theory
sexual reproduction	angiosperm	sex chromosome
spermatogenesis	gametogenesis	X chromosome
oogenesis	sporogenesis	Y chromosome
spermatozoa	gametophyte	homogametic sex
primary spermatogonia	sporophyte	heterogametic sex
primary spermatocytes	stamen	sex linkage
meiocyte	pistil	X linkage
secondary spermatocyte	alternation of generations	hemizygous
spermatid	ovule	homozygous
primary oogonia	pollen	crisscross inheritance
secondary oogonia	anther	reciprocal cross
primary oocyte	gametophyte generation	X-linked recessive trait
unequal cytokinesis	sporophyte generation	X-linked dominant trait
secondary oocyte	spore	Y-linked trait
first polar body		holandric trait
second polar body		
ovum		

7	8
genotypic sex determination	wild type
genic sex determination	mutant allele
Y chromosome sex determination	Mendelian symbolism
X chromosome–autosome balance system	*Drosophila* symbolism
disjunction, nondisjunction	A, a, aa, Aa, AA, aY, AY
segregation	$a^+, a, a^+a^+, a^+a, aa, aY, a^+Y$
chromosome pairing	$A^+, A, AA, A^+A, A^+A^+, AY, A^+Y$
primary nondisjunction	
secondary nondisjunction	
X chromosome nondisjunction	
aneuploidy	
karyotype	
Turner, Klinefelter, XYY syndromes	
dosage compensation	
Barr body	
epigenetic phenomenon	
genetic mosaic	
Lyon hypothesis, lyonization	
X-inactivation center	
XIC, Xic, XIST, Xist	
testis-determining factor gene	
sex reversal individual	
sex determining region Y (*SRY*)	
SRY, Sry genes	
transgene, transgenic organism	
hermaphroditic	
monoecious, dioecious	
mating type	

the answer in this guide is not equivalent to being able to solve the problem independently. You won't have the guide in an exam. By trying to solve problems independently, you will improve your analytical thinking. This will enable you to do better on an exam.

Take care to represent crosses accurately with appropriate symbols. While you can often use Mendelian or normal/abnormal notation (see Box 3.1) equally well, become fluent with both. Geneticists use each type of notation. As you symbolize traits in crosses, consistently use one type of notation, be careful to use a single type of symbol for each gene (e.g., g and g^+ or g and G for green and yellow seed color, not g and Y), and clearly distinguish between X-linked and autosomal traits, as these follow different inheritance patterns. For example, when assigning members of a pedigree symbols for genetic traits, use a notation that reflects whether inheritance is autosomal or X-linked. This will make it easier for you to spot patterns as well as infer expected types of offspring. Use aa for a female with an X-linked recessive trait, AY or a^+Y for a male with an X-linked dominant trait, aY for a male with a recessive trait, and Aa or a^+a for an individual with an autosomal trait.

The proof that genes lie on chromosomes came from relating abnormal patterns of trait inheritance to abnormal patterns of chromosome inheritance. Understanding how chromosome behavior during meiosis relates to the segregation and independent assortment of traits in genetic crosses is therefore central to this material. Review the material in this chapter on chromosome pairing and segregation during meiosis and relate it to how autosomal alleles in a heterozygote segregate during the formation of gametes. Then diagram what happens if the alleles are on an X chromosome and there is X chromosome nondisjunction. Are the normal crisscross inheritance patterns altered? Are the results of testcrosses with the progeny altered?

QUESTIONS FOR PRACTICE

Multiple Choice Questions

1. What two events occur in leptonema of prophase I?
   a. synapsis and crossing-over
   b. disappearance of the nucleolus and disjunction
   c. synaptonemal complex formation and crossing-over
   d. pairing and chiasma formation
   e. pairing and the beginning of crossing-over

2. Which one of the following is *not* true?
   a. Crossing-over begins in leptonema.
   b. Chiasmata are visible manifestations of crossing-over.
   c. Chiasmata are visible in diplonema.
   d. Chiasmata are visible at the time that crossing-over occurs.

3. Which one of the following statements is true about mitosis?
   a. The nucleolus disappears during metaphase.
   b. Nuclear membranes re-form at the end of telophase.
   c. Centromeres are aligned on the metaphase plate at prophase.
   d. The nucleolus re-forms at the end of anaphase.

4. Sister chromatids are
   a. synonymous with homologous chromosomes
   b. present only in meiosis and not in mitosis
   c. identical products of chromosome duplication held together by a replicated but unseparated centromere
   d. visible in interphase just after S phase

5. The chromosome theory of inheritance holds that
   a. chromosomes are inherited
   b. the chromosomes contain the hereditary material
   c. the genes are inherited
   d. the chromosomes are DNA

6. Proof that genes lie on chromosomes was obtained by
   a. correlating aneuploids resulting from nondisjunction with inheritance patterns
   b. showing that some genes appear to be sex linked
   c. showing that males have an X and a Y, while females have two X's
   d. showing that in *Drosophila*, males are the heterogametic sex

7. An individual that produces both ova and sperm
   a. has the XY genotype
   b. reproduces asexually

c. is said to be hermaphroditic

d. both b and c above

8. What piece of evidence best indicates that two X chromosomes are required for a normal human female, even though one of the X chromosomes is inactivated into a Barr body?

a. Turner females are usually infertile and have morphological abnormalities.

b. XXY individuals show a Klinefelter syndrome.

c. 47,XXX individuals are almost completely normal females.

d. Lyonization occurs in 46,XX individuals.

9. The somatic cells of an individual having the karyotype 48,XXXY would have

a. four Barr bodies

b. three Barr bodies

c. two Barr bodies

d. no Barr bodies

10. Pedigrees showing X-linked recessive traits typically have

a. only females being affected

b. all daughters of an affected father being affected

c. all sons of an unaffected mother being normal

d. the trait appearing more frequently in males

11. In humans, what can be said about genes on a normal male's X chromosome?

a. They are hemizygous.

b. They are inactivated when the male's cells form Barr bodies.

c. They are inherited as autosomal recessive traits.

d. They are responsible for male sex type.

12. During spermatogenesis in a male fly, there is X chromosome nondisjunction during meiosis I. What types of sperm are produced?

a. X- and Y-bearing sperm

b. XX- and YY-bearing sperm

c. XY- and no-sex-chromosome-bearing sperm

d. XX-, XY-, X-, and Y-bearing sperm

13. Which of the following is true about organisms where sex type is determined by an X chromosome–autosome balance system?

a. Sex type is determined by the number of X chromosomes. Organisms with two X chromosomes are female, while organisms with one X chromosome are male, regardless of the number of autosomes.

b. Sex type is determined by the ratio of X chromosomes to autosomes. The Y chromosome plays no role in sex determination.

c. A key gene called *SRY* on the Y chromosome determines sex type.

d. The number of autosomes determines sex type. Organisms with one set of autosomes are male, while organisms with two sets of autosomes are female.

14. What unusual feature is associated with the *Xist* gene?

a. It is active on both X chromosomes of a normal female.

b. It is inactive on both X chromosomes of a normal female.

c. It is active only on the nonlyonized X chromosome of a normal female.

d. It is active only on the lyonized X chromosome of a normal female.

15. Which of the following statements is *not* true about sex determination in mammals?

a. It is equivalent to testis determination.

b. In males, it results from the action of the testis-determining factor.

c. Sex reversal occurs when the *SRY* gene is present in males.

d. The *SRY* gene encodes the testis-determining factor.

Answers: 1e, 2d, 3b, 4c, 5b, 6a, 7c, 8a, 9c, 10d, 11a, 12c, 13b, 14d, 15c

Thought Questions

1. When during meiosis does the chromosomal behavior underlying Mendel's laws occur?

2. What is meant by crisscross inheritance, and what clinical significance does it have in pedigree analysis?

3. Develop a flowchart illustrating the decision-making path you would employ to ascertain whether a pedigree showed autosomal dominant, autosomal recessive, X-linked dominant, X-linked recessive, or Y-linked inheritance.

4. Define the term *lyonization*, give an example of it, and discuss its functional significance.

5. Why do you think most flowering plants are monoecious, whereas in most animals, the sexes are separate?

6. If all mammals have evolved from a common ancestor, how do you account for the variety of chromosome numbers they possess? By what mechanism do you think changes in chromosome number could have occurred? (Consider chromosome number variation resulting from nondisjunction and other accidental changes.)

7. Except for the loci involved with sex determination in mammalian males, there is little clear evidence for genetic loci on the Y chromosome. Why might this be the case, and what might this tell you about the Y chromosome in mammals?

8. What are some of the different mechanisms by which sex type can be determined? Do you think it is significant that so many different mechanisms can be employed?

9. Even though X inactivation occurs in XXY individuals, they do not have the same phenotype as XY males. Similarly, even though X inactivation occurs in XX individuals, they do not have the same phenotype as XO individuals. Why might this be the case?

10. An individual who phenotypically appears to have Klinefelter syndrome has a karyotype of XX,46. An individual who phenotypically appears to have Turner syndrome has a karyotype of XY,46. What common explanation might underlie each of these anomalies? (Hint: Review the text material on sex reversal.)

Thought Questions for Media

After reviewing the media on the *iGenetics* CD-ROM, try to answer these questions.

1. How are centromeres, centrioles, and centrosomes different in terms of their cellular location and function during mitosis?

2. At what point in the cell cycle are chromosomes replicated? At what point are replicated chromosomes so extended that individual chromatids cannot be seen? When can they first be visualized? When are they maximally condensed?

3. When do centromeres replicate? When do replicated centromeres split apart?

4. During anaphase, chromosomes move to opposite poles and polar and kinetochore microtubules change length. Which type of microtubule shortens, and which type of microtubule elongates?

5. How do the chromosomes of daughter cells produced by mitosis compare with the chromosomes of the parental cell?

6. After which division (meiosis I or meiosis II) are haploid cells produced? At which point are haploid gametes produced?

7. If nondisjunction occurs in meiosis I, how many types of abnormal gametes are produced? Are any normal gametes produced?

8. If nondisjunction occurs in meiosis II, how many types of abnormal gametes are produced? Are any normal gametes produced?

9. The animation illustrates how nondisjunction in either meiosis I or meiosis II can lead to the production of monosomic and trisomic zygotes. How might nondisjunction be involved in generating a nullosomic (none of one type of chromosome) or a tetrasomic (four copies of one kind of chromosome) zygote?

10. In one meiosis in an *Aa Bb* individual, four gametes are produced. How many different genotypes are produced?

11. Based on the animation and considering your answer to question 1, how can four equal frequencies of gametes (¼ *Ab*, ¼ *AB*, ¼ *aB*, ¼ *ab*) be obtained?

12. At what point during meiosis does the chromosome behavior that underlies independent assortment occur?

1. a. Why is it important to determine that none of the individuals that have married into Anna's or Jackson's family have a separate family history of deafness?

   b. How would your analysis of the pedigree of Jackson's family change if you learned that individuals II-3, II-5, and III-4 have a history of deafness?

   c. Since both congenital and progressive hearing loss are relatively uncommon traits, when a pedigree analysis is being performed, can you just assume that individuals marrying into the family have no family history of deafness? If not, how would you proceed?

2. What is the chance that if Anna and Jackson have two boys and two girls, they will all have normal hearing?

3. Anna may or may not be a carrier for deafness. Suppose Anna and Jackson indeed have two boys and two girls with normal hearing. Can you conclude anything about whether Anna is a carrier for deafness? If so, what? If not, why not?

SOLUTIONS TO TEXT PROBLEMS

3.1   Interphase is a period corresponding to the cell cycle phases of
   a. mitosis
   b. S
   c. $G_1 + S + G_2$
   d. $G_1 + S + G_2 + M$

   Answer: c

3.2   Chromatids joined together by a centromere are called
   a. sister chromatids
   b. homologs
   c. alleles
   d. bivalents (tetrads)

   Answer: a

3.3   Mitosis and meiosis always differ in regard to the presence of
   a. chromatids
   b. homologs
   c. bivalents
   d. centromeres
   e. spindles

   Answer: c

3.4   State whether each of the following statements is true or false, and explain your choice.
   a. The chromosomes in a somatic cell of any organism are all morphologically alike.
   b. During mitosis, the chromosomes divide and the resulting sister chromatids separate at anaphase, ending up in two nuclei, each of which has the same number of chromosomes as the parental cell.
   c. At zygonema, a chromosome can synapse with any other chromosome in the same cell.

   Answer:
   a. False. Chromosomes can have different sizes and can have their centromeres positioned differently. For many organisms, this statement will be false. However, in a diploid cell, a pair of homologous chromosomes will be morphologically alike.
   b. True.
   c. False. In meiosis I, only homologous chromosomes will synapse together.

3.5 For each mitotic event described in the table that follows, write the name of the event in the blank provided in front of the description. Then put the events in the correct order (sequence). Start by placing a 1 next to the description of interphase, and continue through 6, which should correspond to the last event in the sequence.

Name of Event		Order of Event
_____	The cytoplasm divides and the cell contents are separated into two separate cells.	_____
_____	Chromosomes become aligned along the equatorial plane of the cell.	_____
_____	Chromosome replication occurs.	_____
_____	The migration of the daughter chromosomes to the two poles is complete.	_____
_____	Replicated chromosomes begin to condense and become visible under the microscope.	_____
_____	Sister chromatids begin to separate and migrate toward opposite poles of the cell.	_____

Answer: Try to visualize the events of mitosis dynamically as you solve this problem.

Name of Event		Order of Event
cytokinesis	The cytoplasm divides and the cell contents are separated into two separate cells.	6
metaphase	Chromosomes become aligned along the equatorial plane of the cell.	3
interphase	Chromosome replication occurs.	1
telophase	The migration of the daughter chromosomes to the two poles is complete.	5
prophase	Replicated chromosomes begin to condense and become visible under the microscope.	2
anaphase	Sister chromatids begin to separate and migrate toward opposite poles of the cell.	4

3.6 Answer these questions with a "yes" or "no," and then explain the reasons for your answer:
a. Can meiosis occur in haploid species?
b. Can meiosis occur in a haploid individual?

Answer:

a. Yes. If a sexual mating system exists in the species, two haploid cells can fuse to produce a diploid cell that can then go through meiosis to produce haploid progeny. The fungi *Neurospora crassa* and *Saccharomyces cerevisiae* exemplify this life cycle.

   b. No. Meiosis can occur only in a diploid cell. A haploid individual cannot form a diploid cell, so meiosis cannot occur.

3.7   Which of the following sequences describes the general life cycle of a eukaryotic organism?
   a. 1N→meiosis→2N→fertilization→N
   b. 2N→meiosis→N→fertilization→2N
   c. N→mitosis→2N→fertilization→N
   d. 2N→mitosis→N→fertilization→2N

   Answer: b. Only diploid (2N) cells can undergo meiosis, and haploid cells (N) fuse at fertilization, regenerating the diploid state.

3.8   Which statement is true?
   a. Gametes are 2N; zygotes are N.
   b. Gametes and zygotes are 2N.
   c. The number of chromosomes can be the same in gamete cells and in somatic cells.
   d. The zygotic and somatic chromosome numbers cannot be the same.
   e. Haploid organisms have haploid zygotes.

   Answer: c. In a haploid organism, gametes and somatic cells are both N. This provides an example of how (c) is true.

3.9   All of the following happen in prophase I of meiosis, except
   a. chromosome condensation
   b. pairing of homologues
   c. chiasma formation
   d. terminalization
   e. segregation

   Answer: e

3.10   Give the name of each stage of mitosis and meiosis at which each of the following events occurs:
   a. Chromosomes are located in a plane at the center of the spindle.
   b. The chromosomes move away from the spindle equator to the poles.

   Answer:
   a. Metaphase: Metaphase in mitosis, metaphase I and metaphase II in meiosis
   b. Anaphase: Anaphase in mitosis, anaphase I and anaphase II in meiosis

3.11   Consider the diploid meiotic mother cell shown at the top of the next page:

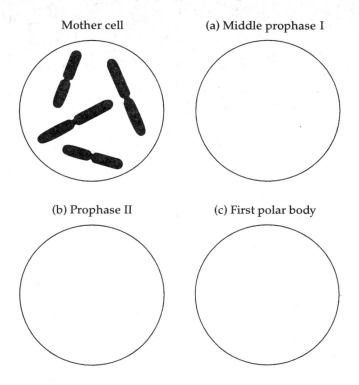

Diagram the chromosomes as they would appear
a. in late pachynema
b. in a nucleus at prophase of the second meiotic division
c. in the first polar body resulting from oogenesis in an animal

Answer:

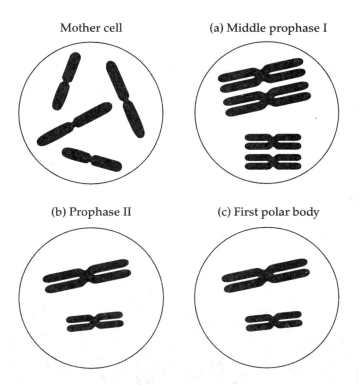

3.12   The cells in the following figure were all taken from the same individual (a mammal).

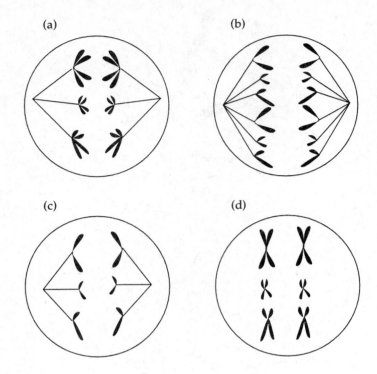

Identify the cell division events occurring in each cell and explain your reasoning. What is the sex of the individual? What is the diploid chromosome number?

Answer: To reconcile how all the cells illustrated could come from one individual, consider that the cells shown could come from either somatic or germline cells. Cell (a) shows three pairs of previously synapsed homologs disjoining and must therefore illustrate anaphase I of meiosis. That three pairs of chromosomes are present indicates that the organism has 2N = 6 chromosomes, so that N = 3. Cell (b) shows the disjoining of chromatids. Since the organism has 2N = 6 chromosomes, and the daughter cells that will form as a result of this cell division will have 6 chromosomes, (b) must illustrate part of mitosis, specifically anaphase. Cell (c) also shows the disjoining of chromatids. Since the daughter cells will receive three chromosomes, this must be anaphase II of meiosis. Cell (d) shows the pairing of homologs and therefore illustrates metaphase I of meiosis. As the individual has three pairs of identically appearing chromosomes, there are two identical sex (= X) chromosomes, indicating that the animal is female.

3.13   Does mitosis or meiosis have greater significance in the study of heredity? Explain your answer.

Answer: Meiosis has greater significance. While mitosis generates progeny cells that are genetically identical to a parent cell, meiosis generates gametes that are genetically diverse. Gamete diversity is obtained when nonparental combinations of genes are obtained in two ways: the random assortment of maternal and paternal chromosomes at anaphase I of meiosis I and crossing-over events in prophase I of meiosis I. When gametes from different parents fuse at fertilization, more diversity is obtained. Hence, meiosis provides a means for genetic variation.

3.14   Consider a diploid organism that has three pairs of chromosomes. Assume that the organism receives chromosomes A, B, and C from the female parent and A', B', and C' from the male parent. Answer the following questions, assuming that crossing-over does not occur:

    a. What proportion of the gametes of this organism would be expected to contain all the chromosomes of maternal origin?

    b. What proportion of the gametes would be expected to contain some chromosomes of both maternal and paternal origin?

Answer:

    a. The chance that a gamete would have a particular maternal chromosome is $\frac{1}{2}$. Apply the product rule to determine the chance of obtaining a gamete with all three maternal chromosomes: $P(A \ and \ B \ and \ C) = P(A) \times P(B) \times P(C) = (\frac{1}{2})^3 = \frac{1}{8}$.

    b. There are two approaches to this question.

        1. List all of the possible options to obtain gametes that satisfy the specified condition—that some maternal and paternal chromosomes are present. Use the product rule to determine the probability of each, and then, using the sum rule, determine the proportion of the gametes having maternal and paternal chromosomes. Gametes that satisfy the condition that both maternal and paternal chromosomes are present are ABC', AB'C, A'BC, AB'C', A'BC', and A'B'C. For each one of these six gamete types, the chance of obtaining it is $\frac{1}{8}$. This is based on applying the product rule and logic similar to that used in part (a) of this question. Apply the sum rule to determine the chance of obtaining one of these types of gametes: $P(ABC' \ or \ AB'C \ or \ A'BC \ or \ AB'C' \ or \ A'BC' \ or \ A'B'C) = \frac{1}{8} + \frac{1}{8} + \frac{1}{8} + \frac{1}{8} + \frac{1}{8} + \frac{1}{8} = \frac{6}{8} = \frac{3}{4}$.

        2. Realize that the set of gametes with some maternal and paternal chromosomes is composed of all gametes *except* those that have only maternal or only paternal chromosomes. That is, P(gamete with *both* maternal *and* paternal chromosomes) $= 1 - P$(gamete with *only* maternal *or* only paternal chromosomes). From part (a), the chance of a gamete having chromosomes from only one parent is $\frac{1}{8}$. Using the sum rule, P(gamete with both maternal and paternal chromosomes) $= 1 - (\frac{1}{8} + \frac{1}{8}) = \frac{3}{4}$.

3.15   Normal diploid cells of a theoretical mammal are examined cytologically at the mitotic metaphase stage for their chromosome complement. One short chromosome, two medium-length chromosomes, and three long chromosomes are present. Explain how the cells might have such a set of chromosomes.

Answer: Since the cells are normal and the cell is diploid, the chromosomes should exist in pairs. The cell could have come from a male mammal with one medium-length pair, one long pair, and a heteromorphic sex chromosome (XY) pair.

3.16   Explain whether the following statement is true or false: "Meiotic chromosomes can be seen after appropriate staining in nuclei from rapidly dividing skin cells."

Answer: False. Skin cells consist entirely of mitotic, not meiotic, cells.

3.17   Explain whether the following statement is true or false: "All the sperm from one human male are genetically identical."

Answer: False. Genetic diversity in the male's sperm is achieved during meiosis. There is crossing-over between nonsister chromatids as well as independent assortment of the male's maternal and paternal chromosomes during spermatogenesis.

3.18   The horse has a diploid set of 64 chromosomes, and the donkey has a diploid set of 62 chromosomes. Mules are the viable, but usually sterile, progeny of a mating between a male donkey and a female horse. How many chromosomes will a mule cell contain?

Answer: A female horse has ova with 32 chromosomes, while a male donkey has sperm with 31 chromosomes. A mule has 32 + 31 = 63 chromosomes.

3.19   The red fox has 17 pairs of large, long chromosomes. The arctic fox has 26 pairs of shorter, smaller chromosomes.
   a. What do you expect to be the chromosome number in somatic tissues of a hybrid between these two foxes?
   b. The first meiotic division in the hybrid fox shows a mixture of paired and single chromosomes. Why do you suppose this occurs? Can you suggest a possible relationship between the mixed chromosomes and the observed sterility of the hybrid?

Answer:
   a. The red fox will have gametes with 17 chromosomes, and the arctic fox will have gametes with 26 chromosomes. The hybrid will have 17 + 26 = 43 chromosomes. Once the fertilized egg is formed, mitotic division will produce somatic cells with 43 chromosomes.
   b. Similar chromosomes pair in meiosis. The pairing pattern seen in the hybrid indicates that some of the chromosomes in the arctic and red foxes share evolutionary similarity, while others do not. Unpaired chromosomes will not segregate in an orderly manner, giving rise to unbalanced meiotic products with either extra or missing chromosomes. This can lead to sterility for two reasons: First, meiotic products that are missing chromosomes may not have genes necessary to form gametes. Second, even if gametes are able to form, a zygote generated from them will not have the chromosome set from either the hybrid, the red, or the arctic fox. The zygote will be an aneuploid with missing or extra genes, causing it to be inviable.

3.20   At the time of synapsis preceding the reduction division in meiosis, the homologous chromosomes align in pairs, and one member of each pair passes to each of the daughter nuclei. In an animal with five pairs of chromosomes, assume that chromosomes 1, 2, 3, 4, and 5 have come from the father, and 1', 2', 3', 4', and 5' have come from the mother. Assuming no crossing-over, in what proportion of the gametes of this animal will all the paternal chromosomes be present together?

Answer: The chance of getting a particular paternal chromosome is $\frac{1}{2}$. Using the product rule, and logic similar to that in the answer presented in question 3.14a, the chance of getting all five paternal chromosomes is $(\frac{1}{2})^5 = \frac{1}{32}$.

3.21   Depict each of the crosses that follow, first using Mendelian and then using *Drosophila* notation (see Box 3.1, p. 57). Give the genotype and phenotype of the F_1 progeny that can be produced.
   a. In humans, a mating between two individuals, each heterozygous for the recessive trait phenylketonuria, whose locus is on chromosome 12.

b. In humans, a mating between a female heterozygous for both phenylketonuria and X-linked color blindness and a male with normal color vision who is heterozygous for phenylketonuria.

c. In *Drosophila*, a mating between a female with white eyes, curled wings, and normal long bristles and a male that has normal red eyes, normal straight wings, and short, stubble bristles. In these individuals, curled wings result from a heterozygous condition at a gene whose locus is on chromosome 2, while the short, stubble bristles result from a heterozygous condition at a gene whose locus is on chromosome 3.

d. In *Drosophila*, a mating between a female from a true-breeding line that has eyes of normal size that are white, black bodies (a recessive trait on chromosome 2), and tiny bristles (a recessive trait called *spineless* on chromosome 3) and a male from a true-breeding line that has normal red eyes, normal grey bodies, normal long bristles, but a reduced eye size (a dominant trait called *eyeless* on chromosome 4).

Answer: The crosses and their progeny are tabulated below:

	Mendelian Notation	Drosophila Notation; Long Version	Drosophila Notation; Short Version
a.			
Allele symbols:	P = normal	p^+ = normal	+ = normal
	p = phenylketonuria	p = phenylketonuria	p = phenylketonuria
Cross:	$Pp \times Pp$	$p^+p \times p^+p$	$+p \times +p$
F_1:	$1\,PP : 2\,Pp : 1\,pp$	$1\,p^+p^+ : 2\,p^+p : 1\,pp$	$1\,++ : 2\,+p : 1\,pp$
F_1 phenotypes:	¾ normal, ¼ phenylketonuria		
b.			
Allele symbols:	C = normal	c^+ = normal	+ = normal
	c = color blind	c = color blind	c = color blind
Cross:	$Cc\,Pp \times CY\,Pp$	$c^+c\,p^+p \times c^+Y\,p^+p$	$+c\,+p \times +Y\,+p$
F_1:	¹⁄₁₆ $CC\,PP$	¹⁄₁₆ $c^+c^+\,p^+p^+$	¹⁄₁₆ $++\,++$
	⅛ $CC\,Pp$	⅛ $c^+c^+\,p^+p$	⅛ $++\,+p$
	¹⁄₁₆ $CC\,pp$	¹⁄₁₆ $c^+c^+\,pp$	¹⁄₁₆ $++\,pp$
	¹⁄₁₆ $Cc\,PP$	¹⁄₁₆ $c^+c\,p^+p^+$	¹⁄₁₆ $+c\,++$
	⅛ $Cc\,Pp$	⅛ $c^+c\,p^+p$	⅛ $+c\,+p$
	¹⁄₁₆ $Cc\,pp$	¹⁄₁₆ $c^+c\,pp$	¹⁄₁₆ $+c\,pp$
	¹⁄₁₆ $CY\,PP$	¹⁄₁₆ $c^+Y\,p^+p^+$	¹⁄₁₆ $+Y\,++$
	⅛ $CY\,Pp$	⅛ $c^+Y\,p^+p$	⅛ $+Y\,+p$
	¹⁄₁₆ $CY\,pp$	¹⁄₁₆ $c^+Y\,pp$	¹⁄₁₆ $+Y\,pp$
	¹⁄₁₆ $cY\,PP$	¹⁄₁₆ $cY\,p^+p^+$	¹⁄₁₆ $cY\,++$
	⅛ $cY\,Pp$	⅛ $cY\,p^+p$	⅛ $cY\,+p$
	¹⁄₁₆ $cY\,pp$	¹⁄₁₆ $cY\,pp$	¹⁄₁₆ $cY\,pp$

(continued)

	Mendelian Notation	Drosophila Notation; Long Version	Drosophila Notation; Short Version
F$_1$ phenotypes:	$\frac{3}{8}$ normal females	$\frac{1}{16}$ phenylketonuria males	
	$\frac{1}{8}$ phenylketonuria females	$\frac{3}{16}$ color-blind males	
	$\frac{3}{16}$ normal males	$\frac{1}{16}$ color-blind phenylketonuria males	

c.

	Mendelian Notation	Drosophila Notation; Long Version	Drosophila Notation; Short Version
Allele symbols:	normal eyes = W	normal eyes = w^+	normal eyes = +
	white eyes = w	white eyes = w	white eyes = w
	normal wings = c	normal wings = C^+	normal wings = +
	curled wings = C	curled wings = C	curled wings = C
	normal bristles = s	normal bristles = S^+	normal bristles = +
	stubble bristles = S	stubble bristles = S	stubble bristles = S
Cross:	$ww\ Cc\ ss$	$ww\ CC^+\ S^+S^+$	$ww\ C++$
	$\times$	$\times$	$\times$
	$WY\ cc\ Ss$	$w^+Y\ C^+C^+\ SS^+$	$+Y\ ++\ S+$
F$_1$ genotypes:	$\frac{1}{8}\ Ww\ Cc\ Ss$	$\frac{1}{8}\ w^+w\ CC^+\ SS^+$	$\frac{1}{8}\ +w\ C+\ S+$
	$\frac{1}{8}\ Ww\ Cc\ ss$	$\frac{1}{8}\ w^+w\ CC^+\ S^+S^+$	$\frac{1}{8}\ +w\ C+++$
	$\frac{1}{8}\ Ww\ cc\ Ss$	$\frac{1}{8}\ w^+w\ C^+C^+\ SS^+$	$\frac{1}{8}\ +w\ ++S+$
	$\frac{1}{8}\ Ww\ Cc\ ss$	$\frac{1}{8}\ w^+w\ C^+C^+\ S^+S^+$	$\frac{1}{8}\ +w\ ++++$
	$\frac{1}{8}\ wY\ Cc\ Ss$	$\frac{1}{8}\ wY\ CC^+\ SS^+$	$\frac{1}{8}\ wY\ C+\ S+$
	$\frac{1}{8}\ wY\ Cc\ ss$	$\frac{1}{8}\ wY\ CC^+\ S^+S^+$	$\frac{1}{8}\ wY\ C+++$
	$\frac{1}{8}\ wY\ cc\ Ss$	$\frac{1}{8}\ wY\ C^+C^+\ SS^+$	$\frac{1}{8}\ wY\ ++S+$
	$\frac{1}{8}\ wY\ cc\ ss$	$\frac{1}{8}\ wY\ C^+C^+\ S^+S^+$	$\frac{1}{8}\ wY\ ++++$
F$_1$ phenotypes:	$\frac{1}{8}$ stubble, curled females	$\frac{1}{8}$ white, stubble, curled males	
	$\frac{1}{8}$ curled females	$\frac{1}{8}$ white, curled males	
	$\frac{1}{8}$ stubble females	$\frac{1}{8}$ white, stubble males	
	$\frac{1}{8}$ normal females	$\frac{1}{8}$ white males	

d.

	Mendelian Notation	Drosophila Notation; Long Version	Drosophila Notation; Short Version
Allele symbols:	white eye = w	white eye = w	white eye = w
	red eye = W	red eye = w^+	red eye = +
	black body = b	black body = b	black body = b
	grey body = B	grey body = b^+	grey body = +
	normal bristles = S	normal bristles = s^+	normal bristles = +
	tiny bristles = s	tiny bristles = s	tiny bristles = s
	normal size eye = e	normal size eye = E^+	normal size eye = +
	reduced size eye = E	reduced size eye = E	reduced size eye = E

	Mendelian Notation	Drosophila Notation; Long Version	Drosophila Notation; Short Version
Cross:	$ww\ bb\ ss\ ee$	$ww\ bb\ ss\ E^+E^+$	$ww\ bb\ ss\ ++$
	$\times$	$\times$	$\times$
	$WY\ BB\ SS\ EE$	$w^+Y\ b^+b^+\ s^+s^+\ EE$	$+Y ++++EE$
F_1 genotypes:	$\frac{1}{2}\ Ww\ Bb\ Ss\ Ee$	$\frac{1}{2}\ w^+w\ b^+b\ s^+s\ EE^+$	$\frac{1}{2}\ +w\ +b\ +s\ +E$
	$\frac{1}{2}\ wY\ Bb\ Ss\ Ee$	$\frac{1}{2}\ wY\ b^+b\ s^+s\ EE^+$	$\frac{1}{2}\ wY\ +b\ +s\ +E$
F_1 phenotypes:	$\frac{1}{2}$ eyeless females,	$\frac{1}{2}$ white, eyeless males	

3.22 In *Drosophila*, white eyes are a sex-linked character. The mutant allele for white eyes (w) is recessive to the wild-type allele for brick-red eye color (w^+). A white-eyed female is crossed with a red-eyed male. An F_1 female from this cross is mated with her father, and an F_1 male is mated with his mother. What will be the eye color of the offspring of these last two crosses?

Answer: The initial cross is $ww \times w^+Y$, so that the F_1 females are ww^+ and the F_1 males are wY. The second set of crosses are therefore $w^+w \times w^+Y$ and $wY \times ww$. The former will give all brick-red females (w^+-) and half white (wY) and half brick-red (w^+Y) males. The latter will give only white-eyed males and females (wY and ww).

3.23 One form of color blindness in humans is caused by a sex-linked recessive mutant gene (c). A woman with normal color vision (c^+) and whose father was color-blind marries a man of normal vision whose father was also color-blind. What proportion of their offspring will be color-blind? (Give your answer separately for males and females.)

Answer: Since fathers always give their X chromosome to their daughters, the woman must be heterozygous for the color-blind trait and is c^+c. As her husband received his X chromosome from his mother and has normal color vision, he is c^+Y. The cross is therefore $c^+c \times c^+Y$. All daughters will receive the paternal X bearing the c^+ allele and have normal color vision. As sons will receive the maternal X half will be cY and be color-blind and half will be c^+Y and have normal color vision.

3.24 In humans, red-green color blindness is recessive and X-linked, whereas albinism is recessive and autosomal. What types of children can be produced as the result of marriages between two homozygous parents—a normal-visioned albino woman and a color-blind, normally pigmented man?

Answer: Let c and c^+ be the color-blind and normal vision alleles, respectively, and let a and a^+ be the albino and normal pigmentation alleles, respectively. Then the cross can be represented as $c^+c^+\ aa \times cY\ a^+a^+$. As all the offspring will be a^+a, all will have normal pigmentation. The offspring will be either c^+c or c^+Y and have normal color vision. The daughters will, however, be carriers for the color-blind trait.

3.25 In *Drosophila*, vestigial (partially formed) wings (vg) are recessive to normal long wings (vg^+), and the gene for this trait is autosomal. The gene for the white-eye trait is on the

X chromosome. Suppose a homozygous white-eyed, long-winged female fly is crossed with a homozygous red-eyed, vestigial-winged male.

a. What will be the genotypes and phenotypes of the F_1 flies?

b. What will be the genotypes and phenotypes of the F_2 flies?

c. What will be the genotypes and phenotypes of the offspring of a cross of the F_1 flies back to each parent?

Answer:

a. The initial cross is $ww\ vg^+vg^+ \times w^+Y\ vgvg$. The F_1 consists of $wY\ vg^+vg$ (white, normal-winged) males and $ww^+\ vg^+vg$ (red, normal-winged) females.

b. The F_2 would be produced by crossing $wY\ vg^+vg$ males and $w^+w\ vg^+vg$ females. In both the male and the female progeny, ⅛ will be white and vestigial, ⅛ will be red and vestigial, ⅜ will be white and normal winged, and ⅜ will be red and normal winged.

c. If the F_1 males are crossed back to the female parent, the cross is $wY\ vg^+vg \times ww\ vg^+vg^+$. All the progeny would be white and normal winged. If the F_1 females are crossed back to the male parent, the cross is $ww^+\ vg^+vg \times w^+Y\ vgvg$. Among the male progeny, there would be ¼ white, vestigial; ¼ red, vestigial; ¼ white, normal winged; and ¼ red, normal winged. Among the female progeny, half would be red and normal winged and half would be red and vestigial.

3.26   In *Drosophila*, two red-eyed, long-winged flies are bred together and produce the offspring listed in the following table:

	Females	Males
red-eyed, long-winged	¾	⅜
red-eyed, vestigial-winged	¼	⅛
white-eyed, long-winged	—	⅜
white-eyed, vestigial-winged	—	⅛

What are the genotypes of the parents?

Answer: From problem 3.25, we know that w is X-linked while vg is autosomal. This can also be determined by considering just one trait at a time and examining the frequency of progeny phenotypes. The ratio of long-winged to vestigial-winged progeny is 3:1 (¾ to ¼) in both sexes, while the ratio of red-eyed to white-eyed progeny is all to none in females and 1:1 in males. This is consistent with vg being autosomal and w being X-linked. The 3:1 ratio of long-winged to vestigial-winged progeny indicates that each parent was heterozygous at the vg locus. Since both parents had red eyes, both had (at least) one w^+ allele. Since half of the sons are white eyed, the mother must have been heterozygous. Therefore, the parents were $w^+w\ vg^+vg$ and $w^+Y\ vg^+vg$.

3.27   In chickens, a dominant sex-linked gene (*B*) produces barred feathers, and the recessive allele (*b*), when homozygous, produces nonbarred (solid-color) feathers. Suppose a nonbarred cock is crossed with a barred hen.

a. What will be the appearance of the F_1 birds?

b. If an F_1 female is mated with her father, what will be the appearance of the offspring?

c. If an F_1 male is mated with his mother, what will be the appearance of the offspring?

Answer:

a. In poultry, sex type is determined by a ZZ (male) and ZW (female) system. The cross can be depicted as *bb* (nonbarred cock) × *BW* (barred hen). The F₁ progeny will be *bW* (nonbarred) hens and *Bb* (barred) cocks.

b. The cross can be represented as *bW* × *bb*. All the progeny will be nonbarred.

c. The cross can be represented as *Bb* × *BW*. The progeny will be ½ barred cocks (¼ *BB*, ¼ *Bb*), ¼ barred hens (*BW*), and ¼ nonbarred hens (*bW*).

3.28 A man (A) suffering from defective tooth enamel, which results in brown-colored teeth, marries a normal woman. All their daughters have brown teeth, but the sons are normal. The sons of man A marry normal women, and all their children are normal. The daughters of man A marry normal men, and 50 percent of their children have brown teeth. Explain these facts genetically.

Answer: Notice that the trait is transmitted from the father to his daughters, indicating crisscross inheritance. This is typical of an X-linked trait. Since the man marries a normal woman and all of their daughters have the trait, the trait must be dominant. The man's X chromosome bearing the defective tooth enamel allele is inherited by all of his daughters and none of his sons. All of his daughters would therefore have defective tooth enamel and be heterozygous for the defective enamel allele. These daughters would transmit the defective enamel allele half of the time, giving rise to 50 percent of their children being affected.

3.29 In humans, differences in the ability to taste phenylthiourea are due to a pair of autosomal alleles. Inability to taste is recessive to ability to taste. A child who is a nontaster is born to a couple who can both taste the substance. What is the probability that their next child will be a taster?

Answer: Since the inability to taste the substance is recessive, the nontaster child must be homozygous for the recessive allele, and each of his parents must have given the child a recessive allele. Since both parents can taste, they must also bear a dominant allele. Let *T* represent the dominant (taster) allele, and *t* represent the recessive (nontaster) allele [Note: Mendelian notation is used here for convenience, but also because there is no value in assigning a normal (+) and abnormal allele.] Then the cross can be written as *Tt* × *Tt*. The chance that their next child will be a taster is the chance that the child will be *TT* or *Tt*, or ¾.

3.30 Cystic fibrosis is inherited as an autosomal recessive. Two parents without cystic fibrosis have two children with cystic fibrosis and three children without. The parents come to you for genetic counseling.

a. What is the probability that their next child will have cystic fibrosis?

b. Their unaffected children are concerned about being heterozygous. What is the probability that a given unaffected child in the family is heterozygous?

Answer:

a. Since the disease is autosomal recessive, and unaffected parents have affected offspring, both parents must be heterozygous. Let c^+ represent the normal allele and *c* represent the disease allele. Then the parental cross can be represented as $c^+c \times c^+c$, and there is a ¼ chance that any conception will produce a *cc* (affected) child.

b. If a child is not affected, the child is either c^+c^+ ($P = \frac{1}{3}$) or c^+c ($P = \frac{2}{3}$). Thus, there is a $\frac{2}{3}$ chance that a nonaffected child is heterozygous.

3.31   Huntington disease is a human disease inherited as a Mendelian autosomal dominant. The disease results in choreic (uncontrolled) movements, progressive mental deterioration, and, eventually, death. In carriers of the trait, the disease appears between 15 and 65 years of age. American folksinger Woody Guthrie died of Huntington disease, as did just one of his parents. Marjorie Mazia, Woody's wife, had no history of this disease in her family. The Guthries had three children. What is the probability that a particular Guthrie child will die of Huntington disease?

Answer: Let H represent the disease allele, and let h represent the normal allele. Since just one of Woody's parents died of the disease, we may assume that only one parent had the disease allele. Thus, Woody must have been Hh. His children are the progeny of a cross that can be represented as $Hh \times hh$. Each will have a 50 percent chance of receiving the H allele.

3.32   Suppose gene A is on the X chromosome, and genes B, C, and D are on three different autosomes. Thus, $A-$ signifies the dominant phenotype in the male or female. An equivalent situation holds for $B-$, $C-$, and $D-$. The cross $AA\,BB\,CC\,DD\,♀ \times aY\,bb\,cc\,dd\,♂$ is made.
   a. What is the probability of obtaining an $A-$ individual in the F_1 progeny?
   b. What is the probability of obtaining an a male in the F_1 progeny?
   c. What is the probability of obtaining an $A-B-C-D-$ female in the F_1 progeny?
   d. How many different F_1 genotypes will there be?
   e. What proportion of F_2 individuals will be heterozygous for the four genes?
   f. Determine the probabilities of obtaining each of the following types in the F_2 individuals: (1) $A-bb\,CC\,dd$ (female), (2) $aY\,BB\,Cc\,Dd$ (male), (3) $AY\,bb\,CC\,dd$ (male), (4) $aa\,bb\,Cc\,Dd$ (female).

Answer:
   a. Since only a single trait is being followed, consider just part of the cross: $AA \times aY$. The progeny will be either AY or Aa, and all are $A-$. Thus, the chance of obtaining an $A-$ individual in the F_1 is 1.
   b. As shown in (a), there is no chance ($P = 0$) of obtaining an aY individual in the F_1.
   c. The F_1 progeny will be $A-Bb\,Cc\,Dd$. Half will be female, so $P = \frac{1}{2}$.
   d. Two, $Aa\,Bb\,Cc\,Dd$ (females) and $AY\,Bb\,Cc\,Dd$ (males).
   e. For the X chromosome trait, the F_1 cross is $AY \times Aa$. Half of the female offspring ($\frac{1}{4}$ of the total) will be heterozygous Aa individuals. For each of the autosomal traits, $\frac{1}{2}$ of the offspring will be heterozygous (e.g., $Bb \times Bb$ gives $\frac{1}{2}$ Bb individuals). Thus, the chance that an F_2 individual will be heterozygous at all four traits is $\frac{1}{4} \times \frac{1}{2} \times \frac{1}{2} \times \frac{1}{2} = \frac{1}{32}$.
   f. Before determining the probabilities, consider that at any autosomal gene, there is a $\frac{1}{4}$ chance of obtaining either type of homozygote (e.g., BB, bb) and a $\frac{1}{2}$ chance of obtaining a heterozygote. At the A gene, the cross is $AY \times Aa$, so there is a $\frac{1}{4}$ chance of obtaining an AY male, a $\frac{1}{4}$ chance of obtaining an aY male, a $\frac{1}{4}$ chance of obtaining an Aa female, and a $\frac{1}{4}$ chance of obtaining an AA female (there will be a $\frac{1}{2}$ chance of

obtaining an *A*–female). Then the chance of obtaining (1) an *A*–*bb CC dd* (female) is $P = (\frac{1}{2} \times \frac{1}{4} \times \frac{1}{4} \times \frac{1}{4}) = \frac{1}{128}$; (2) an *aY BB Cc Dd* (male) is $P = (\frac{1}{4} \times \frac{1}{4} \times \frac{1}{2} \times \frac{1}{2}) = \frac{1}{64}$; (3) an *AY bb CC dd* (male) is $P = (\frac{1}{4} \times \frac{1}{4} \times \frac{1}{4} \times \frac{1}{4}) = \frac{1}{256}$; and (4) an *aa bb Cc Dd* (female) is $P = (0 \times \frac{1}{4} \times \frac{1}{2} \times \frac{1}{2}) = 0$.

3.33   As a famous mad scientist, you have cleverly devised a method to isolate *Drosophila* ova that have undergone primary nondisjunction of the sex chromosomes. In one experiment, you used females homozygous for the sex-linked recessive mutation causing white eyes (*w*) as your source of nondisjunction ova. The ova were collected and fertilized with sperm from red-eyed males. The progeny of this "engineered" cross were then backcrossed separately to the two parental strains. What classes of progeny (genotype and phenotype) would you expect to result from these backcrosses? (The genotype of the original parents may be denoted as *ww* for the females and w^+/Y for the males.)

Answer: First consider what kinds of animals will be produced in the "engineered" cross. Primary nondisjunction of sex chromosomes in the meiosis of a female will result in eggs that have either zero or two X chromosomes. Thus, the eggs will either lack the *w* allele or be *ww*. Sperm from a red-eyed male will either bear a Y chromosome or bear the w^+ allele. Upon fertilization then, four classes of embryos will be obtained: Y only, w^+O, *ww*Y, and w^+ww. Animals with either zero or three X chromosomes are inviable, and so only w^+ (red, XO males) and *ww*Y (white, XXY females) animals will be obtained.

Now consider the backcrosses. The "engineered" red males will be sterile, as they lack a Y chromosome needed for male fertility. Hence, the only backcross that will give progeny is the mating between the *ww*Y females and the w^+Y males. The sex chromosome constitution of the female's gametes is dependent on the pairing and disjunction of the two X chromosomes and the Y chromosome during meiosis. If there is normal pairing and disjunction of the two X chromosomes and the Y chromosome assorts independently of them, XY (*w*Y) and X (*w*) eggs will be produced. If, as described in the text, less frequent secondary nondisjunction occurs, XX-bearing (*ww*) and Y-bearing eggs will be produced. The zygotes produced when these eggs are fertilized by sperm from a w^+Y male are shown in the following Punnett square:

		Gametes of *ww*Y Female			
		Normal Disjunction		Secondary Nondisjunction	
		*w*Y	*w*	*ww*	Y
Gametes of w^+Y Male	w^+	w^+wY (XXY) red female	w^+w (XX) red female	w^+ww (XXX) inviable	w^+Y (XY) red male
	Y	*w*YY (XYY) white male	*w*Y (XY) white male	*ww*Y (XXY) white female	YY inviable

3.34   In *Drosophila*, the bobbed gene (bb^+) is located on the X chromosome: *bb* mutants have shorter, thicker bristles than wild-type flies. Unlike most X-linked genes, however, a bobbed gene is also present on the Y chromosome. The mutant allele *bb* is recessive to bb^+. If a wild-type F$_1$ female that resulted from primary nondisjunction in oogenesis in a cross of a bobbed female with a wild-type male is mated to a bobbed male, what will be the phenotypes and their frequencies in the offspring? List males and females separately in your answer. (Hint: Refer to the chapter for information about the frequency of nondisjunction in *Drosophila*; see p. 59.)

Answer: The cross that gave rise to the wild-type F_1 female can be denoted as $X^{bb}X^{bb} \times X^{bb+}Y^{bb+}$. Nondisjunction in meiosis of the $X^{bb}X^{bb}$ female would produce an $X^{bb}X^{bb}$ egg. Fertilization by a Y^{bb+}-bearing sperm would produce a bb^+ ($X^{bb}X^{bb}Y^{bb+}$) female. We want to know what progeny result when this female is mated to a bb ($X^{bb}Y^{bb}$) male. Normal disjunction (which occurs about 96 percent of the time) in this female will produce $X^{bb}Y^{bb+}$ and X^{bb}-bearing eggs. If these eggs are fertilized by sperm from a bb male (i.e., either X^{bb} or Y^{bb}), half bb^+ ($X^{bb}X^{bb}Y^{bb+}$ and $X^{bb}Y^{bb+}Y^{bb}$) males and females and half bb ($X^{bb}X^{bb}$ and $X^{bb}Y^{bb}$) males and females will be produced. If secondary nondisjunction occurs (about 4 percent of the time), then $X^{bb}X^{bb}$ and Y^{bb+} eggs will be produced. When these eggs are fertilized by sperm from a bb male, the only viable progeny will be bb females ($X^{bb}X^{bb}Y^{bb}$) and bb^+ males ($X^{bb}Y^{bb+}$). This is diagrammed as follows:

		Gametes of $X^{bb}X^{bb}Y^{bb+}$ Female			
		Normal Disjunction—96%		Secondary Nondisjunction—4%	
		48% $X^{bb}Y^{bb+}$	**48% X^{bb}**	**2% $X^{bb}X^{bb}$**	**2% Y^{bb+}**
Gametes of	50% X^{bb}	$X^{bb}X^{bb}Y^{bb+}$ bb^+ female 24%	$X^{bb}X^{bb}$ bb female 24%	$X^{bb}X^{bb}X^{bb}$ inviable 1%	$X^{bb}Y^{bb+}$ bb^+ male 1%
$X^{bb}Y^{bb}$ Male	50% Y^{bb}	$X^{bb}Y^{bb+}Y^{bb}$ bb^+ male 24%	$X^{bb}Y^{bb}$ bb male 24%	$X^{bb}X^{bb}Y^{bb}$ bb female 1%	YY inviable 1%

Among the viable progeny there will be $^{24}/_{98}$ = 24.5% normal females, $^{24}/_{98}$ = 25.5% bobbed females, $^{24}/_{98}$ = 25.5% normal males, and $^{24}/_{98}$ = 24.5% bobbed males.

3.35   An individual with Turner syndrome would be expected to have how many Barr bodies in the majority of cells?

Answer: None. Turner syndrome individuals are XO. They have only one X, so no X is inactivated.

3.36   An XXY individual with Klinefelter syndrome would be expected to have how many Barr bodies in the majority of cells?

Answer: All but one X chromosome is inactivated. An XXY individual, having two X chromosomes, has one inactivated.

3.37   In human genetics, pedigrees are used to analyze inheritance patterns. Females are represented by a circle, males by a square. The figure that follows presents three 2-generation family pedigrees for a trait in humans. Normal individuals are represented by unshaded symbols, people with the trait by shaded symbols. For each pedigree (A, B, and C), state (by answering "yes" or "no" in the appropriate blank space) whether transmission of the trait can be accounted for on the basis of each of the listed simple modes of inheritance:

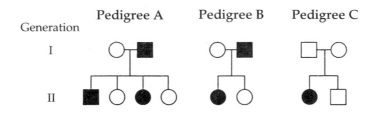

	Pedigree A	Pedigree B	Pedigree C
Autosomal recessive			
Autosomal dominant			
X-linked recessive			
X-linked dominant			

Answer: This problem raises the issue that the precise mode of inheritance of a trait often cannot be determined when a pedigree is small and the frequency of the trait in a population is unknown. For example, Pedigree A could easily fit an autosomal dominant trait (*AA* and *Aa* = affected): The affected father would be heterozygous for the trait (*Aa*), the mother would be unaffected (*aa*), and half of their offspring would be affected. However, if the trait were autosomal recessive (i.e., *aa* = affected individuals) and the mother were heterozygous (*Aa*) and the father homozygous (*aa*), half of the offspring would still be affected. One could also fit the pedigree to an X-linked recessive trait: The mother would be heterozygous *Aa*, the father hemizygous *a*Y, and half of the progeny would be affected (either *aa* or *a*Y). An X-linked dominant trait would not fit the pedigree, as it would require all the daughters of the affected father to be affected (as they all receive their father's X), and not allow a son to be affected (as he does not receive his father's X). Pedigrees B and C can be solved by similar analytical reasoning.

	Pedigree A	Pedigree B	Pedigree C
Autosomal recessive	Yes	Yes	Yes
Autosomal dominant	Yes	Yes	No
X-linked recessive	Yes	Yes	No
X-linked dominant	No	No	No

3.38   Shaded symbols in the following pedigree represent a trait.

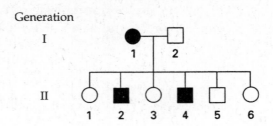

Which of the progeny eliminate X-linked recessiveness as a mode of inheritance for the trait?

a. I-1 and I-2

b. II-4

c. II-5

d. II-2 and II-4

Answer: If the trait were X-linked recessive, then II-5 should be affected, as he would receive his mother's X chromosome. Therefore, the correct answer is (c), II-5.

3.39   When constructing human pedigrees, geneticists often refer to particular individuals by a number. The generations are labeled with Roman numerals, the individuals in each generation with Arabic numerals. For example, in the pedigree in the following figure, the female with the asterisk is I-2:

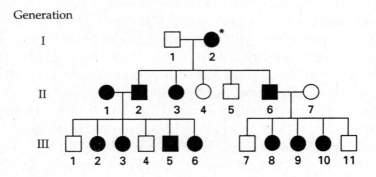

Use this means to designate specific individuals in the pedigree. Determine the probable inheritance mode for the trait shown in the affected individuals (the shaded symbols) by answering the following questions (assume that the condition is caused by a single gene):

a. Y-linked inheritance can be excluded at a glance. What two other mechanisms of inheritance can be definitely excluded? Why can these be excluded?

b. Of the remaining mechanisms of inheritance, which is the most likely? Why?

Answer:

a. Y-linked inheritance can be excluded because females are affected. X-linked recessive inheritance can also be excluded, as an affected mother (I-2) has a normal son (II-5). Autosomal recessive inheritance can also be excluded, as in such a case two affected parents, such as II-1 and II-2, could only have affected offspring, which they do not.

b. The two remaining mechanisms of inheritance are X-linked dominant and autosomal dominant. Genotypes can be written to satisfy both mechanisms of inheritance. Of these two, X-linked dominant inheritance may be more likely, as II-6 and II-7 have only affected daughters, indicating crisscross inheritance. If the trait were autosomal dominant, one would expect half of the daughters and half of the sons to be affected.

3.40   A three-generation pedigree for a particular human trait is shown in the following figure:

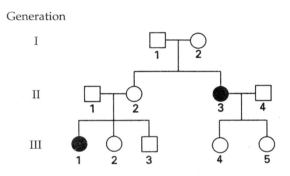

a. What is the mechanism of inheritance for the trait?
b. Which persons in the pedigree are known to be heterozygous for the trait?
c. What is the probability that III-2 is a carrier (heterozygous)?
d. If III-3 and III-4 marry, what is the probability that their first child will have the trait?

Answer:

a. The trait must be recessive, as unaffected parents (I-1 and I-2, II-1 and II-2) have affected offspring. It also must be autosomal, as affected female daughters are born to unaffected fathers.

b. Given the answer to (a), individuals I-1 and I-2, as well as individuals II-1 and II-2, must be heterozygous. Also, individuals III-4 and III-5 must be heterozygous, as they have an affected, homozygous mother.

c. One can denote the parents of III-2 as $Aa \times Aa$. We know individual III-2 is $A-$, so she is either AA ($P = \frac{1}{3}$) or Aa ($P = \frac{2}{3}$).

d. By the same reasoning as described in (c), individual III-3 has a $\frac{2}{3}$ chance of being Aa. The chance that he will contribute an a allele to his offspring is thus $\frac{2}{3} \times \frac{1}{2} = \frac{1}{3}$. The chance that individual III-4 will contribute an a allele to her offspring is $\frac{1}{2}$, as she is heterozygous. Thus, the chance that progeny of III-3 and III-4 will be affected is $\frac{1}{3} \times \frac{1}{2} = \frac{1}{6}$.

3.41   For each of the more complex pedigrees shown in Figure 3.B, determine the probable mechanism of inheritance: autosomal recessive, autosomal dominant, X-linked recessive, X-linked dominant, or Y-linked.

Answer: In complex pedigrees, an appropriate strategy is to ask initially whether the trait is Y-linked, then if it is dominant or recessive, and finally whether it is autosomal or X-linked. Proceeding logically through these choices helps limit the options. Before concluding, it is important to verify that the phenotype of every member of the pedigree fits

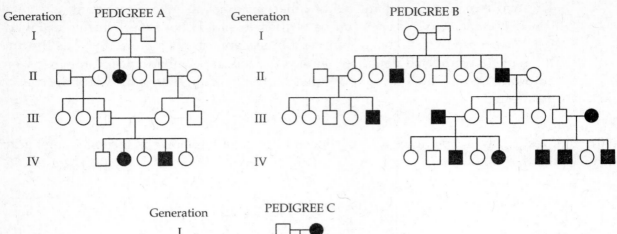

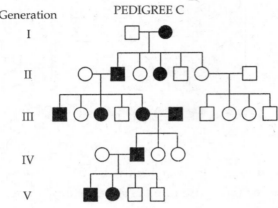

Figure 3.B

the inferred inheritance pattern. Often, there may be only a single individual that can be used to distinguish one type of inheritance pattern from another.

Pedigree A: Clearly, the trait is not Y-linked, as males are not the sole transmitters of the trait. The most striking aspect of this pedigree is that the trait skips generations (i.e., unaffected parents have affected offspring). This indicates that the trait must be recessive. Now, one is left to determine if it is autosomal or X-linked. It does not fit an X-linked pedigree, as two affected females have normal (albeit heterozygous) fathers. For X-linked recessive pedigrees, both parents must contribute an abnormal allele, so that affected daughters must have affected fathers. The trait is therefore autosomal recessive. (Check by assigning appropriate genotypes to each pedigree member.)

Pedigree B: This pedigree also shows that the trait is not Y-linked, and must be recessive, as twice there are unaffected parents who have affected offspring. The preponderance of affected males suggests that it might be X-linked. To determine this for certain, try to assign genotypes to see if such an inheritance mode fits. It does. Could the trait be autosomal recessive? This would be a possibility if the trait were relatively common, as one would have to suppose that the individuals who married into the family in generation II were heterozygotes. This may be the case, as several homozygotes do marry into the family in generation III. What is the best guess then? X-linked recessive is the most likely, partly because of the frequency of affected males, but mostly because the homozygous female in generation III has three affected male offspring.

Pedigree C: The trait is expressed in every generation, making it a good candidate for a dominant trait. It is expressed equally by males and females, making it a good candidate for an autosomal trait. Indeed, it must be autosomal, for two reasons: First, individuals III-5 and III-6 have unaffected daughters, which would not be possible if the trait were X-linked and dominant; and second, individual IV-2, a male, has an affected male offspring, which would not be possible if the trait were X-linked. Thus, the trait is autosomal dominant.

In problems 3.42–3.44, select the correct answer.

3.42   If a genetic disease is inherited on the basis of an autosomal dominant gene, which of the statements that follow are true, and which are false? Why?
   a.   Affected fathers have only affected children.
   b.   Affected mothers never have affected sons.
   c.   If both parents are affected, all of their offspring have the disease.
   d.   If a child has the disease, one of his or her grandparents also had the disease.

Answer: Answer (a) is untrue because if the affected father is heterozygous, only half of his offspring should be affected. Answer (b) is untrue because if the mother is heterozygous, half of her offspring, regardless of sex type, should be affected. Answer (c) is untrue because if the parents are each heterozygous, ¼ of their offspring should be homozygous recessive and normal. Answer (d) is the most likely. Note, however, that if the mutation was new in either the child or his or her parents, his or her grandparent could have been unaffected.

3.43   A genetic disease is inherited as an autosomal recessive. Which of the following statements are true, and which are false? Why?
   a.   Two affected individuals never have an unaffected child.
   b.   Two affected individuals have affected male offspring, but no affected female children.
   c.   If a child has the disease, one of his or her grandparents had it.
   d.   In a marriage between an affected individual and an unaffected one, all the children are unaffected.

Answer: Answer (a) is true because two affected individuals will always have affected children (*aa* × *aa* can give only *aa* offspring). Answer (b) is untrue, as an autosomal trait is inherited independent of sex type. Answer (c) need not be true, as the trait could be masked by normal dominant alleles through many generations before two heterozygotes marry and produce affected, homozygous offspring. Answer (d) could also be true. If the trait is rare, then it is likely that an unaffected individual marrying into the pedigree is homozygous for a normal allele. Since the trait is recessive, and the children receive the dominant, normal allele from the unaffected parent, the children will be normal. Answer (d) would not be true if the unaffected individual was heterozygous. In this case, half of the children would be affected.

3.44   Which of the following statements is not true for a disease that is inherited as a rare X-linked dominant trait?
   a.   All daughters of an affected male will inherit the disease.
   b.   Sons will inherit the disease only if their mothers have it.
   c.   Both affected males and affected females will pass the trait to half the children.
   d.   Daughters will inherit the disease only if their fathers have it.

Answer: The only untrue statement is (d). Since daughters receive an X chromosome from each of their parents, they can inherit an X-linked dominant disease from either their mother or father.

3.45   Women who were known to be carriers of the X-linked recessive hemophilia gene were studied to determine the amount of time required for blood to clot. It was found that the time required for clotting was extremely variable from individual to individual. The values obtained ranged from normal clotting time at one extreme to clinical hemophilia at the other. What is the most probable explanation for these findings?

Answer: Since hemophilia is an X-linked trait, the most likely explanation is that random inactivation of X chromosomes (lyonization) produces individuals with different proportions of cells with a functioning allele. Normal clotting times would be expected in females with a functional h^+ allele, i.e., females whose h-bearing X chromosome was very frequently inactivated. Clinical hemophilia would be expected in females without a functional h^+ allele, i.e., females whose h^+-bearing X chromosome was very frequently inactivated. Intermediate clotting times would be expected to be proportional to the amount of h^+ function, which is related to the frequency of inactivation of the h^+-bearing X chromosome.

3.46   Hurler syndrome is a genetically transmitted disorder of mucopolysaccharide metabolism resulting in short stature, mental retardation, and various bony malformations. Two specific types are described with extensive pedigrees in the medical genetics literature:

Type I: recessive autosomal

Type II: recessive X linked

You are a consultant in a hospital ward with several patients with Hurler syndrome who have asked you for advice about their relatives' offspring. Being aware that both types are extremely rare and that afflicted individuals almost never reproduce, what counsel would you give to a woman with Type I Hurler syndrome, whose normal brother's daughter is planning marriage, about the offspring of that proposed marriage? In your answer, state the probabilities that the offspring will be affected and whether male and female offspring have an equal probability of being affected.

Answer: Draw out the pedigree of the patient and try to assign genotypes to the relevant individuals. Let h represent the Hurler syndrome Type I allele. Then the patient is hh. Since we are told that Hurler syndrome patients virtually never reproduce, neither of her parents is expected to be hh. Still, in order for the patient to be hh, her parents must have both been Hh. (The cross was $Hh \times Hh$.) Since her brother is normal, he is $H-$, with a ⅔ chance of being Hh. The brother's daughter has a half chance of receiving either of the brother's (H or h) alleles. Thus, the chance that the brother's daughter has an h allele is ⅔ × ½ = ⅓. Since the trait is extremely rare, it is likely she will marry an HH individual, and have $H-$ children. Therefore there is no chance the brother's daughter will have affected children, as her husband most likely will provide a dominant, normal H allele. Since Type I is autosomal, there will be no sex differences.

4

EXTENSIONS OF MENDELIAN GENETIC PRINCIPLES

THINKING ANALYTICALLY

While the reasoning methods used in this chapter are not very different from those used in the earlier chapters on Mendelian genetics, there is much more information to consider. Phenotypes are now affected by multiple alleles, gene interactions, and environmental influences. To understand how all of these affect or contribute to a phenotype, *analyze one at a time.* First focus on understanding what each factor contributes to a phenotype. Then consider how several jointly affect a phenotype.

Mutations at different genes can produce the same phenotypes, so it is important to identify how many genes are affected in a set of mutations. This can be determined by using complementation tests. In a complementation test, two mutations with the same phenotype are crossed and the phenotype of the progeny is observed. If the two mutations affect different

REVIEW OF KEY TERMS, SYMBOLS, AND CONCEPTS

In your own words, write a brief, precise definition of each term in the groups below. Check your definitions using the text. Then develop a concept map using the terms in each list.

1	2	3
mutant	dominance	penetrance
multiple alleles	epistasis, epistatic	expressivity
allelic series	hypostasis, hypostatic	population
diploid	dihybrid cross	individual
wild-type allele	dominant epistasis	phenocopy
complete dominance	recessive epistasis	phenocopying agent
complete recessiveness	duplicate dominant epistasis	hereditary trait
incomplete dominance	duplicate recessive epistasis	nature vs. nurture
partial dominance	duplicate gene interaction	norm of reaction
codominance	complementary gene action	sex-limited trait
essential gene	complementation test	sex-influenced trait
lethal allele	cis-trans test	sex-linked gene
recessive lethal allele	9:3:3:1	autosomal gene
dominant lethal allele	9:6:1	
ABO blood group	9:3:4	
glycosyltransferases	9:7	
MN blood group	12:3:1	
Bombay blood type	13:3	
blood-typing	9:6:1	
cellular antigen	15:1	
antibody		
agglutination		
universal donor, recipient		
haplosufficient		

genes, complementation occurs and the progeny are wild type. In this case, the progeny are heterozygous for each mutation and the normal allele at each gene ensures wild-type progeny. If the mutations affect the same gene, no normal allele is present, the mutations will not complement each other, and the progeny will have a mutant phenotype. Complementation tests are elegant and quick ways to organize sets of mutations by the genes they affect: Cross pairs of mutants and observe the phenotype of their progeny.

As geneticists discover additional factors that affect a phenotype, the phenotype is often described in different, more revealing ways. For example, when a disease is first observed, it may be described only in terms of its inheritance pattern—say, as a dominantly inherited trait. As more is learned about the disease, its phenotype will be reassessed in light of the new knowledge. Consequently, as more is understood, different labels will be applied to the disease phenotype. These labels will indicate the kind and degree of interaction that occurs within a set of alleles at one gene (e.g., the types of dominance relationships), between several alleles at more than one gene (e.g., the types of epistasis), and between genetic and environmental influences (e.g., the level of penetrance and/or expressivity). Although complex, these labels are very useful: They quickly convey a coherent, informative description of the phenotype.

The major message of this chapter is that phenotypes can be affected by many factors. These factors include the set of alleles an organism inherits as well as the environment(s) in which it develops and lives.

QUESTIONS FOR PRACTICE

Multiple Choice Questions

1. In clover leaves, chevron pattern (a light-colored triangular leaf pattern) is controlled by seven different alleles at a single gene. From this information alone, what can be said about this trait?
   a. Alleles at this gene show incomplete dominance.
   b. There is a multiple allelic series with seven alleles that controls chevron pattern.
   c. This gene shows epistasis.
   d. One allele at this gene must be completely dominant.

2. The phenotype associated with a wild-type allele of a particular gene is
   a. the phenotype that is always found in nature.
   b. a reference phenotype used by geneticists for comparison.
   c. a special mutant phenotype.
   d. the most common phenotype seen.

3. A phenocopy is different from a particular heritable mutation in that
   a. a phenocopy cannot have the same phenotype as the mutation.
   b. the biochemical anomaly in a phenocopy and a mutation must be different.
   c. the phenocopy is always dominant.
   d. a phenocopy is not a heritable trait.

4. Neurofibromatosis is a dominantly inherited disease that can show mild, moderate, or severe symptoms. Every individual that inherits the dominant allele shows at least mild symptoms. This means that the disease allele shows
   a. variable penetrance and complete expressivity.
   b. variable expressivity and complete penetrance.
   c. variable penetrance and variable expressivity.
   d. complete penetrance and complete expressivity.

5. Two alleles at the *white* locus in *Drosophila melanogaster* are $w^{apricot}$, which has orange-colored eyes, and w^{coral}, which has pink-colored eyes. This is an example of
   a. codominance.
   b. a multiple allelic series.
   c. penetrance.
   d. epistasis.

6. One difference between epistasis and dominance is that
   a. epistasis occurs between two different genes while dominance occurs between alleles at one gene.
   b. only epistasis is influenced by environmental interactions.

c. dominant traits are completely penetrant, while epistatic interactions may not be.

d. dominant traits may show variable penetrance, epistatic interactions may not.

For questions 7, 8, 9, and 10, choose the kind of inheritance that best explains the data. Explain your choices.

    a. recessive lethal allele

    b. dominant lethal allele

    c. incomplete dominance; two allelic pairs

    d. duplicate recessive epistasis; two allelic pairs

    e. dominant epistasis; two allelic pairs

7. When crossed, two pure-breeding varieties of white kernel corn yield progeny with purple kernels. When these purple progeny are selfed, a count of the kernels in the F_2 yields an average of 270 purple kernels : 210 white kernels per ear.

8. When pure-breeding wheat with red kernels is crossed to the pure-breeding white variety, the progeny have pink kernels. One hundred F_2 plants exhibit the following phenotypes: 6 red, 25 dark pink, 38 pink, 25 light pink, 6 white.

9. Yellow mice crossed to pure-breeding agouti mice produce a 50:50 ratio of yellow to agouti offspring. Crosses between yellow mice never produce 100 percent yellow progeny, but rather yield 36 yellow and 15 agouti offspring.

10. At age 52, a man begins to show the symptoms of Huntington disease. Like his mother before him, he eventually dies from the disease. His wife is normal and there is no history of the disease in his wife's family or his father's family. Two of his four children also die from the disease, albeit not until late in their lives.

11. Which one of the following statements is true about complementation?

    a. Two mutants that complement each other affect the same gene.

    b. In a complementation test, two mutants are crossed together and the progeny phenotype is assessed. If the progeny phenotype is normal, the two mutants complement each other.

    c. Complementation tests can be used to determine whether dominant mutations affect the same function.

    d. Alleles at the same gene will complement each other.

12. Six independently isolated mutations ($a-f$) all result in yellow tomatoes and are recessive to the normal red color. When different pairs of homozygous mutants are crossed, all produce plants bearing yellow tomatoes except for pairs involving mutation c. When cc plants are crossed to any of the other strains, progeny plants produce red tomatoes. Based on these observations, which one of the following statements is not true?

    a. The six mutations affect six different genes.

    b. Mutations $a, b, d, e,$ and f are alleles at the same gene.

    c. A complementation test was performed.

    d. Mutation c affects a different gene than do mutations $a, b, d, e,$ and f.

    e. Mutations $a, b, d, e,$ and f are members of a multiple allelic series at one gene.

Answers: 1b, 2b, 3d, 4b, 5b, 6a, 7d, 8c, 9a, 10b, 11b, 12a

Thought Questions

1. Distinguish between dominance and epistasis and between epistasis and multiple alleles.

2. How would you distinguish whether a complex mutant phenotype was caused by one gene or alleles at several genes? (Hint: Review independent assortment.)

3. Tay-Sachs disease is fatal due to a lack of the enzyme hexoaminidase A. Phenotypically normal heterozygotes have about half as much enzyme activity as homozygotes for the normal allele. In what sense do alleles at the gene affected in Tay-Sachs disease show incomplete dominance, codominance, and recessive lethality?

4. A sex-linked gene controls coat color in cats. One allele determines black coat color while another allele determines orange coat color. A separate set of genes determines where color is produced (i.e., where white patches are). Use this information to explain why most calico (black-, orange-, and white-patched) cats are female. How do rare calico males arise? Do you expect them to be fertile?

5. In a very large natural population, could you ever presume to know all the allelic variants in a multiple allelic series? (Hint: Could new variants appear at any time?)

6. Devise a situation in which a multiple allelic series shows elements of codominance as well as dominant lethality.

7. Could a lethal allele, either dominant or recessive, be incompletely penetrant? (Hint: Consider small variations in the timing of events during development.)

8. As you will discover in Chapter 17, single nucleotide polymorphisms (SNPs) are very common, about one per several hundred base pairs of DNA. What does this tell you about the number of alleles that exist at one gene in a population? Do you expect the allelic differences to always show clear patterns of dominance or recessiveness relative to a "wild-type" allele?

9. Some traits are influenced by many genes as well as by the environment. How might you identify genes influencing such a trait? How would you then analyze the epistatic relationships between the genes influencing such a trait?

10. What different explanations are there for a trait to be incompletely penetrant? What approaches can be devised to test these explanations?

Thought Questions for Media

After reviewing the media on the *iGenetics* CD-ROM, try to answer these questions.

1. Based on viewing the animation, give an example where neither of two alleles at a gene is dominant or recessive. Using the alleles of your example, write the genotypes of a cross whose progeny show identical genotypic and phenotypic ratios. What type of dominance relationship is shown by these two alleles?

2. Correct the following false statement: "In incomplete dominance, a heterozygous individual exhibits the phenotype of both homozygotes."

3. Based on viewing the animation, give an example of a situation where two dominant alleles are fully expressed at the same time. Using the alleles of your example, write the genotypes of a cross whose progeny show identical genotypic and phenotypic ratios. What type of dominance relationship is shown by these two alleles?

4. Suppose you are able to show that a cross of two heterozygotes gives progeny in 1:2:1 genotypic and phenotypic ratios. What additional information do you need to distinguish between incomplete dominance and codominance?

1. In the modern courtroom, blood type tests can be used to exclude paternity, but it has become more difficult to use them convincingly to demonstrate paternity.

   a. Should the blood type tests have excluded Charlie Chaplin from paternity of Carol Ann? Why or why not?

b. Suppose a defendant who is the alleged father takes a blood type test that does not exclude him from paternity consideration. What arguments might the lawyer of the defendant make in a modern courtroom to support the exclusion of the test results as "compelling" evidence? (Assume the tests are done correctly.)

c. As mentioned in the iActivity, an individual of type A blood can be either homozygous or heterozygous. Suppose a blood-typing test allowed these two genetic conditions to be distinguished. Based on what you inferred about the ABO blood types of Joan and Carol Ann Berry in the iActivity, and supposing that you could identify their exact blood types, can you now predict the exact blood type of the father?

2. Why is it helpful to use both the M-N and the ABO blood group tests, instead of just one alone? Would a third blood type test be even more useful, or are two enough?

SOLUTIONS TO TEXT PROBLEMS

4.1   In rabbits, C = agouti coat color, c^{ch} = chinchilla, c^h = Himalayan, and c = albino. The four alleles constitute a multiple allelic series. The agouti C is dominant to the other three alleles, c is recessive to the other three alleles, and chinchilla is dominant to Himalayan. Determine the phenotypes of progeny from the following crosses:

   a. $C/C \times c/c$

   b. $C/c^{ch} \times C/c$

   c. $C/c \times C/c$

   d. $C/c^h \times c^h/c$

   e. $C/c^h \times c/c$

   Answer:

   a. All agouti

   b. ¾ agouti, ¼ chinchilla

   c. ¾ agouti, ¼ albino

   d. ½ agouti, ½ Himalayan

   e. ½ agouti, ½ Himalayan

4.2   If a given population of diploid organisms contains only three alleles of a particular gene (say, w^1, w^2, and w^3), how many different diploid genotypes are possible in the populations? List all possible genotypes of diploids. (Consider only the three given alleles.)

   Answer: Six genotypes are possible: w/w, w/w^1, w/w^2, w^1/w^1, w^1/w^2, w^2/w^2.

4.3   The genetic basis of the ABO blood types is

   a. multiple alleles

   b. polyexpressive hemizygotes

   c. allelically excluded alternatives

   d. three independently assorting genes

   Answer: a

4.4   In humans, the three alleles I^A, I^B, and i constitute a multiple allelic series that determines the ABO blood group system, as described in this chapter. For the following problems, state whether the child mentioned can be produced, and explain your answer:

   a. An O child by a mating of two A individuals

   b. An O child by an A and a B mating

   c. An AB child by an A and an O mating

   d. An O child by an AB and an A mating

   e. An A child by an AB and a B mating

   Answer:

   a. Yes, if both parents were $I^A i$.

   b. Yes, if the A parent were $I^A i$ and the B parent were $I^B i$.

    c. No, as neither parent has an I^B allele.

    d. No, as an O child must obtain an i allele from each parent, and one parent is $I^A I^B$.

    e. Yes, providing that the B parent is $I^B i$ (and the child is $I^A i$).

4.5   A man is blood type O, M. A woman is blood type A, M, and her child is type A, MN. The man cannot be the father of this child because

    a. O men cannot have type A children

    b. O men cannot have MN children

    c. an O man and an A woman cannot have an A child

    d. an M man and an M woman cannot have an MN child

Answer: d. If both parents are M, an MN child is impossible because neither has an N allele. An A child is possible if the mother contributes I^A and the father contributes i.

4.6   A woman of blood group AB marries a man of blood group A whose father was of group O. What is the probability that

    a. their two children will both be of group A?

    b. one child will be of group B, the other of group O?

    c. the first child will be a son of group AB and the second child a son of group B?

Answer: First determine the genotypes of the parents. The mother is type AB, so she must have the genotype $I^A I^B$. The father is type A, so he must have at least one I^A allele. The grandfather was type O, so he only had i alleles. Therefore the cross must be $I^A I^B \times I^A i$. The four equally likely genotypes ($I^A I^A$, $I^A i$, $I^B I^A$, $I^B i$) give rise to three phenotypes: A ($P = \frac{1}{2}$), AB ($P = \frac{1}{4}$), or B ($P = \frac{1}{4}$).

    a. The chance that two children will both be group A is $\frac{1}{2} \times \frac{1}{2} = \frac{1}{4}$.

    b. There is no chance of producing an O child, so $P = 0$.

    c. The chance of a male AB child is $\frac{1}{2} \times \frac{1}{4} = \frac{1}{8}$. The chance of a male B child is $\frac{1}{2} \times \frac{1}{4} = \frac{1}{8}$. Therefore the chance of both events happening is $\frac{1}{8} \times \frac{1}{8} = \frac{1}{64}$. The first event is independent of the second, so this is the chance of the events happening in the order specified.

4.7   If a mother and her child belong to blood group O, what blood group could the father not belong to?

Answer: Since the child is group O, its genotype is ii. Therefore the father must have been able to contribute an i allele. The father could not have been type AB, whose genotype is $I^A I^B$.

4.8   A man of what blood group could not be the father of a child of blood type AB?

Answer: A man who has group O blood is genotypically ii and could not contribute either the I^A or I^B to an AB individual.

4.9 In snapdragons, red flower color (C^R) is incompletely dominant to white (C^W); the C^R/C^W heterozygotes are pink. A red-flowered snapdragon is crossed with a white-flowered one. Determine the flower color of
   a. the F_1 snapdragons,
   b. the F_2 snapdragons,
   c. the progeny of a cross of the F_1 snapdragons to the red parent, and
   d. the progeny of a cross of the F_1 snapdragons to the white parent.

   Answer:
   a. The cross is $C^R C^R \times C^W C^W$; the F_1 are all $C^R C^W$ and are pink.
   b. The F_2 will be 1 $C^R C^R$: 2 $C^R C^W$: 1 $C^W C^W$ and be 1 red : 2 pink : 1 white.
   c. $C^R C^W \times C^R C^R$ produces ½ $C^R C^R$ (red) and ½ $C^R C^W$ (pink) progeny.
   d. $C^R C^W \times C^W C^W$ produces ½ $C^R C^W$ (pink) and ½ $C^W C^W$ (white) progeny.

4.10 In shorthorn cattle, the heterozygous condition of the alleles for red coat color (C^R) and white coat color (C^W) is roan coat color. If two roan cattle are mated, what proportion of the progeny will resemble their parents in coat color?

   Answer: A cross between two roan cattle can be denoted as $C^R C^W \times C^R C^W$. The progeny will be ¼ $C^R C^R$, ½ $C^R C^W$, and ¼ $C^W C^W$. Thus, half will be roan progeny.

4.11 What progeny will a roan shorthorn have if bred to
   a. a red shorthorn?
   b. a roan shorthorn?
   c. a white shorthorn?

   Answer: In shorthorn cattle, the color roan indicates a heterozygote ($C^R C^W$) because of incomplete dominance of each allele. Hence, the crosses can be represented as (a) $C^R C^W \times C^R C^R$, (b) $C^R C^W \times C^R C^W$, and (c) $C^R C^W \times C^W C^W$. (a) will give ½ red and ½ roan, (b) will give ¼ red, ½ roan, and ¼ white, and (c) will give ½ roan and ½ white.

4.12 In peaches, fuzzy skin (F) is completely dominant to smooth (nectarine) skin (f), and the heterozygous condition of oval glands at the base of the leaves (G^O) and no glands (G^N) gives round glands. A homozygous fuzzy, no-gland peach variety is bred to a smooth, oval-gland variety.
   a. What will be the appearance of the F_1 peaches?
   b. What will be the appearance of the F_2 peaches?
   c. What will be the appearance of the offspring of a cross of the F_1 peaches back to that smooth, oval-glanded parent?

   Answer:
   a. The parental cross can be written as $FF\ G^N G^N \times ff\ G^O G^O$. The F_1 will be all $Ff\ G^N G^O$ and be fuzzy with round leaf glands.
   b. As the alleles at the G gene show incomplete dominance, there will be a modified 9:3:3:1 ratio in the F_2. The progeny will be ³⁄₁₆ fuzzy, oval-glanded ($F-\ G^O G^O$),

^{6}⁄₁₆ fuzzy, round-glanded ($F-\ G^OG^N$), ^{3}⁄₁₆ fuzzy, no-glanded ($F-\ G^NG^N$), ¹⁄₁₆ smooth, oval-glanded ($ff\ G^OG^O$), ^{2}⁄₁₆ smooth, round-glanded ($ff\ G^OG^N$), and ¹⁄₁₆ smooth, no-glanded ($ff\ G^NG^N$).

c. The cross can be written as $Ff\ G^NG^O \times ff\ G^OG^O$. The progeny will be ¼ fuzzy, oval-glanded ($Ff\ G^OG^O$), ¼ fuzzy, round-glanded ($Ff\ G^NG^O$), ¼ smooth, oval-glanded ($ff\ G^OG^O$), and ¼ smooth, round-glanded ($ff\ G^NG^O$).

4.13   In guinea pigs, short hair (L) is dominant to long hair (l), and the heterozygous condition of yellow coat (C^Y) and white coat (C^W) gives cream coat. A short-haired, cream guinea pig is bred to a long-haired, white guinea pig, and a long-haired, cream baby guinea pig is produced. When the baby grows up, it is bred back to the short-haired, cream parent. What phenotypic classes, and in what proportions, are expected among the offspring?

Answer: The initial cross can be diagrammed as $L-\ C^YC^W \times ll\ C^WC^W$. As a long-haired cream baby ($ll\ C^YC^W$) is produced, one can infer that the short-haired parent must have been Ll. The backcross can be diagrammed as $ll\ C^YC^W \times Ll\ C^YC^W$. There will be ⅛ short-haired yellow, ¼ short-haired cream, ⅛ short-haired white, ⅛ long-haired yellow, ¼ long-haired cream, and ⅛ long-haired white progeny.

4.14   The shape of radishes may be long (S^L/S^L), oval (S^L/S^S), or round (S^S/S^S), and the color of radishes may be red (C^R/C^R), purple (C^R/C^W), or white (C^W/C^W). If a long, red radish plant is crossed with a round, white plant, what will be the appearance of the F_1 and the F_2 plants?

Answer: The cross can be diagrammed as $S^L/S^L\ C^R/C^R \times S^S/S^S\ C^W/C^W$. The F_1 will be oval and purple ($S^L/S^S\ C^R/C^W$). The F_2 consists of ¹⁄₁₆ long red ($S^L/S^L\ C^R/C^R$), ⅛ oval red ($S^L/S^S\ C^R/C^R$), ¹⁄₁₆ round red ($S^S/S^S\ C^R/C^R$), ⅛ long purple ($S^L/S^L\ C^R/C^W$), ¼ oval purple ($S^L/S^S\ C^R/C^W$), ⅛ round purple ($S^S/S^S\ C^R/C^W$), ¹⁄₁₆ long white ($S^L/S^L\ C^W/C^W$), ⅛ oval white ($S^L/S^S\ C^W/C^W$), and ¹⁄₁₆ round white ($S^S/S^S\ C^W/C^W$).

4.15   In poultry, the dominant alleles for a rose comb (R) and a pea comb (P), if present together, give a walnut comb. The recessive alleles of each gene, when present together in a homozygous state, give a single comb. What will be the comb characters of the offspring of the following crosses?

a. $R/R\ P/p \times r/r\ P/p$

b. $r/r\ P/P \times R/r\ P/p$

c. $R/r\ P/p \times r/r\ P/p$

Answer:

a. ¾ walnut ($R/r\ P/-$), ¼ rose ($R/r\ p/p$)

b. ½ walnut ($R/r\ P/-$), ½ pea ($r/r\ P/-$)

c. ¼ walnut ($R/r\ P/p$), ¼ rose ($R/r\ p/p$), ¼ pea ($r/r\ P/p$), ¼ single ($r/r\ p/p$)

4.16   For the following crosses involving the comb character in poultry, determine the genotypes of the two parents:

a. A walnut crossed with a single produces offspring that are ¼ walnut, ¼ rose, ¼ pea, and ¼ single.

b. A rose crossed with a walnut produces offspring that are $3/8$ walnut, $3/8$ rose, $1/8$ pea, and $1/8$ single.

c. A rose crossed with a pea produces five walnut and six rose offspring.

d. A walnut crossed with a walnut produces one rose, two walnut, and one single offspring.

Answer: To approach solving this problem, first consider the possible genotypes of the parents and their offspring from the phenotypes they present. Then consider the phenotypic ratios in the offspring and determine the precise genotype.
a. *R/r P/p* (walnut) × *r/r p/p* (single).
b. *R/r p/p* (rose) × *R/r P/p* (walnut).
c. *R/− p/p* (rose) × *r/r P/p* (pea).
d. *R/r P/p* × *R/r P/p* (note that since a single *r/r p/p* offspring is produced, both parents must have an *r* and a *p* allele).

4.17   In poultry, feathered (*F*) shanks (part of the legs) are dominant to clean (*f*), and white plumage of white leghorns (*I*) is dominant to black (*i*). Comb phenotypes and genotypes are given in Figure 4.5.
a. A feathered-shanked, white, rose-combed bird crossed with a clean-shanked, white, walnut-combed bird produces these offspring: 2 feathered, white, rose; 4 clean, white, walnut; 3 feathered, black, pea; 1 clean, black, single; 1 feathered, white, single; and 2 clean, white, rose. What are the genotypes of the parents?
b. A feathered-shanked, white, walnut-combed bird crossed with a clean-shanked, white, pea-combed bird produces a single offspring that is clean shanked, black, and single combed. In additional offspring from this cross, what proportion may be expected to resemble each parent?

Answer: Approach this problem by considering each set of traits separately.
a. A cross between a feathered and a clean bird produces 6 feathered : 7 clean offspring (approximately 1 : 1), indicating the parents are *F/f* × *f/f*. A cross between two white birds produces 9 white and 4 black offspring (approximately 3 : 1), indicating the parents are *I/i* × *I/i*. Since a cross between a rose and a walnut bird produces single-combed (as well as walnut, rose, and pea) offspring, each parent must have an *r* and a *p* allele. Thus, the parents must have been *R/r p/p* × *R/r P/p*. The complete genotypes of the parents were *F/f I/i R/r p/p* × *f/f I/i R/r P/p*.
b. The single-combed, clean-shanked, black offspring displays the phenotypes associated with the recessive alleles at the *F/f*, *I/i*, *R/r*, and *P/p* genes. Since its genotype is *f/f i/i r/r p/p*, each parent was able to give it recessive *f*, *i*, *r*, and *p* alleles. Considering this and the parents' phenotypes, the parents' genotypes must be *F/f I/i R/r P/p* and *f/f I/i r/r P/p*. To resemble the feathered-shanked, white, walnut-combed parent, the offspring must be *F/− I/− R/− P/−*. The chance of this is $1/2 × 3/4 × 1/2 × 3/4 = 9/64$. To resemble the clean-shanked, white, pea-combed parent, the offspring must be *f/f I/− r/r P/−*. The chance of this is $1/2 × 3/4 × 1/2 × 3/4 = 9/64$.

4.18 F_2 plants segregate $9/16$ colored : $7/16$ colorless. If just one colored plant from the F_2 generation is chosen at random and selfed, what is the probability that there will be *no* segregation of the two phenotypes among its progeny?

Answer: Since there is a modified 9:3:3:1 ratio in the F_2, the F_1 cross must have been of the form *Aa Bb* × *Aa Bb*. An F_2 colored plant must have the genotype *A– B–*, where *A* and *B* are dominant alleles that must be present for colored pigment to be produced. In order for no segregation of the two phenotypes to be observed in its progeny, the colored plant must be true breeding (i.e., *AA BB*). Thus, the question can be rephrased as What is the chance that if an F_2 colored plant is picked, it is true breeding? Of *all* the plants in the F_2, only $\frac{1}{16}$ are *AA BB*. Among the $\frac{9}{16}$ *A– B–* plants, $\frac{1}{9}$ [$= (\frac{1}{16})/(\frac{9}{16})$] are true-breeding plants, so $P = \frac{1}{9}$.

4.19   In peanuts, a plant may be either "bunch" or "runner." Two different strains of peanut, V4 and G2, in which "bunch" occurred were crossed, with the following results:

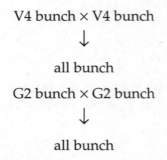

V4 bunch × V4 bunch
↓
all bunch

G2 bunch × G2 bunch
↓
all bunch

The two true-breeding strains of bunch were crossed in the following way:

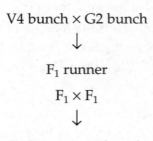

V4 bunch × G2 bunch
↓
F_1 runner

F_1 × F_1
↓

F_2 9 runner : 7 bunch

What is the genetic basis of the inheritance pattern of runner and bunch in the F_2 peanuts?

Answer: The 9:7 ratio in the F_2 is a modified 9:3:3:1 ratio, where the *A– B–* genotypes are runner and the *A– bb*, *aa B–*, and *aa bb* genotypes are bunch. This is an example of needing dominant alleles at each of two genes to observe a phenotype. When only recessive alleles at either of the two genes are present, they block (are epistatic to) the runner phenotype. Thus, this is an example of duplicate recessive epistasis.

4.20   In rabbits, one enzyme (the product of a functional gene *A*) is needed to produce a substance required for hearing. Another enzyme (the product of a functional gene *B*) is needed to produce another substance required for hearing. The genes responsible for the two enzymes are not linked. Individuals homozygous for either one or both of the nonfunctional recessive alleles, *a* or *b*, are deaf.

a. If a large number of matings were made between two double heterozygotes, what phenotypic ratio would be expected in the progeny?

b. The phenotypic ratio found in (a) is a result of what well-known phenomenon?

c. What phenotypic ratio would be expected if rabbits homozygous recessive for trait A and heterozygous for trait B were mated to rabbits heterozygous for both traits?

Answer:

a. Since individuals homozygous for either one or both recessive alleles are deaf, individuals that are *aa B–, A– bb,* or *aa bb* will be deaf. Only *A– B–* individuals are able to hear. Thus, one will get a 9 hearing : 7 deaf phenotypic ratio.

b. This is duplicate recessive epistasis. Homozygous recessive alleles at either of two genes block hearing, and are epistatic to the dominant alleles at the other gene.

c. The cross can be written as: *aa Bb* × *Aa Bb.* Use a branch diagram to show that there would be ⅝ deaf progeny (⅛ *Aa bb* + ½ *aa ––*) and ⅜ hearing (*Aa B–*).

4.21   In doodlewags (hypothetical creatures), the dominant allele *S* causes a solid coat color; the recessive allele *s* results in white spots on a colored background. The black coat color allele *B* is dominant to the brown allele *b,* but these genes are expressed only in the genotype *a/a.* Individuals that are *A/–* are yellow regardless of *B* alleles. Six pups are produced in a mating between a solid yellow male and a solid brown female. Their phenotypes are 2 solid black, 1 spotted yellow, 1 spotted black, and 2 solid brown.

a. What are the genotypes of the male and female parents?

b. What is the probability that the next pup will be spotted brown?

Answer: Two characteristics controlled by three genes are described here. Spotted coat (where pigment is positioned) is controlled by the *S/s* alleles. The pigmentation color itself (brown, black, or yellow) is controlled by the *B/b* and *A/a* genes, where *A–* is epistatic to (i.e., blocks the expression of) alleles at the *B/b* gene. Consider each characteristic separately and remember that when dealing with small numbers of progeny, precise Mendelian ratios are not always obtained.

a. Two solid-colored Doodlewags produce 4 solid and 2 spotted pups. For any spotted Doodlewags to be produced, both parents must have been heterozygous (i.e., *Ss*). The yellow male parent can initially be assigned the genotype *A– – –,* and the brown female must be *aa bb.* Since both black and brown pups are obtained, the yellow parent must be *Aa Bb.* Thus, the parents are *Ss Aa Bb* × *Ss aa bb.*

b. A spotted brown pup has the genotype *ss aa bb.*

P (*ss aa bb* pup)
  = P(*s* from each parent) × P(*a* from each parent) × P(*b* from each parent)
  = (½ × ½) × (½ × 1) × (½ × 1)
  = ¹⁄₁₆.

4.22   The allele *l* in *Drosophila* is recessive, sex linked, and lethal when homozygous or hemizygous (the condition in the male). If a female of genotype *L/l* is crossed with a normal male, what is the probability that the first two surviving progeny will be males?

Answer: The cross can be denoted as *L/l* × *L/Y,* where Y is the Y chromosome. The progeny of this cross are 1 *L/Y* : 1 *l/Y* : 1 *L/L* : 1 *L/l.* Of these, *l/Y* progeny die. Thus, one observes a 1 *L/Y* male : 2 *L/–* female progeny ratio. The chance of observing a surviving male offspring is ⅓. The chance of observing two males as the first two surviving progeny is ⅓ × ⅓ = ⅑.

4.23   A locus in mice is involved in pigment production; when parents heterozygous at this locus are mated, ¾ of the progeny are colored and ¼ are albino. Another phenotype concerns coat color; when two yellow mice are mated, ⅔ of the progeny are yellow and ⅓ are agouti. The albino mice cannot express whatever alleles they may have at the independently assorting agouti locus.

   a.   When yellow mice are crossed with albinos, they produce F_1 mice consisting of ½ albino, ⅓ yellow, and ⅙ agouti. What are the probable genotypes of the parents?

   b.   If yellow F_1 mice are crossed among themselves, what phenotypic ratio would you expect among the progeny? What proportion of the yellow progeny produced here would be expected to be true breeding?

Answer: Two loci are involved here. The first controls pigment production. From the information that a cross of heterozygous parents gives a 3 colored : 1 albino ratio, we can infer that the homozygous recessive condition is that of no pigmentation at all. If we denote this locus as C/c, this also means that $C-$ is required for pigmentation. Another way to state this is that cc blocks the production of pigment.

The second locus determines whether the coat color is yellow or agouti. The 2 yellow : 1 agouti progeny ratio seen in a cross between two yellow mice is a modified 3 : 1 ratio from a monohybrid cross. It indicates recessive lethality. Specifically, it indicates that the yellow allele is dominant for coat color, but that when homozygous, it is lethal. If we denote this locus as Y/y, the YY condition is lethal, Yy individuals are yellow, and yy individuals are agouti.

Now consider the relationship between the two loci. If cc blocks the production of pigment, cc is epistatic to Yy and yy ($cc\ Yy$ and $cc\ yy$ mice will be albino).

   a.   From the above analysis, the phenotype of a yellow mouse indicates that its genotype must be $C-\ Yy$, and the phenotype of an albino mouse indicates that its genotype must be $cc\ y-$. To determine what the unknown alleles are, consider the two traits separately by considering color/no color (the C/c locus) and type of color (the Y/y locus). There are half albino and half colored offspring, indicating that the cross must have been $Cc \times cc$. There is a 2 : 1 ratio of yellow to agouti offspring, indicating that the cross was $Yy \times Yy$. Therefore, the parental genotypes are $Cc\ Yy \times cc\ Yy$.

   b.   The yellow F_1 mice have the genotype $Cc\ Yy$. The progeny that will be obtained from the cross $Cc\ Yy \times Cc\ Yy$ are ¼ albino, ½ yellow, and ¼ agouti (remember that the YY progeny are inviable). None of the yellow mice will be true breeding, as they are all Yy.

4.24   In *Drosophila melanogaster*, a recessive autosomal allele, ebony (e), produces a black body color when homozygous, and an independently assorting autosomal allele, black (b), also produces a black body color when homozygous. Flies with genotypes $e/e\ b^+/-$, $e^+/-\ b/b$, and $e/e\ b/b$ are phenotypically identical with respect to body color. Flies with genotype $e^+/-\ b^+/-$ have a grey body color. True-breeding $e/e\ b^+/b^+$ ebony flies are crossed with true-breeding $e^+/e^+\ b/b$ black flies.

   a.   What will be the phenotype of the F_1 flies?

   b.   What phenotypes and what proportions would occur in the F_2 generation?

   c.   What phenotypic ratios would you expect to find in the progeny of these backcrosses?
      i. $F_1 \times$ true-breeding ebony
      ii. $F_1 \times$ true-breeding black

Answer:

a. The F$_1$ flies will be $b/b^+ e/e^+$ and be wild type in color (grey).

b. The F$_2$ will show a 9 $b^+/- e^+/-$ (grey) : 3 $b^+/- e/e$ (black) : 3 $b/b e^+/-$ (black) : 1 $b/b e/e$ (black) ratio (i.e., $9/16$ grey and $7/16$ black).

c. The F$_1$ × true-breeding ebony can be denoted $b/b^+ e/e^+ \times b^+/b^+ e/e$ and will give a 1 $b^+/- e/e$: 1 $b^+/- e/e^+$ progeny ratio, $1/2$ black and $1/2$ grey. The F$_1$ × true-breeding black can be denoted $b^+/b e/e^+ \times b/b e^+/e^+$ and will give a 1 $b^+/b e^+/-$: 1 $b/b e^+/-$ progeny ratio, $1/2$ grey and $1/2$ black.

4.25 In four-o'clock plants, two genes, Y and R, affect flower color. Neither is completely dominant, and the two interact with each other to produce seven different flower colors:

$Y/Y R/R$ = crimson $Y/y R/R$ = magenta

$Y/Y R/r$ = orange-red $Y/y R/r$ = magenta-rose

$Y/Y r/r$ = yellow $Y/y r/r$ = pale yellow

$y/y R/R$, $y/y R/r$, and $y/y r/r$ = white

a. In a cross of a crimson-flowered plant with a white one ($y/y r/r$), what will be the appearances of the F$_1$ plants, the F$_2$ plants, and the offspring of the F$_1$ plants backcrossed to their crimson parent?

b. What will be the flower colors in the offspring of a cross of orange-red × pale yellow?

c. What will be the flower colors in the offspring of a cross of a yellow with a $y/y R/r$ white?

Answer:

a. $Y/Y R/R \times y/y r/r$ will give a magenta-rose $Y/y R/r$ F$_1$, and an F$_2$ that is $1/16$ crimson ($Y/Y R/R$), $1/8$ orange-red ($Y/Y R/r$), $1/16$ yellow ($Y/Y r/r$), $1/8$ magenta ($Y/y R/R$), $1/4$ magenta-rose ($Y/y R/r$), $1/8$ pale yellow ($Y/y r/r$), and $1/4$ white ($y/y -/-$). A backcross of the F$_1$ to the crimson parent will give $1/4$ crimson ($Y/Y R/R$), $1/4$ magenta-rose ($Y/y R/r$), $1/4$ magenta ($Y/y R/R$), and $1/4$ orange-red ($Y/Y R/r$).

b. The cross can be denoted as $Y/Y R/r \times Y/y r/r$. The progeny will be $1/4$ orange-red ($Y/Y R/r$), $1/4$ magenta-rose ($Y/y R/r$), $1/4$ yellow ($Y/Y r/r$), and $1/4$ pale yellow ($Y/y r/r$).

c. The cross can be denoted as $Y/Y r/r \times y/y R/r$, and the progeny will be $1/2$ $Y/y R/r$ magenta-rose and $1/2$ $Y/y r/r$ pale yellow.

4.26 Two four-o'clock plants were crossed and gave the following offspring: $1/8$ crimson, $1/8$ orange-red, $1/4$ magenta, $1/4$ magenta-rose, and $1/4$ white. Unfortunately, the person who made the crosses was color blind and could not record the flower colors of the parents. From the results of the cross, deduce the genotypes and flower colors of the two parents.

Answer: First determine the genotypes of the progeny. They are $1/8$ $Y/Y R/R$, $1/8$ $Y/Y R/r$, $1/4$ $Y/y R/R$, $1/4$ $Y/y R/r$, and $1/4$ $y/y -/-$. Now consider each gene separately. There are $1/4$ Y/Y, $1/2$ Y/y, and $1/4$ y/y progeny, a 1:2:1 ratio, so the parental cross must have been $Y/y \times Y/y$. Among the $3/4$ of the progeny whose R/r genotypes can be determined, half are R/R and half are R/r, indicating that the parental genotypes were $R/R \times R/r$. Thus the parental cross was $Y/y R/R$ (magenta) × $Y/y R/r$ (magenta-rose).

4.27 Genes *A*, *B*, and *C* are independently assorting and control the production of a black pigment.

a. Suppose that A, B, and C act in the following pathway:

$$\text{colorless} \xrightarrow{A} \xrightarrow{B} \xrightarrow{C} \text{black}$$

The alternative alleles that give abnormal functioning of these genes are designated *a*, *b*, and *c*, respectively. A black *A/A B/B C/C* is crossed with a colorless *a/a b/b c/c* to give a black F_1. The F_1 is selfed. What proportion of the F_2 individuals is colorless? (Assume that the products of each step except the last are colorless, so only colorless and black phenotypes are observed.)

b. Suppose instead that a different pathway is utilized. In it, the *C* allele produces an inhibitor that prevents the formation of black by destroying the ability of *B* to carry out its function. The mechanism is as follows:

$$\text{colorless} \xrightarrow{A} \xrightarrow{B} \text{black}$$
$$\uparrow$$
$$C \text{ (inhibitor)}$$

A colorless *A/A B/B C/C* individual is crossed with a colorless *a/a b/b c/c*, giving a colorless F_1. The F_1 is selfed to give an F_2. What is the ratio of colorless to black in the F_2 individuals? [Only colorless and black phenotypes are observed, as in part (a).]

c. How would you evaluate which of the biochemical pathways hypothesized in parts (a) and (b) is more likely?

Answer:

a. In order for an individual to be black, it must have normal function at each step of the pathway. This is provided by the alleles *A*, *B*, and *C*. Thus, *A/– B/– C/–* individuals will be black, while all others (those having *a/a* and/or *b/b* and/or *c/c*) will be colorless. The chance of obtaining a black *A/– B/– C/–* individual from a cross of *A/a B/b C/c* × *A/a B/b C/c* is $\frac{3}{4} \times \frac{3}{4} \times \frac{3}{4} = \frac{27}{64}$. The proportion of the F_2 that is colorless is $1 - \frac{27}{64} = \frac{37}{64}$. There will be a 37 colorless : 27 black ratio.

b. With this pathway, an individual is black only if it has the first two steps of the pathway (those provided by *A* and *B*) and lacks the inhibitor provided by *C* (i.e., if it is *A/– B/– c/c*). The chance of obtaining this genotype from a cross of *A/a B/b C/c* × *A/a B/b C/c* is $\frac{3}{4} \times \frac{3}{4} \times \frac{1}{4} = \frac{9}{64}$. The proportion of the F_2 that is colorless is $1 - \frac{9}{64} = \frac{55}{64}$. There will be a 55 colorless : 9 black ratio.

c. Use a chi-square test to evaluate how well the ratio of black to colorless observed in the F_2 fits the expectation of each of these two hypothetical pathways.

4.28 In cats, two alleles (*B*, *O*) at an X-linked gene control whether black or orange pigment is deposited. A dominant allele at an autosomal gene *I/i* partially inhibits the deposition of pigment, lightening the coat color from black to grey or from orange to pale orange. A dominant allele at the autosomal gene *T/t* determines whether a tabby, or vertically striped, pattern is present. The tabby pattern depends on a dominant agouti (*A*) allele for its expression, with nonagouti (*a*) epistatic to tabby. The agouti allele also causes a speckled, rather than solid, color coat. Judy, a stray cat with a speckled, tortoiseshell pattern with grey and pale orange spots and no trace of a tabby pattern, gives birth to four kittens. Of the three female offspring, two are solid grey and the third is speckled grey and light orange like her mother, but also shows traces of a tabby pattern. The single male offspring is solid grey.

a. Explain how the tortoiseshell pattern arises in cats. That is, how can a cat have distinct patches of fur with different deposits of pigment?

b. Cats with a tortoiseshell pattern usually are female. Explain why this is the case and also why, when an unusual male tortoiseshell male cat is found, he is atypically large and typically not very swift.

c. What genotype(s) might Judy and her kittens have?

d. Assuming that there is just one father, what phenotype(s) should be considered in assessing the neighborhood males for paternity?

Answer:

a. The tortoiseshell cat has pigment deposited in patches of orange and black, with the pigment color determined by alleles at an X-linked gene. In females, one X chromosome is inactivated, so that only one allele is expressed at a time. The pigment color in each patch depends on which allele is expressed. The color is orange if it is the *O* allele and black if it is the *B* allele. Thus, the tortoiseshell black/orange pattern reflects the pattern of X-chromosome inactivation. When the X chromosome bearing the *B* allele is inactivated, the *O* allele functions to deposit orange pigment. When the X chromosome bearing the *O* allele is inactivated, the *B* allele functions to deposit black pigment.

b. Since the pattern of a tortoiseshell cat results from X inactivation, a process that occurs in animals with more than one X chromosome, it will normally occur only in XX female individuals. Thus, tortoiseshell cats are usually female. The exceptional male tortoiseshell cat probably has a sex-chromosome abnormality, and could be an XXY male. In an XXY male, one of the two X chromosomes would be inactivated. If an XXY male were heterozygous for the *O* and *B* alleles, he could exhibit a tortoiseshell pattern. By analogy with human XXY (Klinefelter) males, one might expect XXY cats to be larger and less intelligent.

c. Consider one trait at a time, and the phenotypic description of each cat. Judy has both black and orange pigments, and so is X^O/X^B. Her pigment deposition is inhibited, as her tortoiseshell patches appear as grey and pale orange spots, and so she is $I/-$. She shows a speckled pattern, and so is agouti and must be $A/-$. Since some of her offspring are not speckled (e.g., her son is solid, and so is *a/a*), she must carry an *a* allele and be *A/a*. She is not tabby, and so is *t/t*. She is X^O/X^B $I/-$ *A/a* *t/t*.

    Her male offspring is solid grey. Since he is grey, he is *B/Y* and also $I/-$. Since he is solid colored, he is not agouti, and must be *a/a*. Since the tabby pattern is dependent on $A/-$ for expression, he may be either *t/t* or *T/t*. Since Judy, his mother, was *t/t*, he cannot be *T/T*. He is either *BY* $I/-$ *a/a* *t/t* or *BY* $I/-$ *a/a* *T/t*.

    Two of the female offspring are, like their brother, solid grey. Since they are females, they are *B/B*. Using the same logic as for their brother, they are either *B/B* $I/-$ *a/a* *t/t* or *B/B* $I/-$ *a/a* *T/t*.

    The third female offspring is speckled, tortoiseshell, light, and tabby. Thus, she is *B/O* $I/-$ *A/-* *T/t*.

d. Consider one trait at a time. Since Judy is not tabby, she is *t/t* and can only contribute *t* alleles. Since one of the offspring is tabby, the father must have been able to contribute a *T* allele, and must have been *T/–*. Since two female offspring are grey and *B/B*, the father must have been *B/Y*. Since Judy is *I/–*, and was able to give an *I* allele to each of her offspring, the father was *i/i, I/i,* or *I/I*. The father could have been either *A/a* or *a/a* (he was not *A/A*, since some of the progeny are solid colored and *a/a*), since Judy could have contributed the *A* allele to her speckled offspring. Thus, the father's genotypes could have been *B/Y* (*I/–* or *i/i*) *T/–* (*A/a* or *a/a*). This means that the father could have been black and tabby (*B/Y i/i T/– A/a*), grey and tabby (*B/Y I/– T/– A/a*), solid black (*B/Y i/i T/– a/a*), or solid grey (*B/Y I/– T/– a/a*).

4.29   In *Drosophila*, a mutant strain has plum-colored eyes. A cross between a plum-eyed male and a plum-eyed female gives ⅔ plum-eyed and ⅓ red-eyed (wild-type) progeny flies. A second mutant strain of *Drosophila*, called stubble, has short bristles instead of the normal long bristles. A cross between a stubble female and a stubble male gives ⅔ stubble and ⅓ normal-bristled flies in the offspring. Assuming that the plum gene assorts independently from the stubble gene, what will be the phenotypes and their relative proportions in the progeny of a cross between two plum-eyed, stubble-bristled flies? (Both genes are autosomal.)

Answer: For either of the stubble or plum mutants, a cross of two mutant individuals gives a 2 mutant : 1 wild-type progeny ratio. This 2 : 1 ratio is a modified 3 : 1 ratio from a monohybrid cross, where 25 percent of the progeny—the homozygous mutant individuals—die. We can denote the cross (using *P/p* for plum, *S/s* for stubble) as *P/p S/s* × *P/p S/s*. Any *PP* and/or *SS* progeny will be inviable. The kinds of progeny expected can be diagrammed as follows:

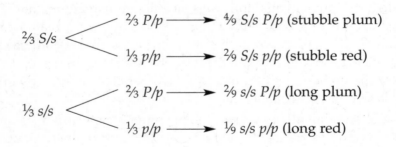

4.30   In *Drosophila*, a recessive, temperature-sensitive mutation in the *transformer-2* (*tra-2*) gene on chromosome 2 causes XX individuals raised at 29°C to be transformed into phenotypic males. At 16°C, these individuals develop as normal females. The sex type of XY individuals is unaffected by the *tra-2* mutation. Suppose you are given three true-breeding, unlabeled vials containing different strains of *Drosophila*, all raised at 16°C. Two of the strains have white eyes, and one has red eyes. You are told that one of the white-eyed strains also carries the *tra-2* mutation. Devise two different methods to determine which white-eyed strain has the *tra-2* mutation. Is there a reason to prefer one method over the other?

Answer: Since the *transformer-2* mutation transforms XX individuals into males at the restrictive temperature of 29°C, one way to decide which white-eyed strain carries the *tra-2* mutation is to examine the progeny produced when the strains are raised at the restrictive temperature. The *tra-2* strain should produce only phenotypic males. Transfer adults from each white-eyed strain to a fresh vial, and collect the eggs that are laid. Raise half of the eggs collected from each strain at the permissive temperature of 16°C, and raise the other half at the restrictive temperature of 29°C. The white-eyed strain carrying the *tra-2* mutation will give only male progeny at 29°C, but both male and female progeny at 16°C. The white-eyed strain carrying a *tra-2*$^+$ allele will give male and female progeny at both temperatures. Raising animals at the permissive temperature serves as a control to assess the relative number of males and females normally produced by each strain.

An alternative method is to employ a set of crosses to assess the chromosomal constitution (XX or XY) of an animal independently of its sexual phenotype ("maleness" or "femaleness"). In such a cross, an XX *tra-2/tra-2* animal raised at the restrictive temperature would appear to be a male. Such crosses can be designed using the X-linked white-eyed gene. Consider the following two crosses between a red-eyed *tra-2*$^+$/*tra-2*$^+$ strain and a white-eyed *tra-2/tra-2* strain. Both crosses are done at 16°C, but the progeny of the second cross are raised at 29°C.

I. *w*/Y *tra-2/tra-2* (white XY male) × *w*$^+$/*w*$^+$ *tra-2*$^+$/*tra-2*$^+$ (red XX female)
Progeny [raise at 16°C]: ½ *w*$^+$/Y *tra-2/tra-2*$^+$ (red XY males)
½ *w*/*w*$^+$ *tra-2/tra-2*$^+$ (red XX females)

II. *w*$^+$/Y *tra-2/tra-2*$^+$ (from cross I) × *w*/*w* *tra-2/tra-2* (white XX female)
Progeny [raise at 29°C]: ¼ *w*$^+$/*w* *tra-2/tra-2* (red XX "males")
¼ *w*$^+$/*w* *tra-2/tra-2*$^+$ (red XX females)
¼ *w*/Y *tra-2/tra-2*$^+$ (white XY males)
¼ *w*/Y *tra-2/tra-2* (white XY males)

In the second cross, XX progeny are red eyed, while XY progeny are white eyed. Since the *tra-2/tra-2* mutation transforms XX animals into males at 29°C, red-eyed "males" will be seen. Thus, if this set of crosses is done with each of the white-eyed strains, the white-eyed strain that also has a *tra-2* mutation can be identified.

Both methods work satisfactorily to identify the *tra-2* strain and each has its own advantage. The first method is faster, as it requires only one cross, not two. The second method has an internal control to identify an animal's chromosomal constitution independent of its sexual phenotype.

4.31 Normal *Drosophila* have straight wings and smooth, well-ordered compound eyes. A strain with curly wings and rough eyes has the following properties: Interbreeding its males and females always gives progeny identical to the parents. An outcross of a male from this strain to a normal female gives 45 curly and 49 rough progeny. An outcross of a female from the same strain to a normal male gives 53 curly and 47 rough progeny. Crossing a curly F$_1$ male and female from the first outcross gives 81 curly and 53 straight progeny. The same curly F$_1$ male mated to a normal female gives 57 curly and 61 normal progeny. Crossing a rough F$_1$ male and female from the first outcross gives 78 rough and 42 smooth progeny. When the same rough F$_1$ male is mated to a normal female, 46 rough and 48 normal progeny are recovered. Develop hypotheses to explain these data, and test them using chi-square tests.

Answer: Since interbreeding males and females from the strain gives progeny identical to the parents, an initial hypothesis is that the strain is true breeding. If this is the case, *and* the rough-eyed and curly-winged traits are dominant, outcrossing a male or a female from the strain should give progeny that are all rough and curly. To demonstrate this, let C be the curly allele, and R be the rough allele. The strain is $C/C\ R/R$, and an outcross is $C/C\ R/R \times +/+\ +/+$, with progeny $C/+\ R/+$. However, we know that outcrossing does not give all curly and rough progeny. It gives about a $1:1$ ratio of curly to rough progeny. (Outcrossing a male gives 45 rough : 49 curly; progeny total = 94, $1:1$ expected = $47:47$, $\chi^2 = 0.17$, df = 1, $0.5 < P < 0.7$, accept as possible. Outcrossing a female gives 53 rough : 47 curly; progeny total = 100, $1:1$ expected = $50:50$, $\chi^2 = 0.36$, df = 1, $0.5 < P < 0.7$, accept as possible.) This suggests that while the curly-winged and rough-eyed traits may be dominant, the strain is not true breeding. How then can inbreeding give progeny phenotypes identical to the parents?

Insight into this question comes from analyzing the other crosses. The cross of two curly F_1 animals gives 81 curly and 53 normal progeny, while the cross of two rough F_1 animals gives 78 rough and 42 normal animals. Each of these progeny ratios are closer to $2:1$ than to $3:1$. (For the 81 curly : 53 normal progeny, progeny total = 134, $2:1$ expected = $89:45$, $\chi^2 = 2.14$, df = 1, $0.1 < P < 0.2$, accept as possible; $3:1$ expected = $100:34$, $\chi^2 = 14.2$, df = 1, $P < 0.001$, reject as unlikely. For the 78 rough : 42 normal, progeny total = 120, $2:1$ expected = $80:40$, $\chi^2 = 0.15$, df = 1, $P = 0.7$, accept as possible; $3:1$ expected = $90:30$, $\chi^2 = 6.4$, df = 1, $0.01 < P < 0.05$, reject as unlikely.) A $2:1$ ratio is a modified $1:2:1$ ratio, where $\frac{1}{4}$ of the progeny die due to recessive lethality. Here, a dominant curly-winged phenotype is seen in about $\frac{2}{3}$ of the viable progeny from one cross, and a dominant rough-eyed phenotype is seen in about $\frac{2}{3}$ of the progeny from the other cross.

When rough or curly F_1 progeny are crossed to normal animals, a 1 mutant : 1 normal progeny ratio is seen. (For curly $F_1 \times$ normal female, 57 curly and 61 normal progeny are seen: progeny total = 118, $1:1$ expected = $59:59$, $\chi^2 = 0.14$, df = 1, $0.70 < P < 0.90$, accept. For rough $F_1 \times$ normal female, 46 rough and 48 normal progeny are seen; progeny total = 94, $1:1$ expected = $47:47$, $\chi^2 = 0.04$, df = 1, $0.70 < P < 0.90$, accept.) These would be the expectations if these crosses were between heterozygotes for a dominant allele and an animal with normal alleles (e.g., $C/+ \times +/+$ and $R/+ \times +/+$).

Thus, the subsequent crosses support the hypotheses that the curly-winged and rough-eyed alleles are dominant alleles that are each recessive lethal. To explain how inbreeding the strain produces identical progeny to parents, consider the possibility that the curly-winged and rough-eyed phenotypes are caused by alleles at two different genes, each on different members of the same homologous pair of chromosomes (i.e., the strain is CR^+/C^+R). Then a cross between two members of the strain is $CR^+/C^+R \times CR^+/C^+R$ and gives a $1\ CR^+/CR^+ : 2\ CR^+/C^+R : 1\ C^+R/C^+R$ progeny ratio. Since R/R and C/C animals are inviable, only CR^+/C^+R animals are recovered. If a CR^+/C^+R animal is crossed to a C^+R^+/C^+R^+, a $1\ CR^+/C^+R^+ : 1\ C^+R/C^+R^+$ ($1\ C : 1\ R$) progeny ratio is obtained. These predictions are consistent with the observed results and supported by the chi-square tests.

Another possibility is that the C and R alleles are mutations at one gene, and that the different mutations represent pleiotropic affects. In this case, one could write the strain as C/R, and a cross between two members of the strain as $C/R \times C/R$. This would give a $1\ C/C : 2\ C/R : 1\ R/R$ progeny ratio. Since R/R and C/C animals are inviable, only C/R animals are recovered. If a C/R animal is crossed to a $+/+$ animal, a $1\ +/C : 1\ +/R$ progeny ratio is obtained. These predictions are also consistent with the observed results and supported by the chi-square tests.

4.32   In sheep, white fleece (W) is dominant over black (w), and horned (H) is dominant over hornless (h) in males, but recessive in females. If a homozygous horned white ram is bred to a homozygous hornless black ewe, what will be the appearances of the F$_1$ and the F$_2$ sheep?

Answer: One can restate the information given in the problem as follows: In males, *H/h* and *H/H* result in horned animals, while *h/h* results in hornless animals. Both the notation and phenotypes are that of a typical dominant trait. In females, the fact that the trait is sex-influenced makes the situation quite different. In females, only *H/H* animals are horned. *H/h* and *h/h* animals are hornless. The white trait shows dominance in both sexes, with *W/–* animals being white and *ww* animals being black.

This information allows one to diagram the cross as *H/H W/W* ♂ × *h/h w/w* ♀. The F$_1$ genotype will be *H/h W/w*. Both sex types will have white fleece, and while males will be horned, females will be hornless. The F$_2$ genotypes will be 9 *H/– W/–* : 3 *H/– w/w* : 3 *h/h W/–* : 1 *h/h w/w*. Because *H/h* and *H/H* show the dominant horned trait in males, the males will be ⁹⁄₁₆ horned white, ³⁄₁₆ horned black, ³⁄₁₆ hornless white and ¹⁄₁₆ hornless black. Female phenotypes can be determined with the following branch diagram:

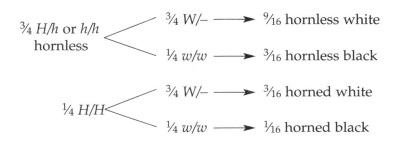

4.33   A horned black ram bred to a hornless white ewe produces the following offspring: Of the males, ¼ are horned, white; ¼ are horned, black; ¼ are hornless, white; and ¼ are hornless, black. Of the females, ½ are hornless and black, and ½ are hornless and white. What are the genotypes of the parents?

Answer: Start by using the symbols of question 4.32 to denote what can be inferred about the genotypes of the parents and the offspring. Remember that in males, the horned condition is dominant, while in females, it is recessive. Hence, *H/–* (= *H/H* or *H/h*) denotes the horned condition in males, while *h/–* (*h/h* or *H/h*) denotes the hornless condition in females. Then the parents are *H/– w/w* (male) × *h/– W/–* (female). The male progeny are ¼ *H/– W/–*, ¼ *H/– w/w*, ¼ *h/h W/–*, and ¼ *h/h w/w*. The female progeny are ½ *h/– w/w* and ½ *h/– W/–*. Since homozygous recessive (male) progeny are recovered, we know that each parent must have had (at least) one recessive *h* and one recessive *w* allele. Since none of the female progeny are *H/H*, only one parent had an *H* allele. Thus, the parents were *H/h w/w* ♂ × *h/h W/w* ♀.

4.34   A horned white ram is bred to the following four ewes and has one offspring by the first three and two by the fourth: Ewe A is hornless and black; the offspring is a horned white female. Ewe B is hornless and white; the offspring is a hornless black female. Ewe C is horned and black; the offspring is a horned white female. Ewe D is hornless and white; the offspring are one hornless black male and one horned white female. What are the genotypes of the five parents?

Answer: Using the information and symbols of question 4.32, start by inferring what you can about genotypes from the phenotypes:

Individual	Phenotype	Initially Inferred Genotype
Male parent	horned white male	*H/– W/–*
Ewe A	hornless black female	*H/h* or *h/h, w/w*
Ewe A offspring	horned white female	*H/H W/–*
Ewe B	hornless white female	*H/h* or *h/h, W/–*
Ewe B offspring	hornless black female	*H/h* or *h/h, w/w*
Ewe C	horned black female	*H/H w/w*
Ewe C offspring	horned white female	*H/H W/–*
Ewe D	hornless white female	*H/h* or *h/h, W/–*
Ewe D offspring #1	hornless black male	*h/h w/w*
Ewe D offspring #2	horned white female	*H/H W/–*

Note that Ewe D's male offspring must be homozygous for the recessive *h* and *w* alleles. This means that both parents must have had at least one recessive allele. Hence the male parent must be *H/h W/w* and Ewe D is *H/h* or *h/h W/w*. That Ewe D's female offspring must be *H/H* means that Ewe D must have an *H* allele. Thus, Ewe D is *H/h W/w*. Since Ewe A has an offspring that is *H/H*, Ewe A necessarily has an *H* allele. Hence, Ewe A must be *H/h w/w*. Since Ewe B has an offspring that is *w/w*, Ewe B must have a *w* allele. We cannot tell if Ewe B is *H/h* or *h/h*. So Ewe B is *H/h W/w* or *h/h W/w*. From the genotype of Ewe C, we know she is *H/H w/w*.

4.35 Pattern baldness is more frequent in males than it is in females. This appreciable difference in frequency is assumed to result from

  a. Y linkage of this trait

  b. X-linked recessive mode of inheritance

  c. sex-influenced autosomal inheritance

  d. excessive beer drinking in males, with the consumption of gin being approximately equal between the sexes

Answer: c

4.36 King George III, who ruled England during the period of the Revolutionary War in the United States, is an ancestor of Queen Elizabeth II. (See Figure 3.28 on p. 68.) Near the end of his life, he exhibited sporadic periods of "madness." In retrospect, it appears that he showed symptoms of porphyria, an autosomal dominant disorder of heme metabolism. In addition to "madness," the symptoms of porphyria, which include a variety of physical ailments that King George III exhibited, are sporadic, variable in severity, can be affected by diet, and, currently, can be treated with medication.

  a. How would you describe this disease in terms of penetrance and expressivity?

  b. If, indeed, King George III had porphyria, what is the chance that the current Prince of Wales (Charles) carries a disease allele? State all of your assumptions.

Answer:

  a. Since the disease varies in its severity in a single individual, it shows variable expressivity. From the description of the disease, it is possible that some individuals in a population would not show the disease at all (e.g., King George III would not have shown the disease had he had a better diet, or had he died before the onset of symptoms.) Therefore, the disease is likely to show reduced penetrance.

b. As shown in text Figure 3.28, the parents of Prince Charles, Elizabeth II and Prince Philip, are related as third cousins and are also both descendants of George III. Since the chance that an $A-$ individual will pass the trait to the next generation is $\frac{1}{2}$, and Elizabeth II and Prince Philip are each six generations removed from George III, there is a $(\frac{1}{2})^6$ chance that they have the disease allele. This assumes that no one else marrying into the lineage contributed a disease allele and that George III was heterozygous. Thus, there is a $[1- (\frac{1}{2})^6]$ chance that either parent of Prince Charles has the normal a allele. The table below shows the possible genotypes of the cross between Elizabeth II and Prince Philip, the probability of obtaining an $A-$ offspring if each cross occurred, the probability of each cross occurring, and the overall probability of obtaining an $A-$ offspring with a particular cross.

Possible Genotypes of Elizabeth II and Prince Phillip	Probability of Cross Producing an A− Offspring	Probability of Cross	Probability of A− Offspring with Specified Cross
$aa \times aa$	0	$[1- (\frac{1}{2})^6] \times [1 - (\frac{1}{2})^6]$	0
$Aa \times aa$	$\frac{1}{2}$	$(\frac{1}{2})^6 \times [1 - (\frac{1}{2})^6]$	$\frac{1}{2} \times (\frac{1}{2})^6 \times [1 - (\frac{1}{2})^6]$
$aa \times Aa$	$\frac{1}{2}$	$(\frac{1}{2})^6 \times [1 - (\frac{1}{2})^6]$	$\frac{1}{2} \times (\frac{1}{2})^6 \times [1 - (\frac{1}{2})^6]$
$Aa \times Aa$	$\frac{3}{4}$	$(\frac{1}{2})^6 \times (\frac{1}{2})^6$	$\frac{3}{4} \times (\frac{1}{2})^6 \times (\frac{1}{2})^6$

The chance that Prince Charles is $A-$ is the sum of the probabilities of obtaining an $A-$ offspring with each possible cross:

$$P = 0 + \{\frac{1}{2} \times (\frac{1}{2})^6 \times [1 - (\frac{1}{2})^6]\} + \{\frac{1}{2} \times (\frac{1}{2})^6 \times [1 - (\frac{1}{2})^6]\} + [\frac{3}{4} \times (\frac{1}{2})^6 \times (\frac{1}{2})^6]$$
$$= (\frac{1}{2})^6 - (\frac{1}{2})^{14}$$
$$= 0.0156.$$

4.37 Jasper Rine and his colleagues at the University of California at Berkeley have launched the Dog Genome Initiative to study canine genes and behavior. They mated Pepper, a vocal, highly affectionate, very social Newfoundland female that is not good at fetching tennis balls but loves water, to Gregor, a quiet, less affectionate, less social border collie that is exceptionally good at fetching tennis balls, but avoids water. They obtained 7 F_1 and 23 F_2 progeny. When the behavioral traits were analyzed, it was found that all 7 F_1 dogs were similar, each showing a mixture of the parents' behavioral traits. When the behaviors of the F_2 dogs were analyzed, differences were more evident. In particular, two of the F_2 dogs (Lucy and Saki) shared Pepper's love of water. (For more information, see Donald McCaig, "California Geneticists Are Going to the Dogs," *Smithsonian* 27 (1996): 126–141.)

a. Develop hypotheses to explain the various observations and, when appropriate, test them using a chi-square test.

b. What practical value might there be in studying the genes of canines?

Answer:

a. That the F_1 showed a similar mixture of traits suggests that none of the traits appear to be controlled by alleles at a single gene that function in a completely dominant/recessive manner. That more differences are more evident in the F_2 suggests that some of the traits may be controlled by a small number of interacting genes. The behavioral differences seen in the F_2 could result from the segregation of alleles at these genes segregate in the F_2.

For the "water-loving" trait, $2/23 = 8.7$ percent of F_2 dogs show the trait associated with Pepper, their grandmother. Analyze this data to model whether one or several genes could "control" the trait.

Hypothesis I: Two alleles at one gene control love/avoidance of water. Let L = love of water, l = avoidance of water. If the parental cross was $LL \times ll$, the F_1 was Ll and the F_2 was 1 LL : 2 Ll : 1 ll. Given the F_1 phenotype, alleles L and l must show partial dominance. About 6 ($\frac{1}{4} \times 23 = 5.75$) F_2 dogs would be expected to love water, and about 17 would be expected to avoid or to be indifferent to water. Using a chi-square test to evaluate this hypothesis, $\chi^2 = 3.6$, df = 1, $0.05 < P < 0.10$, and so this hypothesis cannot be rejected as being unlikely.

Hypothesis II: Two alleles at each of two genes control love/avoidance of water. For convenience in notation let $a/a \ b/b$ represent Pepper's genotype, and $A/A \ B/B$ represent Gregor's genotype. The F_1 is $A/a \ B/b$. As in hypothesis I, given the F_1 phenotypes, A/a and/or B/b must show partial dominance. The F_2 is 9 $A/- \ B/-$: 3 $A/- \ b/b$: 3 $a/a \ B/-$: 1 $a/a \ b/b$. Between 1 and 2 ($\frac{1}{16} \times 23 = 1.44$) dogs would be expected to be $a/a \ b/b$ and love water. This is observed. Using a chi-square test to evaluate this hypothesis, $\chi^2 = 0.23$, df = 1, and $0.50 < P < 0.70$. This hypothesis is accepted as being possible, and more likely than hypothesis I. Although this does not prove that only two genes control this behavior (for example, we do not know that either dog is homozygous for the hypothesized alleles, or anything about the nature of the interactions between alleles at the same or different genes), it does suggest that there may be a small number of genes that are important for this behavior.

b. This approach should lead to an understanding of how many genes are important for these relatively complex behaviors. Coupled with recombinant DNA methods, it will lead to insight into the (biochemical) function of the genes important for these behaviors. Since dogs are mammals, these studies have comparative value for other mammals, including humans. In addition, the Dog Genome Initiative should lead to the identification of genes in canines for other health-related traits. This should lead to advances in treating illness in canines, and perhaps other mammals. This would be of interest to the relatively large number of pet owners, among others.

4.38 Parkinson disease is a progressive neurological disease that causes slowness of movement, stiffness, and shaking, and eventually leads to disability. Actor Michael J. Fox has this disease. Parkinson disease affects about 2 percent of the U.S. adult population over 50 years of age and appears most often in individuals who are between their fifth and seventh decades. There has been much discussion among scientists as to whether the disease is caused by environmental factors, genetic factors, or both. Support for the environmental hypothesis stems from the observation that the disease seems not to have been reported until after the Industrial Revolution and from the discovery that some chemicals can cause symptoms. Support for the genetic hypothesis stems from pedigree analysis.

Consider the pedigree in Figure 4.B (modified to protect patient confidentiality), which shows the incidence of parkinsonism in a family of European descent. The shaded portions of the pedigree indicate family members who reside in the United States. The remaining portions of the pedigree reside in Europe. Members of the U.S. branches of the family have not visited Europe for any extensive period since the initial emigration from Europe.

a. If the disease in this family has a genetic basis, what is its basis? Explain your answer.

b. Why might this pedigree be particularly helpful in distinguishing between an environmental and a genetic cause of Parkinson disease?

c. What reservations, if any, do you have about concluding that the disease has a genetic basis in some individuals?

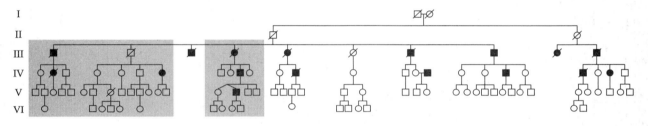

Figure 4.3

Answer:

a. The disease appears to be inherited as an autosomal dominant trait showing reduced penetrance. The trait is transmitted from father to son in several instances, so X-linked inheritance is excluded. The trait shows reduced penetrance, as some unaffected members of the pedigree have affected offspring (e.g., III-2 is unaffected, but his daughter, IV-7, is affected). It also shows delayed expressivity, as the disease does not appear to be present in younger individuals in the pedigree (most of generation V and all of generation VI).

b. If the disease were caused by an agent present in a particular environment (e.g., food, water, pesticide, chemical), the disease should not appear in individuals raised in a completely different environment. That the disease continues to appear in the offspring of pedigree members in generation III who emigrated from Europe is strong evidence that the disease is caused by genetic factors, and not by a particular environmental agent.

c. Some chemicals are known to cause the disease, and there is a relatively high frequency of the disease in adults over 50 years of age. Therefore, it would be very inappropriate to infer that anyone who has the disease has a mutation (notice that in the pedigree shown, individual IV-17, who marries into the family, has the disease). Such an inference can be made only after carefully evaluating an extended family history for the disease and excluding environmental agents as the primary cause of the disease. The late and variable onset of the disease makes such genealogical investigations difficult. It is also difficult to evaluate the potential environmental causes of the disease in just one individual. Since we don't know which environmental factors cause the disease, we also don't know how often disease incidence is related to specific environmental factors. To determine the genetic and environmental contributions to the disease, it is important to study large pedigrees over long periods of time, and undertake epidemiological investigations.

5

QUANTITATIVE GENETICS

CHAPTER OUTLINE

REVIEW OF KEY TERMS, SYMBOLS, AND CONCEPTS

In your own words, write a brief, precise definition of each term in the groups on the next page. Check your definitions using the text. Then develop a concept map using the terms in each list.

1	2	3
quantitative genetics	nature versus nurture	heritability
discontinuous trait	population	broad-sense heritability
continuous trait	sample	narrow-sense heritability
polygenic trait	frequency distribution	phenotypic variance (V_P)
norm of reaction	normal distribution	genetic variance (V_G)
multifactorial trait	binomial distribution	environmental variance (V_E)
familial trait	binomial expansion	genetic-environmental
penetrance	mean	interaction ($V_{G \times E}$)
expressivity	variance	genetic-environmental
pleiotropy	standard deviation	covariation ($COV_{G \times E}$)
epistasis	analysis of variance	additive genetic variance (V_A)
polygene	(ANOVA)	dominance variance (V_D)
contributing alleles	covariance	interaction variance (V_I)
noncontributing alleles	correlation vs. cause and effect	general environmental
polygene or multiple-	correlation vs. identity	effects (V_{Eg})
gene hypothesis for	regression	special environmental effects (V_{Es})
quantitative	regression line	maternal effects (V_{Em})
inheritance	regression coefficient, slope	familial trait
quantitative trait	phenotypic correlation	evolution
locus (QTL)	genetic correlation	artificial selection
marker locus	positive vs. negative	selection response
linkage map	correlation	selection differential

THINKING ANALYTICALLY

Quantitative genetics deals with important but complex issues. Some of the concepts used in the analysis of quantitative traits are subtle. Throughout the chapter, clear and precise thinking is required. To sort out the basis for a quantitative trait, it is important to define the contribution of several different factors. These factors can include the environment as well as multiple genes that themselves may or may not interact. Central to the analysis of a quantitative trait is identifying the degree to which variation in a phenotype can be associated with one or more of these factors. Consequently, understanding how variation is measured and used is of the utmost importance in understanding quantitative genetics.

First, gain a solid understanding of the terms used and the concepts they convey. Geneticists have developed a substantial conceptual framework in which to consider quantitative traits. Some concepts are easy to misunderstand, and to understand what something *is*, it is often important to define what it *is not*. For example, the idea of heritability has important qualifications and limitations. Heritability is a measure of the *proportion of phenotypic variance* that results from genetic differences. It is *not* a measure of the extent to which a trait is genetic, or what proportion of an individual's phenotype is genetic. It is also not fixed for a particular trait or a measurement of genetic differences between populations. The concepts underlying quantitative genetics have significant utility, but only if they are correctly and clearly applied.

Second, concentrate on understanding the statistical methods that are used to analyze data on quantitative traits. Your understanding should consist of more than just knowledge of how to crunch numbers by plugging them into an equation. You need to get a feel for what a statistic tells you about a data set.

Third, relate how measurements are made and analyzed to the concepts that have been developed to explain the inheritance of quantitative traits. For example, relate a measurement of heritability to a selection response. Then explain how you would achieve a particular breeding objective through selection. By doing this, you will obtain a better understanding of the power and utility of quantitative genetics.

QUESTIONS FOR PRACTICE

Multiple Choice Questions

1. Which of the following could be used to describe coat colors in mice?
    a. continuous trait
    b. discontinuous trait
    c. polygenic trait
    d. quantitative trait
    e. both b and c

2. What can generally be said about quantitative traits such as crop yield, rate of weight gain, human birth weight, blood pressure, and the number of eggs laid by *Drosophila*?
    a. They are intractable to molecular genetics.
    b. They are multifactorial.
    c. They are discontinuous.
    d. They are not heritable because they are polygenic.

3. Which statement below best describes the aims of a quantitative genetic analysis of a trait?
    a. to determine whether genes or the environment control a trait
    b. to determine the best breeding strategy to retain a trait
    c. to determine how much of the phenotypic variation associated with a trait in a population is due to genetic variation and how much is due to environmental variation
    d. to determine whether a trait is controlled by nature versus nurture

4. The description of a population in terms of the number of individuals that display varying degrees of expression of a character or range of phenotypes is a
    a. polygraph.
    b. polynomial.
    c. normal distribution.
    d. frequency distribution.

5. The general expression for the binomial expansion is
    a. $(p + q)^n$
    b. $(a^2 + b^2)$
    c. $(a + b^2)$
    d. $a^2 + 2ab + b^2$

6. Which *two* of the following accurately describe the term *variance*?
    a. It is a reflection of the accuracy of an estimated measurement.
    b. It is a measure of how much a set of individual measurements is spread out around the mean.
    c. It is the average value of a set of measurements.
    d. It is equal to $(\Sigma x_i / n)$.
    e. It is equal to $\dfrac{\Sigma(x_i - \bar{x})^2}{n-1}$.

7. What is heritability?
   a. the proportion of a population's phenotype that is attributable to genetic factors
   b. the proportion of a population's phenotypic variation that is attributable to genetic factors
   c. the degree to which family members resemble one another
   d. the degree to which a continuous trait is controlled by genetic factors

8. The proportion of the phenotypic variance that consists of genetic variance, additive or otherwise, is called
   a. heritability.
   b. broad-sense heritability.
   c. narrow-sense heritability.
   d. phenotypic variance derived from genetic-environmental interactions.

9. The narrow-sense heritability of a trait is determined to be very close to 1.0. Which of the following inferences can be made?
   a. The trait will be difficult to select for.
   b. The trait can be readily selected for.
   c. Most of the phenotypic variance results from additive genetic variance.
   d. Most of the phenotypic variance results from environmental variance.
   e. both a and c
   f. both b and c

10. In a population of chickens raised under controlled conditions, body weight is negatively correlated with egg production but positively correlated with egg weight. Because of market conditions, a farmer is interested in producing lots of small eggs. What selection strategy might be beneficial?
    a. Select for smaller hens.
    b. Select for larger hens.

11. Which one of the following statements is true?
    a. Traits shared by family members show high heritability.
    b. Artificial and natural selection can occur in a genetically uniform population.
    c. If heritability is high in each of two populations and the populations differ markedly in a trait, the populations are genetically different.
    d. The selection response depends on (1) the proportion of genetic variance that results from additive genetic variance and (2) the difference between the mean phenotypes of the selected parents and the mean phenotype of the unselected population.

Answers: 1e, 2b, 3c, 4d, 5a, 6b, e, 7b, 8b, 9f, 10a, 11d

Thought Questions

1. Assume that mature fruit weight in pumpkins is a quantitative trait. In the following experiment, environmental factors (weather, soil, etc.) are uniform. Two pumpkin varieties, both of which produce fruit with a mean weight of 20 lb, are crossed. The F_1 produces 20-lb pumpkins. The F_2 plants, however, give the following results:

Mean fruit wt. (lb)	5	12.5	20	27.5	35
Number of plants	19	82	119	79	21

Explain these results: Postulate how many genes are involved and how much each contributes to fruit weight.

2. Distinguish between heritability, broad-sense heritability, and narrow-sense heritability. In what ways can the latter two quantities be measured, and how can they be put to use by plant and animal breeders?

3. What statistics from a regression analysis are used to estimate narrow-sense heritability?

4. What sources contribute to the phenotypic variance associated with a quantitative trait?

5. Distinguish between a genetic correlation and a phenotypic correlation. In particular, address how a phenotypic correlation might exist when the trait is, or is not, influenced by a common set of genes.

6. A positive correlation probably exists between alcohol consumption and the number of Baptist ministers, but these are unlikely to be causally related. What other correlations are likely to be true, but demonstrate that correlations (whether positive or negative) do not imply a cause and effect? Why is a correlation not the same thing as an identity?

7. Defend each of the following statements:
   a. Broad-sense heritability does not indicate the extent to which a trait is genetic.
   b. Heritability does not indicate what proportion of an individual's phenotype is genetic.
   c. Heritability is not fixed for a trait.
   d. If heritability is high in two populations, and the populations differ markedly in a particular trait, one cannot assume that the populations are genetically different.
   e. Familial traits do not necessarily have high heritability.

8. If a population is phenotypically homogeneous for a particular trait, can you predict whether it has a high narrow-sense heritability? If not, what additional information would you need?

9. Why is calculating narrow-sense heritability values for some quantitative human traits especially difficult?

10. Distinguish between a selection differential and a selection response.

11. The mean height in two separate populations A and B is identical and is 5 feet 8 inches. The variance in population A is 14 inches, while the variance in population B is 3 inches. In which population would a 6-foot-tall person be more uncommon?

12. In a single population, is it likely that artificial selection will be equally effective at producing changes in three different traits? Why or why not?

Thought Questions for Media

After reviewing the media on the *iGenetics* CD-ROM, try to answer these questions.

1. In the *Polygene Hypothesis for Wheat Kernel Color* animation, a cross of wheat with red kernels to wheat with white kernels gave wheat with pink kernels. Why does this suggest that this trait shows incomplete dominance?

2. What data given in the animation allow one to rule out incomplete dominance as an explanation for the results illustrated in the animation?

3. An alternate explanation for the results illustrated in the animation was that two independently assorting genes controlled wheat kernel color. Why was this explanation also ruled out?

4. In polygenic inheritance, what is a contributing allele? What is a noncontributing allele? How many contributing alleles can be present at one gene? How many contributing alleles can be present in one organism?

5. How does the explanation of two independently assorting genes with contributing and noncontributing alleles differ from the simpler explanation of two independently assorting genes?

6. Explain why, even though a contributing allele contributes a defined amount of pigment to the red color of the wheat kernel, wheat kernel color still appears to show continuous variation.

SOLUTIONS TO TEXT PROBLEMS

5.1   The following measurements of head width and wing length were made on a series of steamer ducks:

Specimen	Head Width (cm)	Wing Length (cm)
1	2.75	30.3
2	3.20	36.2
3	2.86	31.4
4	3.24	35.7
5	3.16	33.4
6	3.32	34.8
7	2.52	27.2
8	4.16	52.7

a. Calculate the mean and the standard deviation of head width and of wing length for these eight birds.

b. Calculate the correlation coefficient for the relationship between head width and wing length in this series of ducks.

c. What conclusions can you make about the association between head width and wing length in steamer ducks?

Answer:

a. The mean of a sample is obtained by summing all the individual values and dividing by the total number of values. The mean head width is $25.21/8 = 3.15$ cm, and the mean wing length is $281.7/8 = 35.21$ cm.

The standard deviation equals the square root of the variance (s^2). The variance is computed by summing the squares of the differences between each measurement and the mean value, and dividing this sum by the number of measurements minus 1. One has

$$s_{\text{head width}} = \sqrt{s^2} = \sqrt{\frac{\Sigma(x_i - \bar{x})^2}{n-1}} = \sqrt{\frac{1.70}{7}} = \sqrt{0.24} = 0.49 \text{ cm}$$

$$s_{\text{wing length}} = \sqrt{s^2} = \sqrt{\frac{\Sigma(x_i - \bar{x})^2}{n-1}} = \sqrt{\frac{413.35}{7}} = \sqrt{59.05} = 7.68 \text{ cm}$$

b. The correlation coefficient, r, is calculated from the covariance, cov, of two sets of data. Let head width be represented by x and wing length be represented by y. r is defined as

$$r = \frac{\text{cov}_{xy}}{s_x s_y} = \frac{\dfrac{\Sigma x_i y_i - n\bar{x}\bar{y}}{n-1}}{s_x s_y}$$

The first factor ($\Sigma x_i y_i$) is obtained by taking the sum of the products of the individual measurements of head width and wing length for each duck. The next factor is the product of the number of individuals and the means of these two sets of measurements. The difference between these values is then divided by $(n-1)$ and then by the products of the standard deviations of each measurement. One has

$$r = \frac{\dfrac{913.50 - 8 \times 3.15 \times 35.21}{7}}{0.49 \times 7.68} = \frac{3.74}{3.76} = 0.99$$

c. Head width and wing length show a strong positive correlation, nearly 1.0. This means that ducks with larger heads will almost always have longer wings, and ducks with smaller heads will almost always have shorter wings.

5.2   Given the following sets of 30 phenotypic measurements for different traits, decide whether each trait is qualitative or quantitative and explain your answer.

a. Trait 1: 38.9, 47.0, 53.1, 39.1, 62.8, 46.8, 57.5, 54.9, 48.9, 56.3, 52.5, 60.8, 46.7, 48.0, 52.3, 40.7, 50.4, 51.0, 46.5, 47.9, 55.4, 53.1, 58.5, 51.1, 60.2, 50.6, 48.6, 52.5, 54.5, 51.4, 48.1, 49.5, 55.8, 52.9, 42.9, 44.4, 56.4, 38.9, 42.2, 42.2

b. Trait 2: 25.7, 8.8, 11.2, 5.7, 20.6, 34.3, 13.0, 28.8, 20.5, 24.1, 21.2, 14.3, 17.7, 18.7, 24.3, 30.2, 20.2, 25.1, 30.6, 21.2, 31.2, 23.0, 16.9, 10.5, 14.1, 10.2, 30.5, 22.5, 34.1, 10.6, 19.5, 21.0, 20.9, 27.7, 33.0, 7.7, 20.1, 16.9, 18.8, 15.7

c. Trait 3: 31.1, 22.0, 28.1, 14.1, 43.4, 52.8, 32.5, 39.0, 43.1, 52.2, 45.1, 35.8, 36.4, 38.7, 52.8, 42.6, 42.6, 54.8, 43.4, 45.1, 45.1, 49.5, 34.2, 26.1, 35.2, 25.6, 43.1, 48.3, 52.2, 26.4, 40.9, 44.5, 44.3, 36.4, 49.5, 19.4, 42.4, 34.2, 39.0, 31.1

Answer: When given a series of data, the best first step is to graph the data. We can determine the minimum and maximum values, and then create histograms for each series of data using different bin sizes (e.g., by 1, by 2, by 5, etc.) to get a feel for the distribution of the data. Some of these sample histograms follow, along with notes on interpretation and the sources of the original data. One final note is that while many times data from a particular sample do not appear to have a bell-shaped distribution characteristic of a quantitative trait, if we know that we are dealing with a quantitative trait we often assume that they are normally distributed so that we can apply certain statistical techniques in analyzing the data.

a.

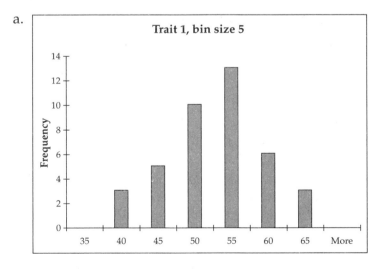

These data appear to be normally distributed, so we could assume that they are representative of phenotypic data we would see for a quantitative trait. In fact, these are 40 sample values taken from a normal distribution with $\mu = 50$ and $s^2 = 5$.

b.

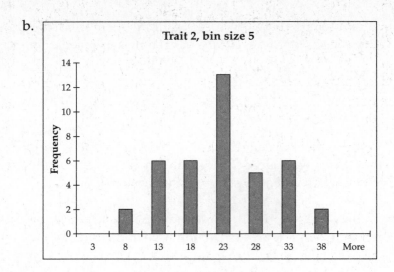

These data appear to have a more pronounced peak in the middle, with no "shoulders" next to the peak like those seen in problem 5.2(a). We could conclude that these data are not representative of a quantitative trait. In fact these data are 10 sample values from a normal distribution with $\mu = 10$ and $s^2 = 2$, 20 values from $\mu = 20$ and $s^2 = 2$, and 10 values from $\mu = 30$ and $s^2 = 2$. This is what we might expect to see from a simple additive Mendelian character with some environmental variance.

c.

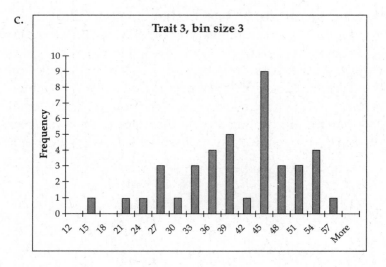

These data have a strong peak, but they do not have the characteristic shape of a normal distribution. You can see that there are no shoulders to the peak and that the peak trails off to the left, but not to the right. In fact, these data include 10 sample values from a normal distribution with $\mu = 25$ and $s^2 = 5$, 20 values from $\mu = 42$ and $s^2 = 5$, and 10 values from $\mu = 55$ and $s^2 = 5$, something we might expect to see from a trait showing simple Mendelian inheritance with a small degree of dominance and substantial environmental variance.

5.3   The F_1 generation from a cross of two pure-breeding parents that differ in a size character usually is no more variable than the parents. Explain.

Answer: The degree of phenotypic variability is related to the degree of genetic variability. Since each pure-breeding parent is homozygous for the genes (however many there are) controlling the size character, the variation seen within parental lines is due only to the environmental variation present. A cross of two pure-breeding strains will generate an F_1 heterozygous for those loci controlling the size trait, but genetically as homogeneous as each of the parents. Therefore, the only variation we would expect to see in the F_1 is from the environment, and it should show no greater variability than the parents.

5.4 Two pairs of independently segregating genes with two alleles each, *A/a* and *B/b*, determine plant height additively in a population. The homozygote *AA BB* is 50 cm tall, and the homozygote *aa bb* is 30 cm tall.

a. What is your prediction of the F_1 height in a cross between the two homozygous stocks?

b. What genotypes in the F_2 will show a height of 40 cm after an $F_1 \times F_1$ cross?

c. What will be the F_2 frequency of the 40-cm plants?

d. What assumptions have you made in answering this question?

Answer:

a. Since the cross is *AA BB* $\times$ *aa bb*, the F_1 genotype will be *Aa Bb*. Since capital alleles determine height additively, and individuals with four capital alleles have a height of 50 cm, while individuals with no capital alleles have a height of 30 cm, each capital allele appears to confer $(50 - 30)/4 = 5$ cm of height over the 30-cm base. *Aa Bb* individuals with two capital alleles should have an intermediate height of 40 cm.

b. Any individuals with two capital alleles will show a height of 40 cm. Thus, *Aa Bb*, *AA bb*, and *aa BB* individuals will be 40 cm high.

c. In the F_2, $\frac{1}{16}$ of the progeny are *AA bb*, $\frac{4}{16}$ are *Aa Bb*, and $\frac{1}{16}$ are *aa BB*. Thus, $\frac{6}{16} = \frac{3}{8}$ of the progeny will be 40 cm high.

d. In answering this question, we have assumed that the *A* and *B* loci assort independently and that each locus and each allele contribute equally to the phenotype.

5.5 Assume that in squashes the difference in fruit weight between a 3-lb type and a 6-lb type results from three independently segregating allelic pairs, *A/a*, *B/b*, and *C/c*. Each capital-letter allele contributes a half pound to the weight of the squash. From a cross of a 3-lb plant (*aa bb cc*) with a 6-lb plant (*AA BB CC*), what will be the phenotypes (weights) of the F_1 and the F_2? What will be their distribution?

Answer: The F_1 will have the genotype *Aa Bb Cc* and so have a weight of $3 + (3 \times 0.5) = 4.5$ lb. The F_2 is expected to show a distribution of different weights. There is a variety of ways to approach this problem. One way to predict the sizes and frequencies of these weights is to use a Punnett square. This is impractical for even four loci, so another is to use the seven coefficients of the binomial expansion $(a + b)^6$ as shown in the first solved problem at the end of the chapter.

These coefficients predict that $\frac{1}{64} = 3$ lb, $\frac{6}{64} = 3.5$ lb, $\frac{15}{64} = 4$ lb, $\frac{20}{64} = 4.5$ lb, $\frac{15}{64} = 5$ lb, $\frac{6}{64} = 5.5$ lb, and $\frac{1}{64} = 6$ lb. Shown here is still another approach, a branch diagram:

$Aa \times Aa$	$Bb \times Bb$	$Cc \times Cc$	Proportion	# Capital-Letter Alleles	Weight
¼AA	¼BB	¼CC	1/64	6	6 lb
		½Cc	2/64	5	5.5 lb
		¼cc	1/64	4	5 lb
	½Bb	¼CC	2/64	5	5.5 lb
		½Cc	4/64	4	5 lb
		¼cc	2/64	3	4.5 lb
	¼bb	¼CC	1/64	4	5 lb
		½Cc	2/64	3	4.5 lb
		¼cc	1/64	2	4 lb
½Aa	¼BB	¼CC	2/64	5	5.5 lb
		½Cc	4/64	4	5 lb
		¼cc	2/64	3	4.5 lb
	½Bb	¼CC	4/64	4	5 lb
		½Cc	8/64	3	4.5 lb
		¼cc	4/64	2	4 lb
	¼bb	¼CC	2/64	3	4.5 lb
		½Cc	4/64	2	4 lb
		¼cc	2/64	1	3.5 lb
¼aa	¼BB	¼CC	1/64	4	5 lb
		½Cc	2/64	3	4.5 lb
		¼cc	1/64	2	4 lb
	½Bb	¼CC	2/64	3	4.5 lb
		½Cc	4/64	2	4 lb
		¼cc	2/64	1	3.5 lb
	¼bb	¼CC	1/64	2	4 lb
		½Cc	2/64	1	3.5 lb
		¼cc	1/64	0	3 lb

Add up the proportion of individuals with the same numbers of capital alleles to obtain:

1/64 → 6 lb

6/64 → 5.5 lb

15/64 → 5 lb

20/64 → 4.5 lb

15/64 → 4 lb

6/64 → 3.5 lb

1/64 → 3 lb

5.6   Refer to the assumptions stated in problem 5.5. Determine the range in fruit weight of the offspring in the following squash crosses:

a. $AA\ Bb\ CC \times aa\ Bb\ Cc$

b. $AA\ bb\ Cc \times Aa\ BB\ cc$

c. $aa\ BB\ cc \times AA\ BB\ cc$.

Answer: In each case, one needs to consider the maximum and minimum number of capital-letter alleles that can be contributed to the progeny, noting that each contributes 0.5 lb above the 3-lb base weight.

a. *AA Bb CC* × *aa Bb Cc*. The gamete from the maternal parent with the most capital alleles would have a genotype of *ABC* and the gamete with the most capital alleles from the paternal parent would have a genotype of *aBC*. Therefore, the heaviest progeny would have squash weights of 5.5 lb. The gamete genotypes with the least numbers of capital alleles are *AbC* and *abc*, making the progeny with the smallest squashes weigh 4 lb.

b. *AA bb Cc* × *Aa BB cc*. The largest offspring squash weights will come from a union of *AbC* and *ABc* gametes, giving the progeny 5-lb squashes. The smallest offspring squash weights will come from a union of *Abc* and *aBc* gametes, giving a progeny with 4-lb squashes.

c. *aa BB cc* × *AA BB cc*. The only gametes that can be produced are *aBc* and *ABc*, so three capital alleles will be contributed to each progeny. The progeny squash weights will all be 4.5 lb.

5.7   Three independently segregating genes (*A*, *B*, *C*), each with two alleles, determine height in a plant. Each capital-letter allele adds 2 cm to a base height of 2 cm.

a. What are the heights expected in the F_1 progeny of a cross between homozygous strains *AA BB CC* and *aa bb cc*?

b. What is the distribution of heights (frequency and phenotype) expected in an $F_1 \times F_1$ cross?

c. What proportion of F_2 plants will have heights equal to the heights of the original two parental strains?

d. What proportion of the F_2 will breed true for height?

Answer:

a. The cross *AA BB CC* × *aa bb cc* will produce all *Aa Bb Cc* progeny. The three capital-letter alleles will add $3 \times 2 = 6$ cm to the 2-cm base height, giving a total of 8 cm.

b. A Punnett square, branch diagram, or binomial expansion can be used to find the following distribution of heights:

$1/64 \rightarrow 14$ cm

$6/64 \rightarrow 12$ cm

$15/64 \rightarrow 10$ cm

$20/64 \rightarrow 8$ cm

$15/64 \rightarrow 6$ cm

$6/64 \rightarrow 4$ cm

$1/64 \rightarrow 2$ cm

c. $1/64$ are 2 cm and $1/64$ are 14 cm, so the answer is $2/64$.

d. Plants must be homozygous for all loci involved in order to breed true. If we do not specify a height, then all homozygous genotypes fulfill this requirement and the answer is just the proportion of homozygous genotypes. There are eight completely homozygous genotypes, and each occurs at a frequency of $1/64$, so the overall proportion of true breeding F_2 individuals is $1/8$.

5.8   Repeat problem 5.7, but assume that one of the loci shows dominance instead of additivity.

Answer: Assume for this problem that the *A* locus shows dominance, and that the *AA* and *Aa* genotypes add 4 cm to height, while the *aa* genotype adds nothing.

a. The cross $AA\,BB\,CC \times aa\,bb\,cc$ will produce all $Aa\,Bb\,Cc$ progeny. The Aa genotype will add 4 cm to height and the other two capital-letter alleles will add $2 \times 2 = 4$ cm to the 2-cm base height, giving a total of 10 cm.

b. A branch diagram is the easiest way to answer this problem. If you construct one as in problem 5.5, you will see that the phenotypes of the groups starting with AA or Aa are the same, but that the Aa group has twice as many. All like phenotypes can be combined to find the following distribution of heights:

$$3/64 \rightarrow 14 \text{ cm}$$
$$12/64 \rightarrow 12 \text{ cm}$$
$$19/64 \rightarrow 10 \text{ cm}$$
$$16/64 \rightarrow 8 \text{ cm}$$
$$9/64 \rightarrow 6 \text{ cm}$$
$$4/64 \rightarrow 4 \text{ cm}$$
$$1/64 \rightarrow 2 \text{ cm}$$

Note the skewed distribution compared to problem 5.7(b).

c. $\frac{1}{64}$ are 2 cm and $\frac{3}{64}$ are 14 cm, so the answer is $\frac{1}{16}$.

d. Even though we have dominance, the answer is still $\frac{1}{8}$, the same as problem 5.7(d), because individuals still need to be homozygous to breed true.

5.9   Assume that three independently segregating, equally and additively contributing pairs of alleles control flower length in nasturtiums. A completely homozygous plant with 10-mm flowers is crossed to a completely homozygous plant with 30-mm flowers. The F_1 plants all have flowers about 20 mm long. The F_2 plants show a range of lengths from 10 to 30 mm, with about $\frac{1}{64}$ of the F_2 having 10-mm flowers and $\frac{1}{64}$ having 30-mm flowers. What distribution of flower length would you expect to see in the offspring of a cross between an F_1 plant and the 30-mm parent?

Answer: Use the clue that $\frac{1}{64}$ of the F_2 progeny show either extreme trait. Note that (as in problems 5.5–5.8) in an F_2 resulting from a trihybrid cross, the proportion of one kind of homozygote is $\frac{1}{64}$. This confirms that three allelic pairs control this quantitative trait. The completely homozygous 10-mm plant would have six lowercase alleles at three loci, $aa\,bb\,cc$, while the completely homozygous 30-mm plant would have six capital-letter alleles at three loci, $AA\,BB\,CC$. If each capital-letter allele contributes $\frac{1}{6}$ of the 20-mm difference between the two extremes, or $\frac{20}{6} = 3.33$ mm, the F_1 trihybrid individual would be 20 mm [= 10 mm base + $3 \times (3.33 \text{ mm/capital-letter allele}) = 20$ mm]. Thus, the trait fits a model in which three capital-letter alleles contribute additively to this quantitative trait.

  If an F_1 plant is crossed to the 30-mm parent, one would have $Aa\,Bb\,Cc \times AA\,BB\,CC$. This cross would produce $\frac{1}{8}$ each of $Aa\,Bb\,Cc$, $Aa\,Bb\,CC$, $AA\,Bb\,Cc$, $Aa\,BB\,Cc$, $AA\,BB\,Cc$, $AA\,Bb\,CC$, $Aa\,BB\,CC$, and $AA\,BB\,CC$. Thus, there will be $\frac{1}{8}$ that are 30 mm long, $\frac{1}{8}$ that are 20 mm long, $\frac{3}{8}$ that are 23.33 mm long, and $\frac{3}{8}$ that are 26.67 mm long.

5.10   An experiment found that the mean internode length in spikes (the floral structures) of the barley variety *asplund* to be 2.12 mm. In the variety *abed binder*, the mean internode length was found to be 3.17 mm. The mean of the F_1 of a cross between the two varieties was approximately 2.7 mm. The F_2 population included individuals similar to both parents, as well as intermediate types. Analysis of the F_3 generation showed that 8 out of the total 125 F_2 individuals of the *asplund* type were true breeding, giving a mean of 2.19 mm.

Nine other F_2 individuals were similar to *abed binder,* and they bred true to type, with a mean internode length of 3.24 mm. Is the internode length in spikes of barley a discontinuous or a quantitative trait? Why?

Answer: Internode length shows the characteristics of a quantitative trait. These characteristics include F_1 progeny that show a phenotype intermediate between the two parental phenotypes, and an F_2 showing a range of phenotypes with extremes in the range of the two parents, some of which have all the original parental alleles.

5.11   Assume that the difference between a corn plant 10 dm (decimeters) high and one 26 dm high results from four pairs of equal and cumulative multiple alleles, with the 26-dm plants being *AA BB CC DD* and the 10-dm plants being *aa bb cc dd.* Make and detail your assumptions, then predict the following:

    a. What will be the size and genotype of an F_1 from across between these two true-breeding types?

    b. Determine the limits of height variation in the offspring from the following crosses:
        i. *Aa BB cc dd × Aa bb Cc dd*
        ii. *aa BB cc dd × Aa Bb Cc dd*
        iii. *AA BB Cc DD × aa BB cc Dd*
        iv. *Aa Bb Cc Dd × Aa bb Cc Dd*

Answer: Assume that the four pairs of alleles are equally responsible for the 16-dm height increase, each allele contributing 2 dm, and that the loci assort independently.

    a. A cross of *AA BB CC DD × aa bb cc dd* will produce *Aa Bb Cc Dd* progeny. With four capital-letter alleles, the F_1 height will be 18 dm.

    b. i. *Aa BB cc dd × Aa bb Cc dd.* The minimum number of capital-letter alleles contributed to the progeny would be one, and the maximum number of capital-letter alleles contributed to the progeny would be four. The height range would be 12 to 18 dm.

       ii. *aa BB cc dd × Aa Bb Cc dd.* At least one and at most four capital-letter alleles would be contributed to the progeny. The height range would be 12 to 18 dm.

       iii. *AA BB Cc DD × aa BB cc Dd.* At least four and at most six capital-letter alleles would be contributed to the progeny. The height range would be 18 to 22 dm.

       iv. *Aa Bb Cc Dd × Aa bb Cc Dd.* A minimum of none and a maximum of seven capital-letter alleles would be contributed to the progeny. The height range would be 10 to 24 dm.

5.12   Refer to the assumptions given in problem 5.11. For this problem, two 14-dm corn plants, when crossed, give nothing but 14-dm offspring (case A). Two other 14-dm plants give one 18-dm, four 16-dm, six 14-dm, four 12-dm, and one 10-dm offspring (case B). Two other 14-dm plants, when crossed, give one 16-dm, two 14-dm, and one 12-dm offspring (case C). What genotypes for each of these 14-dm parents (cases A, B, and C) would explain these results? Would it be possible to get a plant that is taller than 18 dm by selection in any of these families?

Answer: In each of cases A, B, and C, both 14-dm parents must have two capital-letter alleles. Consideration of the possible arrangement of these alleles leads to the solution of each case.

Case A: For two 14-dm plants to give rise only to more 14-dm plants, each of the plants must be homozygous for one of the capital-letter alleles. Both plants could be homozygous for the same capital-letter allele (e.g., *AA bb cc dd* × *AA bb cc dd*), or each could be homozygous for a different capital-letter allele (e.g., *AA bb cc dd* × *aa bb CC dd*). In either case, the progeny plants would have two capital-letter alleles and be 14 dm high. Distinguishing between these two alternatives could be accomplished by intermating the progeny: All 14-dm offspring would be seen with the former, while 10- to 14-dm offspring would be seen with the latter.

Case B: The 1 four capital-letter alleles : 4 three capital-letter alleles : 6 two capital-letter alleles : 4 one capital-letter allele : 1 no capital-letter allele proportions reflect the coefficients in the binomial expansion of $(a + b)^4$. This result suggests that the parents were heterozygous at the two (identical or different) loci (e.g., *Aa Bb cc dd* × *Aa Bb cc dd* or *Aa Bb cc dd* × *aa bb Cc Dd*).

Case C: The 1 one capital-letter allele : 2 two capital-letter alleles : 1 three capital-letter allele proportions are the coefficients in the binomial expansion of $(a + b)^2$. One would expect this ratio if one of the parents were homozygous at one locus and the other heterozygous at two loci. For example, the parents could be *aa BB cc dd* × *Aa Bb cc dd* or *aa BB cc dd* × *Aa bb Cc dd*.

Since each capital-letter allele contributes 2 dm, plants would have to have at least five capital-letter alleles to be taller than 18 dm. To be taller than 18 dm is not possible in case A (the maximum is four), case B (the maximum possible is four), or case C (the maximum possible is four). In order to produce a plant 18 dm tall, we would need to have at least one 16-dm plant and the right 14-dm plant to mate it with.

5.13   Transgressive segregation is the phenomenon in which two pure-breeding strains, differing in a trait, are crossed and produce F_2 individuals with phenotypes that are more extreme than either grandparent (i.e., that are larger than the largest or smaller than the smallest in the original generation). Even if two pure-breeding strains are the same for a quantitative trait, it is possible to see transgressive segregation in an F_2. Propose scenarios with specific assumptions for each of these examples of transgressive segregation.

Answer: In order to see transgressive segregation, at least one of the parents must have some alleles that are "opposite" in effect of the expected direction. For example, imagine we are looking at height. If we assume that there are six loci that contribute to height, and that capital-letter alleles contribute a 5-cm increase over a base height of 1 meter, a cross between an *AA BB CC DD EE FF* individual (160 cm) and an *aa bb cc dd ee ff* individual (100 cm) will produce an F_2 with extreme individuals only as tall and as short as the original parents. If, however, the original parents have the genotypes *AA BB CC DD EE ff* (150 cm) and *aa bb cc dd ee FF* (110 cm), segregation in the F_2 can produce an *AA BB CC DD EE FF* genotype (160 cm) and an *aa bb cc dd ee ff* genotype (100 cm), which are taller and shorter than the original lines used. In this case, the taller parent has "shorter" alleles at one locus and vice versa.

A more extreme case to consider is where the parents have the same phenotype, but produce segregating offspring. Imagine, for example, if an *AA BB CC dd ee ff* (130-cm) individual were crossed with an *aa bb cc DD EE FF* individual (130 cm). Their F_1 offspring would again be 130 cm (*Aa Bb Cc Dd Ee Ff*), but in the F_2, individuals from 160 cm (*AA BB CC DD EE FF*) to 100 cm (*aa bb cc dd ee ff*) could be seen!

5.14 Pigmentation in the imaginary river bottom dweller *Mucus yuccas* is a quantitative character controlled by a set of five independently segregating polygenes with two alleles each: *A/a, B/b, C/c, D/d,* and *E/e.* Pigment is deposited at three different levels, depending on the threshold of gene products produced by the capital-letter alleles. Greyish-brown pigmentation is seen if at least four capital-letter alleles are present; light tan pigmentation is seen if two or three capital-letter alleles are present; and whitish-blue pigmentation is seen if these thresholds are not met. If an *AA BB CC DD EE* animal is crossed to an *aa bb cc dd ee* animal and the progeny are intercrossed, what kinds of phenotypes are expected in the F_1 and F_2?

Answer: The F_1 is *A/a B/b C/c D/d E/e* and so it is greyish-brown. The F_2 phenotypes are determined by the number of capital-letter alleles contributed from each F_1 parent. The easiest way to proceed is to try to find the proportions of individuals with the light tan pigmentation (three or four capital-letter alleles) and the whitish-blue pigmentation (zero or one capital-letter alleles), and the proportion of greyish-brown offspring will be the rest. Start off by determining the chance of obtaining zero, one, two, or three capital-letter alleles in a gamete from each F_1 parent and then ascertain how these combinations make the desired genotypes.

Since the parent is heterozygous at all five loci, the chance of obtaining any specified set of five alleles in one gamete is $(\frac{1}{2})^5$. The chance of obtaining a particular number of capital-letter alleles from an F_1 is the number of ways in which that number of alleles can be obtained, multiplied by $(\frac{1}{2})^5$. There is one way to obtain zero capital-letter alleles, five ways to obtain one capital-letter allele, ten ways of obtaining two capital-letter alleles, and ten ways of obtaining three capital-letter alleles. With this information, we can tabulate the ways in which the progeny classes we are interested in can be formed:

F_1 Gamete #1		F_1 Gamete #2		F_2 Progeny		
Capital-Letter Alleles	Gamete Fraction	Capital-Letter Alleles	Gamete Fraction	Capital-Letter Alleles	F_2 Fraction	Phenotype
0	$(\frac{1}{2})^5$	0	$(\frac{1}{2})^5$	0	$(\frac{1}{2})^{10}$	whitish-blue
0	$(\frac{1}{2})^5$	1	$5(\frac{1}{2})^5$	1	$5(\frac{1}{2})^{10}$	whitish-blue
1	$5(\frac{1}{2})^5$	0	$(\frac{1}{2})^5$	1	$5(\frac{1}{2})^{10}$	whitish-blue
0	$(\frac{1}{2})^5$	2	$10(\frac{1}{2})^5$	2	$10(\frac{1}{2})^{10}$	light tan
1	$5(\frac{1}{2})^5$	1	$5(\frac{1}{2})^5$	2	$25(\frac{1}{2})^{10}$	light tan
2	$10(\frac{1}{2})^5$	0	$(\frac{1}{2})^5$	2	$10(\frac{1}{2})^{10}$	light tan
0	$(\frac{1}{2})^5$	3	$10(\frac{1}{2})^5$	3	$10(\frac{1}{2})^{10}$	light tan
1	$5(\frac{1}{2})^5$	2	$10(\frac{1}{2})^5$	3	$50(\frac{1}{2})^{10}$	light tan
2	$10(\frac{1}{2})^5$	1	$5(\frac{1}{2})^5$	3	$50(\frac{1}{2})^{10}$	light tan
3	$10(\frac{1}{2})^5$	0	$(\frac{1}{2})^5$	3	$10(\frac{1}{2})^{10}$	light tan

Among the F_2, $11(\frac{1}{2})^{10} = 11/1{,}024$ will be whitish-blue and $165(\frac{1}{2})^{10} = 165/1{,}024$ will be light tan. The remaining $[1-176(\frac{1}{2})^{10}] = 848/1{,}024$ will be greyish-brown.

Another method to use to solve this problem is to use the coefficients of the binomial expansion to determine the proportion of progeny with different numbers of capital-letter and lowercase alleles. Let n = total number of alleles, s = number of capital-letter alleles, t = number of lowercase alleles, a = chance of obtaining a capital-letter allele, b = chance of obtaining a lowercase allele, and $x! = (x)(x-1)(x-2) \ldots (1)$, with $0! = 1$. Then the chance P of obtaining progeny with a specified number of each type of allele is given by:

$$P(s,t) = \frac{n!}{s!\,t!} a^s b^t$$

$$\left. \begin{array}{l} P(0,10) = \dfrac{10!}{0!\,10!}\left(\dfrac{1}{2}\right)^0\left(\dfrac{1}{2}\right)^{10} = \dfrac{1}{1,024} \\[3mm] P(1,9) = \dfrac{10!}{1!\,9!}\left(\dfrac{1}{2}\right)^1\left(\dfrac{1}{2}\right)^9 = \dfrac{10}{1,024} \end{array} \right\} \quad \dfrac{11}{1,024} \text{ whitish-blue}$$

$$\left. \begin{array}{l} P(2,8) = \dfrac{10!}{2!\,8!}\left(\dfrac{1}{2}\right)^2\left(\dfrac{1}{2}\right)^8 = \dfrac{45}{1,024} \\[3mm] P(3,7) = \dfrac{10!}{3!\,7!}\left(\dfrac{1}{2}\right)^3\left(\dfrac{1}{2}\right)^7 = \dfrac{120}{1,024} \end{array} \right\} \quad \dfrac{165}{1,024} \text{ light tan}$$

$$1 - \frac{11+165}{1,024} = \frac{848}{1,024} \quad \text{greyish-brown}$$

5.15   Alzheimer's disease (AD) is the leading cause of dementia in older adults. Evidence that genetic alterations are involved in AD comes from three sources: the incidence of AD in first-degree relatives, the incidence in pairs of twins, and pedigree analysis. There is a 24–50 percent risk of AD by age 90 in first-degree relatives of individuals with AD, a 40–50 percent risk of AD in the identical (monozygotic) twin of an individual with AD, and a 10–50 percent risk of AD in the fraternal (dizygotic) twin of an individual with AD. Individuals with AD in a subset of families showing AD have an alteration in the *APP* (amyloid protein) gene on chromosome 21. Individuals with AD in another subset of AD families have a particular allele (*E4*) at the *APOE* (apolipoprotein E) gene on chromosome 19. Individuals homozygous for the *E4* allele have increased risk of AD and earlier disease onset than do heterozygotes. Population studies have shown that 40–50 percent of AD cases are associated with alterations in the *APOE* gene, but less than 1 percent of AD cases are associated with mutations in the *APP* gene.

a. In what sense might AD be considered a polygenic trait?

b. If AD has a genetic basis, why are identical twins not equally affected?

Answer:

a. From the data that are given, it appears that some proportion of cases of AD can be attributed to genetic factors. Multiple genes that increase the risk for AD have been identified, some of which appear to act in a dose-dependent manner. Thus, it could be that a number of different genes contribute to the onset of AD, with some having a greater contribution than others. This is somewhat similar to how polygenic traits control a phenotype, since there, alleles at multiple genes contribute in an additive, dose-dependent fashion to the phenotype.

b. Consider two explanations: First, if AD can be caused by environmental agents, mutation and/or a combination of both environmental agents and mutation, the presence of AD in both twins could be due to the presence of one or more abnormal alleles in both and/or the exposure of both twins to adverse environmental conditions. The presence of AD in only one twin may be due to differences in the exposure of that twin to a contributing or causative environmental agent(s). Second, the presence of a particular allele or a specific mutation may only increase the risk of disease, and not determine its occurrence, since an allele's penetrance may be strongly affected by the environment. In the case of AD, the environmental factors may not be clear-cut or even small in number. There may be multiple environmental factors, some of which may be complex or subtle.

5.16   Since monozygotic twins share all their genetic material and dizygotic twins share, on average, half of their genetic material, twin studies sometimes can be useful for evaluating the genetic contribution to a trait. Consider the following two instances:

An intelligence quotient (IQ) assesses intellectual performance on a standardized test that involves reasoning, ability, memory, and knowledge of an individual language and culture. IQ scores are transformed so that the population mean score is 100 and 95 percent of the individuals have scores in the range between 70 and 130. Observations in the United States and England found that monozygotic twins had an average difference of 6 IQ points, dizygotic twins had an average difference of 11 points, and random pairs of individuals had an average difference of 21 points.

In a large sample of pairs of twins in the United States where one twin was a smoker, 83 percent of monozygotic twins both smoked, whereas 62 percent of dizygotic twins both smoked.

From these data, can you infer the genetic determination of IQ or smoking?

Answer: In neither of these cases is any information provided about the environments in which the twins were raised. While it would initially appear that there is some genetic component in each case, there is no way to evaluate the role of environment. It could be that the identical twins sampled in each case were raised in an identical environment, and this could account for the greater similarity in smoking behavior and IQ results for monozygotic twins than for dizygotic twins.

5.17   A quantitative geneticist determines the following variance components for leaf width in a population of wildflowers growing along a roadside in Kentucky:

Additive genetic variance $(V_A) = 4.2$

Dominance genetic variance $(V_D) = 1.6$

Interaction genetic variance $(V_I) = 0.3$

Environmental variance $(V_E) = 2.7$

Genetic-environmental variance $(V_{G \times E}) = 0.0$

a. Calculate the broad-sense heritability and the narrow-sense heritability for leaf width in this population of wildflowers.

b. What do the heritabilities obtained in (a) indicate about the genetic nature of leaf width variation in this plant?

Answer:

a. The broad-sense heritability of a trait represents the proportion of the phenotypic variance in a particular population that results from genetic differences among individuals, while the narrow-sense heritability measures only the proportion of the phenotypic variance in that population that results from additive genetic variance.

Broad-sense heritability
$$= \text{genetic variance/phenotypic variance} = V_G/V_P$$
$$= (4.2 + 1.6 + 0.3)/(4.2 + 1.6 + 0.3 + 2.7 + 0.0)$$
$$= 6.1/8.8 = 0.69$$

Narrow-sense heritability
$$= \text{additive genetic variance/phenotypic variance}$$
$$= V_A/V_P$$
$$= 4.2/8.8 = 0.48$$

b. About 69 percent of the phenotypic variation in leaf width observed in this population is due to genetic differences among individuals, so we would expect offspring to strongly resemble parents. Only 48 percent of the phenotypic variation is due to additive genetic variation, so there are some loci contributing to the trait that show a degree of dominance. Because the phenotypic variation due to additive genetic variation represents the part of the phenotypic variance that responds to natural selection in a predictable manner, we would expect selection to be able to change the mean phenotype in the population if pressure were applied.

5.18   Members of the inbred rat strain SHR are salt sensitive: They respond to a high-salt environment by developing hypertension. Members of a different inbred rat strain, TIS, are not salt sensitive. Imagine that you placed a population consisting only of SHR rats in an environment that was variable in regard to distribution of salt, so that some rats would be exposed to more salt than others. What would be the heritability of blood pressure in this population?

Answer: SHR rats will continue to respond to salt by developing hypertension. Since the strain is inbred, any variation in blood pressure will result from the amount of exposure to salt, and not from genetic variation. Therefore, heritability for this population will be zero. Similarly, the inbred TIS rats would also have a heritability of zero (and retain a low blood pressure).

5.19   In Kansas, a farmer is growing a variety of wheat called TK138. He calculates the narrow-sense heritability for yield (the amount of wheat produced per acre) and finds that the heritability of yield for TK138 is 0.95. The next year, he visits a farm in Poland and observes that another variety of wheat, UG334, growing there has only about 40 percent as much yield as the TK138 grown on his farm in Kansas. Since he found the heritability of yield in his wheat to be very high, he concludes that the TK138 wheat is genetically superior to the UG334 wheat, and he tells the Polish farmers that they can increase their yield by using TK138. Is his conclusion correct? Why or why not?

Answer: Heritability is a measurement of the genetic variance of a *particular* population in a *specific* environment. Heritability is not fixed for a specific trait. It depends on genetic makeup *as well as* the specific environment of the population in which it is measured. Consequently, heritability cannot be used to make inferences about the basis for the differences between two distinct populations. As the environmental conditions on the two farms differ, the heritability calculated for a population in Kansas cannot be used to infer the future performance of the same strain in another environment. The yield of TK138 grown in Poland would most likely be different than when grown in Kansas, perhaps even lower than the yield of the UG334 variety.

5.20   Dermatoglyphics are the patterns of the ridged skin found on the fingertips, toes, palms, and soles of the feet. (Fingerprints are dermatoglyphics.) Classification of dermatoglyphics frequently is based on the number of triradii: A triradius is a point from which three ridge systems separate at angles of 120°. The number of triradii on all 10 fingers was counted for each member of several families, and the results are tabulated here.

Family	Mean Number of Triradii in the Parents	Mean Number of Triradii in the Offspring
I	14.5	12.5
II	8.5	10.0
III	13.5	12.5
IV	9.0	7.0
V	10.0	9.0
VI	9.5	9.5
VII	11.5	11.0
VIII	9.5	9.5
IX	15.0	17.5
X	10.0	10.0

a. Calculate the narrow-sense heritability for the number of triradii by the regression of the mean phenotype of the parents against the mean phenotype of the offspring.

b. What does your calculated heritability value indicate about the relative contributions of genetic variation and environmental variation to the differences observed in number of triradii?

Answer:

a. The narrow-sense heritability of the number of triradii will equal the slope, b, of the regression line of the mean offspring phenotype on the mean parental phenotype.

$$b = \frac{COV_{xy}}{(s_x)^2} = \frac{\dfrac{\Sigma x_i y_i - n\overline{xy}}{n-1}}{\dfrac{\Sigma(x_i - \overline{x})^2}{n-1}} = \frac{\Sigma x_i y_i - n\overline{xy}}{\Sigma(x_i - \overline{x})^2}$$

For this data set, x is the mean number of triradii in the parents and y is the mean number of triradii in the offspring. Using either a calculator or a spreadsheet or statistics program, you can find the following:

$\Sigma x_i = 111$, $\overline{x} = 11.1$

$\Sigma y_i = 108.5$, $\overline{y} = 10.85$

$$\Sigma(x_i - \bar{x})^2 = 51.4$$
$$\Sigma x_i y_i = 1{,}257.5$$
$$b = \frac{1257.5 - 10 \times 11.1 \times 10.85}{51.4} = \frac{53.15}{51.4} = 1.04$$

b. A slope of 1.04 indicates that additive genetic variation is responsible for essentially all of the observed variation in phenotype. Note that the estimate obtained for h^2 is greater than 1, showing that methods for estimating narrow-sense heritability can overestimate the amount of additive genetic variation among individuals.

5.21 The heights of nine college-age males and the heights of their fathers are presented here.

Height of Son (inches)	Height of Father (inches)
70	70
72	76
71	72
64	70
66	70
70	68
74	78
70	74
73	69

a. Calculate the mean and the variance of height for the sons and for the fathers.

b. Calculate the correlation coefficient for the relationship between the height of father and height of son.

c. Determine the narrow-sense heritability of height in this group by regression of the son's height on the height of father.

Answer: Let x represent the height of the fathers and y represent the height of the sons.

a. Fathers: $\bar{x} = 71.9$, $s^2 = 11.61$
    Sons: $\bar{y} = 70.0$, $s^2 = 10.25$

b. One can calculate $\Sigma x_i y_i = 45{,}333$, so that

$$r = \frac{\text{cov}_{xy}}{s_x s_y} = \frac{\dfrac{\Sigma x_i y_i - n\overline{xy}}{n-1}}{\sqrt{(s_x)^2 (s_y)^2}}$$

$$r = \frac{\dfrac{45{,}333 - 9 \times 71.89 \times 70}{8}}{\sqrt{11.61 \times 10.25}} = \frac{5.29}{10.91} = 0.48$$

c. $b = \dfrac{\text{cov}_{xy}}{(s_x)^2} = \dfrac{5.375}{11.6} = 0.463$

When, as in this case, the mean phenotype of the offspring is regressed against the phenotype of only one parent, the estimate of narrow-sense heritability is $2b = 2(0.46) = 0.92$. Since h^2 is very close to 1, additive effects at the loci involved determine most of the phenotypic variation. Nonadditive factors (genes with dominance, genes with epistasis, environmental factors) appear to contribute little to the phenotypic variation.

5.22   A scientist wants to determine the narrow-sense heritability of tail length in mice. He measures tail length among the mice of a population and finds a mean tail length of 9.7 cm. He then selects the 10 mice in the population with the longest tails: Mean tail length in these selected mice is 14.3 cm. He interbreeds the mice with the long tails and examines tail length in their progeny. The mean tail length in the F_1 progeny of the selected mice is 13 cm.

   Calculate the selection differential, the response to selection, and the narrow-sense heritability for tail length in these mice.

   Answer: The selection differential *(S)* equals $14.3 - 9.7 = 4.6$ cm. The response to selection *(R)* equals $13 - 9.7 = 3.3$ cm. The narrow-sense heritability *(h^2)* equals $R/S = 3.3/4.6 = 0.72$.

5.23   Assume that all phenotypic variance in seed weight in beans is genetically determined and is additive. From a population in which the mean seed weight was 0.88 g, a farmer selected two seeds, each weighing 1.02 g. He planted these and crossed the resulting plants to each other, then collected and weighed their seeds. The mean weight of their seeds was 0.96 g. What is the narrow-sense heritability of seed weight?

   Answer: In this example, there was $0.96 - 0.88 = 0.08$ g of change after one generation of selection. Thus, the response to selection *(R)* was 0.08 g. In this case the selection differential *(S)* is $1.02 - 0.88 = 0.14$ g. The narrow-sense heritability is $h^2 = 0.08/0.14 = 0.57$. Note that (a) under the assumption outlined in the first sentence of the questions that all phenotypic variation is due to additive genetic variation, this would mean there is something wrong because h^2 should be 1; and (b) this is an estimate of h^2 derived from one midparent-offspring comparison—the more families that are included, the better the estimate of h^2!

5.24   The narrow-sense heritability of egg weight in a particular flock of chickens is 0.60. A farmer selects for increased egg weight in this flock. The difference in the mean egg weight of the unselected chickens and the selected chickens is 10 g. How much should egg weight increase in the offspring of the selected chickens?

   Answer:

   Selection response = narrow-sense heritability $\times$ selection differential
   Selection response = 0.60×10 g = 6 g

5.25   Members of a strain of white leghorn chickens are selectively crossed to produce two lines, A and B, that show improved egg production. The progeny from a cross of lines A and B are used for commercial egg production. The selection strategy is shown in Figure 5.A. The mean number of eggs produced in the first egg production year and the mean egg weight (in grams) from hens at an age of 240 days is given for animals at each step of the selection procedure.

a. What is the narrow-sense heritability for the traits at each selection step?

b. Why does the response of the traits to selection change during the selection process?

c. What percentage increase in numbers of eggs produced is obtained when lines A and B are crossed?

d. With the possible exception of dairy cattle, commercial livestock are hybrids produced by crossing breeds, lines, or strains already selected for a set of desirable traits. Why?

Figure 5.A

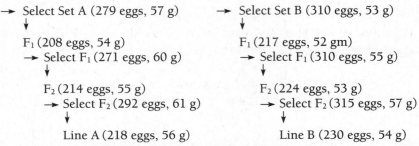

Parental White Leghorn Strain (196 eggs, 51 g)

→ Select Set A (279 eggs, 57 g)   → Select Set B (310 eggs, 53 g)

    F_1 (208 eggs, 54 g) F_1 (217 eggs, 52 gm)
    → Select F_1 (271 eggs, 60 g)   → Select F_1 (310 eggs, 55 g)

        F_2 (214 eggs, 55 g) F_2 (224 eggs, 53 g)
        → Select F_2 (292 eggs, 61 g)   → Select F_2 (315 eggs, 57 g)

           Line A (218 eggs, 56 g)   Line B (230 eggs, 54 g)

Line A × Line B → Hybrid used for commercial production (262 eggs, 57 g)

Answer:

a. The narrow-sense heritability is given in the table here:

Selection Step	Line A		Line B	
	Number of Eggs	**Egg Weight**	**Number of Eggs**	**Egg Weight**
Initial Selection	0.145	0.50	0.184	0.50
F_1 Selection	0.095	0.17	0.075	0.33
F_2 Selection	0.051	0.17	0.066	0.25

b. The response changes because the selection process decreases the amount of genetic variation within the population.

c. Compared to the mean of traits in lines A and B, there is a 17 percent increase in the number of eggs produced, and a 3.6 percent increase in egg weight.

d. When different lines are crossed, an increase due to heterosis, or heterozygote advantage, may be obtained (see Chapter 24). If there are different alleles for a locus that contribute to increased egg production and size, it is possible that alternate alleles were selected for in the two lines. Combining these advantageous alleles across multiple loci could lead to further increases, as well as decrease the effects of recessive, deleterious alleles.

5.26 The following variances were determined for measurements of body length, antenna bristle number, and egg production in a species of moth. Which of these characters would be most rapidly changed by natural selection? Which character would be most slowly affected by natural selection?

Variance	Body Length	Antenna Bristle Number	Egg Production
Phenotypic (V_P)	798	342	145
Additive (V_A)	132	21	21
Dominance (V_D)	122	126	24
Interaction (V_I)	118	136	34
Genetic-environmental ($V_{G \times E}$)	81	23	21
Maternal effects (V_{Em})	345	36	45

Answer: A response to selection depends on (a) variation on which selection can act and (b) a high narrow-sense heritability so that the selected individuals produce similar off-spring. The narrow-sense heritability for each of the traits is V_A/V_P: 0.165 for body length, 0.061 for antenna bristle number, and 0.144 for egg production. The amount of raw variation is also greatest for body length. Thus, body length will respond most to selection, and antenna bristle number will respond least to selection.

5.27   Imagine that you have made the following initial crosses to start your tomato-breeding program. You need to cross a cultivated tomato with a small-fruited, late-flowering wild tomato because the wild tomato has a disease-resistance gene critical for agriculture. The data for each cross below include the lines used as parents (C = cultivated, W = wild), followed by its average fruit weight and days to first flower, along with averages from the progeny (P = F_1 progeny).

Cross Number	Cultivated Parent	Wild Parent	F_1
1	C1 (68 g, 32 d) ×	W1 (6 g, 42 d)	P1 (30 g, 40 d)
2	C1 (68 g, 32 d) ×	W2 (6 g, 41 d)	P2 (38 g, 36 d)
3	C1 (68 g, 32 d) ×	W3 (8 g, 44 d)	P3 (40 g, 41 d)
4	C2 (72 g, 31 d) ×	W1 (6 g, 42 d)	P4 (38 g, 36 d)
5	C2 (72 g, 31 d) ×	W2 (6 g, 41 d)	P5 (34 g, 32 d)
6	C2 (72 g, 31 d) ×	W3 (8 g, 44 d)	P6 (42 g, 42 d)

Which crosses would be your first choices for starting your improvement program? Explain your reasoning.

Answer: Assume that there are multiple loci that contribute equally to fruit weight and days to first flower. In order to recover the cultivated phenotype most quickly from selection after crossing it with the wild genotype, we would like to find the cross where most of the variation is due to additive effects. A quick way to assess this is to look at the phenotype of the F_1: If most of the variation is due to additive effects, the phenotype of the F_1 will be intermediate to both parents. If the F_1 is closer in phenotype to one parent or the other, this can be taken as an indicator that that parent harbors some nonadditive variation. Using this criterion for both traits, crosses 2 and 4 appear to be the best initial crosses to work with.

5.28   Suppose that the narrow-sense heritability of wool length in a breed of sheep is 0.92, and the narrow-sense heritability of body size is 0.87. The genetic correlation between wool length and body size is -0.84. If a breeder selects for sheep with longer wool, what will be the most likely effects on wool length and body size?

Answer: The strong narrow-sense heritability of both wool length and body size indicates that these traits will respond to selection. The negative correlation coefficient between wool length and body size indicates that if longer wool is selected for, you would expect that smaller body size will also be obtained.

6

GENE MAPPING IN EUKARYOTES

CHAPTER OUTLINE

THINKING ANALYTICALLY

When working on a mapping problem, start by systematically organizing the data. This will allow you to extract critical information by inspection. Arrange the data by using symbols to represent the traits, writing out the crosses and their progeny, and tabulating the number of off-spring in each progeny class. After organizing the data, inspect two genes at a time. This often allows you to determine if the two genes are linked, and, if linked, whether they are sex linked. Any trait appearing in progeny in just one sex type should be checked for sex linkage.

Mapping genes using meiotic recombination and testcrosses. It is essential to remember that two genes that show 50 percent recombination in a testcross may either be on different chromosomes or be on the same chromosome but very far apart. The map distances between two genes can exceed 50 map units, but their recombination frequency cannot exceed 50 percent. If two genes showing 50 percent recombination are each linked to a third, intermediate gene (i.e., they show less than 50 percent recombination with the third gene), they will be on the same chromosome but more than 50 map units apart.

In a three-point testcross, a systematic, methodical approach is essential.

1. Assign gene symbols to the phenotypes that are seen and rewrite the crosses.
2. Reorganize the progeny genotypes according to the frequency of each class.

REVIEW OF KEY TERMS, SYMBOLS, AND CONCEPTS

In your own words, write a brief, precise definition of each term in the groups below. Check your definitions using the text. Then develop a concept map using the terms in each list.

1	2
linkage	genetic map
linked genes	meiotic recombination
linkage group	centimorgan (cM), map unit
chromosome	recombination frequency
parental genotypes, classes	coupling vs. repulsion
recombinant genotypes, classes	two-point testcross
genetic recombination	three-point testcross
independent assortment	two-, three-, four-strand, double crossovers
crossing-over	interference
chiasma, chiasmata	coefficient of coincidence
cytological markers	mapping function
genetic, gene markers	
translocation	
reciprocal exchange	

3. Assign parental, single-crossover, and double-crossover classes—parentals will be most frequent, double crossovers will be least frequent.

4. Infer the gene order by comparing the parental and double-crossover classes.

5. Rewrite the testcross; infer which single crossover gives rise to which class.

6. Calculate the recombination frequency and map distance for each gene interval.

7. Calculate the coefficient of coincidence and the interference for the region.

When using the chi-square test to check for linkage, remember that this test, like most statistical tests, serves only as a guide for invalidation of a hypothesis. It does not validate or prove a hypothesis. Carefully consider the hypothesis you are testing and what the values of χ^2 and P mean.

QUESTIONS FOR PRACTICE

Multiple Choice Questions

1. Evidence that genetic recombination is associated with physical exchange between chromosome homolog was found by
   a. observing the existence of chiasmata during meiosis.
   b. observing the pairing of homologs.

    c. using a combined genetic and cytological approach.

    d. showing that genetic recombination was accompanied by an exchange of identifiable chromosomal segments.

    e. both c and d

2. Alleles are considered linked if recombination is less than 50 percent in
    a. a $P \times P \to F_1$ cross.
    b. an $F_1 \times F_1 \to F_2$ cross.
    c. an $F_2 \times F_2 \to F_3$ cross.
    d. an $F_1 \times$ homozygous recessive cross.
    e. any F_2 progeny $\times$ homozygous recessive cross.

3. If two loci *A/a* and *B/b* are 9 map units apart, and an *A b/a B* individual is testcrossed,
    a. there will be 9 percent *A B* individuals.
    b. there will be 91 percent *A b* individuals.
    c. there will be 4.5 percent *a b* individuals.
    d. there will be 91 percent recombinant individuals.

4. If the genes *a* and *b* show 50 percent recombination, genes *a* and *c* show 35 percent recombination, and genes *b* and *c* show 32 percent recombination, then
    a. genes *a* and *b* are on the same chromosome, and 50 map units apart.
    b. genes *a* and *b* are on different chromosomes.
    c. genes *a* and *b* are on the same chromosome, and 67 map units apart.
    d. one cannot tell if genes *a* and *b* are on the same or different chromosomes from the given data.

5. A chi-square test is performed with a hypothesis that two genes are unlinked. A chi-square value of 11.35 is obtained, which in turn gives a value of $P = 0.01$ (i.e., 1 percent). Based on this information, one should
    a. accept the hypothesis as being possible and expect that the genes might be unlinked.
    b. accept the hypothesis as being proven true, and know that the genes are unlinked.
    c. reject the hypothesis as being unlikely and expect that the genes might be linked.
    d. reject the hypothesis as being impossible, and know that the genes are linked.

6. In a three-point cross, interference is a measure of
    a. how often one chiasma physically impedes the occurrence of a second crossover in a particular region.
    b. the accuracy of the map distances between three genes.
    c. whether the data are considered reliable.
    d. whether the genes are linked.

7. In a region in which interference is 0.85,
    a. there are fewer double crossovers than expected.
    b. there are more double crossovers than expected.
    c. one observed the expected number of double crossovers.
    d. 0.85 percent of testcross progeny show double-crossover phenotypes.

8. If two genes *a* and *b* are 8.3 map units apart, one expects to find
    a. 8.3 percent recombinant gametes from a doubly heterozygous parent.
    b. 91.7 percent recombinant gametes from a doubly heterozygous parent.
    c. A chiasma between the *a* and *b* loci in 8.3 percent of the meioses.
    d. A chiasma between the *a* and *b* loci in 91.7 percent of the meioses.

9. Map distances are most accurate when genes are closely linked because as genes become further apart,
   a. they always show independent assortment.
   b. the chance of multiple crossovers increases.
   c. they no longer recombine.
   d. there is chiasma interference.

10. In a three-point testcross, the most frequent classes of progeny were *A B c* and *a b C*. The less frequent classes of progeny were *A b C* and *a B c*. Which gene is in the middle?
    a. *A/a*
    b. *B/b*
    c. *C/c*
    d. It is not possible to tell from just these data.

11. In a three-point testcross with 100 progeny, there were 16 single crossovers between genes *A* and *B*, 29 single crossovers between genes *B* and *C*, and 4 double crossovers. The correct gene order was determined to be *A – B – C*. What is the map distance between *A* and *B*?
    a. 20%
    b. 0.2 mu
    c. 20 mu
    d. 16 mu
    e. 29 mu
    f. 16%

12. In a three-point testcross involving a trihybrid that was *A b C/a B c*, the gene order was *A – B – C* and the value of interference was zero. How do you interpret these results?
    a. There were more single crossovers than double crossovers.
    b. The number of double crossovers observed was exactly the number expected.
    c. A crossover in one interval interfered with a second crossover in the neighboring interval most of the time.
    d. The number of single crossovers between *A* and *B* was the same as the number of single crossovers between *B* and *C*.

Answers: 1e, 2d, 3c, 4c, 5c, 6a, 7a, 8a, 9b, 10a, 11c, 12b.

Thought Questions

1. Do you think that mapping according to the methods described in this chapter will still be useful if the entire DNA sequence of an organism is known? (Hint: Does genetic and/or cytological mapping narrow the field of possibilities?)

2. Why do you think that sex linkage was described before other examples of linkage in animals? (Hint: Is this an example of a correlation of genetic and cytological data?)

3. Does the decision to accept or reject a hypothesis at the 0.05 level mean that one is more likely to discard a correct hypothesis or retain an incorrect one?

4. Do all chromosomes have the same length in terms of map distances?

5. Consider two pairs of genes: *A* and *B*, and *C* and *D*. The number of base pairs of DNA between *A* and *B* is the same as the number of base pairs of DNA between *C* and *D*. Can you infer from this information alone whether these pairs of genes have the same or different map distances between them? Why or why not?

6. In some organisms, correlations have been made between the physical distance between two closely linked loci (i.e., the number of base pairs of DNA) and the map distance between them. If one measures the recombination frequency between two genes in another chromosomal region, can this information be used to infer the physical distance between them as well? (Hint: Consider the basis of interference and the observation that different regions of a chromosome show different amounts of crossing-over.)

7. Compared to a chromosomal region that shows no interference, in a chromosomal region that shows high levels of interference, would you expect there to be the same, greater, or fewer megabases of DNA per map unit?

Thought Questions for Media

After reviewing the media on the *iGenetics* CD-ROM, try to answer these questions.

1. In Curt Stern's cross between a *car B/car⁺ B⁺* female and a *car B⁺/Y* male, how many cytologically different X chromosomes were used? How could each be distinguished, and why was this important for the success of the experiment?

2. Suppose no crossover occurred between the *car/car⁺* and *B/B⁺* loci in the parental female. What eye color, eye shape and cytological phenotypes would be observed? How would your answers differ if a crossover occurred between these two loci?

3. Would the results in the three-point testcross animation be different if the gene order used in the analysis was the one given initially? How?

4. In the chi-square test animation, how, even without assigning genotypes, can you tell that the red-eyed, dumpy-winged and the purple-eyed, normal-winged flies are recombinants?

5. Why is the hypothesis of independent assortment reasonable? When using a chi-square test to evaluate linkage, why not hypothesize that the genes are linked and show a certain recombination frequency?

6. When independent assortment was evaluated using the chi-square test, the data were organized into two classes. Since four different phenotypes were observed, why not organize the data into four classes? If you did this, what parameters in the chi-square test change, and how does it alter the test's results?

7. For a given degree of freedom, the values of P and χ^2 are inversely related. Explain why.

8. What is the map distance between *pr* and *c?*

1. When the initial data generated using testcrosses were analyzed, a hypothesis of "no linkage" was tested.
   a. What was the rationale for testing this hypothesis instead of testing a hypothesis of "linkage"?
   b. Why was this hypothesis not accepted?
   c. Why, in this case, does this suggest that the three genes are linked?

2. When mapping genes using data from trihybrid crosses, it is essential to determine the gene order. This is done by comparing the parental-type and double-crossover-type progeny.
   a. In the trihybrid testcross, how do you know which progeny classes result from double-crossover events?

   b. How does knowing which progeny classes result from a double crossover help you determine the gene order?

3. When mapping the distance between a pair of genes in the trihybrid crosses, the number of recombinants in one of the intervals is equal to the sum of the number of single crossovers in that interval plus the number of double crossovers. Why does the number of double crossovers need to be added?

SOLUTIONS TO TEXT PROBLEMS

6.1   The cross $a^+ a^+ b^+ b^+ \times aa\ bb$ produces an F_1 that is phenotypically $a^+ b^+$. Its F_2 phenotypes appear in the following numbers:

a^+	b^+	110
a^+	b	16
a	b^+	19
a	b	15
Total		160

What F_2 numbers would be expected if the a and b loci assorted independently? Use a chi-square test to evaluate whether they are linked or assort independently.

Answer: If two genes are unlinked in a dihybrid cross, one expects a $9:3:3:1$ ratio in the F_2. For 160 progeny, one would expect $90:30:30:10$. Testing the hypothesis of no linkage using a χ^2 test, one has $\chi^2 = [(110 - 90)^2/90 + (16 - 30)^2/30 + (19 - 30)^2/30 + (15 - 10)^2/30] = 17.5$, df $= 3$, $P < 0.001$. This hypothesis is rejected as being unlikely. It is likely that the genes are linked.

6.2   In corn, a dihybrid for the recessives a and b is testcrossed. The distribution of the phenotypes is as follows:

A	B	122
A	b	118
a	B	81
a	b	79

Test the hypothesis that these genes assort independently using a chi-square test. Explain tentatively any deviation from expected values, and tell how you would test your explanation.

Answer: Use the χ^2 test to evaluate the hypothesis that the genes are unlinked. With this hypothesis, one would expect the cross $Aa\ Bb \times aa\ bb$ to produce four equally frequent classes of progeny (i.e., a $1:1:1:1$ ratio). With 400 total progeny, one would expect 100 progeny in each phenotypic class. One has:

Phenotype	Observed (= *o*)	Expected (= *e*)	$(o - e)^2/e$
$A\ B$	122	100	4.84
$A\ b$	118	100	3.24
$a\ B$	81	100	3.61
$a\ b$	79	100	4.41
		$\chi^2 = 16.1$; df $= 3$, $P < 0.01$	

From the χ^2 test, $P < 0.01$. That is, there is less than a 1 percent likelihood of observing this much deviation from expected values by chance alone. This indicates that one can reject the hypothesis of "no linkage" as being unlikely. Note that the chi-square test neither proves that the genes are unlinked nor proves that they are linked.

Linkage might seem reasonable until examination of the progeny reveals that the minority classes are not reciprocal classes (both carry the *aa* phenotype). If the segregation at each locus is considered, however, the $B/-:b/b$ ratio is about $1:1$ (203:197), while the $A/-:a/a$ ratio is not (240:160). The departure of the observed from expected values is thus due to a reduced number of *a/a* individuals. This departure should be confirmed in other crosses that test the segregation at the *A/a* locus. It would be particularly informative to check if the lack of *a/a* corn plants is associated with ungerminated seeds or seedlings that die before they are able to mature.

6.3   The F_1 from a cross of *A B/A B* $\times$ *a b/a b* is testcrossed, resulting in the following phenotypic ratios:

A B	308
A b	190
a b	292
a B	210

What is the frequency of recombination between genes *a* and *b*?

Answer: Recombination frequency
$$= (\text{\# recombinants/total}) \times 100\%$$
$$= [(190 + 210)/(308 + 190 + 292 + 210)] \times 100\%$$
$$= 40\%$$

6.4   In *Drosophila*, the mutant black (*b*) has a black body, and the wild type has a grey body; the mutant vestigial (*vg*) has wings that are much shorter and crumpled, compared with the long wings of the wild type. In the following cross, the true-breeding parents are listed together with the counts of offspring of F_1 females $\times$ black and vestigial males:

P   black and normal $\times$ grey and vestigial
F_1 females $\times$ black and vestigial males

Progeny:	grey, normal	283
	grey, vestigial	1,294
	black, normal	1,418
	black, vestigial	241

Use these data to calculate the map distance between the black and vestigial genes.

Answer: The cross is $b\,vg^+/b\,vg^+ \times b^+\,vg/b^+\,vg$, which gives an F_1 that is $b\,vg^+/b^+\,vg$. An F_1 female is crossed with a homozygous recessive $b\,vg/b\,vg$. The female is used as the doubly heterozygous parent, as no crossing-over occurs in male *Drosophila*. The classes can be grouped in reciprocal pairs as follows:

Nonrecombinants (parentals):	grey, vestigial	1,294
	black, normal	1,418
Recombinants (nonparentals):	grey, normal	283
	black, vestigial	241
	Total progeny	3,236

Recombination frequency = $[(283 + 241)/3{,}236] \times 100\% = 16.2\%$

There are 16.2 map units between the genes.

6.5   In *Drosophila*, the vestigial (*vg*) gene is located on chromosome 2. Homozygous *vg/vg* animals have incompletely formed vestigial wings; *vg*+/− animals have wild-type long wings. A new eye mutation called maroonlike (*m*) is isolated. Homozygous *m/m* animals have maroon-colored eyes; *m*+/− animals have wild-type bright-red eyes. The location of the *m* gene is unknown, and you are asked to design an experiment to determine whether it is on chromosome 2.

     You cross virgin maroon-eyed females to vestigial males and obtain all wild-type F_1 progeny. Then you allow the F_1 offspring to interbreed. As soon as the F_2 offspring start to hatch, you begin to classify the flies. Among the first six newly hatched flies, you find four wild type, one vestigial-winged red-eyed fly, and one vestigial-winged maroon-eyed fly. You immediately conclude that (1) *m* is not X linked and (2) *m* is not linked to *vg*. How could you tell on the basis of this small sample? On what chromosomes might *m* be located? (Hint: There is no crossing-over in male *Drosophila* flies.)

Answer: To eliminate the possibility that *m* is X linked, consider what kinds of progeny you would expect if it were. If *m* were X linked, one could write the parental cross as: *m/m vg*+/*vg*+ × *m*+/Y *vg/vg*. The F_1 males would necessarily be *m*/Y *vg*+/*vg*, and so be maroon eyed. Since the F_1 are all wild type, *m* cannot be X linked.

     To eliminate the possibility that *vg* is linked to *m*, consider the kinds of progeny that could be obtained if it were. If *vg* and *m* are linked, one could write the crosses as:

$$P \qquad \frac{vg^+\, m}{vg^+\, m}\; \female \; \times \; \frac{vg\, m^+}{vg\, m^+}\; \male$$

$$F_1 \qquad \frac{vg^+\, m}{vg\, m^+}\; \text{(All wild type)}$$

The $F_1 \times F_1$ cross is $\dfrac{vg^+\, m}{vg\, m^+}\; \female \; \times \; \dfrac{vg^+\, m}{vg\, m^+}\; \male$

     Eggs produced by F_1 females may be *vg m*+ (parental), *vg*+ *m* (parental), *vg m* (recombinant), or *vg*+ *m*+ (recombinant). Assuming that the two genes are linked, and remembering that there is no crossing-over in *Drosophila* males, the F_1 males' gametes must be *vg m*+ or *vg*+ *m* (parental types). That is, if the two genes are on the same chromosome, no matter how far apart they are, the F_1 males' gametes must be these parental types. Since none of the F_1 males' gametes are *vg m*, it is impossible for the progeny of the F_1 cross to be homozygous *vg m/vg m*. (Check this by diagramming a Punnett square.) However, since one of the progeny of the F_1 cross is both vestigial winged and maroon eyed, it must have obtained the *m* and *vg* alleles from both parents. Given the genotype of the males, this could occur only through independent assortment (i.e., if the two genes are on different chromosomes). Since the *m* gene is not on the X or chromosome 2, it must be on chromosome 3 or 4.

6.6   Use the following two-point recombination data to map the genes concerned, and show the order and the length of the shortest intervals:

Gene Loci	% Recombination
a,b	50
a,c	15
a,d	38
a,e	8
b,c	50
b,d	13
b,e	50
c,d	50
c,e	7
d,e	45

Answer: Because of the effects of double crossovers, larger recombination frequencies are less accurate measures of map distances than smaller recombination frequencies between close neighbors. As a consequence, the map distances that can be derived from these data are not strictly additive (e.g., $a - d = 38$ and $a - e = 8$, but $d - e = 45$). The distances between nearest neighbors can be used to infer the following map.

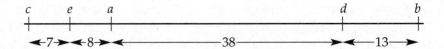

Remember that although recombination frequency cannot exceed 50 percent, map distances can exceed 50 map units. Thus, the map distance between c and b can be inferred to be 66 map units ($= 7 + 8 + 38 + 13$), although c and b show only 50 percent recombination.

6.7   Use the following two-point recombination data to map the genes concerned, and show the order and the length of the shortest intervals:

Gene Loci	% Recombination	Gene Loci	% Recombination
a,b	50	c,d	50
a,c	17	c,e	50
a,d	50	c,f	7
a,e	50	c,g	19
a,f	12	d,e	7
a,g	3	d,f	50
b,c	50	d,g	50
b,d	2	e,f	50
b,e	5	e,g	50
b,f	50	f,g	15
b,g	50		

Answer: As discussed in the solution to problem 6.6, the most accurate map distances are obtained by summing the map distances between nearest neighbors. It is important to remember that while the percent recombination between two genes cannot exceed 50 percent, map distances can exceed 50 map units. When the percent recombination between two genes is 50 percent, either they are far apart on the same chromosome or on different chromosomes.

In some cases, one of these two options can be eliminated: If two genes showing 50 percent recombination both show evidence of linkage to other, intervening loci after sufficient data are gathered, these two genes lie far apart on the same chromosome. The map distances between the two distant loci can be determined by adding up the distances between the intervening loci.

Now consider the other option: The two genes show 50 percent recombination and no data exist to show linkage of each of them to intervening loci. For these genes, it will not be possible to distinguish between the genes lying on different chromosomes and the genes being very far apart on the same chromosome.

Start this problem by constructing a map using the smallest map distances, and then building on it as you check for linkage with other genes. The solution is:

With only the data provided in this problem, it is not possible to tell whether these two linkage groups lie at least 50 map units apart on the same chromosome or are associated with different chromosomes.

6.8   The following data are from Bridges and Morgan's work on recombination between the genes *black* (black body color), *curved* (curved wings), *purple* (purple eyes), *speck* (black specks on wings), and *vestigial* (crumpled wings) in chromosome 2 of *Drosophila*:

Genes in Cross	Total Progeny	Number of Recombinants
black, curved	62,679	14,237
black, purple	48,931	3,026
black, speck	685	326
black, vestigial	20,153	3,578
curved, purple	51,136	10,205
curved, speck	10,042	3,037
curved, vestigial	1,720	141
purple, speck	11,985	5,474
purple, vestigial	13,601	1,609
speck, vestigial	2,054	738

On the basis of the data, map the chromosome for these five genes as accurately as possible. (Remember that determinations for short distances are more accurate than those for long ones.)

Answer: Determine recombination frequencies for each pair of genes by dividing the number of recombinants by the total number of progeny for that pair of genes. One obtains:

Genes in Cross	Percent of Total Progeny
black, curved	22.7
black, purple	6.2
black, speck	47.6
black, vestigial	17.8
curved, purple	20.0
curved, speck	30.2
curved, vestigial	8.2
purple, speck	45.7
purple, vestigial	11.8
speck, vestigial	35.9

Now analyze these data to determine the order of the genes. One approach is to consider that the pair of genes that is farthest apart is the pair with the greatest recombination frequency, and then order genes from one or the other end. The genes *black* and *speck* are farthest apart:

Now fill in the order by finding the sequence of genes that has an increasing order of recombination with *black:*

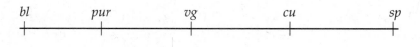

Next, assign map distances based on the recombination frequencies of neighboring genes:

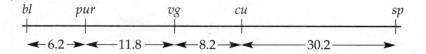

Check gene order by choosing one of the genes (e.g., *curved*) and verifying that the distances between that gene and the others are approximately correct. Because of multiple crossovers, larger recombination frequencies tend to give underestimates of map distance. The most accurate map distances between two genes can be determined by summing the map distances in all intervals between the genes. Thus, the map distance from *bl* to *sp* is most accurately given by 56.4 (= 6.2 + 11.8 + 8.2 + 30.2) map units, and not 47.6 (= RF × 100%) map units.

6.9   A corn plant known to be heterozygous at three loci is testcrossed. The progeny phenotypes and numbers are as follows:

+	+	+	455
a	b	c	470
+	b	c	35
a	+	+	33
+	+	c	37
a	b	+	35
+	b	+	460
a	+	c	475
	Total		2,000

Give the gene arrangement, linkage relationships, and map distances.

Answer: First, note that since the offspring result from a testcross, the phenotypes of the offspring reflect the genotypes of the gametes of the heterozygous parent. Consequently, these phenotypes can be used to determine the gene arrangement in the heterozygous parent.

Second, notice that there are eight genotypic classes that appear in two frequencies. Four of the genotypic classes have about 460 individuals, while the other four have about 35 individuals. This would be expected if two of the genes were linked and the third was not. In this case, the chromosome bearing the two linked genes could undergo crossing-over and give rise to two frequencies of gametes. Because crossing-over between two closely linked genes is infrequent, gametes with noncrossover chromosomes are more frequent. The third gene assorts independently. The independent assortment of the third gene results in four, and not just two, genotypes present in each of two frequencies.

With this information, we can proceed in either of two ways. First, we can tabulate the data considering only two pairs of genes at one time. This allows us to view this cross as a series of two-point crosses.

a	b	c	# Progeny
a	b		470 + 35 = 505
+	b		460 + 35 = 495
a	+		475 + 33 = 508
+	+		455 + 37 = 492
a		c	470 + 475 = 945
+		c	35 + 37 = 72
a		+	33 + 35 = 68
+		+	455 + 460 = 915
	b	c	470 + 35 = 505
	+	c	475 + 37 = 512
	b	+	460 + 35 = 495
	+	+	455 + 33 = 488

Notice that there is a 1 : 1 : 1 : 1 ratio in the offspring when one considers just the *a* and *b* genes or just the *b* and *c* genes. This is what we would expect in a dihybrid cross with two independently assorting genes. Thus, *a* assorts independently of *b*, and *b* assorts independently of c. This is not the case for a and c, however. Here, there are two frequencies of data indicative of two linked genes showing recombination. The recombination frequency between a and c is RF = [(72 + 68)/(72 + 68 + 945 + 915)] × 100% = 7%, and there are 7 map units between a and c.

Alternatively, we can use the reasoning presented above to infer that the four progeny classes with about 460 individuals contain parental types of chromosomes for two linked genes, as well as an independently assorting chromosome with the third gene. By inspection of the most frequent phenotypes, we see that only a c and + + chromosomes are found, while the b/+ gene assorts independently from these noncrossover chromosomes. Thus, the a and c loci are linked, and the b gene is unlinked. The least frequent progeny classes contain a + and + c recombinant chromosomes, so that the percent recombination between a and c is RF = [(35 + 37 + 33 + 35)/2,000] × 100% = 7%. Just as we found in the first approach, the map distance between a and c is 7 map units.

We can diagram these results as

6.10   Genes a and b are linked, with 10 percent recombination. What would be the phenotypes, and the probability of each, among progeny of the following cross?

$$\frac{a \quad b^+}{a^+ \ b} \times \frac{a \quad b}{a \quad b}$$

Answer: The $a\,b^+/a^+\,b$ parent will give 90 percent parental-type ($a\,b^+$ or $a^+\,b$) gametes and 10 percent recombinant-type gametes ($a\,b$ or $a^+\,b^+$). The genotypes of these gametes will determine the offspring phenotypes, as the $a\,b/a\,b$ parent will give only $a\,b$ gametes. Therefore, there will be 45% $a\,b^+$, 45% $a^+\,b$, 5% $a\,b$, and 5% $a^+\,b^+$ offspring.

6.11   Genes a and b are sex linked and are located 7 mu apart on the X chromosome of *Drosophila*. A female of genotype $a^+\,b/a\,b^+$ is mated with a wild-type male ($a^+\,b^+/Y$).
a. What is the probability that one of her sons will be either $a^+\,b^+$ or $a\,b^+$ in phenotype?
b. What is the probability that one of her daughters will be $a^+\,b^+$ in phenotype?

Answer: Since the genes are 7 mu apart, the female will have 93 percent parental-type gametes (46.5% $a^+\,b$ and 46.5% $a\,b^+$) and 7 percent recombinant-type gametes (3.5% $a\,b$ and 3.5% $a^+\,b^+$). The wild-type male will give either an X chromosome bearing $a^+\,b^+$ or a Y chromosome. As the genes are X-linked, the phenotype of the sons will reflect the chromosome they receive from the mother. The phenotype of the daughters will be $a^+\,b^+$, as they receive the $a^+\,b^+$ X chromosome from their father.
a. $P = 3.5\% \ (a^+\,b^+) + 46.5\% \ (a\,b^+) = 50\%$.
b. $P = 100\%$.

6.12   In maize, the dominant genes A and C are both necessary for colored seeds. Homozygous recessive plants give colorless seed, regardless of the genes at the second locus. Genes A and C show independent segregation, and the recessive mutant gene waxy endosperm (wx) is linked with C (20 percent recombination). The dominant Wx allele results in starchy endosperm.

a. What phenotypic ratios would be expected when a plant of constitution *c Wx/C wx A/A* is testcrossed?

b. What phenotypic ratios would be expected when a plant of constitution *c Wx/C wx A/a* is testcrossed?

Answer: Use branch diagrams to assess what kind of gametes will arise from the heterozygous plants.

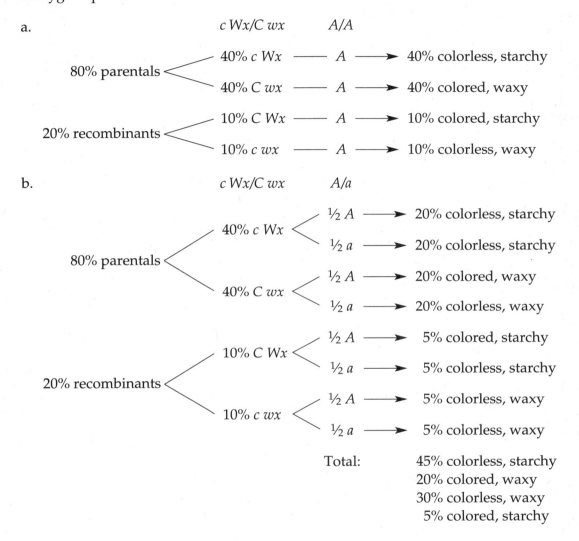

a. *c Wx/C wx*　　*A/A*

80% parentals
- 40% *c Wx* —— *A* ⟶ 40% colorless, starchy
- 40% *C wx* —— *A* ⟶ 40% colored, waxy

20% recombinants
- 10% *C Wx* —— *A* ⟶ 10% colored, starchy
- 10% *c wx* —— *A* ⟶ 10% colorless, waxy

b. *c Wx/C wx*　　*A/a*

80% parentals
- 40% *c Wx*
  - ½ *A* ⟶ 20% colorless, starchy
  - ½ *a* ⟶ 20% colorless, starchy
- 40% *C wx*
  - ½ *A* ⟶ 20% colored, waxy
  - ½ *a* ⟶ 20% colorless, waxy

20% recombinants
- 10% *C Wx*
  - ½ *A* ⟶ 5% colored, starchy
  - ½ *a* ⟶ 5% colorless, starchy
- 10% *c wx*
  - ½ *A* ⟶ 5% colorless, waxy
  - ½ *a* ⟶ 5% colorless, waxy

Total:　45% colorless, starchy
　　　　20% colored, waxy
　　　　30% colorless, waxy
　　　　5% colored, starchy

6.13 In tomatoes, tall vine is dominant over dwarf, and spherical fruit shape is dominant over pear shape. Vine height and fruit shape are linked, showing 20 percent recombination. A certain tall, spherical-fruited tomato plant is crossed with a dwarf, pear-fruited plant. The progeny are 81 tall, spherical; 79 dwarf, pear; 22 tall, pear; and 17 dwarf, spherical. Another tall and spherical plant crossed with a dwarf and pear plant produces 21 tall, pear; 18 dwarf, spherical; 5 tall, spherical; and 4 dwarf, pear. What are the genotypes of the two tall and spherical plants? If they were crossed, what types and frequencies of offspring would they produce?

Answer: Let *T* represent tall vine, *t* represent dwarf, *S* represent spherical, and *s* represent pear. Then, the initial tall, spherical × dwarf, pear cross can be represented as *T S/ − − × t s/t s*. Since this is a testcross, the genotypes of the gametes of the (potentially) heterozygous plant can be inferred from the progeny phenotypes:

Phenotype	Gamete Genotype	#	First Cross Type	#	Second Cross Type
tall, spherical	*T S*	81	parental	5	recombinant
dwarf, pear	*t s*	79	parental	4	recombinant
tall, pear	*T s*	22	recombinant	21	parental
dwarf, spherical	*t S*	17	recombinant	18	parental

In the first cross, the four classes of gametes, their phenotypes and frequencies indicate that the parental genotypes were *T S/t s × t s/t s*. Similarly, the four classes of progeny and their frequencies in the second cross indicate that it is *T s/t S × t s/t s*. If the two tall and spherical plants were crossed, one would have *T S/t s × T s/t S*. One can diagram the expected progeny using a branch diagram:

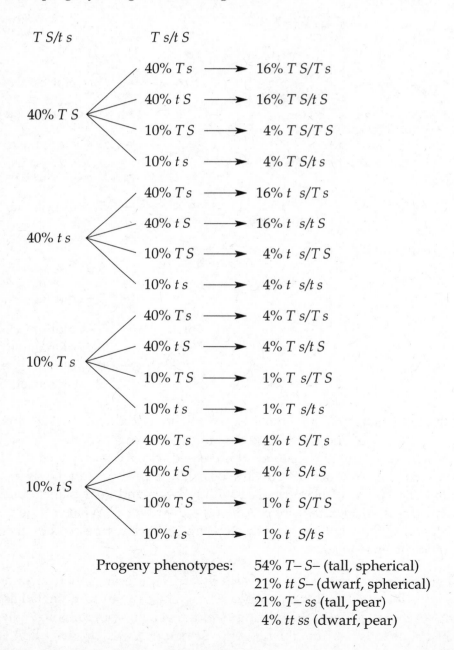

Progeny phenotypes: 54% *T– S–* (tall, spherical)
21% *tt S–* (dwarf, spherical)
21% *T– ss* (tall, pear)
 4% *tt ss* (dwarf, pear)

6.14   Genes *a* and *b* are on one chromosome, 20 mu apart; *c* and *d* are on another chromosome, 10 mu apart. Genes *e* and *f* are on yet another chromosome and are 30 mu apart. A homozygous *A B C D E F* individual is crossed to an *a b c d e f* individual, and the resulting F_1 is crossed back to an *a b c d e f* individual. What are the chances of getting individuals of the following phenotypes in the progeny?

a. *A B C D E F*

b. *A B C d e f*

c. *A b c D E f*

d. *a B C d e f*

e. *a b c D e F*

Answer: The frequency of recombination between two genes can be inferred from the map distance between them. For example, the 20 mu between *a* and *b* indicate that in an *AB/ab* individual, gametes will have 80 percent parental-type chromosomes (40 percent each of *AB, ab*) and 20 percent recombinant-type chromosomes (10 percent each of *Ab, aB*). The cross under consideration is *AB/ab; CD/cd; EF/ef* × *ab/ab; cd/cd; ef/ef*.

a. The chance of obtaining the parental-type *AB, CD,* and *EF* chromosomes is $P = 0.40 \times 0.45 \times 0.35 = 0.063$ or 6.3%.

b. The chance of obtaining *AB, Cd,* and *ef* chromosomes is $P = 0.40 \times 0.05 \times 0.35 = 0.007$ or 0.7%.

c. The chance of obtaining *Ab, cD,* and *Ef* chromosomes is $P = 0.10 \times 0.05 \times 0.15 = 0.00075$ or 0.075%.

d. The chance of obtaining *aB, Cd,* and *ef* chromosomes is $P = 0.10 \times 0.05 \times 0.35 = 0.00175$ or 0.175%.

e. The chance of obtaining the *ab, cD,* and *eF* chromosomes is $P = 0.40 \times 0.05 \times 0.15 = 0.003$ or 0.3%.

6.15   Genes *d* and *p* occupy loci 5 mu apart in the same autosomal linkage group. Gene *h* is in a different autosomal linkage group. What types of offspring are expected, and what is the probability of each, when individuals of the following genotypes are testcrossed?

a. $\dfrac{D\ P}{d\ p}\ \dfrac{h}{h}$ b. $\dfrac{d\ P}{D\ p}\ \dfrac{H}{h}$

Answer:

a. 95% of the progeny will have *D P* or *d p* parental-type chromosomes, and 5% of the progeny will have *D p* or *d P* recombinant-type chromosomes. All of the progeny will be *h*. There will be 47.5% *D P h*, 47.5% *d p h*, 2.5% *D p h*, 2.5% *d P h*.

b. 23.75% *d P H*, 23.75% *d P h*, 23.75% *D p H*, 23.75% *D p h*, 1.25% *D P H*, 1.25% *D P h*, 1.25% *d p H*, 1.25% *d p h*.

6.16   A hairy-winged (*h*) *Drosophila* female is mated with a yellow-bodied (*y*), white-eyed (*w*) male. The F_1 are all wild type. The F_1 progeny are then crossed, and the F_2 that emerge are as follows:

Females:	wild type	757
	hairy	243

Males:	wild type	390
	hairy	130
	yellow	4
	white	3
	hairy, yellow	1
	hairy, white	2
	yellow, white	360
	hairy, yellow, white	110

Give genotypes of the parents and the F_1 and show the linkage relations and distances where possible.

Answer: Careful inspection of the data shows that the F_2 males, but not the F_2 females, show the yellow and white traits. In contrast, the hairy trait is shown in an approximately 3:1 ratio in both the females and males (females = 757 normal : 243 hairy; males = 757 normal : 243 hairy). This indicates that the yellow and white genes are sex linked, while hairy is not.

The normal phenotype of the F_1 heterozygote indicates that the nonhairy, nonyellow, and nonwhite traits are dominant. Therefore, let h and h^+ represent the hairy and nonhairy traits, respectively, y and y^+ represent the yellow and normal body-color traits, respectively, and w and w^+ represent the white- and red-eyed traits, respectively. Then, given the phenotypes of the parents and F_1 and the sex linkage of y and w, the crosses can be written as

P: $y^+ w^+/y^+ w^+ h/h$ females $\times y w/Y h^+/h^+$ males
F_1: $y^+ w^+/y w h^+/h$ females $\times y^+ w^+/Y h^+/h$ males

Now determine the map distance between y and w. This can be inferred from the frequency of recombinants among the F_2 males.

Phenotype of F_2 Males	Genotype	X-Chromosome Recombinant?	Number
wild type	$y^+ w^+/Y h^+/-$	no	390
hairy	$y^+ w^+/Y h/h$	no	130
yellow	$y w^+/Y h^+/-$	yes	4
white	$y^+ w/Y h^+/-$	yes	3
hairy, yellow	$y w^+/Y h/h$	yes	1
hairy, white	$y^+ w/Y h/h$	yes	2
yellow, white	$y w/Y h^+/-$	no	360
hairy, yellow, white	$y w/Y h/h$	no	110

The recombination frequency between y and w is $[(4 + 3 + 1 + 2)/1{,}000] \times 100\% = 1\%$, so there is 1 map unit between these genes. Note that since recombination can be scored only in the F_2 males, the number of recombinants is divided by the number of *male* progeny, and not the total number of progeny.

6.17 In the Maltese bippy, amiable (*A*) is dominant to nasty (*a*), benign (*B*) is dominant to active (*b*), and crazy (*C*) is dominant to sane (*c*). A true-breeding amiable, active, crazy bippy was mated, with some difficulty, to a true-breeding nasty, benign, sane bippy. An F_1 individual from this cross was then used in a testcross (to a nasty, active, sane bippy) and produced, in typical prolific bippy fashion, 4,000 offspring. From an ancient manuscript titled *The Genetics of the Bippy, Maltese and Other,* you discover that all three genes are autosomal, that *a* is linked to *b*, but not to *c*, and that the map distance between *a* and *b* is 20 mu.

   a. Predict all the expected phenotypes and the numbers of each type from this cross.

   b. Which phenotypic classes would be missing had *a* and *b* shown complete linkage?

   c. Which phenotypic classes would be missing if *a* and *b* were unlinked?

   d. Again, assuming *a* and *b* to be unlinked, predict all the expected phenotypes of nasty bippies and the frequencies of each type resulting from a self-cross of the F_1.

Answer: The crosses can be denoted as:

      Parental: *A b/A b C/C* × *a B/a B c/c*

      F_1 testcross: *A b/a B C/c* × *a b/a b c/c*

   a. Since the map distance between *a* and *b* is 20 mu, the F_2 will have 20 percent recombinants (classes indicated by an * in the following table) between *a* and *b*.

Genotype	Phenotype	Percent	Number
A b/a b C/c	amiable, active, crazy	20	800
A b/a b c/c	amiable, active, sane	20	800
a B/a b C/c	nasty, benign, crazy	20	800
a B/a b c/c	nasty, benign, sane	20	800
a b/a b C/c	nasty, active, crazy	5	200*
a b/a b c/c	nasty, active, sane	5	200*
A B/a b C/c	amiable, benign, crazy	5	200*
A B/a b c/c	amiable, benign, sane	5	200*

   b. If *a* and *b* had shown complete linkage, the recombinant (*) classes would be missing.

   c. No phenotypic classes would be missing if *a* and *b* were unlinked. The frequency of each class would be identical (500 individuals, or 12.5 percent), however.

   d. If *a* and *b* were unlinked, the results of selfing a triple heterozygote (the F_1) are as described previously for a trihybrid cross. The F_2 will have 8 phenotypic classes in a ratio of 27:9:9:9:3:3:3:1. The nasty bippies must be *aa* in genotype, and so their distribution is a subset of these classes. If one is considering only the nasty bippies, the other two pairs of phenotypes will be distributed in the 9:3:3:1 ratio expected from a dihybrid cross. The progeny ratio will be 9 nasty, benign, crazy:3 nasty, benign, sane:3 nasty, active, crazy:1 nasty, active, sane.

6.18 In the following table, continuous bars indicate linkage and the order of linked genes is as shown:

Parent Genotypes	Number of Different Possible Gametes	Least-Frequent Classes	
$\dfrac{A}{a}\ \dfrac{b}{B}\ \dfrac{C}{c}$	_____	_____	
$\dfrac{A\ b\ C}{a\ B\ c}$	_____	_____	
$\dfrac{A\ b\ C\ D}{a\ B\ c\ d}$	_____	_____	
$\dfrac{A\ b\ C\ D\ E\ f}{a\ B\ C\ d\ e\ f}$	_____	_____	
$\dfrac{b\ D}{B\ d}$	_____	_____	

Fill in the blanks in the table. In the rightmost column, show two gamete genotypes, unless all types are equally frequent, in which case write "none."

Answer: Note that the number of classes of gametes is a function of how many different heterozygous loci are present. If there are n heterozygous loci, there will be 2^n possible gametes.

Parent Genotypes	Number of Different Possible Gametes	Least-Frequent Classes	
$\dfrac{A}{a}\ \dfrac{b}{B}\ \dfrac{C}{c}$	$2^3 = 8$	none	
$\dfrac{A\ b\ C}{a\ B\ c}$	$2^3 = 8$	$A\ B\ C$	$a\ b\ c$
$\dfrac{A\ b\ C\ D}{a\ B\ c\ d}$	$2^4 = 16$	$A\ B\ C\ d$	$a\ b\ c\ D$
$\dfrac{A\ b\ C\ D\ E\ f}{a\ B\ C\ d\ e\ f}$	$2^4 = 16$	$A\ B\ C\ D\ e\ f$	$a\ b\ C\ d\ E\ f$
$\dfrac{b\ D}{B\ d}$	$2^2 = 4$	$b\ d$	$B\ D$

6.19 Genes at loci f, m, and w are linked, but their order is unknown. The F_1 heterozygotes from a cross of *FF MM WW* × *ff mm ww* are testcrossed. The most frequent phenotypes in testcross progeny will be *F M W* and *f m w*, regardless of what the gene order turns out to be.

a. What classes of testcross progeny (phenotypes) will be least frequent if locus m is in the middle?

b. What classes will be least frequent if locus f is in the middle?

c. What classes will be least frequent if locus w is in the middle?

Answer: The double-crossover class will always be the least frequent.

a. *F m W, f M w*

b. *M f W, m F w*

c. *F w M, f W m*

6.20   The following numbers were obtained for testcross progeny in *Drosophila* (phenotypes):

+	m	+	218
w	+	f	236
+	+	f	168
w	m	+	178
+	m	f	95
w	+	+	101
+	+	+	3
w	m	f	1
	Total		1,000

Construct a genetic map.

Answer: Notice that among the eight phenotypic classes, there are four frequencies of progeny. This would be expected if the three genes were linked and a triply heterozygous parent were testcrossed. Since crossovers between closely linked genes occur less often than do no crossovers, the most frequent classes of progeny will be those having nonrecombinant, or parental-type, chromosomes. Thus, the + m + and w + f chromosomes are parental-type chromosomes and indicate that the triply heterozygous parent was obtained as follows:

P: $w + f/w + f \times + m +/+ m +$

F$_1$: $w + f/+ m + \times w\,m\,f/w\,m\,f$

Since recombination between closely linked genes is infrequent, the least frequent classes result from crossovers between each of the genes in the triple heterozygote — that is, double crossovers (+ + +, w m f). The remaining classes will have resulted from single crossovers.

To determine which gene is in the middle in this and in subsequent problems, consider the example shown here. This shows a double crossover between two nonsister chromatids having the gene order $B/b - A/a - C/c$. Relative to the parental-type chromosomes, the recombinant chromosomes have the middle A/a alleles switched:

In three-point crosses, comparison of one of the double-crossover classes (the least frequent classes) with one of the parental classes (the most frequent classes) allows you to infer which gene is in the middle. Depending on how they are chosen, one of two scenarios will be found. In one scenario, alleles at two genes will be identical while alleles at the third gene will be different. In the other scenario, alleles at two genes will be different while alleles at the third gene will be identical. In each case, the third gene is the gene in the middle. The middle gene is unlike the other two because its alleles have been "switched."

Use this strategy and compare one of the parental-type chromosomes and one of the double-crossover chromosomes. This indicates that the m gene is in the middle:

One parental-type chromosome:	w	+	f
One double-crossover chromosome:	w	m	f
Comparison of alleles:	same	different	same

As *m* is unlike the other two, it is in the middle. While the genes are written in the correct order in this problem, this may not always be the case. (After determining the correct order, always rewrite the arrangement of alleles on the parental chromosomes before proceeding.)

Now, diagram the crossovers in the triple heterozygote to infer the progeny classes that are associated with each crossover type:

single crossover between *w, m:*

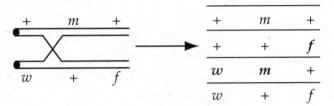

single crossover between *m, f:*

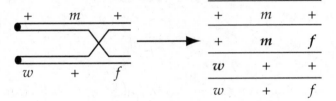

Double crossover:

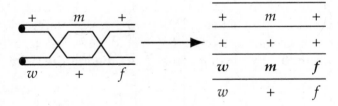

With this information, reevaluate the progeny, and determine how much recombination occurred in each gene interval:

+	m	+	218	parental
w	+	f	236	parental
+	+	f	168	single crossover between *w, m*
w	m	+	178	single crossover between *w, m*
+	m	f	95	single crossover between *m, f*
w	+	+	101	single crossover between *m, f*
+	+	+	3	crossover between *w, m* and between *m, f*
w	m	f	1	crossover between *w, m* and between *m, f*

Total   1,000

Number of recombinants in w, m interval: $168 + 178 + 3 + 1 = 350$
Recombination frequency between $w, m = (350/1,000) \times 100\% = 35\%$

Number of recombinants in m, f interval: $95 + 101 + 3 + 1 = 200$
Recombination frequency between $m, f = (200/1,000) \times 100\% = 20\%$

This information can be represented in the following map:

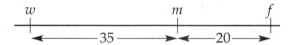

6.21 Three of the many recessive mutations in *Drosophila melanogaster* that affect body color, wing shape, or bristle morphology are black (*b*) body versus grey in the wild type; dumpy (*dp*), obliquely truncated wings versus long wings in the wild type; and hooked (*hk*) bristles at the tip versus nonhooked in the wild type. From a cross of a dumpy female with a black, hooked male, all the F_1 were wild type for all three characters. The testcross of an F_1 female with a dumpy, black, hooked male gave the following results:

wild type	169
black	19
black, hooked	301
dumpy, hooked	21
hooked	8
hooked, dumpy, black	172
dumpy, black	6
dumpy	304
Total	1,000

a. Construct a genetic map of the linkage group (or groups) these genes occupy. If applicable, show the order and give the map distances between the genes.

b. Determine the coefficient of coincidence for the portion of the chromosome involved in the cross. How much interference is there?

Answer:

a. First, use the symbols for the genes to write out the crosses that were done. It does not matter, at this point, what order you assign to the genes. Just be consistent in the order you choose so that you don't get confused.

P: $+ + dp/+ + dp$ (female) $\times b\ hk\ +/b\ hk\ +$ (male)
F_1 testcross: $+ + dp/b\ hk\ +$ (female) $\times b\ hk\ dp/b\ hk\ dp$ (male)

Now rewrite the data to reflect parental-type, single-crossover (SCO) and double-crossover (DCO) classes (based on frequency of each class).

Phenotype	Gamete Genotype			Number	Class
dumpy	+	+	dp	304	parental
black, hooked	b	hk	+	301	parental
hooked, dumpy, black	b	hk	dp	172	SCO
wild type	+	+	+	169	SCO
dumpy, hooked	+	hk	dp	21	SCO
black	b	+	+	19	SCO
hooked	+	hk	+	8	DCO
dumpy, black	b	+	dp	6	DCO

Now determine the gene order by comparing one of the double crossovers to one of the parental types (see solution to problem 6.20, above):

parental-type genotype:	+	+	dp
DCO genotype:	+	hk	+
	same	different	different

The *b* gene is unlike the other two, so it is in the middle. The correct gene order is $dp - b - hk$. As you continue, use this gene order. The F_1 female that was testcrossed was $dp + + / + b\ hk$. Consequently, the $dp\ b\ hk$ and $+++$ progeny are single crossovers between dp and b, while the $dp + hk$ and $+ b +$ progeny are single crossovers between b and hk. Recombination frequencies (RF) can be calculated as:

$$RF(dp - b) = [(172 + 169 + 6 + 8)/1,000] \times 100\% = 35.5\%$$
$$RF(b - hk) = [(21 + 19 + 6 + 8)/1,000] \times 100\% = 5.4\%$$

(Remember to include the double crossovers in this calculation, as each double crossover has a crossover in the interval being considered.) The map distance between dp and b is 35.5 mu and the map distance between b and hk is 5.4 mu. The map distance between dp and hk is $35.5 + 5.4 = 40.9$ mu.

b. The coefficient of coincidence (c.o.c.) is

$$\begin{aligned} \text{c.o.c.} &= \text{frequency observed doubles/frequency expected doubles} \\ &= (14/1,001)/(0.355 \times 0.054) \\ &= 0.014/0.01917 \\ &= 0.73 \end{aligned}$$

$$\begin{aligned} \text{Interference} &= 1 - \text{c.o.c.} \\ &= 1 - 0.73 \\ &= 0.27 \end{aligned}$$

6.22   The frequencies of gametes of different genotypes, determined by testcrossing a triple heterozygote, are as shown in the table on the following page:

a. Which gametes are known to have been involved in double crossovers?

b. Which gamete types have not been involved in any exchanges?

c. The order shown is not necessarily correct. Which gene locus is in the middle?

Gamete Genotype	%
+ + +	12.9
a b c	13.5
+ + c	6.9
a b +	6.5
+ b c	26.4
a + +	27.2
a + c	3.1
+ b +	3.5
Total	100.0

Answer:

a. *a* + *c* and + *b* + (least frequent).

b. + *b c* and *a* + + (most frequent).

c. locus *c* (see solution to problem 6.20 for a method to determine the middle locus).

6.23 Two normal-looking *Drosophila* are crossed and yield the following phenotypes among their progeny:

Females:	+	+	+	2,000
Males:	+	+	+	3
	a	b	c	1
	+	b	c	839
	a	+	+	825
	a	b	+	86
	+	+	c	90
	a	+	c	81
	+	b	+	75
			Total	4,000

Show the parental genotypes, the gene arrangement in the female parent, the map distances, and the coefficient of coincidence.

Answer: The differential appearance of the traits in males and females indicates that the traits are sex linked (i.e., they are X linked). The female progeny are normal, since they received their father's X chromosome, which was + + +. As they are phenotypically normal, they are not helpful in a mapping analysis. The male progeny received an X chromosome from the mother, and therefore can be used to analyze the results of recombination in the mother.

The most frequent progeny classes in the males are + *b c* and *a* + +. These are the parental-type chromosomes. The least frequent classes (the double-crossover classes) are + + + and *a b c*. Comparison of these two classes indicates that gene *a* is in the middle, and the correct order is *b* – *a* – *c* (or *c* – *a* – *b*) (see solution to problem 6.20). The parental cross (with the correct gene order) was *c* + *b*/+ *a* + (female) × + + +/Y (male). Now analyze the data to determine which progeny classes result from crossing-over in each gene interval and calculate the recombination frequencies in each interval (SCO = single crossover; DCO = double crossover).

Gamete Genotype	Number	Class
c + b	839	parental
+ a +	825	parental
+ a b	86	SCO (a – b)
c + +	90	SCO (a – b)
c a +	81	SCO (c – a)
+ + b	75	SCO (c – a)
c a b	1	DCO (a – b, c – a)
+ + +	3	DCO (a – b, c – a)
Total	2,000	

Note that, since recombination cannot be scored in the female progeny, and only the 2,000 male progeny are being considered, the total that is used as the divisor in the calculation of recombination frequency (RF) is 2,000.

$$RF\ (a - b) = [(86 + 90 + 1 + 3)/2{,}000] \times 100\% = 9.0\%$$
$$RF\ (c - a) = [(81 + 75 + 1 + 3)/2{,}000] \times 100\% = 8.0\%$$

With this information, draw the following map:

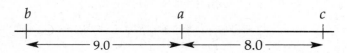

To calculate the coefficient of coincidence (c.o.c.), compare the frequency of actual double crossovers to that expected based on the crossover frequency observed in each interval.

$$c.o.c = observed\ DCO\ frequency/expected\ DCO\ frequency$$
$$= [(3 + 1)/2{,}000]/(0.09 \times 0.08)$$
$$= 0.002/0.0072$$
$$= 0.28$$

6.24   The questions that follow make use of this genetic map:

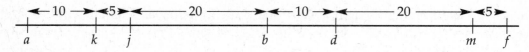

Calculate

a. the frequency of *j b* gametes from a *J B/j b* genotype
b. the frequency of *A M* gametes from an *a M/A m* genotype
c. the frequency of *J B D* gametes from a *j B d/J b D* genotype
d. the frequency of *J B d* gametes from a *j B d/J b D* genotype
e. the frequency of *j b d/j b d* genotypes in a *j B d/J b D × j B d/J b D* mating
f. the frequency of *A k F* gametes from an *A K F/a k f* genotype

Answer:

a. There are 20 map units between *j* and *b,* so there will be 20 percent recombinant-type and 80 percent parental-type gametes. In a *J B/j b* parent, *j b* gametes are half of the parental-type gametes, or 40 percent of the total gametes.

b. There are 65 map units between *A* and *M,* so these loci will show independent assortment, as the frequency of recombinant-type gametes cannot exceed 50 percent. In an *a M/A m* parent, the *A M* gametes are half of the recombinant-type gametes. There will be 25 percent *A M* gametes.

c. In a *j B d/J b D* individual, *J B D* gametes are produced by a double crossover: a crossover in the interval *j – b* and a crossover in the interval *b – d.* As there are 20 map units between *j* and *b* and 10 map units between *b* and *d,* the frequency of double crossovers is expected to be $0.20 \times 0.10 \times 100\% = 2\%$. *J B D* gametes are half of the double crossovers produced, so will be seen 1 percent of the time.

d. In a *j B d/J b D* individual, *J B d* gametes are one-half of the gametes produced by a single crossover in the interval *j – b.* Since there are 20 map units between these genes, 10 percent of the gametes will be *J B d.*

e. By the reasoning in (c), *j b d* gametes will be seen 1 percent of the time in a *j B d/J b D* individual. To obtain a *j b d/j b d* genotype, one must obtain such gametes from both parents. The frequency of this is $(0.01 \times 0.01) \times 100\% = 0.01\%$.

f. Based on the map distances, one expects 10 percent recombination between *a* and *k* and 50 percent recombination between *k* and *f.* (Again, note that even though *k* and *f* are more than 50 map units apart, one observes only 50 percent recombination between these two genes.) In an *A K F/a k f* individual, an *A k F* gamete results from a double crossover: one crossover between *a* and *k* (10 percent chance) and one crossover between *k* and *f* (50 percent chance). The chance of both crossovers occurring simultaneously is $0.10 \times 0.50 = 0.05$, or 5 percent. Since *A k F* gametes are half of the double crossovers produced, they will be 2.5 percent of the total.

6.25   A female *Drosophila* carries the recessive mutations *a* and *b* in repulsion on the X chromosome. (She is heterozygous for both.) She is also heterozygous for an X-linked recessive lethal allele, *l.* When she is mated to a true-breeding, normal male, she yields the following progeny:

Females:	1,000	+	+
Males:	405	*a*	+
	44	+	*b*
	48	+	+
	2	*a*	*b*

Draw a chromosome map of the three genes, in the proper order and with map distances as nearly as you can calculate them.

Answer: First, consider what happens when a female that is heterozygous for an X-linked lethal is crossed to a normal male. One can diagram such a cross as:

P: *l/+* female × *+/Y* male

F$_1$:   females: *l/+* and *+/+* (phenotypically normal)
        males: *+/Y* (normal) and *l/Y* (dead, not recovered)

In such a cross, one-half of the male progeny are not recovered due to the presence of the lethal allele. Only progeny bearing the + allele are recovered. Thus, the 499 males that are recovered in this cross are half of the expected male progeny.

Second, consider that the lethal-bearing chromosome can be recovered in the female progeny. Since the lethal allele is a recessive mutation, it is "rescued" by the normal + allele contributed to the female progeny by the father's X chromosome. However, as the father's X chromosome bears the + alleles of each gene, the females are phenotypically normal and are not helpful in a recombination analysis.

Now, analyze the phenotypes of the male progeny that are recovered to infer map distances and the gene order. Since one-half of the male progeny are not recovered, each of the four classes seen represents one of the two reciprocal classes of progeny recovered in a three-point cross. Include the third ($l/+$) locus and assign each genotype to a progeny class based on its frequency:

$$
\begin{array}{lllll}
\text{Male progeny:} & 405 & a\ \ +\ \ + & \text{parental type} \\
& 44 & +\ \ b\ \ + & \text{single crossover} \\
& 48 & +\ \ +\ \ + & \text{single crossover} \\
& 2 & a\ \ b\ \ + & \text{double crossover}
\end{array}
$$

Comparison of the parental and double-crossover classes indicates that b is in the middle, and the correct order is $a - b - l$ (see solution to problem 6.20 for determining the middle locus). Since one of the parental-type chromosomes was $a + +$, the other must have been $+ b l$, the reciprocal. The heterozygous female was therefore $a + +/+ b l$.

The $44 + b +$ progeny are obtained from gametes arising from single crossovers between b and l. They are half of the total single crossovers in that interval. The other half were not recovered, as they bore the l allele. Similarly, the $48 + + +$ progeny are half of the single crossovers in the interval $a - b$. This information can be used to calculate recombination frequencies (RF) and draw a map.

$$\text{RF}\ (a - b) = [(48 + 2)/499] \times 100\% = 10\%$$
$$\text{RF}\ (b - l) = [(44 + 2)/499] \times 100\% = 9.2\%$$

6.26   The following *Drosophila* cross is done:

$$\frac{a\ +\ b}{+\ c\ +} \times \xrightarrow{a\ c\ b}$$

Predict the numbers of phenotypes of male and female progeny that will emerge if the gene arrangement is as shown, the distance between a and c is 14 mu, the distance between c and b is 12 mu, the coefficient of coincidence is 0.3, and the number of progeny is 2,000.

Answer: Since the male parent is triply recessive, the phenotypes associated with the female parent's gametes will be evident equally in males and females. The map distances between the loci give the frequency of recombinants (i.e., crossovers) in each gene interval. There will be 14 percent recombinants in the $a - c$ interval (7% each of $a\ c$ and $+\ +$), and 12 percent recombinants in the $c - b$ interval (6% each of $+\ +$ and $c\ b$). These recombinants will be distributed between both single- and double-crossover classes.

The coefficient of coincidence gives the percentage of expected double crossovers that are observed. The expected double crossovers can be calculated based on the recombination frequency in each of the two gene intervals, and is $(0.12 \times 0.14) \times 100\% = 1.68\%$. Since the coefficient of coincidence is 0.3, only 30 percent of the expected double crossovers are observed, or $1.68\% \times 30\% = 0.50\%$ (0.25 percent each of $a\ c\ b$ and $+\ +\ +$).

The remaining recombinants will be single-crossover classes. This can be calculated by considering that some of the crossovers in each gene interval contribute to the double-crossover classes, and each double crossover has a crossover in each gene interval. This means the frequency of single crossovers in each gene interval is equal to the difference between the frequency of crossovers in that interval and the frequency of double crossovers. Consequently, there will be $14\% - 0.5\% = 13.5\%$ single crossovers in the $a - c$ interval (6.75 percent each $a\ c\ +$ and $+\ +\ b$), and $12\% - 0.5\% = 11.5\%$ single crossovers in the $c - b$ interval (5.75 percent each $a\ +\ +$ and $+\ c\ b$).

The remaining progeny $[100\% - (13.5\% + 11.5\% + 0.5\%) = 74.5\%]$ will be parental types (37.25% $a\ +\ b$ and 37.25% $+\ c\ +$). The types or progeny are therefore:

Genotype	Percent	Number
$a + b$	37.25	745
$+ c +$	37.25	745
$a\ c +$	6.75	135
$+ + b$	6.75	135
$a + +$	5.75	115
$+ b\ c$	5.75	115
$a\ c\ b$	0.25	5
$+ + +$	0.25	5

6.27 A farmer who raises rabbits wants to break into the Easter market. He has stocks of two true-breeding lines. One is hollow and long eared, but not chocolate, and the second is solid, short eared, and chocolate. Hollow (h), long ears (le), and chocolate (ch) are all recessive and autosomal and are linked as shown in the following map:

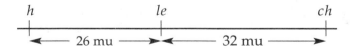

The farmer can generate a trihybrid by crossing his two lines, and at great expense he is able to obtain the services of a male who is homozygous recessive at all three loci to cross with his F_1 females. The farmer has buyers for both solid and hollow bunnies; however, all must be chocolate and long eared. Assuming that interference is zero, if he needs 25 percent of the progeny of the desired phenotypes to be profitable, should he continue with his breeding? Calculate the percentage of the total progeny that will be the desired phenotypes.

Answer: The crosses under consideration are:

P: *h le Ch/h le Ch* × *H Le ch/H Le ch*

F_1:   *h le Ch/H Le ch* × *h le ch/h le ch*

To be profitable, the farmer needs 25 percent of the progeny to be either *h le ch/h le ch* or *H le ch/h le ch*. Hence, he is concerned with the probability of obtaining an *h le ch* or an *H le ch* gamete from an *h le Ch/H Le ch* individual. Gametes that are *h le ch* arise from a single crossover between *le* and *ch*. Since there are 32 mu between *le* and *ch*, there will be 16 percent *h le ch* individuals (one-half of the recombinants between *le* and *ch*). Gametes that are *H le ch* arise from a crossover in each interval. Since there is no interference, their frequency will be half of the expected double-crossover frequency, or $(0.5 \times 0.32 \times 0.26) \times 100\% = 4.16\%$. The expected frequency of *h le ch* and *H le ch* gametes is $4.16\% + 16\% = 20.16\%$. The farmer should stop breeding and cut his losses.

6.28   Three different semidominant mutations affect the tails of mice. These mutations are alleles of linked genes, and all three are lethal in the embryo when homozygous. Fused-tail (*Fu*) and kinky-tail (*Ki*) mice have kinky-appearing tails, whereas brachyury (*T*) mice have short tails. A fourth gene, histocompatibility-2 (*H-2*), is linked to the three tail genes and is concerned with tissue transplantation. Mice that are *H-2/+* will accept tissue grafts, whereas *+/+* mice will not. The phenotypes of the progeny are as follows for four crosses, with the normal allele represented by +:

(1) $\dfrac{Fu\ +}{+\ Ki} \times \dfrac{+\ +}{+\ +}$

Fused tail	106
Kinky tail	92
Normal tail	1
Fused-kinky tail	1

(2) $\dfrac{Fu\ H\text{-}2}{+\ +} \times \dfrac{+\ +}{+\ +}$

Fused tail, accepts graft	88
Normal tail, rejects graft	104
Normal tail, accepts graft	5
Fused tail, rejects graft	3

(3) $\dfrac{T\ H\text{-}2}{+\ +} \times \dfrac{+\ +}{+\ +}$

Brachy tail, accepts graft	1,048
Normal tail, rejects graft	1,152
Brachy tail, rejects graft	138
Normal tail, accepts graft	162

(4) $\dfrac{Fu\ +}{+\ T} \times \dfrac{+\ +}{+\ +}$

Fused tail	146
Brachy tail	130
Normal tail	14
Fused-brachy tail	10

Make a map of the four genes involved in these crosses, giving gene order and map distances between the genes. If more than one map is possible, draw all possible maps.

Answer: Since all of the genes have a dominant phenotype, and crosses are made to a homozygous recessive strain, the data can be analyzed as a series of two-point testcrosses as shown in the following table:

Cross	Genes Involved	# Recombinants / Total	Recombination Frequency (%)	Apparent Map Distance
1	*Fu, Ki*	2/200	1	1 mu
2	*Fu, H-2*	8/200	4	4 mu
3	*T, H-2*	300/2,500	12	12 mu
4	*Fu, T*	24/300	8	8 mu

Two maps are consistent with these data.

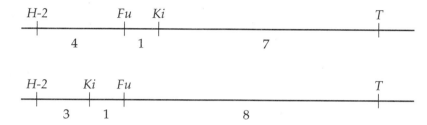

6.29 In *Drosophila*, a cross of

$$\frac{a^+ \ b^+ \ c \ \ d \ \ e}{a \ \ b \ \ c^+ \ d^+ \ e^+} \times \frac{a \ b \ c \ d \ e}{a \ b \ c \ d \ e}$$

gave 1,000 progeny of the following 16 phenotypes:

Genotype	Number
(1) $a^+ \ b^+ \ c \ \ d \ \ e$	220
(2) $a^+ \ b^+ \ c \ \ d \ \ e^+$	230
(3) $a \ \ b \ \ c^+ \ d^+ \ e$	210
(4) $a \ \ b \ \ c^+ \ d^+ \ e^+$	215
(5) $a \ \ b^+ \ c^+ \ d^+ \ e$	12
(6) $a \ \ b^+ \ c^+ \ d^+ \ e^+$	13
(7) $a^+ \ b \ \ c \ \ d \ \ e^+$	16
(8) $a^+ \ b \ \ c \ \ d \ \ e$	14
(9) $a \ \ b^+ \ c^+ \ d \ \ e^+$	14
(10) $a \ \ b^+ \ c^+ \ d \ \ e$	13
(11) $a^+ \ b \ \ c \ \ d^+ \ e^+$	8
(12) $a^+ \ b \ \ c \ \ d^+ \ e$	8
(13) $a^+ \ b^+ \ c^+ \ d \ \ e^+$	7
(14) $a^+ \ b^+ \ c^+ \ d \ \ e$	7
(15) $a \ \ b \ \ c \ \ d^+ \ e^+$	6
(16) $a \ \ b \ \ c \ \ d^+ \ e$	7

a. Draw a genetic map of the chromosome, indicating the linkage of the five genes and the number of map units separating each.

b. From the single-crossover frequencies, what would be the expected frequency of $a^+ \ b^+ \ c^+ \ d^+ \ e^+$ flies?

Answer:

a. Approach this problem by considering two genes at a time. Then one has:

Gene Pair	# Parental-Type Progeny	# Recombinant-Type Progeny	Recombination Frequency (%)	Linked?
a,b	902	98	9.8	yes
a,c	973	27	2.7	yes
a,d	957	43	4.3	yes
a,e	497	503	50.0	no
b,c	875	125	12.5	yes
b,d	945	55	5.5	yes
b,e	497	503	50.0	no
c,d	930	70	7.0	yes
c,e	498	502	50.0	no
d,e	496	504	50.0	no

The genes *a*, *b*, *c*, and *d* are linked, while *e* is unlinked to the other four genes. Use the smallest distances as the most accurate map distances. One has:

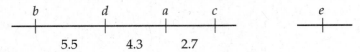

b. Rewrite the parental cross using the correct gene order:

$$\frac{b^+\ d\ \ a^+\ c}{b\ \ d^+\ a\ \ c^+}\ \ \frac{e}{e^+} \times \frac{b\ d\ a\ c}{b\ d\ a\ c}\ \ \frac{e}{e}$$

To obtain a $b^+\ d^+\ a^+\ c^+$ fly, there must be crossovers between *b* and *d*, *d* and *a*, and *a* and *c*. Two reciprocal crossovers will be found: one-half of the progeny will be $b^+\ d^+\ a^+\ c^+$ and one-half of the progeny will be *b d a c*. Among these progeny, one-half will be e^+ and one-half will be *e*.

$$P(b^+\ d^+\ a^+\ c^+\ e^+) = P(\text{triple crossover}) \times 1/2 \times 1/2$$
$$= (0.055 \times 0.043 \times 0.027) \times 0.25$$
$$= 1.6 \times 10^{-5}.$$

6.30   In *Drosophila*, many different mutations have been isolated that affect a normally brick-red eye color caused by the deposition of brown and bright-red pigments. Two X-linked recessive mutations are *w* (white eyes, map position 1.5) and *cho* (chocolate-brown eyes, map position 13.0), with *w* epistatic to *cho*.

a. A white-eyed female is crossed with a chocolate-eyed male, and the normal, red-eyed F_1 females are crossed with either wild-type or white-eyed males. Determine the frequency of the progeny types produced in each cross.

b. The recessive mutation *st* causes scarlet (bright-red) eyes and maps to the third chromosome at position 44. Mutant flies with only *st* and *cho* alleles have white eyes, and *w* is epistatic to *st*. Suppose a true-breeding *w* male is crossed with a true-breeding *cho*, *st* female. Determine the frequency of the progeny types you would expect if the F_1 females are crossed to true-breeding scarlet-eyed males.

Answer:

a. The cross is $w +/w +$ (white female) $\times + cho/Y$ (chocolate male), and the F_1 females are $w +/+ cho$ (red). Since the *w* and *cho* genes are 11.5 map units apart, the F_1 females will have 11.5 percent recombinant-type gametes (5.75% each $+ +$ and *w cho*) and 88.5

percent parental-type gametes (44.25% each w + and + cho). When these gametes are combined with those of wild-type (50% each + + and Y) or white-eyed (50% each w + and Y) males, one obtains the progeny shown in the following branch diagrams.

Cross: w +/+ cho × + +/Y

Gametes of		Progeny
w +/+ cho	+ +/Y	

0.4425 w +
  0.5 + + ⟶ 0.22125 w +/+ + (red female)
  0.5 Y ⟶ 0.22125 w +/Y (white male)

0.4425 + cho
  0.5 + + ⟶ 0.22125 + cho/+ + (red female)
  0.5 Y ⟶ 0.22125 + cho/Y (chocolate male)

0.0575 + +
  0.5 + + ⟶ 0.02875 + +/+ + (red female)
  0.5 Y ⟶ 0.02875 + +/Y (red male)

0.0575 w cho
  0.5 + + ⟶ 0.02875 w cho/+ + (red female)
  0.5 Y ⟶ 0.02875 w cho/Y (white male)

Result: 50% red females

2.875% red males
25% white males
22.125% chocolate males

Cross: w +/+ cho × w+/Y

Gametes of		Progeny
w +/+ cho	w +/Y	

0.4425 w +
  0.5 w + ⟶ 0.22125 w +/w + (white female)
  0.5 Y ⟶ 0.22125 w +/Y (white male)

0.4425 + cho
  0.5 w + ⟶ 0.22125 + cho/w + (red female)
  0.5 Y ⟶ 0.22125 + cho/Y (chocolate male)

0.0575 + +
  0.5 w + ⟶ 0.02875 + +/w + (red female)
  0.5 Y ⟶ 0.02875 + +/Y (red male)

0.0575 w cho
  0.5 w + ⟶ 0.02875 w cho/w + (white female)
  0.5 Y ⟶ 0.02875 w cho/Y (white male)

Result: 25% red females

25% white females

2.875% red males
25% white males
22.125% chocolate males

b. The initial cross is $w\ +/Y$; $+/+ \times +\ cho/+\ cho$; st/st with F_1 females being $w\ +/+\ cho$; $+/st$. These trihybrid females will have eight types of gametes: 88.5 percent will be parental types [22.125% each of ($w\ +$; $+$), ($w\ +$; st), ($+\ cho$; $+$), ($+\ cho$; st)], and 11.5 percent will be recombinants between w and cho [2.875% each of ($+\ +$; $+$), ($+\ +$; st), ($w\ cho$; $+$), ($w\ cho$; st)]. The results obtained when the F_1 female is crossed to a true-breeding scarlet male [with ($+\ +$; st) and (Y; st) gametes] are shown in the following branch diagrams:

Cross: $w\ +/+\ cho$; $st/+ \times +\ +/Y$; st/st

Gametes of		
$w\ +/+\ cho$; $st/+$	$+\ +/Y;st/st$	Progeny

0.2215 w +; +
- 0.5 + +; st → 0.110625 w +/+ +; +/st (red female)
- 0.5 Y; st → 0.110625 w +/Y; +/st (white male)

0.2215 w +; st
- 0.5 + +; st → 0.110625 w +/+ +; st/st (scarlet female)
- 0.5 Y; st → 0.110625 w +/Y; st/st (white male)

0.2215 + cho; +
- 0.5 + +; st → 0.110625 + cho/+ +; +/st (red female)
- 0.5 Y; st → 0.110625 + cho/Y; +/st (chocolate male)

0.2215 + cho; st
- 0.5 + +; st → 0.110625 + cho/+ +; st/st (scarlet femal
- 0.5 Y; st → 0.110625 + cho/Y; st/st (white male)

0.02875 + +; +
- 0.5 + +; st → 0.014375 + +/+ +; +/st (red female)
- 0.5 Y; st → 0.014375 + +/Y; +/st (red male)

0.02875 + +; st
- 0.5 + +; st → 0.014375 + +/+ +; st/st (scarlet female)
- 0.5 Y; st → 0.014375 + +/Y; st/st (scarlet male)

00.02875 w cho; +
- 0.5 + +; st → 0.014375 w cho/+ +; +/st (red female)
- 0.5 Y; st → 0.014375 w cho/Y; +/st (white male)

0.02875 w cho; st
- 0.5 + +; st → 0.014375 w cho/+ +; st/st (scarlet femal
- 0.5 Y; st → 0.014375 w cho/Y; st/st (white male)

Result: 25% red females   1.4375% red males
         25% scarlet females   1.4375% scarlet males
                                 36.0625% white males
                                 11.0625% chocolate males

6.31   Breeders of thoroughbred horses used for racing keep extensive information on pedigrees. Such information can be useful in determining simple inheritance patterns (e.g., chestnut coat color has been determined to be recessive to bay coat color) and in speculating whether racehorses that win competitive races ("classy" horses) share genetic traits. Sharpen Up was a chestnut stallion that was only somewhat successful as a racehorse. At age 4, he was retired from horse racing and put out to stud. His progeny were very successful: Of 367 foals fathered in the United States, 43 were prizewinners in highly competitive races, and of 200 foals fathered in England, 40 were prizewinners in highly competitive races. A commentator who analyzed Sharpen Up's progeny (and that of other chestnut prizewinners) has suggested that whatever gene combinations produced class (winning horses) were tied to the horses' chestnut coat. Indeed, of the 83 progeny that have shown class (won highly competitive races), about 45 were also chestnut in color. Use a chi-square test to assess whether there is any reason to believe that if there is a gene (or genes) for class, it is linked to the gene for chestnut coat color. Examine this issue using two different assumptions: (1) Sharpen Up was mated equally frequently to homozygous bay, heterozygous bay/chestnut, and homozygous chestnut mares, and (2) Sharpen Up was mated equally frequently to heterozygous bay/chestnut and homozygous chestnut horses. State carefully any additional assumptions and your hypothesis.

Answer: First, use the chi-square test to evaluate the hypothesis that there is no relationship between chestnut coat color and class. Further investigation into the potential linkage of this coat color gene and a hypothetical class gene is warranted if we can reject this hypothesis. The assumptions in this initial chi-square test involve the genotypes of the horses bred to Sharpen Up, as stated in the problem, and the hypothesis of the chi-square test, that chestnut coat color and class are unrelated.

From the initial hypothesis of *no relationship between class and chestnut coat color*, the likelihood of Sharpen Up siring a classy horse is uniform with regard to its coat color. There were 83 classy horses produced from a total of 627 (367 + 260) progeny, so the chance of Sharpen Up siring a classy horse, independent of its coat color, is 83/627 = 13.24%. To perform the chi-square test, we need to compare the expected and observed number of classy chestnut and classy bay horses.

Assumption I: Sharpen Up is mated equally frequently to homozygous bay, heterozygous bay/chestnut, and homozygous chestnut mares. Chestnut is recessive to bay, so let c represent chestnut and C represent bay. Sharpen Up is chestnut (cc), so the crosses and their progeny are (1) $cc \times CC \rightarrow$ all $C-$ (bay); (2) $cc \times Cc \rightarrow \frac{1}{2} Cc$ (bay), $\frac{1}{2} cc$ (chestnut); and (3) $cc \times cc \rightarrow$ all cc (chestnut). If each cross is equally likely ($P = \frac{1}{3}$), the expected number of chestnut offspring, rounding up or down to the nearest whole horse, is $627 \times [(\frac{1}{3} \times 0) + (\frac{1}{3} \times \frac{1}{2}) + (\frac{1}{3} \times 1)] = 314$. The remaining $627 - 314 = 313$ offspring are expected to be bay. Using assumption I and assuming that the frequency of classy offspring is uniform (13.24%) with respect to their coat color, the expected number of classy chestnut progeny is $314 \times 0.1324 = 42$, and the expected number of classy bay progeny is $313 \times 0.1324 = 41$. The observed numbers of classy horses were 45 chestnut and 38 bay. For these values, $\chi^2 = [(45 - 42)^2/42 + (38 - 41)^2/41] = 0.43$, df = 1, $0.50 < P < 0.70$. Under assumption I, then, the hypothesis that chestnut coat color and class are unrelated is accepted as possible.

Assumption II: Sharpen Up is mated equally frequently to heterozygous bay/chestnut and chestnut mares. These crosses and their progeny are (1) $cc \times Cc \rightarrow \frac{1}{2} Cc$ (bay), $\frac{1}{2} cc$ (chestnut); and (2) $cc \times cc \rightarrow$ all cc (chestnut). If each cross is equally likely ($P = \frac{1}{2}$), the expected number of chestnut offspring is $627 \times [(\frac{1}{2} \times \frac{1}{2}) + (\frac{1}{2} \times 1)] = 470$. The expected number of bay offspring is $627 - 470 = 157$. Using assumption II and assuming that the frequency of classy offspring is uniform (13.24%) with respect to their coat color, the classy progeny are expected to be $470 \times 0.1324 = 62$ chestnut and $157 \times 0.1324 = 21$ bay. The observed numbers of classy horses were 45 chestnut and 38 bay. For these values, $\chi^2 = [(45 - 62)^2/62 + (38 - 21)^2/21] = 18.4$, df = 1, $P < 0.001$. Under assumption II, then, the hypothesis that chestnut coat color and class are unrelated is rejected as being unlikely. It would be reasonable to consider the hypothesis that a gene closely linked to chestnut/bay coat color might contribute to class.

Notice that the evidence for a relationship between chestnut coat color and class hinges on knowing what alleles at the chestnut/bay gene were present in the mares bred to Sharpen Up. This information is available (although not in this problem). Additional assumptions required to specifically test for linkage to a class gene might include assumptions about the number of alleles in the population of horses, the dominance relationships between them, and which alleles reside on the same homolog with the chestnut allele.

7

ADVANCED GENE MAPPING IN EUKARYOTES

CHAPTER OUTLINE

TETRAD ANALYSIS IN CERTAIN HAPLOID EUKARYOTES
Using Random-Spore Analysis to Map Genes in Haploid Eukaryotes
Calculating Gene-Centromere Distance in Organisms by Using Ordered Tetrads
Using Tetrad Analysis to Map Two Linked Genes

MITOTIC RECOMBINATION
Discovery of Mitotic Recombination
Mitotic Recombination in the Fungus *Aspergillus nidulans*
Retinoblastoma, a Human Tumor That Can Be Caused by Mitotic Recombination

MAPPING HUMAN GENES
Mapping Human Genes by Recombination Analysis
lod Score Method for Analyzing Linkage of Human Genes
High-Density Genetic Maps of the Human Genome
Physical Mapping of Human Genes

THINKING ANALYTICALLY

Mapping genes using tetrad analysis. Many students find the analysis of tetrads quite challenging. Step back for a minute, though, and ask yourself, "Can I recognize the *principles* behind these methods?" Tetrad analysis is based on the normal events of meiosis in organisms in which all four products of a meiosis remain together. A solid understanding of chromosome behavior during meiosis will enable you to master tetrad analysis. Review how chromosomes behave during meiosis. Consider how T, PD, and NPD tetrads are produced. Ask yourself what chromosomal events underlie first- and second-division segregation. Approaching the material this way will provide you with a solid, *mechanistic* understanding of how genes are mapped using tetrad analysis.

Once you understand how meiotic events lead to different types of ordered and unordered tetrads, you will discover that analyzing the results of crosses in which tetrads are produced is reasonably straightforward. Try employing the following systematic approach:

1. Assign gene symbols to the traits and write out the cross. Consider just two loci at a time.
2. Decide which are parental-type spores and which are recombinant-type spores.

REVIEW OF KEY TERMS, SYMBOLS, AND CONCEPTS

In your own words, write a brief, precise definition of each term in the groups below. Check your definitions using the text. Then develop a concept map using the terms in each list.

1	2	3
tetrad analysis	mitotic recombination	meiotic recombination
ordered tetrads	meiotic recombination	pedigree analysis
unordered tetrads	independent assortment	lod score method
linear tetrads	mitotic crossing-over	DNA marker
ascus, asci	twin spot	molecular phenotype
Neurospora crassa	mosaic	mapping panel
Saccharomyces cerevisiae	heterokaryon	genetic map
Chlamydomonas reinhardtii	*Aspergillus nidulans*	physical map
ascospore	haploidization	fluorescent in situ
vegetative cell	parasexual system	hybridization (FISH)
mycelium	sporadic retinoblastoma	radiation hybrid (RH)
mating type	hereditary retinoblastoma	
random-spore analysis	unilateral tumor	
parental-ditype tetrad (PD)	bilateral tumor	
nonparental-ditype tetrad (NPD)		
tetratype tetrad (T)		
first-division segregation		
second-division segregation		
gene-gene distance		
gene-centromere distance		

3. If you have unordered asci, calculate the recombination frequency (RF):

   RF = (# recombinant spores/total # spores) × 100%

4. If you have ordered tetrads, identify PD, T, and NPD tetrads, and then calculate RF.
   a. If PD = NPD (or is close), the genes are not linked.
   b. If PD >> NPD, the genes are linked and RF = ($\frac{1}{2}$ T + NPD)/(total # asci) × 100%.

5. If you are analyzing linear (ordered) tetrads, consider each locus separately, determine the number of first-division (MI, 2:2 or 4:4) and second-division (MII, 1:1:1:1 or 2:2:2:2) segregation patterns, and then calculate RF.

   RF(gene–centromere) = ($\frac{1}{2}$ MII patterns)/(total # asci) × 100%

6. Draw a map.

Understanding mitotic recombination. Just as a solid understanding of chromosomal behavior during meiosis will enable you to have a better understanding of tetrad analysis, a solid understanding of chromosome behavior during mitosis will enable you to master mitotic recombination. In problems involving mitotic recombination, it often helps to remind yourself of one of the outcomes of mitosis: Normally, parent cells give rise to genetically identical daughter cells. In the absence of a mitotic crossover, each daughter cell receives chromosomes identical to those present in its parent. This is because in mitosis chromatids segregate according to their centromere, not according to the loci that are attached to that centromere.

To see how a mitotic crossover leads to a single or a twin spot, follow the normal segregation of centromeres (e.g., in text Figure 7.8) when there is a mitotic crossover. Normally, the movement of chromosomes to opposite poles in mitosis ensures that two identical daughter cells will be produced. A rare mitotic crossover leads to the production of a clone of cells that are homozygous at all loci distal to the crossover event. Depending on the arrangement of alleles in the heterozygote and the location of the crossover relative to the position of the centromere and the alleles, single or twin spots can be seen. Note that the tissue surrounding a spot will remain heterozygous. Since the organism is now composed of tissues having different genotypes, it is referred to as a mosaic.

When solving mapping problems in organisms such as *Aspergillus nidulans,* it is important to place chromosome segregation in the context of a parasexual cycle. Analyze whether cells have undergone diploidization, haploidization, and/or mitotic recombination, and consider how these processes can lead to different types of haploid or diploid segregants. Apply the principles of chromosome segregation when analyzing the linkage relationships between genes: Alleles attached to different centromeres will segregate to different haploid segregants during haploidization, genes on different chromosomes will assort independently during haploidization, a rare mitotic crossover in a heterozygous diploid will make all genes distal to the crossing-over point (on the same chromosome arm) homozygous.

Mapping human genes. It is important to understand how two fundamentally different approaches are used to map genes in humans. The first approach to map genes constructs maps that reflect meiotic recombination frequencies between genes or markers. While the frequency of recombinants for X-linked loci can be inferred from the analysis of large pedigrees, statistically based lod score methods can evaluate linkage between any two markers. A lod score is the $\log_{10}$ of an odds ratio, the ratio of the probability that two markers are linked with a particular recombination frequency to the probability that two markers are unlinked. Scores over 3 (odds of 1,000 to 1) are generally accepted as evidence of linkage. To estimate the distance between two linked markers, identify the recombination frequency between 0 and 0.5 that gives the highest lod score.

The second approach to map genes constructs maps that reflect the physical distance between genes. This approach includes the use of fluorescent in situ hybridization (FISH) and radiation hybrids. FISH can provide an approximate chromosomal location, while radiation hybrids can provide finer resolution. Radiation hybrid maps also rely on statistical methods to evaluate the distance between two markers. Consider why this is the case. Radiation hybrids contain fragments of chromosomes. Two genes that are close together are more likely to be together on one fragment and so are more likely to be found within one radiation hybrid. However, radiation hybrids contain many different chromosomal fragments, so two genes that are not close together also may be found in one radiation hybrid as a consequence of the random introduction of two different chromosomal fragments into the same radiation hybrid. Therefore, statistical methods are necessary to distinguish markers that are linked from markers that are found together by chance alone, and to evaluate the degree of linkage.

QUESTIONS FOR PRACTICE

Multiple Choice Questions

1. Tetrad analysis is a technique for
   a. determining base sequences in DNA
   b. measuring the number of chromatids at meiosis
   c. separating *Neurospora* mycelia
   d. mapping genes in organisms retaining the four meiotic products in one structure

2. Ordered tetrads occur in the asci of
   a. *Neurospora*
   b. *Clamydomonas*
   c. *Saccharomyces*
   d. *Drosophila*

3. A single crossover between two gene loci during meiosis in *Neurospora* spore formation produces an ascus with a genetic configuration known as
   a. tetratype
   b. parental ditype
   c. nonparental ditype
   d. double recombinant

4. The formula ($\frac{1}{2}$ T + NPD)/(total # of tetrads) $\times$ 100% is used to calculate
   a. the length of a chromosome
   b. the distance between a gene and its centromere
   c. the number of crossovers between two genes
   d. the percentage of recombinants between two genes

5. 156 tetrads were analyzed in a series of two-point crosses in *Neurospora*. One hundred and two of the tetrads are PD, 40 are T, and 14 are NPD. What is the frequency of recombination between the two genes?
   a. 14%
   b. 22%
   c. 34%
   d. 50%

6. Two hundred tetrads were analyzed in a *Neurospora* cross involving two genes. There were no T tetrads, 96 PD tetrads, and 104 NPD tetrads. What do you infer about these two genes?
   a. The genes are 50 mu apart and lie on the same chromosome.
   b. The genes are very close together on the same chromosome.
   c. The genes are 100 mu apart and lie on the same chromosome.
   d. The genes are either far apart on one chromosome or on different chromosomes.

7. In yeast, the cross *a b* $\times$ + + produces a tetrad having two *a* + spores and two + *b* spores. How would you classify this tetrad?
   a. nonparental ditype
   b. parental ditype
   c. tetratype
   d. mating type α

8. Which type of tetrads have half recombinant-type spores?
   a. NPD
   b. PD
   c. T
   d. R

9. In a cross involving two linked genes, the frequency of parental ditype tetrads will be
   a. equal to the frequency of tetratype tetrads
   b. greater than the frequency of nonparental ditype tetrads
   c. equal to the frequency of nonparental ditype tetrads
   d. less than the frequency of tetratype tetrads

For questions 10 and 11, refer to the following ordered tetrad from the *Neurospora* cross $a + \times + b$.

   $a +$
   $a +$
   $a\ b$
   $a\ b$
   $+ b$
   $+ b$
   $+ +$
   $+ +$

10. What configuration does this tetrad represent?
    a. NPD
    b. PD
    c. T
    d. NPD + PD

11. What segregation pattern is shown for the *a* locus?
    a. equal
    b. random
    c. first division
    d. second division

For questions 12–14, refer to text Figure 7.8 and the following information. In *Drosophila,* the *y* mutation is recessive and causes a yellow body color and yellow bristles. The y^+ allele results in grey body color and black bristles. The *sn* mutation causes short, twisted bristles, while the sn^+ allele results in long, straight bristles. A $sn^+ y/sn\ y^+$ heterozygote is produced by crossing a $sn\ y^+/Y$ male to a $sn^+ y/sn^+ y$ female.

12. What is normally the phenotype of the $sn^+ y/sn\ y^+$ heterozygote?
    a. grey body and black, long bristles
    b. yellow body and yellow, long, straight bristles
    c. yellow body and yellow, short, twisted bristles
    d. grey body and black, short, twisted bristles

13. On the surface of some animals, twin spots are seen. What phenotype do these have?
    a. One spot has short, twisted, yellow bristles. The other has long, straight, yellow bristles.
    b. One spot has short, twisted, black bristles. The other has long, straight, yellow bristles.
    c. One spot has short, twisted, black bristles. The other has long, straight, black bristles.
    d. One spot has short, twisted, yellow bristles. The other has long, straight, black bristles.

14. How did the twin spots arise?
    a. from chiasma interference
    b. from second-division segregation
    c. from crossing-over during meiosis
    d. from crossing-over between the most proximal gene and the centromere during mitosis
    e. from crossing-over between the two genes during mitosis
    f. from crossing-over between the most distal gene and the telomere during mitosis

15. Two markers are evaluated for linkage using the lod score method with a range of recombination frequencies. A maximal lod score of 4.0 is seen for the two markers at a distance of 20 percent recombination. All other recombination frequencies show lower lod scores (e.g., the lod score for the same two markers at 5 percent recombination is 1.0). How do you interpret these data?
    a. The two markers exhibit a 10,000:1 odds in favor of linkage and are about 20 map units apart.
    b. The two markers are not linked.
    c. The two markers are linked, but we can't infer how far apart they are.
    d. There are about 4 million base pairs of DNA between the two markers.

Answers: 1d, 2a, 3a, 4d, 5b, 6d, 7a, 8c, 9b, 10c, 11c, 12a, 13b, 14d, 15a

Thought Questions

1. In what different ways can genetic maps be constructed in haploid organisms?

2. *Drosophila* normally have deep-red eyes. A large number of alleles at the X-linked *white* locus cause lighter-colored eyes, with w^1 causing a completely white eye, w^a causing apricot (light orange) eyes, and w^c causing coral (pink) eyes. In females heterozygous for these two *white* alleles, the amount of pigmentation contributed by each allele is additive, so that w^1/w^a females have apricot eyes and w^a/w^c females have reddish eyes. Diagram the appearance of a twin spot that results from a single mitotic crossover between the centromere and the *white* locus in a w^a/w^c *Drosophila* female.

3. We know that crossing-over occurs during leptonema of prophase I of meiosis. How could ordered tetrad analysis be used to prove that meiotic recombination occurs at the four-strand stage of meiosis, and not before chromosomes are replicated into chromatids? (Hint: Use diagrams to compare the results of a single crossover between a and b in a cross of $a\,b \times ++$ that occur before and after DNA replication.)

4. How could you use mitotic recombination to map linked loci relative to each other? Do you expect the map distances obtained from mitotic recombination to be the same as those derived using meiotic recombination?

5. Recall that in *Drosophila*, meiotic recombination does not occur during spermatogenesis. Might mitotic recombination occur in *Drosophila* males? For all chromosomes? If the answer is yes, what might this tell you about the "normal" role of mitotic recombination in cells?

6. How might you determine the order, relative to the centromere, of two cloned segments of DNA using FISH?

7. How are statistical methods helpful to evaluate linkage relationships in humans? Why are they needed?

8. In what ways is a parasexual cycle different than a sexual cycle? In what different ways can recombinants be generated in each cycle?

Thought Questions for Media

After reviewing the media on the *iGenetics* CD-ROM, try to answer these questions.

1. Explain why, for two unlinked genes, PD and NPD tetrads are about equal in frequency, while for linked genes, PD tetrads are much greater in frequency than NPD tetrads.

2. How do T tetrads arise when two genes are unlinked?

3. What are the different ways that PD and T tetrads can be produced when two genes are linked?

4. Without consulting the text, derive the tetrad mapping formula from the basic mapping formula. Explain the logic involved in each step of your derivation.

1. What characteristics of the data immediately allowed you to infer that *b* and *c* are unlinked and that *b* and *d* are linked?

2. Can two genes be linked to each other but not to their centromeres? Can two genes be linked to their centromeres but not to each other? Explain.

SOLUTIONS TO TEXT PROBLEMS

7.1   A cross was made between a pantothenate-requiring (*pan*) strain and a lysine-requiring (*lys*) strain of *Neurospora crassa,* and 750 random ascospores were analyzed for their ability to grow on a minimal medium (a medium lacking pantothenate and lysine). Thirty colonies subsequently grew. Map the *pan* and *lys* loci.

Answer: First, write out the strains that were crossed and the kinds of progeny that are possible.

$$pan + \times + lys \rightarrow$$

pan	+	parental-type spore
+	*lys*	parental-type spore
pan	*lys*	recombinant-type spore
+	+	recombinant-type spore

Notice that, since spores that are *pan* or *lys* require supplements to grow, the only spores that can grow on minimal medium are wild type, or + + spores. All parental-type spores and half of the recombinants cannot grow. Thus, the 30 spores that grew are half of the recombinants. Therefore, there were a total of 60 recombinants in 750 progeny tested, giving a map distance of (60/750) $\times$ 100% = 8 mu.

7.2   Four different albino strains of *Neurospora* were each crossed to the wild type. All crosses resulted in half wild-type and half albino progeny. Crosses were made between the first strain and the other three, with the following results:

1×2:   975 albino, 25 wild type
1×3:   1,000 albino
1×4:   750 albino, 250 wild type

Which mutations represent different genes, and which genes are linked? How did you arrive at your conclusions?

Answer: First, consider the results of the set of crosses to the wild type.

albino $\times$ + $\rightarrow$ 50% albino, 50% +

The observation of equal frequencies of parental-type spores in the progeny indicates that the albino strains have an allele that behaves in a Mendelian fashion. That is, for each of the four strains, the albino and + alleles segregate from each other as would be expected of alleles at one gene. The question now becomes whether each of the albino strains results from a mutation in the same or a different gene. This issue can be resolved by considering the second set of crosses and the possibility that the different strains are mutant at different genes. If different strains have mutations at different genes, the wild-type spores that are found represent half of the recombinant types of spores, as shown below.

1×2: $a^1 + \times + a^2$ $\rightarrow$ 975 albino ($a^1 a^2$, $a^1 +$, or $+ a^2$)
                                                     25 wild type (+ +)

1×3: $a^1 + \times + a^3$ $\rightarrow$ 1,000 albino ($a^1 a^3$, $a^1 +$, or $+ a^3$)
                                                     0 wild type (+ +)

1×4: $a^1 + \times + a^4$ $\rightarrow$ 750 albino ($a^1 a^4$, $a^1 +$, or $+ a^4$)
                                                     250 wild type (+ +)

In the cross of 1 × 2, the appearance of 25 wild-type spores indicates that there are a total of 50 recombinant spores (5 percent). Hence, genes 1 and 2 are linked, and five map units apart. In the cross of 1 × 3, there are no recombinant spores, and so these albino alleles appear to be at the same gene, and may be identical. In the cross of 1 × 4, there are a total of 500 recombinant spores (50 percent) and so genes 1 and 4 appear to be unlinked. There are therefore three different genes, with genes 1 (=3) and 2 being linked.

7.3   Genes *met* and *thi* are linked in *Neurospora crassa*; we want to locate *arg* with respect to *met* and *thi*. From the cross *arg* × *thi met*, the following random ascospore isolates were obtained.

arg thi met	26	*arg* + +	51
arg thi +	17	+ *thi* +	4
arg + *met*	3	+ + *met*	14
+ *thi met*	56	+ + +	29

Map these three genes.

Answer: Notice that there are eight classes of progeny that appear in four frequencies, reminiscent of a three-point cross with three linked loci. Therefore, treat this as you would a three-point cross. Comparison of one of the parental types (+ *thi met*) to one of the double recombinants (*arg* + *met*) tells us that *met* is in the middle, between *arg* and *thi*. Rewrite the progeny types as (SCO = single crossover):

parental	+ *met thi*	56
parental	*arg* + +	51
SCO between *arg* – *met*	*arg met thi*	26
SCO between *arg* – *met*	+ + +	29
SCO between *met* – *thi*	*arg* + *thi*	17
SCO between *met* – *thi*	+ *met* +	14
double crossover	*arg met* +	3
double crossover	+ + *thi*	4
	Total	200

The map distance between *arg* and *met* is [(26 + 29 + 3 + 4)/200] × 100% = 31 mu, and the map distance between *met* and *thi* is (17 + 14 + 3 + 4)/200 = 19 mu. Therefore, we have the following map:

7.4   Double exchanges between two loci can be of several types, called two-strand, three-strand, and four-strand doubles.

a. Four recombination gametes would be produced from a tetrad in which the first of two exchanges is depicted in the following figure:

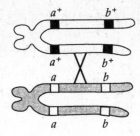

Draw in the second exchange.

b. In the following figure, draw in the second exchange so that four nonrecombination gametes would result:

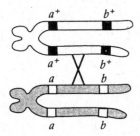

Answer: A solution to (a) and (b) can follow from trying to diagram a number of different crossovers and their resolution (by following a chromosome "through" the crossover from left to right). It is also possible to reason through these as follows: In (a), four recombinant gametes are needed. The crossover shown occurs between two nonsister chromatids, and gives two of these. A crossover between the remaining two nonsister chromatids would give two more, making a total of four. In (b), four nonrecombinant gametes are needed. The crossover shown gives two recombinant gametes, and so a second crossover must be drawn between these same two nonsister chromatids to change them back to nonrecombinant-type chromosomes. The following diagram is of a four-strand double crossover, the solution to (a), and a two-strand double crossover, the solution to (b), with resolution for the crossover events. Note that the diagram shown in (b) is equivalent to a two-strand double crossover between any two nonsister chromatids.

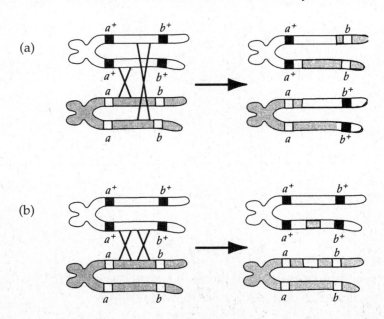

7.5 A cross between a pink (p^-) yeast strain of mating type **a** and a cream strain (p^+) of mating type α produced the following tetrads:

18	p^+ a	p^+ a	p^- α	p^- α
8	p^+ a	p^- a	p^+ α	p^- α
20	p^+ α	p^+ α	p^- a	p^- a

On the basis of these results, are the p and the mating-type genes on separate chromosomes?

Answer: The initial cross is p^- a $\times$ p^+ α, so there are 20 PD tetrads, 18 NPD tetrads, and 8 T tetrads. The number of PD and NPD tetrads is approximately equal, so these two genes assort independently and therefore are on different chromosomes. (Note that once tetrad types are assigned, determination of linkage is much easier than in nontetrad analysis. Simply ask if PD = NPD.)

7.6 The following asci were obtained from the cross *leu* + $\times$ + *rib* in yeast.

110	45	6	39
leu +	*leu rib*	+ +	*leu* +
+ *rib*	*leu* +	*leu rib*	+ *rib*
leu +	+ +	*leu rib*	+ +
+ *rib*	+ *rib*	+ +	*leu rib*

Draw the linkage map and determine the map distance.

Answer: Categorizing these unordered asci shows that there are 110 PD, 6 NPD, and 45 + 39 = 84 T tetrads. Since PD >> NPD, the loci are linked. The percent recombination between them is given by

$$RF = [(\tfrac{1}{2}\,T + NPD)/(total\ \#\ asci)] \times 100\%$$
$$= [[(\tfrac{1}{2} \times 84) + 6]/(110 + 6 + 45 + 39)] \times 100\%$$
$$= 24\%$$

The map distance between *leu* and *rib* is 24 mu.

7.7 The genes *a*, *b*, and *c* are on the same chromosome arm in *Neurospora crassa*. The following ordered asci were obtained from the cross *a b* + $\times$ + + *c*:

45	5	146	1	10	20	15	58
a b +	a b +	a b +	a b +	a b +	a b +	a b +	a b +
+ b c	a + +	a b +	+ + +	a + c	+ + c	a b c	+ b +
a + +	+ b c	+ + c	a b c	+ b +	a b +	+ + +	a + c
+ + c	+ + c	+ + c	+ + c	+ + c	+ + c	+ + c	+ + c

Determine the correct gene order and calculate all gene-gene and gene-centromere distances.

Answer: To solve this problem, consider only one pair of loci at a time, determine the gene-gene and gene-centromere distances for that pair, and then relate the findings for both pairs of loci. For the *a* and *b* loci, one can assign the type of tetrad and the segregation pattern (MI = alleles segregated during meiosis I, MII = alleles segregated during meiosis II) as shown below:

45		5		146		1		10		20		15		58	
T		**T**		**PD**		**PD**		**T**		**PD**		**PD**		**T**	
a	*b*	*a*	*b*	*a*	*b*	*a*	*b*	*a*	*b*	*a*	*b*	*a*	*b*	*a*	*b*
+	*b*	*a*	+	*a*	*b*	+	+	*a*	+	+	+	*a*	*b*	+	*b*
a	+	+	*b*	+	+	*a*	*b*	+	*b*	*a*	*b*	+	+	*a*	+
+	+	+	+	+	+	+	+	+	+	+	+	+	+	+	+
MII	MI	MI	MII	MI	MI	MII	MII	MI	MII	MII	MII	MI	MI	MII	MI

For the cross *a b* × + +, then, there are 146 + 1 + 20 + 15 = 182 PD, 45 + 5 + 10 + 58 = 118 T, and 0 NPD tetrads. PD >> NPD, so the loci are linked. The recombination frequency between them is given by

$$RF = [(½ T + NPD)/\text{total \# asci}] \times 100\%$$
$$= [(½ \times 118)/300] \times 100\%$$
$$= 19.7\%$$

The recombination frequency of *a* and *b* with the centromere is given by

$$RF \text{ (gene–centromere)} = (½ \text{ MII-type patterns/total}) \times 100\%$$
$$RF \text{ (}a\text{–centromere)} = [½(45 + 1 + 20 + 58)/300] \times 100\%$$
$$= 20.7\%$$
$$RF \text{ (}b\text{–centromere)} = [½(5 + 1 + 10 + 20)/300] \times 100\%$$
$$= 6\%$$

For the cross *b* + × + *c*, we have

45		5		146		1		10		20		15		58	
T		**T**		**PD**		**T**		**PD**		**PD**		**T**		**PD**	
b	+	*b*	+	*b*	+	*b*	+	*b*	+	*b*	+	*b*	+	*b*	+
b	*c*	+	+	*b*	+	+	+	+	*c*	+	*c*	*b*	*c*	*b*	+
+	+	*b*	*c*	+	*c*	*b*	*c*	*b*	+	*b*	+	+	+	+	*c*
+	*c*	+	*c*	+	*c*	+	*c*	+	*c*	+	*c*	+	*c*	+	*c*
MI	MII	MII	MI	MI	MI	MII	MI	MII	MII	MII	MII	MI	MII	MI	MI

There are 146 + 10 + 20 + 58 = 234 PD, 45 + 5 + 1 + 15 = 66 T, and 0 NPD tetrads. PD >> NPD, so *b* and *c* are linked. The recombination frequency between them is given by

$$RF = [(½ \times 66)/300] \times 100\%$$
$$= 11\%$$

The recombination frequency of *c* with the centromere is given by

$$RF \text{ (}c\text{–centromere)} = [½(45 + 10 + 20 + 15)/300] \times 100\%$$
$$= 15\%$$

For the cross $a + \times + c$, we have

45	5	146	1	10	20	15	58
PD	**PD**	**PD**	**T**	**T**	**PD**	**T**	**T**
$a +$	$a +$	$a +$	$a +$	$a +$	$a +$	$a +$	$a +$
$+ c$	$a +$	$a +$	$+ +$	$a\ c$	$+ c$	$a\ c$	$+ +$
$a +$	$+ c$	$+ c$	$a\ c$	$+ +$	$a +$	$+ +$	$a\ c$
$+ c$	$+ c$	$+ c$	$+ c$	$+ c$	$+ c$	$+ c$	$+ c$

There are $45 + 5 + 146 + 20 = 216$ PD, $1 + 10 + 15 + 58 = 84$ T, and 0 NPD tetrads. PD >> NPD, so a and c are linked. The recombination frequency between them is given by

$$RF = [(\tfrac{1}{2} \times 84)/300] \times 100\%$$
$$= 14\%$$

Using the calculations above, $a - b$ is 19.7 map units, $a - c$ is 14 map units, $b - c$ is 11 map units, and $b -$ centromere is 6 map units. Thus, the gene order is centromere $- b - c - a$, and one has the following map (taking into consideration that, due to increasing numbers of multiple crossovers, more accurate RFs are obtained over smaller intervals):

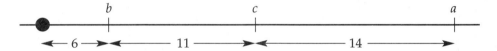

7.8 Under transmitted light, spores of wild-type (+) *Neurospora* appear black, while spores of an albino mutant (*al*) appear white.

    a. Assume that there is no chromatid interference—that is, that crossing-over occurs equally frequently between any of the four chromatids during meiosis. What patterns of ordered asci do you expect to see, and in what frequencies, if there is exactly one crossover between *al* and its centromere in every meiosis of an *al* / +?

    b. Under the preceding conditions, what is the map distance between *al* and its centromere? Are *al* and its centromere linked or unlinked?

Answer:

    a. Draw out the possible single crossovers between the *al* locus and its centromere in the meioses of an *al* / + diploid. Single crossovers occur between nonsister chromatids and with no chromatid interference, there are four equally frequent types of meioses. As shown in the following figure, all are MII patterns, half 2:4:2 and half 2:2:2:2:

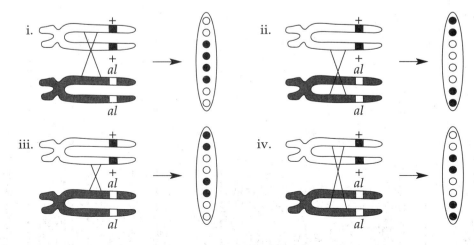

b. RF(*al* – centromere) = ½(MII pattern percentage) = 50%, so *al* is unlinked to its centromere.

7.9   The frequency of mitotic recombination in experimental organisms can be increased by exposing them to low levels of ionizing radiation (such as X-rays) during development. Hans Becker used this method to examine the patterns of clones produced by mitotic recombination in the *Drosophila* retina. (*Drosophila* has a compound eye consisting of many repetitive units called ommatidia.) What type of spots would be produced in the *Drosophila* retina if you irradiated a developing *Drosophila* female obtained from crossing a white-eyed male with a cherry-eyed female? (See Table 4.2 [p. 84] for a description of the *w* and *w^{ch}* alleles.)

Answer: The cross is *w*/Y (white ♂) × *w^{ch}*/*w^{ch}* (cherry ♀), so the female progeny are *w*/*w^{ch}* heterozygotes. Normal mitotic division in a *w*/*w^{ch}* cell, diagrammed in the top half of the figure here, generates two identical *w*/*w^{ch}* daughter cells. As shown in the bottom half of the figure, when a single mitotic crossover occurs between the *w* locus and its centromere, two genetically different daughter cells are produced after cell division: *w*/*w* and *w^{ch}*/*w^{ch}*.

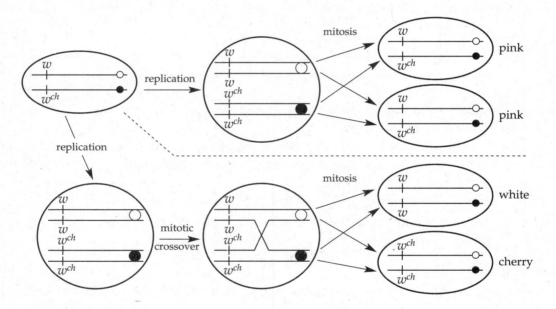

The alleles at the *white* locus control the amount of pigment deposition in the cells of the eye (Chapter 4). Phenotypically white homozygotes (*w*/*w*) have 0.44 percent of wild-type pigment levels, whereas cherry-colored homozygotes (*w^{ch}*/*w^{ch}*) have 4.1 percent of wild-type pigment levels. The *white* alleles show partial dominance, so *w*/*w^{ch}* heterozygotes should have about 2 percent of wild-type pigment levels and show an intermediate pink color. If mitotic recombination occurs during the development of the eye, neighboring *w*/*w* and *w^{ch}*/*w^{ch}* daughter cells will be produced. After these and the surrounding *w*/*w^{ch}* cells divide by mitosis to form the adult eye, twin spots will be seen: In a pink background (*w*/*w^{ch}*), there will be neighboring white (*w*/*w*) and cherry (*w^{ch}*/*w^{ch}*) twin spots.

7.10   A diploid strain of *Aspergillus nidulans* (forced between wild type and a multiple mutant) that was heterozygous for the recessive mutations *y* (yellow), *w* (white), *ad* (adenine), *sm* (small), *phe* (phenylalanine), and *pu* (putrescine) produced haploid segregants. Forty-one haploid white and yellow segregants were tested and found to have the following genotypes and numbers:

white:	*y w pu ad sm phe*	7
	y w pu ad + +	11
yellow:	*y + + + sm phe*	16
	y + + + +	7

What are the linkage relationships of these genes?

Answer: Haploid segregants are produced from the diploid by haploidization. In this process, the two homologs of each chromosome assort independently. Here, three blocks of genes are assorting independently: *y; w, pu,* and *ad;* and *sm* and *phe.* Therefore, *y* is on one linkage group, *w, pu,* and *ad* are on a second linkage group, and *sm* and *phe* are on a third linkage group.

7.11   A heterokaryon was established in the fungus *Aspergillus nidulans* between a *met⁻ trp⁻* auxotroph and a *leu⁻ nic⁻* auxotroph. A diploid strain was selected from this heterokaryon. From the diploid strain, the following eight haploid strains were obtained from conidial isolates, via the parasexual cycle, in approximately equal frequencies:

1.	$nic^+ leu^+ met^- trp^-$	5.	$leu^+ nic^- met^+ trp^+$
2.	$met^+ leu^+ nic^+ trp^+$	6.	$met^- nic^- leu^- trp^-$
3.	$trp^- met^- leu^+ nic^-$	7.	$trp^+ leu^- met^+ nic^+$
4.	$leu^- nic^- trp^+ met^+$	8.	$nic^+ met^- trp^- leu^-$

Which, if any, of these four marker genes are linked, and which are unlinked?

Answer: In a parasexual cycle, haploid nuclei fuse to form a diploid nucleus. Mitotic crossing-over may occur in the diploid nucleus before haploidization. Here, the diploid nucleus is formed from haploid cells that are *met⁻ trp⁻ leu⁺ nic⁺* and *met⁺ trp⁺ leu⁻ nic⁻.* A mitotic crossover will make all genes distal to the crossing-over point (on the same arm) homozygous, while during haploidization, unlinked genes assort independently. Tabulate the genotypes, assess what patterns are seen, and evaluate how these patterns can be explained:

Isolate	Gene			
Type	nic	leu	met	trp
1	+	+	−	−
2	+	+	+	+
3	−	+	−	−
4	−	−	+	+
5	−	+	+	+
6	−	−	−	−
7	+	−	+	+
8	+	−	−	−

Inspection of the data reveals that only parental combinations of the *met* and *trp* genes are obtained (50 percent each *met⁻ trp⁻* and *met⁺ trp⁺*) and that these genes assort independently of *nic* and *leu*. In addition, the *leu* and *nic* genes are found in four combinations (25% each *leu⁺ nic⁺*, *leu⁺ nic⁻*, *leu⁻ nic⁻*, and *leu⁻ nic⁺*) that are distributed among the equally frequent eight genotypes. Therefore, *met* and *trp* are in one linkage group, *leu* is in a second linkage group, and *nic* is in a third linkage group.

It is possible that mitotic crossing-over could have occurred prior to haploidization, and so it is useful to consider how the data would be affected if it had occurred. For the *met* and *trp* loci, mitotic crossing-over between the most proximal gene and the centromere followed by haploidization would produce a parental genotype (and so not be detected), while a mitotic crossover between *met* and *trp* followed by haploidization would produce *met⁺ trp* and *trp met⁺* haploids. The latter are not seen, so mitotic crossing-over did not occur between these loci. This could be because mitotic crossing-over is rare and not enough conidial isolates were sampled. It could also indicate that the loci are closely linked. For either of the *nic* and *leu* loci, rare mitotic crossovers between a locus and its centromere would produce the same genotypes that are observed, and so not be detected.

7.12 A (green) diploid of *Aspergillus nidulans* is heterozygous for *each* of the following recessive mutant genes: *sm, pu, phe, bi, w* (white), *y* (yellow), and *ad*. Analysis of white and yellow haploid segregants from this diploid indicated several classes with the following genotypes:

			Genotype			
sm	**pu**	**phe**	**bi**	**w**	**y**	**ad**
sm	pu	phe	+	w	y	ad
+	pu	+	+	w	y	ad
+	pu	+	bi	w	+	ad
sm	+	phe	+	+	y	+
+	+	+	+	+	y	+
sm	pu	phe	bi	w	+	ad

How many linkage groups are involved, and which genes are on which linkage group?

Answer: The two homologs of each chromosome assort independently during haploidization. Here, three blocks of genes segregate independently, identifying three linkage groups: *ad w pu* and *ad⁺ w⁺ pu⁺* segregate independently of *y bi⁺* and *y⁺ bi*, which segregate independently from *sm phe* and *sm⁺ phe⁺*.

7.13 A (green) diploid of *Aspergillus nidulans* is homozygous for the recessive mutant gene *ad* and heterozygous for the following recessive mutant genes: *paba, ribo, y* (yellow), *an, bi, pro*, and *su−ad*. Those recessive alleles which are on the same chromosome are in coupling. The *su−ad* allele is a recessive suppressor of the *ad* allele: The +/*su−ad* genotype does not suppress the adenine requirement of the *ad/ad* diploid, whereas the *su−ad/su−ad* genotype does suppress that requirement. Therefore the parental diploid requires adenine for growth. From this diploid two classes of segregants were selected: yellow and adenine independent. The following table lists the types of segregants obtained:

Segregant Type Selected								Phenotype
Adenine-independent	+							
	ribo							
	ribo	an						
	ribo	an	pro	paba	y		bi	
Yellow					y	ad	bi	
				paba	y	ad	bi	
			pro	paba	y	ad	bi	
	ribo	an	pro	paba	y		bi	

Analyze these results as completely as possible to determine the location of the centromere and the relative locations of the genes.

Answer: Consider the adenine-independent segregants first. The original strain is homozygous for *ad*, so the adenine-independent segregants must be *su−ad/su−ad*. A mitotic crossover will make all genes distal to the crossing-over point (on the same arm) homozygous. Therefore, *su−ad* homozygotes were generated by a mitotic crossover between the centromere and *su−ad*. For the four *su−ad/su−ad* segregants to exhibit different combinations of the other genes, the *su−ad* gene must either be distal to them on the same chromosome arm or on a different chromosome arm. Segregants that are *su−ad/su−ad* and *an* are always *ribo*. This is consistent with the order centromere−*an−ribo−su−ad*: The *an ribo su−ad* segregants were generated by mitotic crossovers between the centromere and *an*, the *ribo su−ad* segregants were generated by mitotic crossovers between *an* and *ribo*, and the *su−ad* (+) segregants were generated by mitotic crossovers between *ribo* and *su−ad*. The *pro, paba, y,* and *bi* genes are on a different chromosome arm, as they segregate as a block and are separated from the *an ribo su−ad* genes by these mitotic crossovers. The *ribo an pro paba y bi* segregant results if no mitotic crossovers occur.

Now consider the *y* segregants. Since the original strain was heterozygous for *y*, these were generated by a mitotic crossover between the centromere and *y*. Segregants that are *y* and *pro* are always *paba* and *bi*, and segregants that are *y* or *paba* are always *bi*. This is consistent with the gene order centromere−*pro−paba−y−bi*: The *y bi* segregants were generated by a mitotic crossover between *paba* and *y*, the *paba y bi* segregants were generated by a mitotic crossover between *pro* and *paba*, and the *pro paba y bi* segregants were generated by a mitotic crossover between the centromere and *pro*. All the *y bi, y bi paba,* and *y bi paba pro* segregants are *ad* (*su−ad/*+) and not *ribo* or *an*. This is because the *su−ad, ribo,* and *an* genes lie on a different chromosome arm and these genes segregate as a block when a mitotic crossover occurs on the chromosome arm bearing the *y, bi, paba,* and *pro* genes.

Therefore, the relative locations of these genes and their centromere is *su−ad−ribo−an*−centromere−*pro−paba−y−bi*. The *ad* locus is homozygous in the original strain, so it cannot be mapped by these methods.

7.14   High-density genetic maps can be generated through the use of mapping panels with a set of DNA markers and lod score methods. The same DNA markers can be mapped by means of radiation hybrid methods.

a. In what ways will the maps generated by these two methods be identical, and in what ways will they differ?

b. Much of the entire genomic sequence has been obtained for a number of complex eukaryotes, including humans, mice, and the plant *Arabidopsis thaliana*. What is the value of high-density genetic maps in the genetic analysis of organisms whose genome has been sequenced?

Answer:

a. Mapping panels and lod score methods generate maps based on recombination frequency, while radiation hybrid methods generate maps based on physical distance. The order of genes should be the same using the two methods, but the distances are estimated differently.

b. High-density genetic maps are useful to define the location of markers and genes within the DNA sequence and to predict how often recombinants between two loci will be observed.

7.15   Two panels of radiation hybrids were produced by irradiating human tissue culture cells and then fusing them with hamster tissue culture cells. The differing properties of the two panels are shown in the following table (1 Mb = 10^6 bp DNA):

	Panel GB4	Panel G3
X-ray dosage used to generate cell hybrids	3,000 rad	10,000 rad
Number of cell lines established	93	83
Average retention of human genome per hybrid	32%	16%
Average human DNA fragment size	25 Mb	2.4 Mb
Effective map resolution	1 Mb	0.25 Mb

a. A haploid human genome has about 3×10^9 bp of DNA. About how many different human DNA segments are present, on average, in the hybrid cells of each panel?

b. Two human markers are found together in some, but not all, cell hybrids. Are they necessarily linked?

c. How do these panels differ in their advantages with respect to mapping genes and markers?

d. DNA markers *A*, *B*, and *C* derive from a single chromosomal region. Their presence or absence is assessed in DNA isolated from the hybrids of each panel, with the following results:

	Number of Hybrids Testing Positive For Markers	
Marker(s) Present	**Panel GB4**	**Panel G3**
A only	0	4
B only	1	6
C only	2	15
A and *B* only	2	11
A and *C* only	1	2
B and *C* only	0	0
A, *B*, and *C*	27	0

Why do the two panels give such different results? What reasonable hypothesis can you generate concerning the arrangement of these three markers?

Answer:

a. GB4 has $[(3 \times 10^9 \text{ bp/genome} \times 0.32 \text{ genome})/(25 \times 10^6 \text{ bp/segment})] \approx 38$ segments. G3 has ~200 segments.

b. The two markers may reside on segments derived from different chromosomes that, by chance alone, reside in the same hybrid. Their coexistence in one or even several hybrids is insufficient to indicate linkage, without calculating the likelihood that this occurred by chance.

c. To resolve the relative order of a set of markers, overlapping DNA segments bearing marker subsets must be employed. GB4 has larger DNA segments, so it is useful for mapping markers that span larger distances, at least 1 MB. However, it will not resolve the order of closer-spaced markers, as they will most often lie within the same DNA segment. In contrast, since it has smaller DNA segments, G3 will resolve markers that are at least 250,000 bp apart. The differing properties of the two panels make them complementary: GB4 provides long-range map continuity, while G3 gives higher local resolution.

d. The two panels have human DNA segments with different average lengths, and this determines their capacity to resolve close markers. Even though the GB4 panel has a similar total number of hybrids containing each marker, it fails to resolve the markers. All three markers are simultaneously present in the majority of positive hybrids and so these markers are relatively close together, probably within a 1-MB interval. The G3 panel is able to resolve them as hybrids with just one or two markers are obtained. There are more hybrids positive for only *A* and *B* than *A* and *C*, and there are no hybrids positive for *B* and *C* or *A*, *B*, and *C*. This is consistent with a hypothesis that *A* and *B* are closer together than *A* and *C*, and their order is *B–A–C*.

7.16 As discussed in Chapter 3, XO individuals have Turner syndrome. Some individuals who display a Turner phenotype are mosaic individuals with 45,X/46,XX or 46,XY/45,X karyotypes. It is clinically important to address mosaicism in Turner individuals, as some types of mosaics have an increased risk of gonadal cancer.

a. What chromosomal events could lead to a mosaic Turner individual? When do such events occur?

b. How might physical mapping methods be adapted to determine, reliably and readily, whether a Turner individual is mosaic?

Answer:

a. Nondisjunction (see Chapter 3) of the X chromosome during mitosis in a 46,XX cell would produce 45,X and 47,XXX cells. If this occurred in a cell produced after the first mitotic division and the 47,XXX cell did not survive, a 45,X/46,XX individual would be formed. Similarly, nondisjunction of the Y chromosome during a mitotic division in an XY cell could lead to 45,X/46,XY individuals.

b. Obtain a skin (fibroblast) or blood sample from a Turner individual and culture the cells. Label probes specific for the X and Y chromosomes with different flurochromes. Then use fluorescent in situ hybridization (FISH) to characterize the X and Y chromosomes present in the cultured cells. Mosaics will have cells with just one X (45,X) and cells with either two Xs (46,XX) or XY (46,XY) chromosomes.

7.17   Some dogs love water, while others avoid it. A dog that loved water was mated with a dog that avoided it, and their F_1 progeny were interbred to give an F_2. The parental, F_1, and F_2 dogs were evaluated by DNA typing, and the lod score method was used to assess linkage between DNA markers and genes for water affection (*waf* genes). Suppose that the following data were obtained for one marker, where θ gives the value of the recombination frequency between the marker and a *waf* gene used in calculating the lod score:

θ	lod Score
0	$-\infty$
0.05	-12.51
0.10	-2.34
0.15	-1.32
0.20	2.66
0.25	4.01
0.30	3.21
0.35	2.14
0.40	1.56
0.50	0

Graph these lod scores and evaluate whether the marker is linked to a *waf* gene. If it is, estimate the physical distance between the marker and the gene.

Answer:

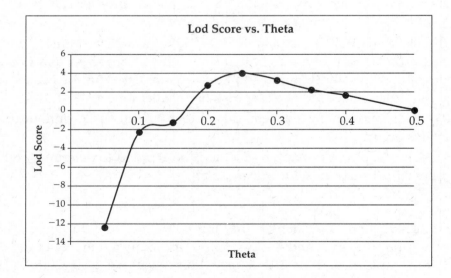

A graph of θ vs. lod score reveals a maximum lod score of 4.01 at a theta of 0.25, a map distance of about 25 mu. Since the lod score is greater than 3, the marker is linked to a *waf* gene. However, the marker is not closely linked. Using the estimate from humans that 1 mu corresponds on average to 1 Mb, the marker is about 25 Mb away from a *waf* gene.

8

VARIATIONS IN CHROMOSOME STRUCTURE AND NUMBER

THINKING ANALYTICALLY

In this chapter, it is crucial to be able to visualize the physical structure of chromosomes in the context of cellular processes such as meiosis and gene expression. Start by developing mental images of chromosomal aberrations. Practice drawing inversions, transpositions, translocations, duplications, and deletions.

To understand fully the impact of chromosomal mutations on gene expression and inheritance of traits, you will need a solid grasp of cellular processes involving chromosome movement. Review how chromosomes move and align during meiosis. Then consider how chromosomal aberrations and mutations affecting chromosome number alter the behavior of chromosomes during meiosis and thereby affect the inheritance of traits. The disruptions in meiosis that result from a particular chromosomal aberration are best understood by drawing out the chromosomal mutation as it proceeds through meiosis. As you do the problems, take the time to draw out the consequences of crossing-over in inversion and translocation heterozygotes. To understand how chromosome aberrations affect gene expression and gene

169

REVIEW OF KEY TERMS, SYMBOLS, AND CONCEPTS

In your own words, write a brief, precise definition of each term in the groups below. Check your definitions using the text. Then develop a concept map using the terms in each list.

1	2
chromosomal mutation, aberration	euploid
deletion	aneuploid
duplication	polyploid
translocation	monoploid
inversion	diploid
position effect	haploid
karyotype	nullisomy
polytene chromosomes, bands and interbands	monosomy
endoreduplication	trisomy
pseudodominance	tetrasomy
terminal tandem, reverse tandem duplication	double monosomic, tetrasomic
multigene family	autopolyploidy
unequal crossing-over	allopolyploidy
pericentric vs. paracentric inversion	allotetraploid
dicentric bridge, acentric fragment	odd-number polyploidy
intra-, interchromosomal translocation	even-number polyploidy
reciprocal vs. nonreciprocal translocation	seedless fruit
alternate, adjacent 1, adjacent 2 segregation	chromosome
semisterility	centromere
fragile sites	distal to/from
triplet repeat amplification	promixal to/from
premutation	p, q arms
normal transmitting male	Down syndrome
cri-du-chat syndrome	Patau syndrome
Prader-Willi syndrome	Edwards syndrome
Drosophila Bar mutation	Turner syndrome
Philadelphia chromosome	Klinefelter syndrome
Burkitt lymphoma	familial Down syndrome
chronic myelogenous leukemia	Robertsonian translocation
proto-oncogene	
oncogenes, *c-abl, c-myc*	
fragile X syndrome, mental retardation	
FMR-1	

function, consider what happens when the structure of a gene is rearranged and broken into two separate parts by a chromosomal aberration. Ask yourself, What are the possible consequences if a rearrangement breakpoint is in the amino acid–coding region of the gene, if it is in the promoter or other regulatory region, if it rearranges the gene nearby the enhancer of a different gene, or if it fuses two unrelated genes in their intronic regions?

Aneuploidy and polyploidy have direct effects on gene dosage: A normal phenotype requires proper gene dosage. As you examine the consequences of different types of aneuploidy and polyploidy, consider how the different phenotypes that arise result from altering gene dosage.

QUESTIONS FOR PRACTICE

Multiple Choice and True-False Questions

1. The number of polytene chromosomes in dipterans, such as *Drosophila*, is characteristically
   a. twice the diploid number
   b. hundreds or more times the diploid number
   c. half the diploid number
   d. the same as the diploid number

2. The chromosomes that become polytene chromosomes are
   a. somatically paired
   b. genetically inert
   c. uniformly pycnotic
   d. acentric

3. Nonhomologous chromosomes that have exchanged segments are the products of a
   a. double deletion
   b. reciprocal translocation
   c. pericentric inversion
   d. paracentric inversion

4. The abbreviated karyotype, $2N - 1$, describes
   a. nullisomy
   b. monosomy
   c. trisomy
   d. haploidy

5. The abbreviated karyotype $2N + 1 + 1$ describes
   a. double trisomy
   b. tetrasomy
   c. double monosomy
   d. none of these

6. All known monosomics in humans have been
   a. lethal
   b. semilethal
   c. treatable
   d. deletion heterozygotes

7. The cultivated bread wheat *Triticum aestivum*, although a polyploid, is fertile because it
   a. has an odd number of chromosome sets
   b. has an even multiple of chromosome sets
   c. is a double hybrid
   d. is propagated by grafting

8. Cultivated bananas and Baldwin apples are
   a. double diploids
   b. tetraploids

    c. double haploids

    d. triploids

9. Down syndrome is associated with

    a. an inversion

    b. a Robertsonian translocation

    c. trisomy-21

    d. both b and c

10. Gene expression can be altered by

    a. an inversion

    b. a duplication

    c. a transposition

    d. any of the above

Decide whether each of the following statements is true or false.

11. In terms of severity of symptoms found in infants, trisomies for chromosomes 13, 18, and 21 are all about equally severe.

12. All deletions are cytologically visible.

13. A dicentric chromosome has two arms.

14. Crossing-over within the inverted region of a pericentric inversion, in an inversion heterozygote, produces gametes with duplications and deletions.

15. The Philadelphia chromosome (a specific translocation between chromosomes 9 and 22) found in chronic myelogenous leukemia fuses two genes, resulting in an oncogene.

16. The semi-sterility seen in individuals heterozygous for chromosomal rearrangements such as inversions and translocations always results from the production of aneuploid gametes following crossing-over.

17. Duplications can result in dominant phenotypes if the duplication includes a dose-sensitive region.

18. Pericentric inversions occur on one arm of a chromosome.

19. Polytene chromosomes are chromosomes that have replicated without nuclear division.

20. In contrast to triploid animals, which usually die, triploid plants usually live (but are sterile).

Answers: 1c, 2a, 3b, 4b, 5a, 6a, 7b, 8d, 9d, 10d, 11F, 12F, 13F, 14T, 15T, 16F, 17T, 18F, 19T, 20T

Thought Questions

1. Identify and describe five examples of changes in chromosome structure that alter gene expression.

2. A chromosomal inversion in the heterozygous condition seems to act as a suppressor of crossing-over. Why might this be the case? How might such chromosomes be important for keeping a laboratory stock of three linked recessive mutations?

3. What are two modes by which Down syndrome can be caused? Why might one be more frequent than the other? If a normal female has a sibling with Down syndrome, what information would be critical in the assessment of the chances that she might have a Down syndrome child? How would the assessment be different if the assessed individual were a normal male with a Down syndrome sibling?

4. Consider the data presented in Table 8.2. Given these data, how do you account for the fact that most children with Down syndrome are born to women under 25 years of age? Does this alter the need for caution and prenatal diagnosis in women over 35 years of age?

5. How do you account for the fact that polyploidy is significantly more common in plants than in animals? (Hint: What basic difference between sexual and asexual reproduction might be significant here?)

6. What might be the significance of the following in the evolutionary process of speciation? (1) autopolyploidy, (2) allopolyploidy, (3) inversions, (4) translocations. Provide an explanation that considers how animals might be affected differently than plants.

7. Many small deletions, when heterozygous, display a normal phenotype. Others display a mutant phenotype. For example, a small deletion that removes just the *Notch* gene in *Drosophila*, when heterozygous with a normal chromosome, displays a mutant phenotype of notched (nicked) wings. A nearby deletion that removes just the *white* gene, when heterozygous with a normal chromosome, displays the wild-type phenotype of deep-red eyes. Speculate why some heterozygous deletions might show a phenotype while others do not.

8. Which, if any, of the chromosomal aberrations discussed in this chapter are able to revert to a normal phenotype?

Thought Questions for Media

After reviewing the media on the *iGenetics* CD-ROM, try to answer these questions.

1. After viewing the *Crossing-over in an Inversion Heterozygote* animation, consider the following two situations.

   I. No crossovers occur in the inverted region of a pericentric inversion in an inversion heterozygote.

   II. A single crossover occurs in the inverted region of a pericentric inversion in an inversion heterozygote.

   a. What different types of *viable* gametes are produced in these two situations?

   b. Are the ratios of different types of *viable* gametes are produced in these situations the same or different?

   c. Are *inviable* gametes produced in either of these situations? If so, how are they produced and why are they inviable?

2. In the *Meiosis in a Translocation Heterozygote* animation, the chromosomes in a reciprocal translocation heterozygote were depicted as N_1, N_2, T_1, and T_2. N_1 and N_2 were the normal ordered, nonhomologous chromosomes. T_1 and T_2 were the chromosomes in which a reciprocal translocation occurred. N_1 and T_1 have homologous centromeres, as do N_2 and T_2. In the animation, homologous regions of these chromosomes paired during meiosis I and aligned in a crosslike figure. They were arranged with N_1 in the upper left-hand corner, N_2 in the lower right-hand corner, T_1 in the lower left-hand corner, and T_2 in the upper right-hand corner, as follows:

   $N_1 \rightarrow T_2$

   $T_1 \rightarrow N_2$

   Use this illustration as you answer the following questions:

a. The animation illustrated three different patterns of segregation of these chromosomes during meiosis I called alternate, adjacent-1, and adjacent-2 segregation. Which chromosomes go to each pole in the different types of segregation?

b. Which two types of segregation are most frequent? Which one type of segregation is rare?

c. Which type(s) of segregation pattern produces balanced gametes? Which type(s) produces inviable gametes?

d. Why are individuals with a reciprocal translocation semi-sterile?

1. Suppose Sonia and Ramon conceive a fetus that has the T(14;21) translocation. Can you tell, from this information alone, whether the child will have Down syndrome or be healthy? If not, what additional information about the karyotype would you need to know? If the child has Down syndrome, what can you infer about the chromosomes carried by the sperm used in fertilization?

2. Suppose Sonia and Ramon, having had one normal child, are interested in having a second child. They seek genetic counseling for advice. What advice should the counselor give them, and how, if at all, would it differ from any advice they might have received when attempting to have their first child?

3. What is the chance that Sonia and Ramon have a child with the same karyotype as Ramon? With the same karyotype as Sonia?

SOLUTIONS TO TEXT PROBLEMS

8.1   A normal chromosome has the following gene sequence:

$$A \; B \; C \; D \; E \; F \; G \; H$$

Determine the chromosomal mutation illustrated by each of the following chromosomes:

a. $A \; B \; C \; F \; E \; D \; G \; H$
b. $A \; D \; E \; F \; B \; C \; G \; H$
c. $A \; B \; C \; D \; E \; F \; E \; F \; G \; H$
d. $A \; B \; C \; D \; E \; F \; F \; E \; G \; H$
e. $A \; B \; D \; E \; F \; G \; H$

Answer:

a. pericentric inversion (inversion of D–o–E–F)
b. nonreciprocal translocation (B–C moved from left to right arm)
c. tandem duplication (E–F duplicated)
d. reverse tandem duplication (E–F duplicated)
e. deletion (C deleted)

8.2   Distinguish between pericentric and paracentric inversions.

Answer: A pericentric inversion is an inversion that includes the centromere, while a paracentric inversion lies wholly within one chromosomal arm. See text Figure 8.8.

8.3   In some instances, very small deletions behave like recessive mutations. Why are some recessive mutations known not to be deletions?

Answer: Deletions, whether just a few bases or large segments of DNA, are unable to be reverted. Point mutations that affect only a single base can be reverted. If a recessive mutation is able to be reverted to a wild-type phenotype, it cannot be a deletion.

8.4   Inversions are known to affect crossing-over. The following homologs have the indicated gene order (the filled and open circles are homologous centromeres):

$$\bullet A \; B \; C \; D \; E$$

$$\circ A \; D \; C \; B \; E$$

a. Considering the position of the centromere, what is this sort of inversion called?
b. Diagram the alignment of these chromosomes during meiosis.
c. Diagram the results of a single crossover between homologous genes B and C in the inversion.

Answer:

a. Paracentric inversion, because the centromere is not included in the inverted DNA segment

b.

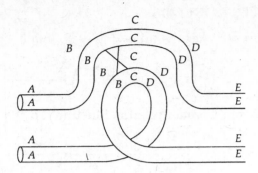

c. A crossover between B and C results in the following chromosomes:

A B C D E   (normal order)

A B C D A   (dicentric, duplication for A, deletion for E)

E B C D E   (acentric, duplication for E, deletion for A)

A D C B E   (inverted order)

8.5   Single crossovers within the inversion loop of inversion heterozygotes give rise to chromatids with duplications and deletions. What happens if, within the loop, there is a two-strand double crossover in such an inversion heterozygote when the centromere is outside the loop?

Answer: If a two-strand double crossover occurs within a paracentric inversion, the four products of meiosis have a complete set of genes, without duplications or deletions. No acentric or dicentric fragments are formed. Consequently, all four meiotic products are viable. This is illustrated in the following example:

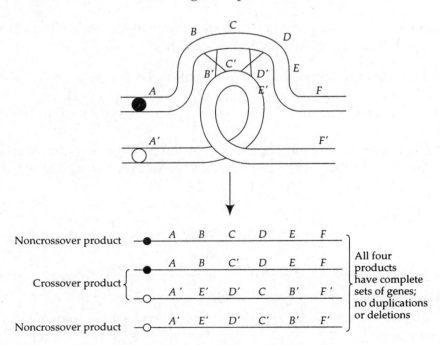

8.6   An inversion heterozygote possesses one chromosome with genes in the normal order:

a b c d e f g h

It also contains one chromosome with genes in the inverted order:

$$\underset{\circ}{\underline{a\ b\ f\ e\ d\ c\ g\ h}}$$

A four-strand double crossover occurs in the areas *e–f* and *c–d*. Diagram and label the four strands at synapsis (showing the crossovers) and at the first meiotic anaphase.

Answer: Diagrammed below is a four-strand double crossover, where the crossover between *c* and *d* involves strands 2 and 4, and the crossover between *e* and *f* involves strands 1 and 3. The bridge strands will break at anaphase I as the centromeres move toward opposite poles of the cell.

Synapsis:

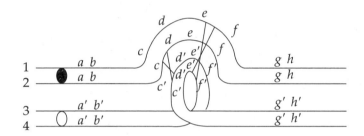

First Anaphase:

1
2   a   b   c   d   e   f'   b'   a'   3
4   **double dicentric**

a   b   c   d'   e'   f'   b'   a'

h   g   f   e   d   c'   g'   h'

acentric fragments

h   g   f   e'   d'   c'   g'   h'

8.7   The following gene arrangements in a particular chromosome are found in *Drosophila* populations in different geographic regions:

a. *A B C D E F G H I*

b. *H E F B A G C D I*

c. *A B F E D C G H I*

d. *A B F C G H E D I*

e. *A B F E H G C D I*

Assuming that the arrangement in (a) is the original arrangement, in what sequence did the various inversion types probably arise?

Answer: One series of sequential inversions is:

a → c → e → d
        ↓
        b

The regions inverted in each step are illustrated here.

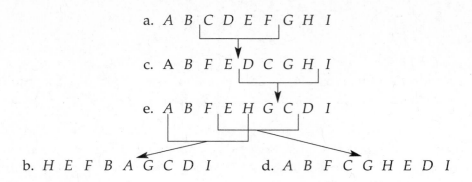

8.8   The following figure shows map distances observed for genes in one chromosomal region (the open circle represents a centromere):

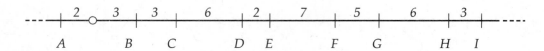

A new paracentric inversion bears a recessive mutation *d* at the *D* locus. Its proximal breakpoint lies between *B* and *C*, 1 mu from *B*, and its distal breakpoint lies between *G* and *H*, 1 mu from *G*. A heterozygote for this inversion and the wild-type arrangement mates with a *dd* individual homozygous for the inversion.

a. In the absence of multiple crossovers, what is the chance that a *dd* offspring will have a homolog with a wild-type arrangement?

b. What type of event is required to produce a *dd* offspring having a homolog with a wild-type arrangement? What is the likelihood of such an event?

c. On the basis of your answers to (a) and (b), how might spontaneously arising inversions contribute to the maintenance of genetic differences between subpopulations of a species?

Answer: The following figure, in which the inverted region is depicted by an arrow, depicts the chromosomes in the heterozygote.

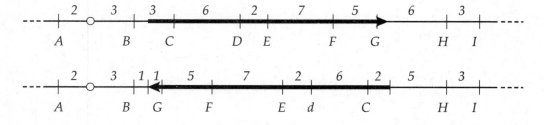

a. If no crossovers occur in the inversion heterozygote or if a crossover occurs outside of the inverted region, the *d*-bearing chromosome contributed by the inversion heterozygote will be inverted. If one crossover occurs, inviable deletion products would result (see text Figure 8.9). Therefore, for zero or one crossover, no *dd* offspring will have chromosomes with only wild-type arrangements.

b. A two-strand double crossover could produce a normal-ordered, *d*-bearing homolog (see the solution to problem 8.5). To produce a *d*-bearing homolog, one crossover must occur between the proximal (toward the centromere) breakpoint of the inversion and the *d* locus, and a second crossover involving the same pair of chromatids must occur between the *d* locus and the distal (toward the telomere) breakpoint of the inversion. There are 15 map units between the proximal breakpoint and *d*, and 8 map units between the *d* and the distal breakpoint. Assuming that there is no crossover interference, P(double crossover) $= 0.15 \times 0.08 = 0.012$. Two-strand double crossovers are ¼ of the possible double crossovers (See text Figure 6.5), so P (desired event) $= ¼ \times 0.012 = 0.003$.

c. Most of the time, mutations within an inversion will be transmitted with that inversion and not with the normal ordered chromosome. Over time, additional genetic differences will accumulate in the inverted region. These will not usually be shared with individuals having normal ordered chromosomes. Therefore, the inversion contributes to the maintenance of genetic differences between subpopulations of a species.

8.9 The deep-red eye color of normal *Drosophila* results from pigment deposition controlled by the *white* gene, which lies on the X chromosome at map position 1.5, far from centromeric heterochromatin (which starts at about map position 66). Hermann Muller screened for new *white* mutants by irradiating wild-type *Drosophila* males (w^+/Y) and mating them to white-eyed (*w/w*) females. He isolated several mutant females bearing mottled red eyes (red eyes with varying amounts of white spotting). One mutant, w^{M5}, was associated with a reciprocal translocation with breakpoints near the *white* locus and centromeric heterochromatin of chromosome 4. A different mutant, w^{M4}, was associated with an X-chromosome inversion with breakpoints near the *white* locus and centromeric heterochromatin. Kenneth Tartof screened for revertants of the mottled-eye phenotype of w^{M4} by crossing irradiated w^{M4}/Y males with *w/w* females. He recovered three different normal-eyed female revertants, each associated with a new X-chromosome inversion. In addition to having the original w^{M4} breakpoints near the *white* locus and in centromeric heterochromatin, each had a third euchromatic breakpoint.

a. On the basis of these results, is the mottled-eye phenotype in Muller's mutants due to a mutation within the *white* gene? If not, what is its most likely cause?

b. How can the w^{M4} mutation be reverted by an additional inversion with a euchromatic breakpoint?

Answer:

a. The following diagram shows a normal chromosome bearing w^+, the w^{M4}-associated inversion, and a second inversion found in a w^+ revertant. The genes *a, b, c,* and *d* are inserted nearby the breakpoints of the different inversions to help visualize the inverted regions. Euchromatin is represented by a thin line, centromeric heterochromatin by a thick line, and the centromere by an open circle. The brackets delineate inverted regions.

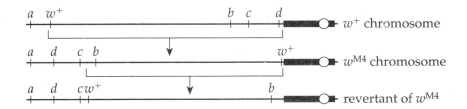

The mottled-eye phenotype is associated with chromosomal rearrangements induced on a w^+-bearing chromosome that place the w^+ gene near heterochromatin. When a rearrangement is heterozygous with a *w* allele, only the w^+ gene on the rearranged chromosome can provide for normal eye pigmentation. The mottled appearance of the eye indicates that it functions in some but not all cells. This suggests that the *white* gene's DNA sequence is unaltered. It is more likely an epigenetic phenomenon caused by a position effect, a phenotypic change due to inactivation of the w^+ allele by neighboring heterochromatin.

b. The second inversion that occurs on the w^{M4} chromosome repositions the w^+ gene to a euchromatic location. This supports the view that the mottled-eye phenotype is caused by a position effect.

8.10   Human abnormalities associated with chromosomal mutations often exhibit a range of symptoms, of which only some subsets appear in a particular individual. Recombinant 8 [Rec(8)] syndrome is an inherited chromosomal abnormality found primarily in individuals of Hispanic origin. Phenotypic characteristics associated with the syndrome include congenital heart disease, urinary system abnormalities, eye abnormalities, hearing loss, and abnormal muscle tone. Most reported cases of Rec(8) have been found in the offspring of phenotypically normal parents who are heterozygous for an inversion of chromosome 8 with breakpoints at p23.1 and q22.1. Individuals with Rec(8) syndrome typically have a duplication of part of 8q (from q22.1 to the terminus of the q arm) and a deletion of 8p (from p23.1 to the terminus of the p arm).

a. Using diagrams, explain why individuals with Rec(8) syndrome typically have a duplication and a deletion for part of chromosome 8.

b. An individual is heterozygous for an inversion on chromosome 8 with breakpoints at p23.1 and q22.1. If a crossover occurs within the inverted region during a particular meiosis, what is the chance that the resulting offspring will have Rec(8) syndrome?

c. Why might the phenotypes of Rec(8) individuals vary?

d. A child with some of the symptoms of Rec(8) syndrome is referred to a human geneticist. The karyotype of the child reveals heterozygosity for a large pericentric inversion in chromosome 8 with breakpoints at p23.1 and q22.1. Cytogenetic analysis of her phenotypically normal mother and phenotypically normal maternal grandmother reveals a similar karyotype. According to the child's mother, the father has a normal phenotype, but he is unavailable for examination. Propose at least two explanations for why the child, but not her mother or maternal grandmother, is affected with some of the symptoms of Rec(8) syndrome. (Hint: Consider the limitations of karyotype analysis using G-banding methods [see Chapter 3, p. 42] and also consider what is unknown about the father.)

Answer:

a. Parents of Rec(8) individuals are heterozygous for a pericentric inversion with breakpoints at 8p23.1 and 8q22.1. Rec(8) offspring with 8q-duplication and 8p-deletion probably arose from a single crossover within the pericentric inversion. Such an event is diagrammed in text Figure 8.10.

b. As shown in text Figure 8.10, a single crossover between two nonsister chromatids in an inversion heterozygote results in four products: Two have the noncrossover chromosomes (one is normal ordered and one is inverted) and two are duplication/deletion products. Here, the product with 8q-duplication and 8p-deletion contribute to a viable zygote with Rec(8) syndrome. The product with 8q-deletion and 8p-duplication is not discussed in the problem. It may be that zygotes with this product do not survive. In this case, 1/3 of the surviving zygotes have Rec(8) syndrome. Of the 2/3 normal zygotes, 1/2 carry the chromosome 8 inversion.

   c. The phenotypes of Rec(8) individuals could vary for one or a combination of reasons. (1) There could be several different chromosome 8 inversions in the population that vary slightly in their inversion breakpoints. The Rec(8) individuals resulting from single crossovers in inversion heterozygotes would differ symptomatically due to variation in genes that are duplicated and deleted or due to differences in gene activation or gene inactivation. (2) There may be a position effect. (3) The genetic background could vary. The phenotypic effects of gene deletion or duplication could depend on genetic interactions with other genes in the genome. In this case, alleles inherited from the father that are different from those inherited from the mother and grandmother could contribute to the phenotype. (4) Environmental effects could exacerbate the effects of the deleted and duplicated region. These effects may not be uniform, and so could contribute to the observed phenotypic variability. Since many of the symptoms associated with Rec(8) syndrome are developmental abnormalities, variation of the environment during fetal development may contribute to phenotypic variability. (5) There may be other, cytologically invisible mutations associated with the Rec(8) individuals that could strongly affect their phenotype.

   d. The child has the chromosome 8 inversion, but not the duplication/deletion chromosome that results from a single crossover in an inversion heterozygote; she is an atypical Rec(8) individual. There are several explanations for why some of her symptoms overlap with those of Rec(8) syndrome. She may have an additional mutation near one of the Rec(8) breakpoints, in a region that is duplicated or deleted in Rec(8) syndrome, or in a gene that interacts with genes in the duplicated or deleted regions. Alternatively, it is possible that the inversion disrupts the function of a gene or genes at one or both breakpoints, and that normally, the inversion is an asymptomatic condition. In this case, the inversion chromosome (in her mother and grandmother) would bear a recessive mutation. If she had a new allelic mutation, or her paternally contributed chromosome had an allelic mutation, she would be affected. This could also explain why she has only some of the symptoms of Rec(8) syndrome; she would have fewer genes affected than most Rec(8) individuals.

     Small deletions would be cytologically invisible, as would point mutations. Thus, the explanations given above could not be evaluated solely by karyotype analysis. FISH, DNA marker, and/or DNA sequence analyses (see text Chapter 7) could be used to evaluate the integrity of the chromosomal regions near the breakpoints.

8.11 A particular plant species that had been subjected to radiation for a long time in order to produce chromosomal mutations was then inbred for many generations until it was homozygous for all of these mutations. It was then crossed to the original unirradiated plant, and the meiotic process of the F_1 hybrids was examined. It was noticed that a cell with a dicentric chromosome (bridge) and a fragment occurred at a low frequency in anaphase I of the hybrid.

   a. What kind of chromosomal mutation occurred in the irradiated plant? In your answer, indicate where the centromeres are.

   b. Explain, in words and with a clear diagram, where crossover(s) occurred and how the bridge chromosome of the cell arose.

Answer:

   a. The irradiated chromosome has a paracentric inversion (i.e., an inversion within one of its arms). Dicentric chromosomes and fragments arise as the result of single crossovers within paracentric inversions, and dicentric chromosomes with two bridges and two fragments result from a four-strand double crossover within paracentric inversions. An example of such a chromosome is diagrammed here.

$$\underline{a \quad b \quad c \bullet \underline{d \quad e \quad f \quad g \quad h \quad i}} \quad \text{normal order}$$

$$\underline{a \quad b \quad c \bullet \underline{d \quad h \quad g \quad f \quad e \quad i}} \quad \text{paracentric inversion}$$

b. The bridge chromosome would arise by a single crossover within an inversion loop during meiosis. In the above example, the crossover would occur between d and i. See text Figure 8.9 for an illustration of such a crossover.

8.12 On a normal-ordered chromosome, two loci, a and b, lie 15 map units apart on the left arm of a metacentric chromosome. A third locus, c, lies 10 map units to the right of b on the right arm of the chromosome. What frequency of progeny phenotypes do you expect to see in a testcross of an $a\, b\, c/a^+\, b^+\, c^+$ individual if the $a^+\, b^+\, c^+$ chromosome

a. has a normal order?

b. has an inversion with breakpoints just proximal (toward the centromere) to a and just distal (away from the centromere) to b?

c. has an inversion with breakpoints just proximal to a and just proximal to c?

d. has an inversion with breakpoints just distal to a and just distal to c?

Answer: Since the crosses are three-point testcrosses, the progeny phenotypes are specified by the genotypes of the gametes of the trihybrid parent. In each case, it helps to diagram the two chromosomes of the trihybrid and then consider the different types of gametes that can be produced by crossovers in the trihybrid parent. Since an inversion is present in parts b, c, and d of this question, some recombinant gametes will not be viable due to the aneuploidy that results when crossovers of this question occur within the inverted region. As discussed below, this will alter the proportions of gamete genotypes relative to those seen in trihybrids with normal-ordered chromosomes.

a. The chromosomes can be diagrammed as:

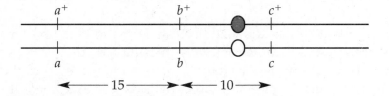

The trihybrid has no inversion, and its gametes will be exactly those expected from a trihybrid testcross. Since there are 15 mu between a and b, and 10 mu between b and c, 15 percent of the gametes will be single or double recombinants with crossovers between a and b, and 10 percent of the gametes will be single or double recombinants with crossovers between b and c. The remaining $[100 - (10 + 15)] = 75$ percent of the gametes will be parental types. This gives a frequency of 37.5% each of $a^+\, b^+\, c^+$ and $a\, b\, c$. Assuming that there is no interference, the expected frequency of double recombinants in the a–c interval is $0.10 \times 0.15 = 0.015$, or 1.5%. This gives a frequency of 0.75 percent each of $a^+\, b\, c^+$ and $a\, b^+\, c$. The remaining recombinants will result from single crossovers. There will be $15 - (1.5/2) = 14.25\%$ single recombinants in the a–b interval, giving 7.125% each of a^+ b c and a b^+ c^+. There will be $10 - (1.5/2) = 9.25\%$ single recombinants in the $b-c$ interval, giving 4.625% each of a^+ b^+ c and a b c^+.

b. Here, nearly the entire region between a^+ and b^+ is inverted. The chromosomes can be diagrammed as indicated here, using a thick line with an open-headed arrowhead to represent the inverted region.

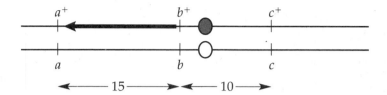

The inverted region is a paracentric inversion. As shown in text Figure 8.9, a single crossover within a paracentric inversion results in a dicentric bridge that produces deletion-bearing chromosomes. Gametes that receive these chromosomes are not viable. If this region were not inverted, nearly 15 percent of the gametes would be recombinant for the $a-b$ interval. However, because of the inversion, these recombinants will not be recovered. As a consequence, the frequency of the remaining recombinant and parental gametes will not be the same as these gamete types in a trihybrid parent with normal-ordered chromosomes. Relative to a trihybrid parent with normal-ordered chromosomes, the frequency of the remaining recombinant and parental gametes will be increased by a factor of $[1/(1.00 - 0.15)] = 1.176$.

The only viable recombinant gametes from this trihybrid parent are those that result from crossovers between b and c. In a normal-ordered chromosome, since there are 10 mu between b and c, 10% of the gametes will be recombinant in this interval. Since 15 percent of the gametes are not seen in this trihybrid parent, there will be $1.176 \times 10\% = 11.76\%$ recombinants in the $b-c$ interval. The remaining gametes $[1.176 \times 90\% = 88.24\%]$ will be parental types. This gives 5.88% a^+ b^+ c, 5.88% a b c^+, 44.12% a^+ b^+ c^+, and 44.12% a b c.

c. Here, nearly the entire region between a^+ and c^+ has been inverted. The chromosomes can be diagrammed as shown below, using a thick line with an open-headed arrowhead to show the inverted region.

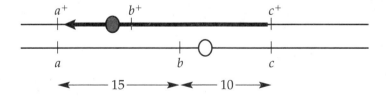

The inverted region is a pericentric inversion. As shown in text Figure 8.10, the only viable gametes that are produced from a meiosis with a single crossover within the inverted region are parental types (either normal-ordered or inversion-bearing chromosomes). Gametes with either of the two types of duplication/deletion-bearing chromosomes resulting from a single crossover are not viable. Thus, single crossovers occurring in nearly the entire $a-c$ interval will not produce recombinant gametes.

Crossovers that occur simultaneously in each of the $a-b$ and $b-c$ intervals (double crossovers) may produce viable recombinants, depending on whether they are two-strand, three-strand, or four-strand double crossovers. As shown in the answer to problem 8.5, a two-strand double crossover will produce viable gametes. Half of the gametes from such a meiosis are double recombinants ($a\ b^+\ c$ and $a^+\ b\ c^+$). As shown in the answer to problem 8.6, a four-strand double crossover will not produce any viable gametes. There are two possible types of three-strand double crossovers (see text Figure 6.5). These are diagrammed for an inversion heterozygote in the figure here.

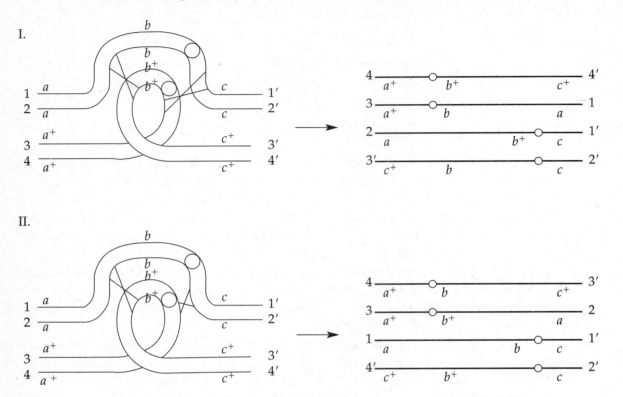

The three-strand double crossover shown in part I will produce the doubly recombinant chromosome $a\ b^+\ c$, two duplication/deletion chromosomes, and the parental-type chromosome $a^+\ b^+\ c^+$. The three-strand double crossover shown in part II will produce the doubly recombinant chromosome $a^+\ b\ c^+$, two duplication/deletion chromosomes, and the parental-type chromosome $a\ b\ c$. Thus, in a meiosis with either type of three-strand double crossover, two gametes will not be viable (the duplication/deletion-bearing gametes) and two gametes will be viable (one parental and one recombinant gamete). The results of meioses with each type of double crossover are summarized in the table here:

		Viable Gametes		
Type of Double Crossover	Nonviable Gametes	Recombinants	Parentals	Total
Two-strand	0	2	2	4
Three-strand (I)	2	1	1	2
Three-strand (II)	2	1	1	2
Four-strand	4	0	0	0

If each of the four types of double-crossover meioses occur equally frequently in a trihybrid with a heterozygous inversion, on average these meioses will produce 1/2 viable gametes and 1/4 doubly recombinant gametes. By comparison, double-crossover meioses in a trihybrid with normal-ordered chromosomes will produce on average all viable gametes and half doubly recombinant gametes. Thus, the pericentric inversion reduces both the number of gametes and the number of double recombinants by half. Since this pericentric inversion spans nearly the entire $a-c$ interval, the observed number of double recombinants will be very close to half of the expected number of double recombinants in a normal-ordered trihybrid parent. From (a), a normal-ordered trihybrid parent has 1.25% doubly recombinant gametes, so here there will be $1/2 \times 1.25\% = 0.625\%$. This gives $(100 - 0.625)/2 = 49.6875\%$ each of $a^+ b^+ c^+$ and $a\ b\ c$, and 0.3125% each of $a^+ b\ c^+$ and $a\ b^+ c$.

d. Here, the inversion is only slightly larger than that in (c), as the inverted region extends distally just past a and c. The trihybrid chromosomes can be diagrammed as shown here, using a thick line with an open-headed arrowhead to show the inverted region.

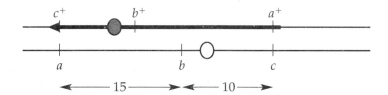

The frequency of progeny phenotypes will be very close to those observed in (c), since the inversion is very similar in size.

8.13 Mr. and Mrs. Lambert have not yet been able to produce a viable child. They have had two miscarriages and one severely defective child who died soon after birth. Studies of banded chromosomes of father, mother, and child showed that all chromosomes were normal except for pair number 6. The number 6 chromosomes of mother, father, and child are shown in the following figure:

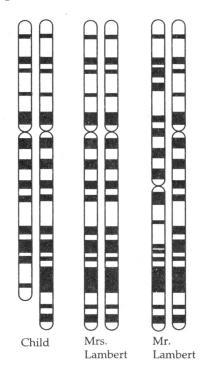

a. Does either parent have an abnormal chromosome? If so, what is the abnormality?

b. How did the chromosomes of the child arise? Be specific as to what events in the parents gave rise to those chromosomes.

c. Why is the child not phenotypically normal?

d. What can be predicted about future conceptions by this couple?

Answer:

a. Mr. Lambert is heterozygous for a pericentric inversion of chromosome 6. Relative to the centromere of the normal chromosome 6, one of the breakpoints is within the fourth light band up from the centromere, while the other is in the sixth dark band below the centromere. Mrs. Lambert's chromosomes are normal.

b. When Mr. Lambert's number 6 chromosomes paired during meiosis, they formed an inversion loop that included the centromere. Crossing-over occurred within the loop and gave rise to the partially duplicated, partially deficient chromosome 6 that the child received.

c. The child's abnormalities stem from having three copies of some, and only one copy of other, chromosome 6 regions. The top part of the short arm is duplicated, and there is a deficiency of the distal part of the long arm in this case.

d. The inversion appears to cover more than half of the length of chromosome 6, so crossing-over will occur in this region in the majority of meioses. In the minority of meioses where there is a two-strand double crossover inside the loop, or where crossing-over occurs outside the loop, and in the cases where a crossover has occurred within the loop but the child receives a noncrossover chromosome, the child can be normal. There is significant risk of abnormality, so fetal chromosomes should be monitored.

8.14   Mr. and Mrs. Simpson have been trying for years to have a child, but have been unable to conceive. They consulted a physician, and tests revealed that Mr. Simpson had a markedly low sperm count. His chromosomes were studied, and a testicular biopsy was done. His chromosomes proved to be normal, except for pair 12. The following figure shows Mrs. Simpson's normal pair of number 12 chromosomes and Mr. Simpson's number 12 chromosomes:

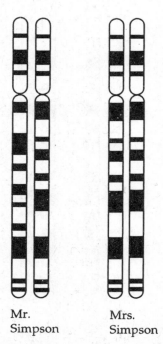

Mr.
Simpson

Mrs.
Simpson

a. What is the nature of the abnormality in pair number 12 of Mr. Simpson's chromosomes?

b. What abnormal feature would you expect to see in the testicular biopsy? (Cells in various stages of meiosis can be seen.)

c. Why is Mr. Simpson's sperm count low?

d. What can be done about Mr. Simpson's low sperm count?

Answer:

a. Mr. Simpson has a paracentric inversion in the long arm of one of his number 12 chromosomes. Moving downward (distally) from the centromere of the normal chromosome, the breakpoints are in the first dark band and in the sixth light band.

b. Crossing-over within the inversion loop will produce dicentric chromatids that will form anaphase bridges. These chromatin bridges joining the two chromatin masses at the beginning of anaphase I will be visible in the testicular biopsy.

c. The inversion is large, so the frequency of crossing-over within it will be significant. Consequently, bridges will be formed in the majority of meioses. Cells in which bridges form do not complete meiosis or form sperm in mammals.

d. Nothing can be done to increase Mr. Simpson's sperm count. This might be an instance to consider in vitro fertilization.

8.15 Chromosome I in maize has the gene sequence *ABCDEF*, whereas chromosome II has the sequence *MNOPQR*. A reciprocal translocation resulted in *ABCPQR* and *MNODEF*. Diagram the expected pachytene (see Chapter 3, p. 48) configuration in the F_1 of a cross of homozygotes of these two arrangements.

Answer:

```
        F        F
        E        E
        D        D
  A B C           O N M


  A B C           O N M
        P        P
        Q        Q
        R        R
```

8.16 Diagram the pairing behavior at prophase of meiosis I (see Chapter 3, pp. 46–48) of a translocation heterozygote that has normal chromosomes of gene order *abcdefg* and *tuvwxyz* and has the translocated chromosomes *abcdvwxyz* and *tuefg*. Assume that the centromere is at the left end of all chromosomes.

Answer:

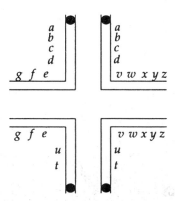

8.17   Mr. and Mrs. Denton have been trying for several years to have a child. They have experienced a series of miscarriages, and last year they had a child with multiple congenital defects. The child died within days of birth. The birth of this child prompted the Dentons' physician to order a chromosome study of parents and child. The results of the study are shown in the accompanying figure. Chromosome banding was done, and all chromosomes were normal in these individuals except some copies of number 6 and number 12. The number 6 and number 12 chromosomes of mother, father, and child are shown in the figure (the number 6 chromosomes are the larger pair):

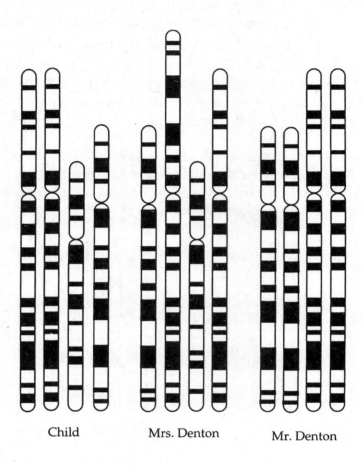

Child   Mrs. Denton   Mr. Denton

a. Does either parent have an abnormal karyotype? If so, which parent has it, and what is the nature of the abnormality?

b. How did the child's karyotype arise? (What pairing and segregation events took place in the parents?)

c. Why is the child phenotypically defective?

d. What can this couple expect to happen in subsequent conceptions?

e. What medical help, if any, can be offered to the couple?

Answer:

a. Mr. Denton has normal chromosomes. Mrs. Denton is heterozygous for a balanced reciprocal translocation between chromosomes 6 and 12. Most of the short arm of chromosome 6 has been reciprocally translocated onto the long arm of chromosome 12. The breakpoints appear to be in the first thick, dark band just above the centromere of 6 and in the third dark band below the centromere of 12.

b. The child received a normal 6 and a normal 12 from his father. In prophase I of meiosis in Mrs. Denton, chromosomes 6 and 12 and the reciprocally translocated 6 and 12 paired to form a cruciform-like structure. Segregation of adjacent, nonhomologous centromeres to the same pole ensued, so that the child received a gamete containing a normal 6 and one of the translocation chromosomes. See text Figure 8.12 for an illustration of adjacent-1 segregation.

c. The child has a normal 6 and a normal 12 chromosome from Mr. Denton. The child also has a normal 6 chromosome from Mrs. Denton. However, the child also has one of the translocation chromosomes from Mrs. Denton. With this chromosome, the child is partially trisomic as well as partially monosomic. It has three copies of part of the short arm of chromosome 6 and only one copy of most of the long arm of chromosome 12. This abnormality in gene dosage is the cause of its physical abnormality.

d. Segregation of adjacent homologous centromeres to the same pole is relatively rare. The segregation pattern seen in this child (adjacent-1 segregation) and alternate segregation (see text Figure 8.12) are more common. About half of the time, when alternate segregation occurs, the gamete will have a complete haploid set of genes, and the embryo should be normal. However, half of the gametes resulting from alternate segregation will be translocation heterozygotes.

e. Prenatal monitoring of fetal chromosomes could be done, and given the severity of the abnormalities (high probability of miscarriage and multiple congenital abnormalities), therapeutic abortion of chromosomally unbalanced fetuses would be a consideration.

8.18   Irradiation of *Drosophila* sperm produces translocations between the X chromosome and autosomes, between the Y chromosome and autosomes, and between different autosomes. Translocations between the X and Y chromosomes are not produced. Explain the absence of X-Y translocations.

Answer: Mature sperm bear either an X or a Y chromosome, not both. For X-Y translocations to be obtained, both chromosomes must be present in the same sperm cell.

8.19   Define the terms *aneuploidy, monoploidy,* and *polyploidy.*

Answer: An aneuploid cell or organism is one in which there is not an exact multiple of a haploid set of chromosomes, or one in which part or parts of chromosomes have been duplicated or deleted. It is one that does not have a *euploid* number of chromosomes. It is a general term used to describe a typically abnormal individual with an "unbalanced" chromosomal set.

A monoploid cell or individual has only one set of chromosomes. In humans, a monoploid cell would have 23 chromosomes instead of the normal diploid number of 46. In this case, a monoploid cell is also a haploid cell. A haploid number of chromosomes is typically the number of chromosomes in a gamete. Thus, in diploid individuals, a haploid gamete is also a monoploid cell. This is not always the case for nondiploid individuals. For example, in a hexaploid plant that has 36 chromosomes, a gamete will have 18 chromosomes (haploid number = 18), but the monoploid number will be 6.

A polyploid cell or individual has multiple sets of chromosomes. It is euploid, having multiple *complete* sets of chromosomes.

8.20   If a normal diploid cell is 2N, what is the chromosome content of the following?

    a. a nullisomic

    b. a monosomic

    c. a double monosomic

    d. a tetrasomic

    e. a double trisomic

    f. a tetraploid

    g. a hexaploid

Answer:

    a. $2N - 2$ (two copies of the same chromosome are missing)

    b. $2N - 1$ (missing one chromosome)

    c. $2N - 1 - 1$ (one copy of each of two different chromosomes are missing)

    d. $2N + 2$ (two extra copies of one chromosome)

    e. $2N + 1 + 1$ (an extra copy of each of two different chromosomes)

    f. $4N$

    g. $6N$

8.21   In humans, how many chromosomes would be typical of nuclei of cells that are

    a. monosomic?

    b. trisomic?

    c. monoploid?

    d. triploid?

    e. tetrasomic?

Answer:

    a. 45

    b. 47

    c. 23

    d. 69

    e. 48

8.22   An individual with 47 chromosomes, including an additional chromosome 15, is said to be

    a. triplet

    b. trisomic

    c. triploid

    d. tricycle

Answer: b

8.23   A color-blind man marries a homozygous normal woman, and after four joyful years of marriage they have two children. Unfortunately, both children have Turner syndrome, although one has normal vision and one is color blind. The type of color blindness involved is a sex-linked recessive trait.

a. For the color-blind child, did nondisjunction occur in the mother or in the father? Explain your answer.

b. For the child with normal vision, in which parent did nondisjunction occur? Explain your answer.

Answer:

a. The cross can be written as $++ \times c$Y. Turner syndrome children are XO. If the child is color blind, it received its father's c allele via a chromosomally normal X-bearing sperm. Therefore, the egg that was fertilized must have lacked an X, and nondisjunction occurred in the mother.

b. If the child has normal vision, it must have received its mother's normal X. To be XO, this must be the only X the embryo received, so that the egg must have been fertilized by a nullo-X, nullo-Y sperm. Nondisjunction occurred in the father.

8.24   The frequency of chromosome loss in *Drosophila* can be increased by a recessive chromosome-2 mutation called *pal*. The mutation causes the preferential loss of chromosomes contributed to a zygote by *pal/pal* fathers. The paternally contributed chromosomes are lost during the first few mitotic divisions after fertilization. What phenotypic consequences do you expect in offspring of the following crosses? (Keep in mind how sex is determined in *Drosophila*, that loss of an entire chromosome 2 or chromosome 3 is lethal, and that the loss of one copy of the small chromosome 4 is tolerated.)

a. X chromosome loss at the first mitotic division in a cross between a true-breeding *yellow* (recessive, X linked mutation causing yellow body color) female and a *pal/pal* father.

b. Chromosome 4 loss at the first mitotic division in a cross between a true-breeding *eyeless* (recessive mutation on chromosome 4 causing reduced eye size) female and a *pal/pal* father.

c. Chromosome 3 loss at the first mitotic division in a cross between a true-breeding *ebony* (recessive mutation on chromosome 3 causing black body color) female and a *pal/pal* father.

Answer: The *pal* mutation causes the loss of paternally contributed chromosomes. Following chromosome loss, one daughter nucleus has a normal diploid set of chromosomes, while the other is monosomic for one chromosome. This problem considers what happens when a chromosome is lost at the very first mitotic division (and only at that division).

a. The cross is y/y $+/+$ ♀ $\times$ $+/$Y *pal/pal* ♂, with progeny $y/+$ *pal/+* (daughters) and $y/$Y *pal/+* (sons). The paternally contributed X is only found in the $y/+$ *pal/+* daughter, so we only need to consider the consequence of its loss in daughters. If a paternally contributed X chromosome (+) is lost during the first mitotic division in a $y/+$ *pal/+* zygote, one daughter cell will lose an X chromosome and be y *pal/+*. The other daughter cell will have two X chromosomes and be $y/+$ *pal/+*. The cell with two X chromosomes would be female (XX) and produce nonyellow cells ($y/+$), while the cell with one X chromosome would be male (XO) and produce yellow cells (y). The animal will be a mosaic with cells of two sex types that are marked by yellow (male) or grey (female) cuticle.

b. The cross is $+/+$ *eye/eye* ♀ $\times$ *pal/pal* $+/+$ ♂, with progeny *pal/+* *eye/+* (daughters and sons). The paternally contributed fourth chromosome is +. If it is lost during the first mitotic division in a *pal/+* *eye/+* zygote, one daughter cell will lose a fourth chromosome and be *pal/+* *eye*. The other daughter cell will have two fourth chromosomes and be normal, while the cell with one fourth chromosome will be *eye*. The animal will be a mosaic with some cells that are haploid for the fourth chromosome and some cells that are diploid for the fourth chromosome. If a patch of haplo-4 cells forms an eye during development, the eye will be reduced in size.

c. The cross is $+/+$ e/e ♀ × pal/pal $+/+$ ♂, with progeny $pal/+$ $e/+$. The paternally contributed third chromosome is $+$. If it is lost during the first mitotic division in a $pal/+$ $e/+$ zygote, one daughter cell will lose a third chromosome, and be $pal/+$ e. This cell is inviable, and so will not be recovered in the organism, should the organism survive. Consequently, if the organism survives, it will be phenotypically normal ($pal/+$ $e/+$).

8.25   Assume that x is a new mutant gene in corn. A female x/x plant is crossed with a triplo-10 individual (trisomic for chromosome 10) carrying only dominant alleles at the x locus. Trisomic progeny are recovered and crossed back to the x/x female plant.

a. What ratio of dominant to recessive phenotypes is expected if the x locus is not on chromosome 10?

b. What ratio of dominant to recessive phenotypes is expected if the x locus is on chromosome 10?

Answer:

a. If the new mutant is not on chromosome 10, the cross can be written as x/x × $+/+$ (considering only the chromosome carrying the x gene). The progeny will be $x/+$, and the backcross will be $x/+$ × x/x. A $1:1$ ratio of x/x (recessive mutant) to $x/+$ (normal) individuals will be seen.

b. If the new mutant is on chromosome 10, the cross can be written as x/x × $+/+/+$. Trisomic progeny will be $x/+/+$, and the backcross can be written as $x/+/+$ × x/x. In the trisomic $x/+/+$ individual, there will be four kinds of gametes produced depending on how chromosome 10 segregates during meiosis. In $\frac{1}{3}$ of the meioses, x and $+/+$ gametes will be produced, giving $\frac{1}{6}x$ and $\frac{1}{6}$ $+/+$ gametes. In $\frac{2}{3}$ of the meioses, $+$ and $+/x$ gametes will be produced, giving $\frac{1}{3}$ $+/x$ and $\frac{1}{3}$ $+$ gametes. When such gametes fuse with an x-bearing gamete (from the xx parent) at fertilization, $\frac{1}{6}$ of the progeny will have the mutant phenotype (i.e., be x/x) and $\frac{5}{6}$ will be normal (i.e., $+/+/x$, $+/x/x$, $+/x$). The phenotypic ratio would be 5 dominant:1 recessive mutant.

8.26   Why are polyploids with even multiples of the chromosome set generally more fertile than polyploids with odd multiples of the chromosome set?

Answer: Polyploids with even multiples of the chromosome set can better form chromosome pairs in meiosis I than can polyploids with odd multiples. Triploids, for example, will generate an unpaired chromatid pair for each chromosome type in the genome, so that chromosome segregation to the gametes is irregular and the resulting zygotes will not be euploid.

8.27   One plant species (N = 11) and another plant species (N = 19) produced an allotetraploid. For the following statements, select the correct answer from the key:

I. The chromosome number of this allotetraploid is 30.

II. The number of linkage groups of this allotetraploid is 30.

Key:

a. Statement I is true and Statement II is true.

b. Statement I is true but Statement II is false.

c. Statement I is false but Statement II is true.

d. Statement I is false and Statement II is false.

Answer: To form the initial alloploid, gametes from each species fused, so that the initial alloploid had 11 + 19 = 30 chromosomes. A fertile, allotetraploid plant was produced by the doubling of this chromosome set, so that the allotetraploid plant has 60 chromosomes, two sets of 11 and two sets of 19. Thus Statement II is true, but Statement I is not. The allotetraploid has 60 chromosomes and 30 linkage groups (30 pairs of homologs). Answer (c) is correct.

8.28   According to Mendel's first law, genes *A* and *a* segregate from each other and appear in equal numbers among the gametes. But Mendel did not know that his plants were diploid. In fact, because plants are frequently tetraploid, he could have been unlucky enough to have started with peas that were 4N rather than 2N. Let us assume that Mendel's peas were tetraploid, that every gamete contains two alleles, and that the distribution of alleles to the gamete is random. Suppose we have a cross of *AAAA* × *aaaa*, where *A* is dominant, regardless of the number of *a* alleles present in an individual.

a. What will be the genotype of the F_1 peas?

b. If the F_1 peas are selfed, what will be the phenotypic ratios in the F_2 peas?

Answer:

a. The F_1 will be *AA aa*.

b. If we label the four alleles in the F_1 as A^1, A^2, a^1, and a^2, there are six possible gamete genotypes: A^1A^2, A^1a^1, A^1a^2, A^2a^1, A^2a^2, a^1a^2. This the F_1 gametes will be ⅙ *AA*, 4/6 *Aa*, and ⅙ *aa*. As shown in the following Punnett square, selfing the F_1 gives a phenotypic ratio of 35 *A*− : 1 *aa*.

	⅙ *AA*	4/6 *Aa*	⅙ *aa*
⅙ *AA*	1/36 *AAAA*	4/36 *AAAa*	1/36 *AAaa*
4/6 *Aa*	4/36 *AAAa*	16/36 *AAaa*	4/36 *Aaaa*
⅙ *aa*	1/36 *AAaa*	4/36 *Aaaa*	1/36 *aaaa*

8.29   What phenotypic ratio of *A* to *a* is expected if *AAaa* plants are testcrossed against *aaaa* individuals? (Assume that the dominant phenotype is expressed whenever at least one *A* is present, no crossing-over occurs, and each gamete receives two chromosomes.)

Answer: Label the four alleles of the *AAaa* plant as A^1, A^2, a^1, and a^2. As in problem 8.28, these four alleles can segregate into gametes in six ways: A^1A^2, A^1a^1, A^1a^2, A^2a^1, A^2a^2, a^1a^2 giving ⅙ *AA*, 4/6 *Aa*, and ⅙ *aa* gametes. When these gametes fuse with *aa* gametes at fertilization, the testcross progeny will be ⅙ *AAaa*, 4/6 *Aaaa*, and ⅙ *aaaa*. The phenotypic ratio will be 5 *A* :1 *a*.

8.30   The root-tip cells of an autotetraploid plant contain 48 chromosomes. How many chromosomes were contained by the gametes of the diploid from which this plant was derived?

Answer: The somatic cells of an autotetraploid plant have four identical sets of chromosomes. As root tip cells are mitotically dividing somatic cells, they, too, have four identical sets of chromosomes. Therefore, the gametes of the diploid from which this plant was derived had 12 chromosomes.

8.31   A number of species of the birch genus have a somatic chromosome number of 28. The paper birch is reported as occurring with several different chromosome numbers; *fertile* individuals with the somatic numbers 56, 70, and 84 are known. How should the 28 chromosome individuals be designated with regard to chromosome number?

Answer: Plants with 56, 70, and 84 chromosomes have gametes that have 28, 35, and 42 chromosomes, respectively. If these plants are fertile polyploids, they should have an even number of chromosome sets. This would be the case if the monoploid number of these plants is 7, and the 56, 70, and 84 chromosome plants have eight, ten, and twelve times the monoploid number of chromosomes. A plant with 28 chromosomes would therefore be tetraploid, and should be fertile.

8.32   How many chromosomes would be found in somatic cells of an allotetraploid derived from two plants, one with $N = 7$ and the other with $N = 10$?

Answer: The initial allopolyploid will have 17 chromosomes. After doubling, the somatic cells will have 34 chromosomes.

8.33   Plant species A has a haploid complement of four chromosomes. A related species, B, has five. In a geographic region where A and B are both present, C plants are found that have some characters of both species and somatic cells with 18 chromosomes. What is the chromosome constitution of the C plants likely to be? With what plants would they have to be crossed to produce fertile seed?

Answer: The C plants are allotetraploids, containing a diploid chromosome set from each of species A and species B. These plants should be fertile, as they will have no abnormal chromosome pairing or unpaired chromosomes. They should be able to produce gametes with 9 chromosomes (the four of species A and the five of species B) and can either be selfed or crossed to other C plants to produce fertile seed.

9

GENETICS OF BACTERIA AND BACTERIOPHAGES

CHAPTER OUTLINE

THINKING ANALYTICALLY

While this chapter focuses on the genetics of bacteria and bacteriophages, many of the core concepts introduced here are used throughout the study of genetics. Of particular importance are the genetic principles that underlie selection, recombination, and complementation. These have had a major impact on how we think about gene structure and function: Geneticists working in many different research areas are concerned with how to *select* for or against specific phenotypes; analysis of recombination within a gene (*intragenic* recombination) formed a foundation for our current understanding of gene structure; the concept of a gene as a unit of function was developed by using complementation tests to measure whether two mutants affect the same function. It is therefore essential to thoroughly understand these genetic principles as they are developed using bacterial and phage systems in this chapter.

REVIEW OF KEY TERMS, SYMBOLS, AND CONCEPTS

In your own words, write a brief, precise definition of each term in the groups below. Check your definitions using the text. Then develop a concept map using the terms in each list.

1	2	3
transfer of genetic material	transduction, cotransduction	bacteriophage cross
conjugation	phage vectors	turbid, clear plaques
plating	transductants	permissive host
colony	phage lysate	nonpermissive host
titer	virulent vs. temperate phage	bacterial lawn
minimal medium	lysogenic pathway	host range property
complete medium	prophage	intergenic mapping
auxotroph	lysogeny	"beads-on-a-string" view
prototroph	lytic cycle	intragenic mapping
replica plating	generalized transduction	fine-structure mapping
transconjugants, exconjugant	specialized transduction	homoallelic mutations
sex factor	transducing phage	heteroallelic mutations
F^-, F', Hfr, F^+ strains	selected, unselected marker	deletion mapping
plasmid, episome	transformation	point, deletion, revertant
F-duction (sexduction)	competent cells	mutations
merodiploid	natural, engineered	hot spot
interrupted mating	transformation	cis-trans test
donor vs. recipient	electroporation	complementation test
F-pili (sex-pili)	heteroduplex DNA	cistron
F factor origin	cotransformation	gene as a unit of function

Having stressed the importance of the concepts presented in this chapter, it is also important to be forewarned that much of the material used to develop these concepts is complicated and will not necessarily be intuitive. For example, there are many refinements of bacteria and phage types and special characteristics that are associated with one or another aspect of bacterial or phage growth and mating. Carefully read the chapter sections that describe these characteristics until you thoroughly understand them. The names are quite descriptive, and once you use them a few times in solving problems, you will become adept at using them. Approach using the terms with more than rote memorization. Use them in context until you fix upon what is practically a visual image of them.

As with the preceding chapters, carefully read the statements and conditions of a particular problem so that you gain a clear understanding of what it asks and what data it gives you to work with. Then organize and reorganize the data, and sometimes reorganize again, using scratch paper and pencil until you can see how to use the data to solve the problem. If you still get stuck, try representing the processes that are used in the problem with diagrams, and ask yourself how the data fit into the biological process you've drawn out.

The problems in this chapter are especially well organized to help you build up your understanding of the material. Work through each of them sequentially to fully develop your understanding and analytical skills.

QUESTIONS FOR PRACTICE

Multiple Choice Questions

1. A strain, such as a strain of *E. coli*, that requires nutritional or other kinds of supplements for growth and/or survival is known as
   a. a prototroph
   b. a heterotroph
   c. a pleiotroph
   d. an auxotroph

2. A conjugating bacterial cell that typically transfers only part of its *F* factor and some chromosomal genes is
   a. F^+
   b. F'
   c. *Hfr*
   d. F^-

3. The process by which cells take up genetic material from the extracellular environment and incorporate it into their genetic complement is called
   a. transduction
   b. transformation
   c. translocation
   d. DNA transfusion

4. In bacteria, conjugation involves
   a. the union of two bacterial genomes
   b. the fusion of two cells of opposite mating types
   c. a virus-mediated exchange of DNA
   d. the transfer of DNA from one cell into another

5. A self-replicating genetic element found in the cytoplasm of bacteria is a
   a. sex-pilus
   b. plasmid
   c. viral capsid
   d. contransductant

6. Which of the following is true about cells having *F* factors?
   a. In both F^+ and *Hfr* cells, fertility genes are transferred first.
   b. In both F^+ and *Hfr* cells, genes are transferred through sex-pili.
   c. In *Hfr* cells, bacterial genes closest to the origin are transferred first.
   d. In F' cells, only *F* factor genes are transferred.

7. Phage that can grow using either a lytic or lysogenic pathway are called
   a. virulent
   b. temperate
   c. intemperate
   d. transductant

8. Recombination that occurs between two alleles at one gene is
   a. unheard of
   b. intergenic recombination
   c. intragenic recombination
   d. more frequent than recombination between two closely linked genes

9. A complementation test measures
   a. whether two mutants undergo intergenic recombination
   b. whether two mutants undergo intragenic recombination
   c. whether two mutants are linked
   d. whether two mutants affect the same function

10. *E. coli* K12(λ) cells will only support the growth of T4 phage that are r^+. If two different mutant *rII* phage coinfect an *E. coli* K12(λ) cell and lysis occurs,
    a. intragenic recombination must have occurred
    b. complementation must have occurred
    c. the two *rII* phage must have mutations in different cistrons
    d. both a and b are true
    e. both b and c are true
    f. a, b, and c are true

11. How can you distinguish between a point mutation and a deletion mutation?
    a. Point mutations can't be seen cytologically, deletion mutations can be seen.
    b. Point mutations can be reverted, deletion mutations can't be reverted.
    c. A deletion mutation will complement a point mutation in the same gene.
    d. Point mutations can be transduced, deletion mutations can't be transduced.

12. Which of the statements accurately describe an episome and a plasmid?
    a. An episome is a plasmid that can integrate into the host chromosome.
    b. A plasmid is a circular extrachromosomal element.
    c. An *F* factor is a plasmid that is also an episome.
    d. All episomes are *F* factors, but not all *F* factors are plasmids.
    e. a, b, and c, but not d, are accurate
    f. a, b, c, and d are accurate

Answers: 1d, 2c, 3b, 4d, 5b, 6c, 7b, 8c, 9d, 10e, 11b, 12e

Thought Questions

1. Explain why two genes that are close together on a chromosome show a high frequency of cotransduction and cotransformation.
2. Compare the life cycles of T4, P1, and λ. Which is (are) temperate? Which is (are) virulent? Which can undergo lysogeny?
3. Contrast the features and/or conditions of generalized transduction and specialized transduction.
4. What is the significance of bacterial transformation and conjugation in relation to the rapidity with which certain infectious organisms adapt to changing environmental conditions or factors, such as antibiotics? (Hint: Sexual reproduction promotes genetic variability. How do parasexual systems compare?)
5. Explain how intragenic recombination is different from complementation.
6. Deletion mapping as used by Benzer in T4 was important in defining the location of point mutations at the *rII* locus. What might it mean for recessive mutations to be "uncovered" by

deletions in eukaryotes. Diagram how you might use a set of nested deletions to localize a *Drosophila* point mutant. (*Drosophila* has polytene chromosomes in which deletion breakpoints can be mapped cytologically.)

7. Consider how we may select for prototrophic and drug-resistant bacterial strains. Can you think of clever (or even not so clever) ways to attempt to select for (a) a pesticide-resistant strain of beetles, (b) bacteria that are able to degrade a toxic chemical, (c) very sweet sweetcorn, and (d) a strain of cats that doesn't respond to catnip?

8. When two true-breeding white-flowered sweet peas are crossed, a purple F_1 is produced. When the F_1 is selfed, the F_2 is 9 purple : 7 white. As described in Chapter 4, the 9 : 7 ratio is characteristic of duplicate recessive epistasis. Why is duplicate recessive epistasis also called complementary gene action?

9. Suppose a linear piece of DNA with genes (in order) $a^+ b^+ c^+$ enters a bacterial cell that is a^- b^- and c^-. How many crossovers are required to convert the bacterial cell to (a) $a^+ b^+ c^+$, (b) $a^+ b^+ c^-$, (c) $a^+ b^- c^+$, (d) $a^- b^+ c^-$? How would your answers change if the DNA was circular? In each case, which would be the most frequent and the least frequent event?

10. P1 and λ phage have been modified to serve as cloning vectors by removing selected genes (thereby making room for DNA inserts) and introducing specific mutations to remove some functions. Carefully consider their life cycles, and speculate what features of their life cycles it might be useful to retain in a cloning vector and what features it might be necessary to remove?

Thought Questions for Media

After reviewing the media on the *iGenetics* CD-ROM, try to answer these questions.

1. The *Mapping Bacterial Genes by Conjugation* animation shows that the leu^+ and thr^+ genes are transferred about 8 minutes after conjugation begins but that we can't tell which one is transferred first. What does this tell you about the limits of resolution of mapping by using the interrupted mating technique?

2. What are the units used to relate the distance between two bacterial genes mapped by interrupted mating? Do you expect these units to be proportional to the number of DNA base pairs between the two loci?

3. Could you perform an interrupted mating experiment between an F^+ strain and an F^- strain? Why or why not?

4. Why is it necessary to use diploids to perform a complementation test?

5. Suppose you had two true-breeding mutant lines of deaf rabbits. How would you determine if the cause of deafness in each line was the same?

6. To assess if two genes complement each other, is it important to obtain recombinants between them?

1. Suppose that in the initial cross between the prototrophic strain to the F^- strain that was leu^-, arg^-, his^-, and trp^-, the media used for selection of leu^+ recombinants was prepared incorrectly. On plate A, streptomycin was mistakenly left out of the medium, and only supplemental arginine, histidine, and tryptophan were added. On plate B, streptomycin was added, but leucine was mistakenly added along with the other supplements (arginine, histidine, and tryptophan). Compared to all of the other plates used to select for recombinants at the other loci, you observe a much larger number of colonies on plates A and B.

    a. Why are there so many more colonies on plates A and B, compared to the others?

    b. Suppose you picked several dozen individual colonies from each of plates A and B. Would you expect the genotypes of all of the colonies from both plates to be the same? Would you expect the genotypes of all of the colonies from just one of the plates to be the same?

    c. Could any of the colonies on plates A and B be recombinants?

    d. What would you expect their genotype(s) to be?

    e. How could you experimentally address your answers to (c) and (d)?

2. As presented in the iActivity, multiple *Hfr* strains are used to map the relative location of genes distributed throughout the *E. coli* chromosome.

    a. Does each *Hfr* strain have one or more than one *F* factor inserted in its *E. coli* chromosome?

    b. If, in an *Hfr* strain, a bacterial gene is far from the *F* factor site of insertion, will it be transferred frequently to an *F⁻* strain?

    c. Why are multiple *Hfr* strains needed to map genes distributed throughout the *E. coli* chromosome?

    d. When maps are being constructed, why doesn't it matter if the *F* factors in the different *Hfr* strains are inserted in opposite directions (relative to each other) in the *E. coli* chromosome?

SOLUTIONS TO TEXT PROBLEMS

9.1   In $F^+ \times F^-$ crosses, the F^- recipient is converted to a donor with very high frequency. However, it is rare for a recipient to become a donor in $Hfr \times F^-$ crosses. Explain why.

Answer: The frequency with which a recipient is converted to a donor reflects the frequency with which a complete F factor is transferred. In $F^+ \times F^-$ crosses, only the F factor is transferred, and this occurs relatively quickly. In $Hfr \times F^-$ crosses, transfer starts at the origin within the F element and then must proceed through the bacterial chromosome before reaching the F factor. For the entire F factor to be transferred, the whole chromosome would have to be transferred. This would take about 100 minutes, and usually the conjugal unions break apart before then.

9.2   With the technique of interrupted mating, four *Hfr* strains were tested for the sequence in which they transmitted a number of different genes to an F^- strain. Each *Hfr* strain was found to transmit its genes in a unique order, as shown in the accompanying table. (Only the first six genes transmitted were scored for each strain.)

Order of Transmission	*Hfr* Strain			
	1	2	3	4
First	O	R	E	O
	F	H	M	G
	B	M	H	X
	A	E	R	C
	E	A	C	R
Last	M	B	X	H

What is the gene sequence in the original strain from which these *Hfr* strains derive? On a diagram, indicate the origin and polarity of each of the four *Hfr* strains.

Answer: The diagram here shows the gene sequence in the original F^+ strain and the relative location and orientation of the four different F factor insertions. (Note that only one insertion exists in a particular *Hfr* strain.)

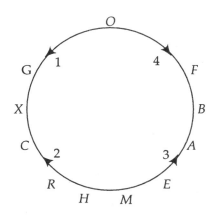

9.3   At time zero, an *Hfr* strain (*Hfr* 1) was mixed with an *F⁻* strain, and at various times after mixing, samples were removed and agitated to separate conjugating cells. The cross may be written as

$$Hfr\ 1:\quad a^+\ \ b^+\ \ c^+\ \ d^+\ \ e^+\ \ f^+\ \ g^+\ \ h^+\ \ str^S$$
$$F^-:\quad a\ \ \ b\ \ \ c\ \ \ d\ \ \ e\ \ \ f\ \ \ g\ \ \ h\ \ \ str^R$$

(No order is implied in listing the markers.) The samples were then plated onto selective media to measure the frequency of $h^+\ str^R$ recombinants that had received certain genes from the *Hfr* cell. A graph of the number of recombinants against time is shown in the accompanying figure.

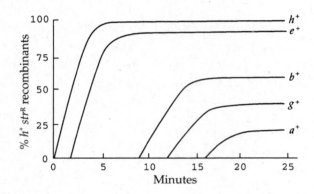

a. Indicate whether each of the following statements is true or false, and why.
i.   All F^+ cells that received a^+ from the *Hfr* in the chromosome transfer process must also have received b^+.
ii. The order of gene transfer from *Hfr* to F^- was a^+ (first), then g^+, then b^+, then e^+, and finally, h^+.
iii. Most $e^+\ str^R$ recombinants are likely to be *Hfr* cells.
iv. None of the $b^+\ str^R$ recombinants plated at 15 minutes are also a^+.
b. Draw a linear map of the *Hfr* chromosome, indicating
i.   the point of nicking (the origin) and the direction of DNA transfer
ii. the order of the genes a^+, b^+, e^+, g^+, and h^+
iii. the shortest distance between consecutive genes on the chromosomes

Answer:

a. i.   True. The graph indicates that a^+ is transferred last, starting about 16 minutes after conjugal pairing. The genes are transferred linearly in the order h^+ (starting at 0 minutes), e^+ (starting at 2 minutes), b^+ (starting at 9 minutes), g^+ (starting at 12 minutes), and then a^+.
ii. False. The data show that a^+ took the longest time to be transferred and so was the last gene to be received. The correct order of gene transfer was h^+ (first), e^+, b^+, g^+, and a^+ (last).
iii. False. In order for the recipient cell to be converted to F^+ or *Hfr*, it must receive a complete copy of the F factor. Since only part of the F factor is transferred at the beginning of conjugation, and e^+ is transferred quite quickly after conjugal pairing, most $e^+\ str^R$ recombinants will have only a small part of the donor chromosome. The remaining part of the F factor is transferred only after the entire bacterial

chromosome is transferred. Because of turbulence, the conjugal pairing is almost always disrupted before transfer of an entire chromosome is completed. Hence, the chance that complete transfer will occur is quite remote.

iv. True. The graph shows that a^+ is not transferred until after about 16 minutes has elapsed.

b.

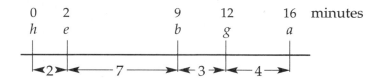

9.4   What steps would you take to selectively grow each of the bacterial cell types found in the following mixtures A through D?

Mixture	Genotypes Present	Phenotypes
A	*his, his$^+$*	*his* cells require supplemental histidine; *his$^+$* cells are able to grow without supplemental histidine.
B	*azi^R, azi^S*	*azi^R* cells are able to grow even in the presence of the poison sodium azide; *azi^S* cells die in the presence of sodium azide.
C	*lac$^-$, lac$^+$*	*lac$^+$* cells can grow even if lactose is the only sugar present; *lac$^-$* cells cannot utilize lactose for growth, they require a sugar other than lactose for growth whether or not lactose is present
D	*pcsA$^+$, pcsA*	*pcsA* cells are cold sensitive and grow at 37°C, but not at 30°C; *pcsA$^+$* cells can grow at both 37°C and 30°C.

Answer:

Mixture A: Leaving out histidine from an otherwise complete medium will select for *his$^+$* cells, as they can synthesize their own histidine. To select for *his* cells, plate cells on medium with histidine where both *his* and *his$^+$* cells grow. Then check individual colonies for their ability to grow on medium without histidine (use replica plating). Colonies able to grow on medium with histidine but not on medium without histidine are *his.*

Mixture B: Adding sodium azide to complete medium will select for *azi^R* cells. To select for *azi^S* cells, plate cells on medium with sodium azide and then check individual colonies for their ability to grow on medium with sodium azide (use replica plating). The *azi^S* colonies will not grow on medium with sodium azide.

Mixture C: Using lactose as the sole sugar in the medium will select for *lac$^+$* cells. To select for *lac* cells, plate cells on medium with a sugar other than lactose (e.g., glucose), where both *lac* and *lac$^+$* cells can grow. Then assess whether individual colonies can grow on medium with lactose as the sole sugar. Colonies unable to grow on this medium are *lac.*

Mixture D: Since *pcsA* cells are cold sensitive, incubating plated cells at 30°C will select for *pcsA⁺* cells. To select for *pcsA* cells, incubate plated cells at 37°C, and then test individual colonies for their ability to grow at 30°C. Cells unable to grow at 30°C but able to grow at 37°C are *pcsA*.

9.5   If an *E. coli* auxotroph *A* could grow only on a medium containing thymine, and an auxotroph *B* could grow only on a medium containing leucine, how would you test whether DNA from *A* could transform *B*?

Answer: Strain *A* is *thy⁻ leu⁺*, while strain *B* is *thy⁺ leu⁻*. To test if DNA from *A* can transform *B*, determine whether the *leu⁻* allele of *B* can be transformed to *leu⁺* by DNA from *A*. Add DNA from strain *A* to a leucine-fortified culture of *B*. Incubate long enough for transformation to occur, then plate out the potentially transformed *B* cells on minimal medium or on medium supplemented only with thymine. As shown in the table below, one can select for transformants by plating on such media, as growth on such media requires *leu⁺*.

| | | Medium Type | | |
Strain	Minimal	Plus Leucine	Plus Thymine	Plus Thymine and Leucine
leu⁺ thy⁺	+	+	+	+
leu⁺ thy⁻	−	−	+	+
leu⁻ thy⁺	−	+	−	+
leu⁻ thy⁻	−	−	−	+

9.6   Three different prototrophic strains (*1, 2,* and *3*) that are all sensitive to the antibiotic streptomycin are isolated. Each is individually mixed with an auxotrophic *F⁻* strain that is *a b c d e f g h* (and therefore requires compounds A, B, C, D, E, F, G, and H to grow) and that is also resistant to the antibiotic streptomycin. At 1-minute intervals after the initial mixing, a sample of the mixture is removed, shaken violently, and plated on media to select for *c⁺ str^R* recombinants. Recombinants are then tested for the presence of other genes. The following results are obtained:

Strain *1* × *F⁻*: No *c⁺* recombinants are ever obtained, even after 25 minutes.

Strain *2* × *F⁻*: *c⁺* recombinants are obtained at 6 minutes, *g⁺* at 8 minutes, *h⁺* at 11 minutes, *a⁺* at 14 minutes, *b⁺* at 16 minutes. No *d⁺, e⁺,* or *f⁺* recombinants are obtained.

Strain *3* × *F⁻*: *c⁺* recombinants are obtained at 1 minute, and *c⁺ g⁺* recombinants are obtained at or after 3 minutes. No *a⁺, b⁺, d⁺, e⁺, f⁺,* or *h⁺* recombinants are obtained.

If *c⁺* recombinants obtained at 16 minutes from the cross involving strain 2 are mixed with an *amp^R* (ampicillin-resistant) *F⁻* strain, no *c⁺ amp^R* recombinants are ever recovered. However, if *c⁺* recombinants obtained at 16 minutes from the cross involving strain 3 are mixed with an *amp^R F⁻* strain, *amp^R c⁺* recombinants can be recovered after 1 minute of mating.

a. How was the initial selection for $c^+\, str^R$ recombinants done? How were the subsequent selections done?

b. Use the given data to ascertain, as best you can, whether each strain is F^-, *Hfr*, F^+, or F'. If these data do not allow you to make an unambiguous determination, indicate the possibilities.

c. To the extent you can, draw a map of the chromosomes that might be present in each of strains *1, 2,* and *3*. Indicate the location and distance between genes a^+, b^+, d^+, e^+, f^+, and h^+ as best you can.

Answer:

a. To initially select for $c^+\, str^R$ recombinants, plate the progeny on minimal medium without compound C, but supplemented with streptomycin and compounds A, B, D, E, F, G, and H. To assess the complete genotype of the $c^+\, str^R$ recombinants, replica plate them onto different minimal media supplemented with streptomycin and all but two of the compounds (compound C and one other). For example, to test if a $c^+\, str^R$ colony was also a^+, replica plate it onto a medium that lacked compound A, but was supplemented with streptomycin and B, D, E, F, G, and H. If the colony were able to grow on this medium, it would be $a^+\, c^+\, str^R$. If it were unable to grow, it would be $a\, c^+\, str^R$.

b. *Strain 1:* Since no c^+ recombinants are ever obtained, strain *1* is unable to transfer c^+. This means it is either (1) F^-; (2) *Hfr*, but with the F factor inserted either far from c^+, or close to it but in an orientation so that genes are transferred in a direction opposite to c^+; (3) F', with c^+ in the bacterial chromosome. It should not be F^+, as then, at a very low frequency, some c^+ recombinants would be obtained.

　　Strain 2: Since c^+ recombinants are obtained at 6 minutes, and g^+, h^+, a^+, and b^+ recombinants are obtained at subsequent time intervals, strain *2* is *Hfr*. The genes are transferred in the following order: c^+, g^+, h^+, a^+, and b^+. From the times of their transfer, the map position of the genes is: origin (0) $-\, c^+$ (6) $-\, g^+$ (8) $-\, h^+$ (11) $-\, a^+$ (14) $-\, b^+$ (16). The location of genes d^+, e^+, and f^+ cannot be precisely determined; since they are not transferred in an *Hfr* × F^- cross, they are either far away from the F factor insertion site, or close to it but near the fertility genes, which are only rarely transferred by an *Hfr* strain. When the recombinants obtained from the strain *2* × F^- mating at the 16-minute time period are crossed to an $amp^R\, F^-$ strain, c^+ is not transferred. If these recombinants cannot conjugate with F^-, this indicates that although strain *2* is fertile, it did not transfer a complete F factor. It therefore must be *Hfr*.

　　Strain 3: Strain *3* transfers c^+ within 1 minute and g^+ by three minutes. From analysis of the strain *2* × F^- cross, we knew that these genes are 2 minutes apart. This data supports this conclusion. Since no other recombinants are obtained, no other genes are transferred. This suggests that strain *3* is F', and that the segment of DNA containing c^+ and g^+ is in the F' factor. If this is the case, the complete F factor will be transferred in a strain *3* × F^- cross if the mating is allowed to proceed long enough. This is observed: c^+ recombinants from the strain *3* × F^- cross obtained at 16 minutes are able to transfer c^+ to an $F^-\, amp^R$ strain. Therefore, strain *3* is F'.

c. Information known with certainty is diagrammed below. The location of genes in strains *1* and *3* is inferred from crosses with strain *2*. The location of genes d^+, e^+, and f^+ is unknown.

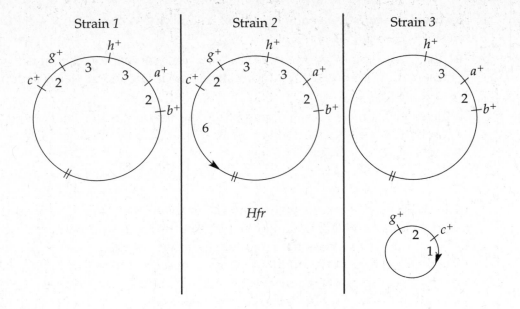

9.7   You are given a prototrophic str^R (streptomycin-resistant) *Hfr* strain and an amp^R (ampicillin-resistant) F^- auxotrophic strain that requires leucine (*leu*), arginine (*arg*), lysine (*lys*), purine (*pur*), and biotin (*bio*).

  a. Devise a strategy to determine quickly which gene (leu^+, arg^+, lys^+, pur^+, or bio^+) lies closest to the *F*-factor origin of replication.

  b. Even when the prototrophic *Hfr* strain is mixed with the F^- strain for very long periods, streptomycin resistance is not transferred. State two hypotheses that explain this finding.

Answer:

  a. Prepare a set of bacterial plates with five different media, each containing minimal media supplemented with ampicillin and all but one of leucine, arginine, lysine, purine, and biotin. Mix the *Hfr* strain with the F^- strain. After, say, 2 minutes and 5 minutes, remove some of the bacteria from the mating, violently shake them, and plate them on each of the media. This will select for growth of amp^R recombinants that are able to grow in the absence of one supplement. For example, plating on medium with ampicillin, leucine, arginine, lysine, and purine will select for $bio^+ amp^R$ recombinants. Incubate the plates to let colonies grow, and identify the plate that has the most colonies after the shortest mating interval. This will identify the prototrophic gene transferred first, and so identify the gene closest to the origin of replication of the *F* factor. If, say, only plates lacking leucine have growth after 2 minutes, then leu^+ is closest to the *F* factor origin of replication.

  b. The str^R gene could be far from the origin of replication of the *F* factor in the *Hfr* chromosome. Alternatively, the str^R gene might be on a separate plasmid unable to be transferred to the F^- cell (perhaps it is a defective *F* factor).

9.8   Indicate whether each of the following lettered items

  Occurs or is a characteristic of generalized transduction (GT)

  Occurs or is a characteristic of specialized transduction (ST)

Occurs in both (B)

Occurs in neither (N)

a. Phage carries DNA of bacterial or viral DNA origin, never both.

b. Phage carries viral DNA covalently linked to bacterial DNA.

c. Phage integrates into a specific site on the host chromosome.

d. Phage integrates at a random site on the host chromosome.

e. "Headful" of bacterial DNA is packaged into phage.

f. Host is lysogenized.

g. Prophage state exists.

h. Temperate phage is involved.

i. Virulent phage is involved.

Answer:

a. GT

b. ST

c. ST

d. GT

e. GT

f. B

g. B

h. B

i. N

9.9   Consider the following transduction data:

Donor	Recipient	Selected Marker	Unselected Marker	%
aceF⁺ dhl	*aceF dhl⁺*	*aceF⁺*	*dhl*	88
aceF⁺ leu	*aceF leu⁺*	*aceF⁺*	*leu*	34

Is *dhl* or *leu* closer to *aceF*?

Answer: The notation *aceF* represents a specific insertion site for an *F* factor. This table shows that cells selected for transduction of F^+ were isolated and then tested for the cotransduction of *dhl* or *leu*. It shows that in one experiment, of the *aceF⁺* transductants that were isolated, 88 percent were also transduced with the *dhl* marker. In another experiment, of the *aceF⁺* transductants that were isolated, only 34 percent were also transduced with the *leu* marker.

In order to obtain a transductant, a double crossover must occur between the trans-duced, donor DNA, and the host chromosome. The closer two loci are together on the same chromosome, the greater the probability that they will both be included within the limits of a double crossover. Therefore, two loci that are closer together will show a greater frequency of cotransduction: *dhl* is closer to *aceF⁺* than *leu*.

9.10   Consider the following data pertaining to P1 transduction:

Donor	Recipient	Selected Marker	Unselected Marker	%
aroA pyrD+	aroA+ pyrD	pyrD+	aroA	5
aroA+ cmlB	aroA cmlB+	aroA+	cmlB	26
cmlB pyrD+	cmlB+ pyrD	pyrD+	cmlB	54

Choose the correct order:

a. *aroA cmlB pyrD*
b. *aroA pyrD cmlB*
c. *cmlB aroA pyrD*

Answer: The frequency of cotransduction gives an indication of the closeness of each pair of genes: the higher their cotransduction frequency, the closer the two genes. The *pryD* and *cmlB* genes show the highest cotransduction frequency, and are the closest together. This eliminates (c) as a possibility. The genes *aroA* and *pyrD* show the lowest cotransduction frequency, and so they are the farthest apart. This eliminates (b) as a possibility. The *aroA* and *cmlB* genes show an intermediate cotransduction frequency, as would be expected if *cmlB* were between *aroA* and *pyrD*. Thus, the correct answer is (a).

9.11   Order the mutants *trp*, *pyrF*, and *qts* on the basis of the following three-factor transduction cross:

Donor	trp^+ pyr^+ qts
Recipient	trp pyr qts^+
Selected Marker	trp^+

Unselected Markers	Number
pyr^+ qts^+	22
pyr^+ qts	10
pyr qts^+	68
pyr qts	0

Answer: Notice that in this cross, by selecting for trp^+ one selects for recombinants. The frequency of the different classes of recombinants can be affected by two factors: the distance between genes and the number of crossovers that are needed to obtain a particular genotype. Since the bacterial chromosome is circular, recombinants are obtained only by either two or some multiple of two crossovers. Obtaining four crossovers will be less common than obtaining two crossovers, and so a genotype that requires four crossovers to be produced is likely to be the least common. The parents are trp^+ pry^+ qts and trp pyr qts^+. To produce the least frequent trp^+ pyr qts class by a quadruple crossover, the gene order must be *trp–pyr–qts*. The other three transductant classes can be generated by double crossovers. Try drawing this out.

9.12   Order *cheA*, *cheB*, *eda*, and *supD* from the following data:

Markers	% Cotransduction
cheA−eda	15
cheA−supD	5
cheB−eda	28
cheB−supD	2.7
eda−supD	0

Answer: The higher the frequency of cotransduction, the closer the loci. Thus, the relative proximity of the loci to each other is:

cheB−eda	closest together
cheA−eda	↓
cheA−supD	↓
cheB−supD	↓
eda−supD	farthest apart

A gene order that is consistent with these relationships is *eda−cheB−cheA−supD*.

9.13   Wild-type phage T4 grows on both *E. coli B* and *E. coli K12(λ)*, producing turbid plaques. The *rII* mutants of T4 grow on *E. coli B*, producing clear plaques, but do not grow on *E. coli K12(λ)*. This host range property permits the detection of a very low number of r^+ phages among a large number of *rII* phages. With such a sensitive system, it is possible to determine the genetic distance between two mutations within the same gene—in this case the *rII* locus. Suppose *E. coli B* is mixedly infected with *rIIx* and *rIIy*, two separate mutants in the *rII* locus. Suitable dilutions of progeny phages are plated on *E. coli B* and *E. coli K12(λ)*. A 0.1-mL sample of a thousandfold dilution plated on *E. coli B* produces 672 plaques. A 0.2-mL sample of undiluted phage plated on *E. coli K12(λ)* produces 470 turbid plaques. What is the genetic distance between the two *rII* mutations?

Answer: The plaques produced on *E. coli K12(λ)* are r^+, while those on *E. coli B* may be either r^+ or r^-. Thus, the total number of progeny can be inferred from the number of plaques formed on *E. coli B*. Since *E. coli B* is coinfected with *rIIx* and *rIIy*, the only way to obtain an r^+ progeny phage is to have a crossover within the *rII* locus. The progeny resulting from a crossover would be half r^+ and half *rIIxy* recombinants. The number of recombinant phage is twice the number of r^+ phage, which can be assayed for by growth on *E. coli K12(λ)*.

$$\text{\# recombinant progeny in 1 mL} = 2 \times (\text{number of } r^+ \text{ phage/mL})$$
$$= 2 \times (470/0.2)$$
$$= 4,700/\text{mL}$$

$$\text{total \# of progeny in 1 mL} = (\text{dilution factor}) \times (\text{\# progeny phage/mL})$$
$$= 1,000 \times (672/0.1)$$
$$= 6.72 \times 10^6/\text{mL}$$

$$RF = [4,700/(6.72 \times 10^6)] \times 100\%$$
$$= 0.07\%$$

The map distance between *rIIx* and *rIIy* is 0.07 mu.

9.14   Construct a map from the following two-factor phage cross data (show the map distances):

Cross	% Recombination
$r1 \times r2$	0.10
$r1 \times r3$	0.05
$r1 \times r4$	0.19
$r2 \times r3$	0.15
$r2 \times r4$	0.10
$r3 \times r4$	0.23

Answer: Recall that, as discussed in Chapter 6, the most accurate map distances are those obtained over the shortest intervals.

9.15   The following two-factor crosses were made to analyze the genetic linkage between four genes in phage λ: *c, mi, s,* and *co.*

Parents	Progeny
$c + \times + mi$	1,213 c +, 1,205 + mi, 84 + +, 75 c mi
$c + \times + s$	566 c +, 808 + s, 19 + +, 20 c s
$co + \times + mi$	5,162 co +, 6,510 + mi, 311 + +, 341 co mi
$mi + \times + s$	502 mi +, 647 + s, 65 + +, 56 mi s

Construct a genetic map of the four genes.

Answer: Analyze the data as you would a set of two-factor crosses:

Cross	Number of Progeny	# Recombinants	mu
$c + \times + mi$	2,577	159	6.2
$c + \times + s$	1,413	39	2.8
$co + \times + mi$	12,324	652	5.3
$mi + \times + s$	1,270	121	9.5

Two maps are compatible with these data:

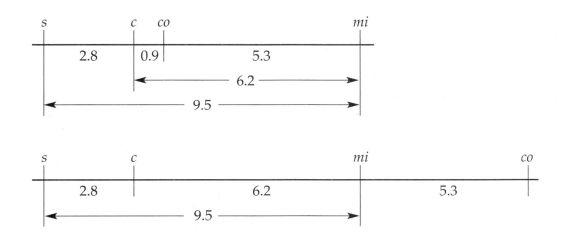

9.16   Wild-type (r^+) strains of T4 produce turbid plaques, whereas *rII* mutant strains produce larger, clearer plaques. Five *rII* mutations (*a*–*e*) in the A cistron of the *rII* region of T4 give the following percentages of wild-type recombinants in two-point crosses:

Cross	% of Wild-Type Recombinants	Cross	% of Wild-Type Recombinants
$a \times b$	0.2	$e \times d$	0.7
$a \times c$	0.9	$e \times c$	1.2
$a \times d$	0.4	$e \times b$	0.5
$b \times c$	0.7	$b \times d$	0.2
$e \times a$	0.3	$d \times c$	0.5

What is the order of the mutational sites, and what are the map distances between the sites?

Answer: Between any two *rII* mutants *rIIx and rIIy* there are two products of intragenic recombination: *rII++* and *rIIxy*. Thus, the frequency of wild-type recombinants is half the total recombinant frequency. These data give the following map:

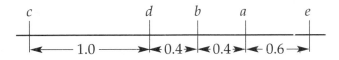

9.17   Given the following map with point mutants and given the data in the following table, draw a topological representation of deletion mutants *r21*, *r22*, *r23*, *r24*, and *r25*. (Be sure to show the endpoints of the deletions; = r^+ recombinants are obtained, 0 = r^+ recombinants are not obtained.)

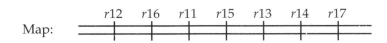

Deletion Mutants	Point Mutants						
	r11	r12	r13	r14	r15	r16	r17
r21	0	+	0	+	0	+	+
r22	+	+	0	0	+	+	0
r23	0	0	0	+	0	0	+
r24	+	+	0	0	+	+	+
r25	+	+	0	0	0	+	+

Answer: If no r^+ recombinants are obtained, the deletion removes the site of the point mutant. If r^+ recombinants are obtained, the site of the point mutation is not within the boundaries of the deletion. To determine the deleted region then, define the region that includes all of the point mutations unable to recombine with the deletion. Check your answer by verifying that all of the point mutations outside of this region do recombine with the deletion.

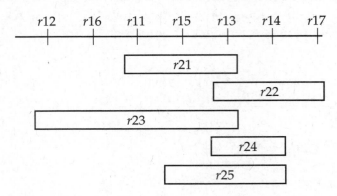

9.18   Given the following deletion map with deletions r31, r32, r33, r34, r35, and r36, place the point mutants r41, r42, etc., on the map. Be sure to show where they lie with respect to end points of the deletions.

r31

r32

r33

r34

r35   r36

	Deletion Mutants (+ = recombinants produced; 0 = no r^+ recombinants produced)					
Point Mutants	r31	r32	r33	r34	r35	r36
r41	0	0	0	0	+	0
r42	0	0	0	+	0	+
r43	0	0	+	+	+	0
r44	0	0	0	0	+	+
r45	0	+	0	+	+	+
r46	0	0	+	0	+	0

Show the dividing line between the *A* cistron and the *B* cistron on your map from the following data [+ growth on strain *K12(λ)*, 0 = no growth on strain *K12(λ)*]:

Mutant	Complementation with	
	rIIA	*rIIB*
*r*31	0	0
*r*32	0	0
*r*33	0	+
*r*34	0	0
*r*35	0	+
*r*36	0	0
*r*41	0	+
*r*42	0	+
*r*43	+	0
*r*44	0	+
*r*45	0	+
*r*46	0	+

Answer: First define where the point mutants lie using a systematic approach that utilizes two facts: (1) If a point mutant is *able* to recombine with a deletion mutant, it must lie *outside* of the deleted region; and (2) if a point mutant is *unable* to recombine with a deletion mutant, it must lie *within* the deleted region. This means that a point mutant must lie within a region remaining intact in each of the deletions with which it does recombine, and, conversely, it must lie in a deleted region that is shared by each of the deletions with which it does not recombine. Employ both of these inferences to define the region in which the point mutant lies.

Then determine the positions of the *A* and *B* cistrons. Use the fact that if two mutants are unable to grow on *E. coli K12(λ)* ("0"), the mutants cannot together provide the functions to complete the *rII* pathway. Either the *rIIA* or *rIIB* function is missing. Consequently, both mutants must be defective in the same function. In other words, the mutants do not complement each other. For example, if a mutant does not complement a known *rIIA* mutant (i.e., there is no growth on *E. coli K12(λ)*, ("0"), then the mutation is in the *A* cistron. Note that deletions can affect both the *A* and *B* cistrons, while point mutations can only affect one or the other cistron.

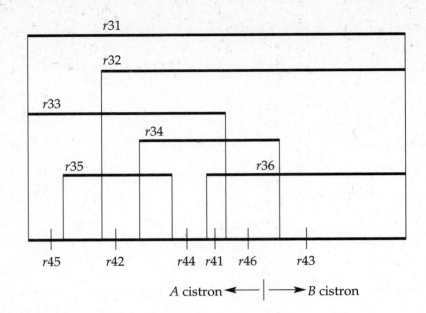

9.19   Some adenine-requiring mutants of yeast are pink because of the intracellular accumulation of a red pigment. Diploid strains were made by mating haploid mutant strains. The diploids exhibited the following phenotypes:

Cross	Diploid Phenotypes
1 × 2	pink, adenine-requiring
1 × 3	white, prototrophic
1 × 4	white, prototrophic
3 × 4	pink, adenine-requiring

How many genes are defined by the four different mutants? Explain.

Answer: When two haploid strains mate to form a diploid cell, each contributes its genome. Consequently, there are two sets of genes in a diploid cell, and complementation between two mutants can be observed. If two mutations are in the same unit of function (i.e., the same gene) then they will not complement when together in a diploid cell and the mutant phenotype will be exhibited. If two mutations are in different genes, they will complement each other in a diploid cell and the wild-type phenotype will result.

In the data presented here, mutations 1 and 3 and mutations 1 and 4 complement each other, indicating that mutations 1 and 3 are in different genes and mutations 1 and 4 are in different genes. On the other hand, mutations 1 and 2 and mutations 3 and 4 fail to complement each other. Thus, mutations 1 and 2 are in one gene, while mutations 3 and 4 are in a different gene. There are two genes.

9.20   In *Drosophila* mutants *A, B, C, D, E, F,* and *G,* all have the same phenotype: the absence of red pigment in the eyes. In pairwise combinations in complementation tests, the following results were produced, where + = complementation and − = no complementation.

	A	B	C	D	E	F	G
G	+	−	+	+	+	+	−
F	−	+	+	−	+	−	
E	+	+	−	+	−		
D	−	+	+	−			
C	+	+	−				
B	+	−					
A	−						

a. How many genes are present?

b. Which mutants have defects in the same gene?

Answer: Mutants that fail to complement each other are in the same gene. Mutants G and B fail to complement each other, mutants C and E fail to complement each other, and mutants A, F, and D fail to complement each other. Therefore, there are three genes, with mutants G and B in one, mutants A, F, and D in a second, and mutants C and E in a third.

9.21 Specialized transduction can be used to develop fine-structure maps. Five different λd *gal* phage were isolated (1, 2, 3, 4, 5). Each was infected into five different *gal* point mutants (*a, b, c, d, e*) and *gal*$^+$ recombinants were selected by plating the cells on media containing galactose as the sole carbon source. The results are shown in the following table (+ indicates that *gal*$^+$ recombinants were obtained; − indicates no *gal*$^+$ recombinants were obtained):

E. coli *gal* mutant	λd *gal* phage				
	1	2	3	4	5
a	−	+	−	−	−
b	−	+	−	+	−
c	+	+	+	+	+
d	+	+	+	+	−
e	+	+	−	+	−

Each of the λd *gal* phage was then coinfected into *E. coli* with each of five lambda point mutants (*j, k, l, m, n*), and a selection performed for wild-type lambda progeny. The following table shows the results (+ indicates that wild-type λ recombinants were obtained; − indicates no wild-type λ recombinants were obtained):

λ mutant	λd *gal* phage				
	1	2	3	4	5
j	+	−	+	+	+
k	+	−	−	−	+
l	+	−	+	−	+
m	+	−	−	−	−
n	−	−	−	−	−

Draw the *gal-bio* region of an *E. coli* λ lysogen and label the location of the *att* site, the *gal* and *bio* genes and λ. Below your drawing, indicate the relative map positions of the five *gal* mutations, the relative map positions of the five λ mutations, and the regions of the *gal* gene and λ genome that are present in each of the λ*d gal* phage.

Answer: Specialized transduction by λ occurs when a portion of the *E. coli* chromosome near the site of the λ chromosome insertion (*att*) is transduced into the recipient cell. See text Figure 9.16. The *gal* and *bio* genes are near the *att* site, so portions of these genes can be transduced. If a transducing phage can recombine with a mutant to give a wild-type phenotype, the transducing phage must carry normal DNA that can recombine with and replace the mutant site. This information lets you deduce the relative order of the mutants *a−e* and identify the regions of the *gal* locus that are transduced by the five λ*d gal* phage. For example, λ*d gal* phage 5 can recombine only with mutant *c*, so must contain DNA for the mutant *c* site, but not any of the other mutant sites. Therefore, λ*d gal* phage 5 carries the least amount of DNA for the *gal* locus, and mutant *c* is the closest to the *att* site where the λ chromosome recombines into *E. coli*. In contrast, λ*d gal* phage 2 can recombine with all of the mutants, so it must contain DNA for all of their sites. Therefore, λ*d gal* phage 2 contains the most DNA of the *gal* locus and extends furthest into the *gal* locus. By similar reasoning, λ*d gal* phage 3 covers the *c* and *d* mutants, λ*d gal* phage 1 covers the *c*, *d*, and *e* mutants, and λ*d gal* phage 4 covers the *c*, *d*, *e*, and *b* mutants. This gives the mutant order: (*gal*) *a−b−e−d−c−att−bio*.

During specialized transduction, an abnormal outlooping occurs in which λ gains part of the *E. coli* chromosome and loses part of its own chromosome. The second set of data shed light on how much of the λ chromosome is lost. Using the same reasoning as before, a λ*d gal* phage will recombine with a λ mutant if it has DNA for the mutant site in λ. No λ*d gal* phage recombines with λ mutant *n* and so none of the phage have DNA containing this site. In contrast, λ*d gal* phage 1 can recombine with λ mutants *j*, *l*, *k*, and *m*, and so has DNA for all of these sites. By similar reasoning, λ*d gal* phage 5 covers mutant sites *k*, *l*, and *j*, λ*d gal* phage 3 covers mutant sites *l* and *j*, λ*d gal* phage 4 covers mutant site *j*, and λ*d gal* phage 2 covers no mutant sites. This gives the order of λ mutant sites as: *j−l−k−m−n*.

These data can be summarized in the following map:

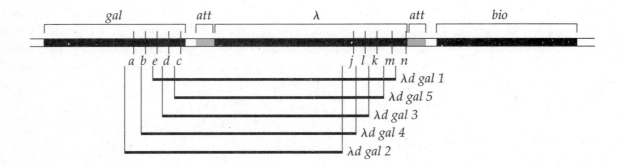

9.22   In *E. coli*, eight spontaneously and independently arising *leu* mutants were isolated from a parental F^- str^R strain. Each mutant requires supplemental leucine to grow, but is resistant to streptomycin. Interrupted mating experiments were performed with each of the eight *leu* mutants and a prototrophic *E. coli* Hfr strain sensitive to streptomycin. In each cross, str^R leu^+ recombinants were recovered just after 4 minutes of mating. The fine structure of the *leu* region was then evaluated using generalized transduction. Each of the eight mutants was individually infected with a generalized transducing phage. The resulting lysate was used to infect the other mutants, and *leu⁺* recombinants were selected. The table here shows the results (+ indicates that *leu⁺* recombinants were recovered, – indicates that *leu⁺* recombinants were not recovered).

leu Mutant	1	2	3	4	5	6	7	8
1	–	–	+	–	+	+	+	+
2		–	–	–	+	–	–	+
3			–	+	+	+	–	+
4				–	–	+	+	+
5					–	+	+	+
6						–	+	+
7							–	–
8								–

a. Draw a map showing the relative order and locations of the mutant sites in this region. (Hint: first identify the deletions.)

b. Can you infer if any of these mutations are point mutants? If not, how would you address this issue?

c. Explain whether you can infer how many cistrons involved in leucine biosynthesis are present in this region.

Answer:

a. If DNA transduced into a particular *leu* mutant recombines with its chromosome to produce a *leu⁺* recombinant, the transduced DNA must contain a wild-type site that can replace the mutated *leu* site. Therefore, pairs of mutants that produce *leu⁺* recombinants affect different sites, while pairs of mutants unable to produce *leu⁺* recombinants affect one or more common sites. The sites may be single or multiple base-pair regions. Mutants that affect more than one site must be deletions. Deletions can be recognized by identifying mutants unable to produce *leu⁺* recombinants with pairs of mutants that lie in different sites. For example, mutants 3 and 8 can recombine to produce *leu⁺* recombinants, so they lie in different sites. Mutant 7 cannot recombine with either mutant 3 or mutant 8, so it must delete both sites. Using this logic, mutants 2, 4, and 7 must be deletions. In the map below, these deletions are shown by open boxes.

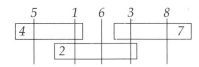

b. A site may be one or more base pairs. To address if a site is a point mutation, check if a mutant can be reverted. Point mutants, but not deletions, can be reverted.

c. This analysis does not address how many cistrons are present in this region. For this, complementation tests are needed.

9.23   A homozygous white-eyed Martian fly (w_1/w_1) is crossed with a homozygous white-eyed fly from a different stock (w_2/w_2). It is well known that wild-type Martian flies have red eyes. This cross produces all white-eyed progeny. State whether each of the following is true or false. Explain your answer.

a. w_1 and w_2 are allelic genes.
b. w_1 and w_2 are nonallelic.
c. w_1 and w_2 affect the same function.
d. The cross was a complementation test.
e. The cross was a *cis-trans* test.
f. w_1 and w_2 are allelic by terms of the functional test.

The F_1 white-eyed flies are allowed to interbreed, and when you classify the F_2 you find 20,000 white-eyed flies and 10 red-eyed progeny. Concerned about contamination, you repeat the experiment and get exactly the same results. How can you best account for the presence of the red-eyed progeny? As part of your explanation, give the genotypes of the F_1 and F_2 generation flies.

Answer:

a. True. In the cross $w_1/w_1 \times w_2/w_2$, the offspring are w_1w_2. That they are white-eyed indicates that w_1 and w_2 are mutations in the same function (i.e., the same gene). Thus, w_1 and w_2 are allelic genes.

b. False. If the genes were nonallelic, each mutation would be in a different function. The two mutations would each be in different steps of a pathway, with each mutant retaining the function of a different step. Since the progeny obtain a set of genes from each mutant parent, they would have the function of both steps and have red eyes. Since they have white eyes, this statement is false.

c. True.

d. True. A complementation test examines whether two mutations affect the same function.

e. True. The F_1 is a *trans*-heterozygote.

f. True. If two genes affect the same function, they are allelic.

The complementation test indicates that the two mutations affect the same function and are therefore alleles at the same gene. However, it does not indicate whether they are *identical* alleles or not. To evaluate if two allelic mutations lie in exactly the same position in a gene, one needs to assess whether recombination can occur between them. If two alleles are identical and lie in the same site, *intragenic* recombination between them cannot occur. If they lie in different sites, intragenic recombination can occur, giving rise to wild-type recombinant chromosomes. The appearance of wild-type F_2 progeny results from intragenic recombination, as illustrated here.

If w_1 and w_2 lie in different sites in the same gene, the cross and the F_1 progeny can be written as

$$\text{P:} \quad \frac{w_1 \ +}{w_1 \ +} \ \times \ \frac{+ \ w_2}{+ \ w_2} \qquad \text{F}_1\text{:} \quad \frac{w_1 \ +}{+ \ w_2}$$

A rare intragenic crossover between the two mutant sites can be diagrammed as follows:

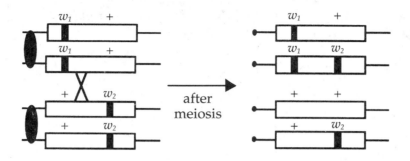

This crossover gives rise to gametes that are $w_1 w_2$ and $+ +$. The $w_1 w_2$ gamete would not be detected ($w_1 w_2 / w_1 +$ and $w_1 w_2 / + w_2$ flies are white eyed, as are their noncrossover siblings), but a $+ +$ gamete would give rise to a red-eyed offspring that has a genotype of $+ + / w_1 +$ or $+ + / + w_2$. It is interesting to note that from the frequency of the white-eyed progeny, one can determine the recombination frequency (RF) between the two point mutants, just as in Benzer's *cis-trans* tests in the *rII* region of T4.

RF $(w_1, w_2) = [(2 \times 10)/20{,}010] \times 100\% = 0.1\%$.

The w_1 and w_2 mutation sites are 0.1 mu apart.

9.24   Propose a genetic explanation for the ugly duckling phenomenon: Two white parents have a rare black offspring amid a prolific number of white offspring.

Answer: First think about why the parents might be white and have mostly white offspring. Presumably, the parents are white because they lack a functional gene that makes a product (an enzyme perhaps) that produces black pigment. If the parents were white because they lacked the function of two *different* genes needed for black pigment production, a cross between them would produce black, and not white, offspring. Since most of their offspring are white, it is likely that they each lack an identical function needed for black pigment production. Consider the options for how this could occur and then consider how a black offspring might arise under each option.

Option 1: Each parent is homoallelic at a gene that results in white and not black pigmentation. In this case, denote the cross as $a^1a^1 \times a^2a^2$, all progeny are a^1/a^2. Since the parents are homoallelic, no intragenic recombination will occur. In this case, an a^+ allele could arise by a rare, new mutation (a reversion mutation), which would restore the function of the gene in the black pigment pathway. The offspring would be $a^{++}/(a^1$ or $a^2)$, and be black.

Option 2: At least one parent is heteroallelic at a gene that results in white and not black pigmentation. In this case, one can denote the cross as $a^1/a^2 \times a^-/a^-$, where a^- represents any mutant, nonfunctional allele at this gene (the two a^- alleles need not be identical). In the heteroallelic parent a^1/a^2, a rare intragenic crossover between the sites of the a^1 and a^2 mutations will result in a^+ and $a^{1,2}$ alleles, so that an a^+/a^- offspring will be produced. This offspring will have a^+ gene function and so it will be black.

9.25   Both *trpA* and *trpB* mutants of *E. coli* lack tryptophan synthetase activity. All *trpA* mutants complement all *trpB* mutants. Explain how two different complementing mutants (*trpA* and *trpB*) can affect the activity of the same enzyme.

Answer: There are at least two options. First, the enzyme could be composed of multiple polypeptide subunits. The mutants are at different genes, each of which encodes a polypeptide that is part of the multimeric enzyme. Second, the enzyme is composed of just one type of polypeptide but the polypeptide is modified before it becomes active as an enzyme. One mutant is at a gene that encodes the polypeptide that will be modified to become the enzyme; the second mutant is at a gene that encodes a protein that modifies the enzymatic polypeptide and that is required to make it functional (e.g., it could be a protease that cleaves a proenzyme, making it active as an enzyme; it could be a kinase that phosphorylates a nonactive form of the enzyme, making it active).

9.26   Herpes simplex virus type 1 (HSV-1) is a large eukaryotic virus whose growth proceeds sequentially. Progression from one stage to the next requires completion of the earlier stage. Understanding how different viral genes are used at each stage should aid in the development of therapies for viral infection. Nine different mutations (*B2, B21, B27, B28, B32, 901, LB2, D, c75*) block viral growth at a very early stage. Each mutant grows at the permissive temperature of 34°C and fails to grow at the restrictive temperature of 39°C. All of the mutants except for *c75* spontaneously revert to wild type at about the same low frequency; *c75* reverts to wild type, but much less frequently than others. Complementation analysis of these mutants and a tenth temperature-sensitive mutation that blocks growth at a later stage, *J12g*, was performed by coinfecting pairs of mutants into cells at 39°C, collecting the cell culture media, and assaying its virus content by infecting cells at 34°C. Virus production was quantified using an index *I*. For two mutants *A* and *B*, *I* = [yield(coinfection of *A* and *B*)]/[yield(infection of *A*) + yield(infection of *B*)]. *I* must be over 2 to be considered positive. The following table gives the values of *I* for pairs of coinfected mutants.

Virus	B21	B27	B28	B32	901	LB2	D	c75	J12g
B2	1.5	0.70	0.81	0.37	0.40	0.55	0.48	0.70	170
B21		0.31	0.27	0.86	0.76	0.88	0.33	0.32	4.9
B27			0.28	0.18	1.9	1.5	0.61	0.68	19
B28				0.20	0.42	0.68	0.13	0.50	72
B32					1.4	0.84	0.28	0.38	580
901						0.45	0.10	0.20	570
LB2							0.44	0.91	22
D								0.35	444
c75									30

Pairwise recombination frequencies of the eight mutants that block very early growth were determined. The recombination frequency (RF) for two mutants *A* and *B* was calculated as RF = [yield(coinfection of *A* and *B*) at 39°C]/[yield(coinfection of *A* and *B*) at 34°C] × 2 × 100. The following table gives RF values for pairs of mutants.

Mutant	B21	B27	B28	B32	901	LB2	D	c75
B2	0.36	0.51	2.6	2.5	6.0	1.7	4.0	0.87
B21		0.91	3.0	2.8	6.4	2.3	4.5	1.6
B27			2.1	2.2	5.5	1.4	3.6	1.8
B28				0.11	3.8	0.71	1.55	0.31
B32					3.4	0.73	1.45	0.55
901						4.1	1.9	0.89
LB2							2.2	0.40
D								0.0

Analyze these results and answer the following questions about the nine mutants that block HSV-1 growth at a very early stage.

a. Are the mutants point mutations or deletions?

b. How many functions are affected by the mutants?

c. Do any of the mutants affect the same site?

d. Do any of the mutants affect multiple sites?

e. What is the rationale behind the calculations of *I* and RF?

f. What are the relative map positions of *B2, B21, B27, B28, B32, 901, LB2,* and *D*?

g. RF values for the *c75* mutant are inconsistent with those of the other mutants. Assuming that there are no technical errors, what might explain this?

Answer:

a. Since all nine very early mutants can be reverted, all are point mutants.

b. All nine very early mutants fail to complement each other (none produce enough virus in pairwise coinfections to be considered positive), so they affect one function.

c. Eight of the mutants (*B2, B21, B27, B28, B32, 901, LB2, D*) are able to recombine with each other and so affect different sites. Mutant *c75* fails to recombine with mutant *D*, so these two mutants may affect the same site.

d. Mutant *c75* may be a mutant affecting multiple sites. It reverts at a lower frequency than the others and also shows inconsistent recombination relative to the other mutants.

e. The mutants incompletely block viral growth, so some virus is produced by each mutant. *I* compares the amount of virus produced by coinfection of two mutants to the sum of the amounts produced by individual infections. If two viruses are blocked in the same function, a coinfection should produce low amounts of virus similar to the sum of two separate infections and *I* will be about 1.0. If two mutants are blocked in different functions, coinfection allows for complementation, as the function blocked in one mutant is provided by the other mutant. Substantial viral growth will occur and *I* should be much larger than 1.

At the restrictive temperature of 39°C, neither parent nor doubly mutant recombinants can grow. Only wild-type recombinants can grow. These are half of the recombinants, so doubling the amount of virus produced at 39°C will estimate the number of recombinants between the two mutant sites. At the permissive temperature of 34°C, mutant and wild-type virus can grow, so the amount of virus produced at 34°C measures the total amount of virus produced by coinfecting the two mutant strains. RF is calculated by doubling the amount of virus produced by coinfecting two mutants at 39°C and then dividing this number by the amount of virus produced by coinfecting the mutants at 34°C, so RF estimates recombination between two mutants.

f.

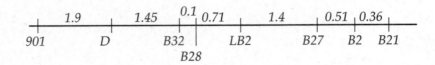

g. Mutants *c75*'s reversion rate is less than the other mutants, suggesting that it is more complex than a simple point mutant. If it affected multiple sites, it would have recombination data inconsistent with the other mutants.

9.27 A large number of mutations in *Drosophila* alter the normal deep-red eye color. As we discussed in Chapter 4, even alleles at one gene (*white, w*) can display a variety of phenotypes.

You are given a wild-type, deep-red strain and six independently isolated, true-breeding mutant strains that have varying shades of brown eyes, with the assurance that each mutant strain has only a single mutation. How would you determine

a. whether the mutation in each strain is dominant or recessive?

b. how many different genes are affected in the six mutant strains?

c. which mutants, if any, are allelic?

d. whether any of these mutants are alleles of genes already known to affect eye color?

Answer:

a. Cross each mutation to the wild-type, deep-red strain and observe the phenotypes of the progeny. A dominant mutation, by definition, is one that appears in a heterozygote. If the progeny show deep-red eyes, the mutation must be recessive. If the progeny show brownish eyes, the mutation is dominant.

b. Set up pairs of crosses between the mutants to perform complementation tests. Mutations that affect the same gene function will produce brown-eyed progeny when crossed and belong to the same complementation group. Mutations that affect different functions will produce deep-red-eyed progeny when crossed and belong to different complementation groups. Counting the number of different complementation groups will give the number of genes that are affected.

c. Allelic mutations are those that are members of the same complementation group, as determined in (b).

d. One could assess if a particular mutant is allelic to a known eye color gene by performing complementation tests between it and mutants at all known eye color genes. This would involve crossing the mutant to mutants from a collection of strains with known eye color mutations and observing the progeny of each cross. If the progeny have a mutant eye color, one would infer that the mutations carried in the two strains are allelic. However, this would be a tremendous amount of work. There are many eye color mutations, and this would require a large number of crosses. It would be faster to first determine which of the six mutations are allelic and then choose a representative allele from each complementation group and identify its chromosomal location using a set of two- and/or three-point mapping crosses. Once this is done, examine published genetic maps of the *Drosophila* genome (e.g., at http://www.flybase.org) and ask if any known eye color mutations lie in the same region. Then obtain strains with these eye color mutations and perform complementation tests between these mutant strains and a representative mutant from each complementation group identified with the new eye color mutations.

9.28   In *Drosophila*, the *kar*, *ry*, and *l(3)26* loci are located on chromosome 3 at map positions 51.7, 52.0, and 52.2, respectively. Mutants at *kar* have karmoisin (bright red) eyes. Mutants at *l(3)26* are recessive lethal. Mutants at *ry* lack the enzyme xanthine dehydrogenase. They survive and have rosy eyes if their dietary purine is limited but die if it is not. Wild-type ry^+ animals have deep-red eyes and survive if fed a diet rich in purine. You want to test whether Benzer's findings at the *rII* locus in T4 phage can be replicated in eukaryotes, so embark on a fine-structure analysis of the *ry* locus. Over the years, hundreds of mutants with rosy eyes have been identified by different researchers, and you have obtained many of them. Describe your experimental design, and address each of the following concerns.

a. There are many loci that affect eye color in *Drosophila*. What methods will you use to ensure that a rosy-eyed mutant is caused by a mutation at the *ry* locus?

b. What sets of crosses would you perform? In general terms, what progeny and frequencies do you expect to see in each cross?

c. How will you efficiently select for intragenic recombinants at the *ry* locus?

d. If you undertake both fine-structure recombination and complementation analyses, what results do you expect to see if Benzer's findings are replicated?

Answer:

a. Start by performing complementation tests. Cross a rosy-eyed mutant (call it *x*) to a known mutation, *ry*, at the *rosy* locus, and observe the phenotype of the offspring. If the progeny have wild-type (deep-red) eyes, the mutations complement each other, suggesting that *x* is not an allele at the *rosy* locus. This can be verified by mapping mutant *x* using two- and three-point testcrosses. If mutant *x* maps to a different locus, it can be excluded from further consideration. However, if *x* is tightly linked to the *rosy* locus, *x* and the known *ry* mutation may lie in different cistrons at the *ry* locus. If *x* does not complement *ry*, they *x* affects the same cistron as the known *ry* mutant.

b. Intragenic recombination can be evaluated by observing whether the progeny of heterozygotes bearing different *ry* alleles (e.g., ry^1/ry^2) receive an ry^+ allele. Using a logic similar to that of Benzer, intragenic recombination in ry^1/ry^2 heterozygotes will generate ry^+ and $ry^{1,2}$ chromosomes. If the heterozygotes are crossed to *ry/ry* males, recombinant offspring that are $ry^{1,2}/ry$ will have rosy eyes, while recombinants that are ry^+/ry will have deep-red eyes. Therefore, set up these types of crosses (females heterozygous for different *ry* alleles crossed to *ry/ry* males) and count the number of offspring with deep-red eyes. The recombination frequency is given by (2 × # deep-red eyed progeny)/(total progeny). Since recombination within a gene is rare, very few recombinants are expected (0.01% < RF < 0.3% based on Benzer's findings) and so this could be tedious.

   Since intragenic recombination is rare, it is important to verify that the deep-red-eyed progeny result from intragenic recombination and not reversion of one of the *ry* mutations. If the crosses were set up so that one of the *ry* chromosomes had flanking markers, intragenic recombination could be verified by checking for flanking marker exchange. For example, suppose the cross was $kar\ ry^1\ l(3)26/kar^+\ ry^2\ l(3)26^+$ (females) $\times\ ry^3/ry^3$ (males), that there was intragenic recombination between the sites of ry^1 and ry^2, and that the relative order of the mutant sites was $kar—ry^1—ry^2—l(3)26$. Then the resulting ry^+ recombinant would be $kar^+\ ry^+\ l(3)26/ry^3$ (demonstrate this by drawing it out). Flanking marker exchange associated with crossing over between ry^1 and ry^2 could be verified by testcrossing the deep-red-eyed fly to a $kar\ ry^+\ l(3)26/kar^+\ ry\ l(3)26^+$ individual and counting the number of progeny phenotypes. In this case, flanking marker exchange should be associated with a 2 ry^+ : 1 ry ratio due to the recessive lethality of the $l(3)26$ mutation. If the order of the mutant sites was $kar—ry^2—ry^1—l(3)26$, then the resulting ry^+ recombinant would be $kar\ ry^+\ l(3)26^+$, and flanking marker exchange would be associated with a 3 kar^+ : 1 kar progeny ratio.

c. One way to efficiently select for intragenic recombinants is to use supplemental purines in the diet to eliminate all ry progeny. Estimate the number of progeny (by counting the number of eggs laid, for example) and then raise the offspring on media rich in purines. Only ry^+ progeny would survive and so this would select for intragenic recombinants (and unintentionally, ry^+ revertants). This would make the process much less tedious as it wouldn't be necessary to look at the eye color of tens of thousands of flies.

d. Based on Benzer's work, one would expect that intragenic recombination is rare, that mutations can be mapped to sites within the ry locus, that some "hot spots" (sites where mutations are more prevalent) may exist, that some mutations will be unable to recombine with others (because they are deletions), and that there will be a minimum map distance between the closest ry alleles (corresponding to the recombination frequency between mutants affecting neighboring DNA base pairs). A complementation analysis will reveal if the ry locus has multiple cistrons (functional units).

10

DNA: THE GENETIC MATERIAL

CHAPTER OUTLINE

THINKING ANALYTICALLY

This chapter presents a large amount of complex, related information. As such, its information and problems require attention to detail. To comprehend and retain this information, organize it and develop strategies to place it in context. In addition to concept maps, it will help to construct labeled diagrams that display information contextually. Sketch labeled diagrams of nucleic acid structure, chromosome packaging, and sequence organization of chromosomes.

This chapter starts with a presentation of the key experimental evidence for the nature and structure of the genetic material. What we now accept as fact once was hotly debated. These insightful experiments resolved fundamental questions in biology. To organize this information and place it in context, pay close attention to the subtleties of the experiments that are presented, and strive to see *how* they provided evidence for or against a particular view.

225

REVIEW OF KEY TERMS, SYMBOLS, AND CONCEPTS

In your own words, write a brief, precise definition of each term in the groups below. Check your definitions using the text. Then develop a concept map using the terms in each list.

1	2	3
macromolecule	genetic material	Chargaff's rules
chromosome	DNA	% GC
genome	RNA	X-ray diffraction
nucleic acid	TMV	angstrom unit
DNA, RNA	transformation	3'-OH, 5'-P
ribose, deoxyribose	transforming principle	polarity
phosphodiester bond	nuclease	antiparallel strands
nucleotide	ribonuclease (RNase)	complementary base pair
nucleoside	deoxyribonuclease (DNase)	double helix
polynucleotide	Avery's experiment	major groove
nitrogenous base	Griffith's experiment	minor groove
adenine, guanine	Hershey-Chase experiment	right-handed helix
cytosine, thymine		left-handed helix
uracil		A-, B-, Z-DNA
purine		oligomer, oligo
pyrimidine		polarity
		Watson-Crick model

4	5	6
viral genome	histones	C-value paradox
prokaryotic genome	nonhistones	unique-sequence DNA
single-, double-stranded	H1, H2A, H2B, H3, H4	moderately repetitive DNA
circular chromosome	linker DNA	highly repetitive DNA
linear chromosome	nucleosome	dispersed repeated DNA
T-even phage	"beads-on-a-string"	interspersed repeated DNA
λ, ΦX174	solenoid model	tandemly repeated DNA
nucleoid	C-value	LINEs
supercoiling	chromatin	SINEs
topoisomerases	10-nm chromatin fiber	transposons
looped domains	30-nm chromatin fiber	
	euchromatin	
	heterochromatin	
	facultative heterochromatin	
	constitutive heterochromatin	
	CEN sequences	

4	5	6
	chromosome scaffold scaffold-associated regions (SARs) simple telomeric sequences telomere-associated sequences	

It is essential to know the chemical structures of nucleic acids to fully understand the material in upcoming chapters. Start by drawing out the structures using paper and pencil. Copy the component parts of a nucleotide (the sugar, the bases, the phosphodiester linkages) from the text figures, paying close attention to their polarity. Practice them for a few days in a row so that you can draw them easily from memory.

Having a good sense of the dimensions involved in nucleic acid structure and chromosomal packaging is helpful in grasping a mental image of these physical processes. For this reason, it is important to learn (or review) units of measurement and to be able to make conversions between different dimension units. In solving problems, it is helpful to convert values presented in many different units to a smaller set of units. Particularly important unit conversions are:

UNIT/CONVERSION	EXAMPLE
$1 \text{ Å} = 1 \text{ angstrom} = 1 \times 10^{-10} \text{ meter}$	distance between 2 protons in $H_2 = 0.74 \text{ Å}$
$1 \text{ nm} = 1 \text{ nanometer} = 1 \times 10^{-9} \text{ meter}$	width of a DNA double helix $= 2 \text{ nm}$
$1 \text{ μm} = 1 \text{ micrometer} = 1 \times 10^{-6} \text{ meter}$	typical red blood cell diameter $= 7 \text{ μm}$
1 dalton (unit of mass) $= 1.66 \times 10^{-24} \text{ gram}$	1 H atom = 1 dalton

The chromosomal organization and packaging of DNA varies in different organisms. Consequently, while there are general principles that underlie DNA packaging, there are also organism-specific requirements. As you study this chapter, first identify the general principles used to organize and package genetic material. Then relate these principles to specific examples of organization and packaging in viruses, prokaryotes, and eukaryotes.

DNA packaging occurs in three dimensions, and so it helps to visualize DNA packaging by making sketches. You do not have to be a great artist; the point is to convey the different strategies that organisms use to fit a DNA molecule, which if left unpackaged, would be larger than most cells, into a cell and have it remain functional.

Not everyone visualizes well in three dimensions and so it can be especially challenging to visualize supercoiling in a circular DNA molecule. The following exercise might help: Take a spiral telephone cord and hold the ends together to circularize it. Sequentially introduce one, two, three or four additional twists that *tighten* the spirals. For each complete twist that is introduced, a supercoil is added. Since the twist tightens the spiral, it introduces a positive supercoil. If you let the cord "relax" and then introduce additional twists that *loosen* the spirals, you will be introducing negative supercoils. To determine whether your phone cord is a right- or left-handed helix, wrap your right hand around the spiral, and trace the direction of the spiral with your right index finger. If, while you hold the spiral with your right hand, your index finger traces up the spiral in the same direction that your right thumb points, you are holding a right-handed helix. If it traces down the spiral in an opposite direction, you are holding a left-handed helix. Notice that the orientation of the helix remains the same regardless of whether your right thumb points up or down.

QUESTIONS FOR PRACTICE

Multiple Choice Questions

1. DNA and RNA are polymers of
   a. nucleosides
   b. nucleotides
   c. pentose sugars connected by phosphodiester bonds
   d. ribonucleotides

2. A molecule consisting of ribose covalently bonded to a purine or pyrimidine base is a
   a. ribonucleoside
   b. ribonucleotide
   c. nuclease
   d. deoxyribonucleotide

3. The transforming principle was found to be
   a. a cellular material that could alter a cell's heritable characteristic
   b. a substance derived from killed viruses
   c. modified RNA that could change a living cell
   d. a transmissible substance that revives dead cells

4. When Griffith injected mice with a mixture of live *R* pneumococcus derived from a *IIS* strain and heat-killed *IIIS* bacteria,
   a. the mice survived and he recovered live type *IIIR* organisms
   b. the mice died, but he recovered live type *IIIR* cells
   c. the mice died, but he recovered live type *IIS* cells
   d. the mice died, but he recovered live type *IIIS* cells

5. In the Hershey-Chase experiment, T2 phage were radioactively labeled with either ^{35}S or ^{32}P, and allowed to infect *E. coli*. What results proved that DNA was the genetic material?
   a. The ^{35}S was found in progeny phage, and the ^{32}P was found in phage ghosts.
   b. The ^{35}S was found in phage ghosts, and the ^{32}P was found in progeny phage.
   c. The ^{35}S was found in both progeny phage and in phage ghosts.
   d. The ^{32}P was found in both progeny phage and in phage ghosts.

6. Analysis of the bases of a sample of nucleic acid yielded these percentages: A, 20 percent; G, 30 percent; C, 20 percent; T, 30 percent. The sample must be
   a. double-stranded RNA
   b. single-stranded RNA
   c. double-stranded DNA
   d. single-stranded DNA

7. Which of the following is *not* true about a linear molecule of double-stranded DNA?
   a. It is a double helix composed of antiparallel strands.
   b. Bases are paired via hydrogen bonds.

   c. At one end, two 5' phosphate groups can be found.

   d. Pentose sugars are linked via covalent phosphodiester bonds.

8. Which kind of DNA is *not* likely to be found in cells?
   a. A-form DNA
   b. B-form DNA
   c. Z-form DNA
   d. none of the above

9. Two double-stranded 25-base-pair DNA fragments are heated in solution. Fragment A has 60 percent GC, and fragment B has 40 percent GC. Which observation(s) might be made as the solution temperature increases?
   a. At a low enough temperature, both fragments will remain double-stranded.
   b. A will separate into single strands at a lower temperature than B.
   c. B will separate into single strands at a lower temperature than A.
   d. At a high enough temperature, both A and B will separate into single strands.
   e. a, c, and d

10. The chromosome of *E. coli* is packaged in the nucleoid region in a
    a. nuclear membrane
    b. semicircular form
    c. relaxed form
    d. supercoiled form

11. Topoisomerases are enzymes that do all of the following except
    a. untwist relaxed DNA
    b. introduce negative supercoils into relaxed DNA
    c. introduce positive supercoils into relaxed DNA
    d. change the topological form of the DNA, but not the DNA sequence

12. The total amount of DNA in the haploid genome of any organism is
    a. its karyotype
    b. its C value
    c. twice its GC content
    d. an indication of its organizational and structural complexity

13. Facultative heterochromatin
    a. is always inactive
    b. is inactive only in certain cells
    c. contains only moderately repetitive DNA
    d. can contain unique-sequence DNA
    e. both b and d
    f. both a and c

14. Which of the following is *not* true about both centromeres and telomeres?
    a. They are characteristically heterochromatic.
    b. They are associated with consensus sequence elements.
    c. They are essential for eukaryotic chromosome function.
    d. They contain a short, species-specific sequence that is repeated hundreds to thousands of times.

15. The fundamental unit of chromatin packaging in eukaryotes is
    a. a histone protein
    b. a nucleosome

c. a 30-nm chromatin fiber

d. a looped domain

16. Examples of tandemly repeated DNA sequences include all of the following except

    a. genes for ribosomal RNA

    b. LINEs

    c. genes for histones

    d. simple telomeric sequences

17. Which type of chromatin contains expressed genes?

    a. constitutive heterochromatin

    b. facultative heterochromatin

    c. euchromatin

    d. Barr bodies

Answers: 1b, 2a, 3a, 4d, 5b, 6d, 7c, 8a, 9e, 10d, 11c, 12b, 13e, 14d, 15b, 16b, 17c

Thought Questions

1. The histones, H1, H2A, H2B, H3, and H4, are the most highly conserved of all proteins. It has been proposed that they serve a basic function in all eukaryotes. What are the implications of this assertion for evolutionary theory?

2. The histones are highly basic proteins that interact with DNA (specifically, with the acidic sugar-phosphate backbone). What kinds of proteins are the nonhistones? How might they differ from the histones in their interactions with DNA?

3. How do you account for the presence of both unique and repetitive sequences in the genomes of eukaryotes? Why are the latter mostly lacking in prokaryotes?

4. Explain one of the fundamental puzzles of life: An amphibian that lives in a swamp has 30-fold more DNA than a human that lives on the 48th floor of a luxury apartment complex. In what way is the amphibian richer?

5. Summarize the structural composition and organization of the chromosomes of eukaryotic cells. How do they compare with the chromosomes of viruses and of prokaryotes?

6. Describe the logic behind the series of experiments that conclusively demonstrated that DNA was the genetic material in some cells. Why was this hotly debated for so long? When RNA was found to be the genetic material for some viruses, was the debate rekindled, or was this the exception that proved the rule?

7. In a double-stranded DNA molecule, how many strands of DNA exist in a single chromosome during the following stages of the cell cycle (mitosis and meiosis). G_1, G_2, prophase, metaphase, anaphase, pachynema, diplonema, anaphase I, and anaphase II?

8. Describe the various forms of DNA (A, B, Z) that have been identified. What are their most significant differences? Are any of these differences likely to have functional significance? If so, what are they?

9. Watson and Crick reputedly deduced the structure of DNA by applying observations made by others to molecular models, without any direct experimental observations of their own. Is this legitimate scientific procedure? If so, why? If not, why not?

10. Draw out the chemical structure of a double-stranded DNA molecule so that one strand has the sequence 5'-ATG-3'. Indicate the polarity of each strand, the location and the kinds of bonds that exist, and the approximate dimensions of the molecule.

11. Double-stranded DNA molecules have negatively charged sugar-phosphate backbones and a major and minor groove into which the chemical groups of the bases project. Some proteins that interact with DNA do so in a highly sequence-specific manner, while others interact in a largely sequence-nonspecific manner. What different features of the DNA molecule might these two classes of protein be recognizing?

Thought Questions for Media

After reviewing the media on the *iGenetics* CD-ROM, try to answer these questions.

1. In the animation depicting the experiment by Avery and his colleagues that showed that bacteria could be transformed with DNA, how did they show that polysaccharides were not the transforming principle?
2. In the animation depicting the Hershey-Chase experiment, how were the T2 phage initially labeled with either ^{35}S or ^{32}P?
3. The T2 phage labeled with ^{32}P only had their DNA labeled. Why doesn't this information, by itself, provide evidence that DNA is the genetic material of T2?
4. What experiment was done to provide evidence that DNA is the genetic material?
5. In the animation depicting how the *E. coli* chromosome is supercoiled, is the DNA of *E. coli* compacted by adding negative or positive supercoils?
6. Which enzyme is responsible for adding negative supercoils?
7. Which enzyme is responsible for removing negative supercoils?

1. We now know that the genetic material of viruses can be either RNA or DNA and that it can be either single or double stranded. Why is it still important to determine the type of nucleic acid associated with a new virus?
2. Once you determine the type of nucleic acid of a new virus, how can you determine which one of several known viruses is most closely related to it?

SOLUTIONS TO TEXT PROBLEMS

10.1 Griffith's experiment injecting a mixture of dead and live bacteria into mice demonstrated that (choose the correct answer):

a. DNA is double-stranded.

b. mRNA of eukaryotes differs from mRNA of prokaryotes.

c. a factor was capable of transforming one bacterial cell type to another.

d. bacteria can recover from heat treatment if live helper cells are present.

Answer: c

10.2 In the 1920s, while working with *Streptococcus pneumoniae* (the agent that causes pneumonia), Griffith injected mice with different types of bacteria. For each of the following bacteria types injected, indicate whether the mice lived or died:

a. type *IIR*

b. type *IIIS*

c. heat-killed *IIIS*

d. type *IIR* + heat-killed *IIIS*

Answer:

a. lived

b. died

c. lived

d. died (DNA from the *IIIS* bacteria transformed the *IIR* bacteria to a virulent form.)

10.3 Several years after Griffith described the transforming principle, Avery, MacLeod, and McCarty investigated the same phenomenon.

a. List the steps they used to show that DNA from dead *S. pneumoniae* cells was responsible for the change from a nonvirulent to a virulent state.

b. Did their work confirm or disconfirm Griffith's work, and how?

c. What was the role of the enzymes used in their experiments?

Answer:

a. They made extracts from *S* cells and showed that they could transform *R* cells by exposing them to these extracts. These extracts contained different cellular macromolecules, such as lipids, polysaccharides, proteins, and nucleic acids. They first showed that nucleic acids were the only macromolecular component that could transform the *R* cells. Then, by treating the nucleic acid component with either ribonuclease or deoxyribonuclease, they showed that the transforming principle copurified with DNA. This strongly suggested that the genetic material was DNA. Although the work suggested that the genetic material was DNA, it could be criticized because the nucleic acids were contaminated by proteins.

b. Their work affirmed and extended Griffith's work in that it identified the probable nature of Griffith's transforming principle as being DNA.

c. Enzymes were used to destroy one of the two components of the nucleic acid mixture, and served to identify which of two nucleic acids was the genetic material. Ribonuclease did not destroy the transforming principle, but deoxyribonuclease did abolish transforming activity. This result supported the contention that DNA was the genetic material.

10.4   Hershey and Chase showed that when phages were labeled with ^{32}P and ^{35}S, the ^{35}S remained outside the cell and could be removed without affecting the course of infection, whereas the ^{32}P entered the cell and could be recovered in progeny phages. What distribution of isotopes would you expect to see if parental phages were labeled with isotopes of

a. C

b. N

c. H

Explain your answer.

Answer: a, b, and c. Phage ghosts (supernatant) and progeny would have isotope. Both amino acids and nucleic acids have carbon, nitrogen, and hydrogen. If a parental phage was labeled with isotopes of C, N, or H, one would expect the phage to have a labeled protein coat as well as labeled DNA. As a consequence, these isotopes would be recovered in the DNA of the progeny phage, as well as in the phage ghosts left behind in the supernatant after phage infection. This experiment would not be very helpful to determine whether DNA or protein was the genetic material, as neither material would be selectively labeled. The elegance of the Hershey-Chase experiment was that the isotopes used *selectively* labeled only one component of the phage, so that it could be followed from one generation to the next.

10.5   What is the evidence that the genetic material of tobacco mosaic virus (TMV) is RNA?

Answer: Initial evidence that RNA is the genetic material of TMV came from the experiments of Gierer and Schramm. They showed that tobacco plants inoculated with RNA purified from TMV developed typical virus-induced lesions, and that these effects were specifically due to RNA, as they were not seen following RNase treatment. Fraenkel-Conrat and Singer confirmed this when they showed that reconstituted hybrid viruses, which had RNA from one strain and protein from another strain, produced progeny viruses that had protein subunits specified by the RNA component of the hybrid.

10.6   The X-ray diffraction data obtained by Rosalind Franklin suggested (choose the correct answer)

a. DNA is a helix with a pattern that repeats every 3.4 nm

b. purines are hydrogen bonded to pyrimidines

c. DNA is a left-handed helix

d. DNA is organized into nucleosomes

Answer: a

10.7   What evidence do we have that, in the helical form of the DNA molecule, the base pairs are composed of one purine and one pyrimidine?

Answer: Two different lines of evidence support the view that a base pair is composed of one purine and one pyrimidine.

1. When the chemical components of double-stranded DNA from a wide variety of organisms were analyzed quantitatively by Chargaff, it was found that the amount of purines equaled the amount of pyrimidines. More specifically, it was found that the amount of adenine equaled the amount of thymine and that the amount of cytosine equaled the amount of guanine. The simplest hypothesis to explain these observations was the existence of complementary base pairing, A on one strand paired with T on the other strand, and G paired with C.

2. More direct physical evidence was provided by X-ray diffraction studies. These established the dimensions of the DNA double helix and allowed for comparison with the known sizes of the bases. The diameter of the double helix is constant throughout its length at 2 nm. This is the right size to accommodate a purine paired with a pyrimidine, but too small for a purine-purine pair and too large for a pyrimidine-pyrimidine pair.

10.8   What exactly is a deoxyribonucleotide made up of, and how many different deoxyribonucleotides are there in DNA? Describe the structure of DNA, and describe the bonding mechanism of the molecule (i.e., the kind of bonds are on the sides of the "ladder" and the kind of bonds hold the two complementary strands together). Base pairing in DNA consists of purine-pyrimidine pairs, so why can't A-C and G-T pairs form?

Answer: A deoxynucleotide consists of 2'-deoxyribose plus a phosphate group (PO_4^{-2}) attached to its 5'-carbon, plus a nitrogenous base attached to its 1'-carbon. Since there are four different nitrogenous bases in DNA, adenine (A), thymine (T), guanine (G), and cytosine (C), there are four different nucleotides. In a DNA molecule, one finds monophosphate nucleotides, so that there is deoxyadenosine monophosphate (dAMP), thymidine monophosphate (TMP), deoxyguanosine monophosphate (dGMP), and deoxycytidine monophosphate (dCMP). Along the sides of the "ladder," the 5'-carbon of one deoxyribose is connected by a covalent phosphodiester (O–P–O) bond to the 3'-carbon of another. A phosphate group is found at the 5' end of a DNA polynucleotide chain, and a hydroxyl (OH) group is found at the 3' end of a DNA polynucleotide chain. Weaker hydrogen bonds between complementary A-T and G-C bases hold the complementary strands together. A-T pairs have two hydrogen bonds, while G-C pairs have three hydrogen bonds. See text Figure 10.12.

Although A-C and G-T pairs would be purine-pyrimidine base pairs, they do not form because of their inability to pair using hydrogen bonding. Note that A-T base pairs have two hydrogen bonds, while G-C base pairs have three hydrogen bonds. This is consistent with the evidence provided by Chargaff. Quantitative measurements of the four bases in double-stranded DNA isolated from a wide variety of organisms indicated that in all cases, the amount of A equaled the amount of T, and the amount of G equaled the amount of C. Moreover, different DNAs exhibited different base ratios so that while (A) = (T) and (G) = (C), in most organisms (A + T) ≠ (G + C) . Put another way, the percent GC content of different DNA samples varies. The simplest hypothesis is that there are two base pairs in DNA, A-T and G-C, and the proportion of the two base pairs varies from organism to organism.

10.9   What is the base sequence of the DNA strand that would be complementary to the following single-stranded DNA molecules?

a. 5'-AGTTACCTGATCGTA-3'
b. 5'-TTCTCAAGAATTCCA-3'

Answer:

a. 3'-TCAATGGACTAGCAT-5' (or 5'-TACGATCAGGTAACT-3').
b. 3'-AAGAGTTCTTAAGGT-5' (or 5'-TGGAATTCTTGAGAA-3').

10.10   Describe the bonding properties of G-C and T-A. Which would be the hardest to break? Why?

Answer: The adenine-thymine base pair has two hydrogen bonds, while the guanine-cytosine base pair has three hydrogen bonds. Thus, the guanine-cytosine base pair requires more energy to break apart and so is harder to break apart.

10.11   The double-helix model of DNA, as suggested by Watson and Crick, was based on data gathered on DNA by other researchers. The facts fell into the following two general categories:
a. chemical composition
b. physical structure
Give two examples of each.

Answer:
a. The double helix model of DNA suggested by Watson and Crick had to incorporate existing information about its chemical composition and physical structure. In terms of its chemical composition, it was known that DNA is composed of polynucleotides, that (A) = (T) and (G) = (C) (Chargaff's rules), and that while the percent GC varies between organisms, the A/T and G/C ratios do not.
b. The structure and molecular dimensions of the component molecules (the bases, sugars, and phosphates) were known. It was also known from studies of Franklin and Wilkins that the molecule is organized in a highly ordered, helical structure, and that there are two distinctive regularities at 0.34 and 3.4 nm along the molecule's axis.

10.12   For double-stranded DNA, which of the following base ratios always equals 1?
a. (A + T) / (G + C)
b. (A + G) / (C + T)
c. C / G
d. (G + T) / (A + C)
e. A / G

Answer: Since (G) = (C) and (A) = (T), it follows that (G + A) = (C + T) and (G + T) = (A + C). Thus, b, c, and d are all equal to 1.

10.13   The ratio of (A + T) to (G + C) in a particular DNA is 1.0. Does this indicate that the DNA is probably composed of two complementary strands of DNA or a single strand of DNA, or is more information necessary?

Answer: More information is needed. That (A + T) / (G + C) = 1 indicates only that (A + T) = (G + C). If the DNA were double stranded, (G) = (C), and (A) = (T). For the observed ratio, there would need to be 25 percent A, 25 percent T, 25 percent C, and 25 percent G. However, if the DNA were single stranded, one could still observe this ratio. In single-stranded DNA, there are no restrictions on the relative amounts of the different bases. There are many ways in which one could observe $(A + T)/(G + C) = 1$. For example, suppose there were 35 percent A, 15 percent T, 20 percent G, and 30 percent C.

10.14   The percentage of cytosine in a double-stranded DNA is 17. What is the percentage of adenine in that DNA?

Answer: In double-stranded DNA, if $(C) = 17$ percent, then $(G) = 17$ percent. This means that the DNA has 34 percent GC, and 66 percent AT base pairs. Hence, the DNA will have $(66/2)(100\%) = 33\%$ A.

10.15   A double-stranded DNA polynucleotide contains 80 thymidylic acid and 110 deoxyguanylic acid residues. What is the total nucleotide number in this DNA fragment?

Answer: Since the DNA molecule is double-stranded, $(A) = (T)$ and $(G) = (C)$. If there are 80 T residues, there must be 80 A residues. If there are 110 G residues, there must be 110 C residues. The molecule has $(110 + 110 + 80 + 80) = 380$ nucleotides, or 190 base pairs.

10.16   Analysis of DNA from a bacterial virus indicates that it contains 33 percent A, 26 percent T, 18 percent G, and 23 percent C. Interpret these data.

Answer: First, notice that $(A) \neq (T)$ and $(G) \neq (C)$. Thus, the DNA is not double stranded. The bacterial virus appears to have a single-stranded DNA genome.

10.17   The following are melting temperatures for different double-stranded DNA molecules:
a.   73°C   d.   78°C
b.   69°C   e.   82°C
c.   84°C

Arrange these molecules from lower to higher content of G-C pairs.

Answer: G-C base pairs have three hydrogen bonds, while A-T base pairs have two. As a consequence, G-C base pairs are stronger than A-T base pairs. If a double-stranded molecule in solution is heated, the thermal energy will "melt" the hydrogen bonds, denaturing the double-stranded molecule into single strands. Double-stranded molecules with more G-C base pairs require more thermal energy to break the hydrogen bonds, and so dissociate into single strands at higher temperatures. Put another way, the higher the GC content of a double-stranded DNA molecule, the higher its melting temperature. Reordering the molecules from lowest to highest percent GC, one has (b) (69°), then (a) (73°), (d) (78°), (e) (82°), and (c) (84°).

10.18   What is a DNA oligomer?

Answer: A DNA oligomer is a short stretch of DNA containing a small number of bases. It is often referred to as an oligonucleotide, or an oligo, for short. Sometimes, when the exact size of the oligomer can be specified, this information might be included. For example, a single-stranded oligomer having 20 bases might be referred to as a 20-base oligomer, a 20-base oligo, or a 20-mer.

10.19 The genetic material of bacteriophage ΦX174 is single-stranded DNA. What base equalities or inequalities might we expect for single-stranded DNA?

Answer: No predictions can be made regarding the base content of single-stranded DNA. Any base pair equality would depend on the overall sequence of the chromosome: `(A)` might be equal to `(T)`, but that need not be the case, as that is only one of many possibilities.

10.20 Through X-ray diffraction analysis of crystallized DNA oligomers, different forms of DNA have been identified. These forms include A-DNA, B-DNA, and Z-DNA, and each has unique molecular attributes.

 a. Which form is the most common form in living cells?
 b. Z-DNA has an unusual conformation resulting in more base pairs per helical turn than B-DNA has. What is the conformation? Does this molecule have any function in living cells?
 c. Which of the given forms is never found in living cells?

Answer:

 a. B-DNA is the form most common to living cells.
 b. To understand why Z-DNA is considered to have an unusual conformation, it helps to compare it with B-DNA. Z-DNA is a long and thin (about 2 nm wide) double helix, like B-DNA. Unlike the right-handed B-DNA, however, Z-DNA is left-handed. It has an axis that runs through the minor groove and has 12 base pairs per helical turn that are inclined 8.8° from a plane perpendicular to the axis. In contrast, B-DNA has an axis that runs through the base pairs, and has 10 base pairs per helical turn that are inclined 2° from a plane perpendicular to the axis. The major groove of Z-DNA is not very distinct, as it is thin and flattened out along the helix surface, while the major groove of B-DNA is wide and intermediate in depth (between that of A-DNA and Z-DNA). The minor groove of Z-DNA is extremely narrow and very deep, while that of B-DNA is narrow and of intermediate depth. It has been proposed that regions with Z-DNA provide a stretch of left-handed helical turns involved in replication, recombination, and transcription. Stretches of left-handed turns may aid in unwinding right-handed helical turns in B-DNA during these processes. Z-DNA may also be more stable under extreme environmental conditions.
 c. A-DNA is found only when the DNA is dehydrated, so it is unlikely that lengthy sections of A-DNA would be found in living cells.

10.21 If a virus particle contains double-stranded DNA with 200,000 bp, how many complete 360° turns occur in this molecule?

Answer: With 10 base pairs per complete 360° turn of a B-form double-stranded DNA molecule, there are 200,000/10 = 20,000 complete turns in the viral DNA.

10.22 A double-stranded DNA molecule is 100,000 bp (100 kb) long.

 a. How many nucleotides does it contain?
 b. How many complete turns are there in the molecule?
 c. How long is the DNA molecule?

Answer:

a. Each base pair has two nucleotides, so the molecule has 200,000 nucleotides.

b. There are 10 base pairs per complete 360° turn, so there will be 100,000/10 = 10,000 complete turns in the molecule.

c. There is 0.34 nm between the centers of adjacent base pairs. There will be 100,000 × 0.34 nm = 3.4×10^4 nm = 34 μm.

10.23 Different organisms have vastly different amounts of genetic material. *E. coli* has about 4.6 million bp of DNA in one circular chromosome, the haploid budding yeast (*S. cerevisiae*) has 12,057,500 bp of DNA in 16 chromosomes, and the gametes of humans have about 2.75 billion bp of DNA in 23 chromosomes.

a. For each of these organism's cells, if all of the DNA were B-DNA, what would be the average length of a chromosome in the cell?

b. On average, how many complete turns would be in each chromosome?

c. Would your answers to (a) and (b) be significantly different if the DNA were composed of, say, 20 percent Z-DNA and 80 percent B-DNA?

d. What implications do your answers to these questions have for the packaging of DNA in cells?

Answer:

a. For *E. coli*, the length is $(4.6 \times 10^6 \text{ bp}) \times (0.34 \text{ nm/bp}) = 1.6 \times 10^6$ nm = 1,600 μm (microns). Since *E. coli* has a circular chromosome, this would be a chromosome with a diameter of about 510 μm. For yeast, the average chromosome length is (12,057,500 bp/16 chromosomes) × (0.34 nm/bp) = 2.56×10^5 nm = 256 μm. For humans, the average chromosome length is $(2.75 \times 10^9 \text{ bp}/23 \text{ chromosomes}) \times (0.34 \text{ nm/bp}) = 4.07 \times 10^7$ nm = 40,700 μm.

b. For B-DNA, there are 10 bp per helical turn. Therefore, the *E. coli* chromosome would have $(4.6 \times 10^6 \text{ bp/chromosome}) \times (1 \text{ turn}/10 \text{ bp}) = 4.6 \times 10^5$ turns. The average yeast chromosome would have (12,057,500 bp/16 chromosomes) × (1 turn/10 bp) = 7.54×10^4 turns. The average human chromosome would have $(2.75 \times 10^9 \text{ bp}/23$ chromosomes) × (1 turn/10 bp) = 1.2×10^7 turns.

c. While Z-DNA and B-DNA differ in the distance between successive base pairs and the number of base pairs per turn, these are not large compared to chromosome length. Thus, the answers to (a) and (b) will not be very different.

Z-DNA has more space between successive base pairs, as there is about 0.57 nm between successive base pairs instead of the 0.34 nm found in B-DNA. Thus, if 20% of each chromosome were Z-DNA, each chromosome would be 20% × [(0.57 − 0.34)/0.34] = 13.5% longer. A full turn of the double helix in Z-DNA utilizes more base pairs, as there are 12 bp/turn in Z-DNA instead of the 10 bp/turn in B-DNA. Thus, if 20% of each chromosome were Z-DNA, each chromosome would have 20% × [(12 − 10)/12] = 3.33% fewer turns.

d. These answers point out the need for DNA to be flexible so that it can be packaged and "fit" into a cell. Other forms of DNA, including Z-DNA, do not necessarily shorten a chromosome but they can affect its flexibility. In the examples considered in this problem, the cells are hundreds to tens of thousands of times smaller than the length of their (average) chromosome. *E. coli* is about 2 μm × 0.5 μm, with a 510-μm diameter chromosome; a yeast cell is about 20 μm wide, with 16 chromosomes that average 256 μm; and a "typical" human cell is about 10 μm wide, with 46 chromosomes that average 40,700 μm. Thus, chromosomes cannot remain as unpackaged molecules inside cells.

10.24   If nucleotides were arranged at random in a piece of single-stranded RNA 10^6 nucleotides long, and if the base composition of this RNA was 20 percent A, 25 percent C, 25 percent U, and 30 percent G, how many times would you expect the specific sequence 5'-GUUA-3' to occur?

Answer: The chance of finding the sequence 5'-GUUA-3' is $(0.30 \times 0.25 \times 0.25 \times 0.20)$ = 0.00375. In a molecule 10^6 nucleotides long, there are nearly 10^6 groups of four bases: The first group of four is bases 1, 2, 3, and 4, the second group is bases 2, 3, 4, and 5, etc. Thus, the number of times this sequence is expected to appear is 0.00375×10^6 = 3,750.

10.25   Two double-stranded DNA molecules from a population of T2 phages were denatured to single strands by heat treatment. The result was the following four single-stranded DNAs:

1   T A G C T C C →   3   G C T C C T A →

and

2   ← A T C G A G G   4   ← C G A G G A T

These separated strands were then allowed to renature. Diagram the structures of the renatured molecules most likely to appear when (a) strand 2 renatures with strand 3 and (b) strand 3 renatures with strand 4. Label the strands and indicate sequences and polarity.

Answer:

a. The sequence C G A G G in molecule 2 is complementary to the sequence G C T C C in molecule 3. When these pair up, one has:

3   ← G C T C C T A →

2   A T C G A G G

Each strand has two unpaired bases sticking out. These bases are complementary to each other, so that if the molecule bends, one has:

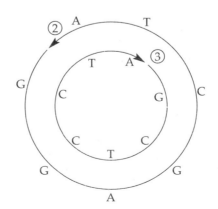

b. The sequence in molecule 3 is complementary to the sequence in molecule 4. It also has opposite polarity, so that the two strands can pair up. One has:

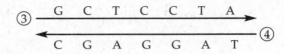

10.26   Define topoisomerases, and list the functions of these enzymes.

Answer: Topoisomerases are a class of enzymes that convert one topological form of DNA to another. Topoisomerases untwist relaxed DNA to produce negative supercoils or twist negatively supercoiled DNA to convert it to a more relaxed state. These enzymes allow DNA to be converted between negatively supercoiled, compact DNA and a more relaxed, less compact state.

10.27   What is the relationship between cellular DNA content and the structural or organizational complexity of the organism?

Answer: Paradoxically, there is no simple relationship between the haploid DNA content of a cell and its structural or organizational complexity. This is the "C value paradox." While organisms in some taxa show little variation, organisms in other taxa show as much as tenfold variation in their C values. In addition, organisms do not have C values corresponding to their organizational or structural complexity. At least one reason for this is that there is considerable variation in the amount of repetitive-sequence DNA in the genome.

10.28   In a particular eukaryotic chromosome (choose the best answer),
   a. heterochromatin and euchromatin are regions where genes make functional gene products (that is, where they are active)
   b. heterochromatin is active, but euchromatin is inactive
   c. heterochromatin is inactive, but euchromatin is active
   d. both heterochromatin and euchromatin are inactive

Answer: c

10.29   Compare and contrast eukaryotic chromosomes and bacterial chromosomes with respect to the following features:
   a. centromeres   f. nonhistone proteins scaffolds
   b. hexose sugars   g. DNA
   c. amino acids   h. nucleosomes
   d. supercoiling   i. circular chromosome
   e. telomeres   j. looping

Answer:
   a. Only eukaryotic chromosomes have centromeres, the sections of the chromosome found near the point of attachment of mitotic or meiotic spindle fibers. In some organisms, such as *S. cerevisiae,* they are associated with specific CEN sequences. In other organisms, they have a more complex repetitive structure.

b. Hexose sugars are not found in either eukaryotic or bacterial chromosomes, as *pentose* sugars are found in DNA.

c. Amino acids are found in proteins that are involved in chromosome compaction, such as the proteins that hold the ends of looped domains in prokaryotic chromosomes, and the histone and nonhistone proteins in eukaryotic chromatin.

d. Both eukaryotic and bacterial chromosomes are supercoiled.

e. Telomeres are found only at the ends of eukaryotic chromosomes, and are required for replication and chromosome stability. They are associated with specific types of sequences: simple telomeric sequences and telomere-associated sequences.

f. Nonhistone protein scaffolds are found only in eukaryotic chromosomes, and have structural (higher-order packaging) and possibly other functions.

g. DNA is found in both prokaryotic and eukaryotic chromosomes. (Some viral chromosomes have RNA as their genetic material.)

h. Nucleosomes are the fundamental unit of packaging of DNA in eukaryotic chromosomes, and are not found in prokaryotic chromosomes.

i. Circular chromosomes are found only in prokaryotes, not in eukaryotes.

j. Looping is found in both eukaryotic and prokaryotic chromosomes. In eukaryotic chromosomes, the 30-nm nucleofilament is packed into looped domains by nonhistone chromosomal proteins. In bacterial chromosomes such as that of *E. coli*, there are about 100 looped domains, each containing about 40 kilobases of supercoiled DNA.

10.30 Discuss the structure of nucleosomes, their hierarchical packaging, and the composition of a core particle of a nucleosome. Also discuss the functions of nucleosomes.

Answer: Nucleosomes are the fundamental unit of DNA packaging in eukaryotic chromosomes. In a nucleosome core particle, a short segment of DNA is wrapped about one-and-three-quarters times around a protein core. This protein core is an octamer consisting of two copies of each of four histone proteins: H2A, H2B, H3, H4. The nucleosome core particle packs the DNA into a flattened disk about 5.7×11 nm. Packaging into nucleosomes effectively condenses the DNA by a factor of about seven. The interactions between the histones and the DNA are not sequence specific, but rather are based on the basic, positively charged histone proteins interacting with the acidic, negatively charged sugar-phosphate backbone of the DNA.

Nucleosome core particles serve to package double-stranded DNA at regular intervals, with short stretches of linker DNA between particles. At this level of packaging, DNA is packed into a "beads-on-a-string" form of chromatin known as the 10-nm nucleofilament. This condensation of DNA also forms the basis for subsequent packaging, first through associations between nucleosomes and then by associations with nonhistone chromosomal proteins. Associations between nucleosomes result in further compaction of the 10-nm nucleofilament, producing a 30-nm chromatin fiber that condenses the DNA another sixfold. Histone H1 plays an important role in this coiling. The 30-nm chromatin fiber is further condensed into looped domains, with each loop containing between ten and hundreds of kilobases of DNA. The looped domains are anchored inside the nuclear envelope to a filamentous structural framework of protein, the nuclear matrix.

10.31   Set up the following "rope trick": Start with a belt (representing a DNA molecule; imagine the phosphodiester backbones lying along the top and bottom edges of the belt) and a soda can. Holding the belt buckle at the bottom of the can, wrap the belt flat against the side of the can, counterclockwise three times around the can. Now remove the "core" soda can and, holding the ends of the belt, pull the ends of the belt taut. After some reflection, answer the following questions:

a. Did you make a left- or a right-handed helix?

b. How many helical turns were present in the coiled belt before it was pulled taut?

c. How many helical turns were present in the coiled belt after it was pulled taut?

d. Why does the belt appear more twisted when pulled taut?

e. About what percentage of the length of the belt was decreased by this packaging?

f. Is the DNA of a linear chromosome that is coiled around histones supercoiled?

g. Why are topoisomerases necessary to package linear chromosomes?

Answer:

a. The belt forms a right-handed helix. Although you wrapped the belt around the can axis in a counterclockwise direction from your orientation (looking down at the can), the belt was winding up and around the side of the can in a clockwise direction from its orientation. With the belt wrapped around the can, curve the fingers of your right hand over it, and use your index finger to trace the direction of the belt's spiral. Your right index finger will trace the spiral upward, the same direction your thumb points when you wrap your hand around the can. Therefore the belt has formed a right-handed helix. If you do this with your left hand, your thumb will point away from the direction the helix twists, indicating that it has not formed a left-handed helix.

b. Three.

c. Three. The number of helical turns is unchanged, although the twist in the belt is changed.

d. The belt appears more twisted because the pitch of the helix was altered, and the edges of the belt (positioned much like the complementary base pairs of a double helix) are twisted more tightly.

e. While twisted around the can, the length of the belt decreases by about 70 to 80 percent, depending on the initial length of the belt and the belt diameter.

f. Yes. The DNA of linear chromosomes becomes supercoiled as it is wrapped around histones to form the 10-nm nucleofilament. Just as you must add twists to the belt in order for it to lie flat on the surface of the can, supercoils must be introduced into the DNA for it to wrap around the histones.

g. Topoisomerases increase or reduce the level of supercoiling in DNA. In order for linear DNA to be packaged, negative supercoils must be added to it.

10.32   What are the main molecular features of yeast centromeres?

Answer: The centromeres of *S. cerevisiae* have similar, though not identical, DNA sequences, called *CEN* sequences. Each contains three sequence domains, called centromere DNA elements (CDEs). CDEII, a 76–86-bp region that is >90% AT, is flanked by CDEI, a conserved RTCACRTG sequence (R = A or G), and CDEIII, a 26-bp AT-rich conserved domain. The CDEs interact with kinetochore proteins that mediate the attachment of centromeres to spindle fiber microtubules. In one model, CDEI binds centromere-binding factor I (CBFI), CDEII is wrapped once around a histone octamer, and CDEIII binds the protein complex CBF3. In turn, a linker protein may attach CBF3 to the spindle microtubule. Some yeast centromeres are quite different. For example, those of the fission yeast *S. pombe* are 40 to 80 kb long. They contain complex rearrangements of repeated sequences. Nonetheless, all centromeres provide the same function: They ensure accurate chromosome segregation during mitosis and meiosis.

10.33   Telomeres are unique repeated sequences. Where on the DNA strand are they found? Do they serve a function?

Answer: Telomeres are characteristically heterochromatic sequences found at the ends of a linear, eukaryotic chromosome. For most organisms, the telomeric sequences at the extreme end of a chromosome are simple, highly repeated, and species specific. Nearby, but not at the very end of a chromosome, are telomere-associated sequences. These are repeated, complex sequences and extend for thousands of bases from the simple telomeric sequences. These sequences are replicated, along with other chromosomal DNA sequences, by a set of enzymes that include DNA polymerases. Telomere organization is quite different in some organisms. In organisms such as *Drosophila*, telomeres consist of transposons belonging to the LINE family of repeated sequences. Telomeres function in DNA replication and provide chromosome stability.

10.34   Would you expect to find most protein-coding genes in unique-sequence DNA, in moderately repetitive DNA, or in highly repetitive DNA?

Answer: in unique-sequence DNA

10.35   Both histone and nonhistone proteins are essential for DNA packaging in eukaryotic cells. However, these classes of proteins are fundamentally dissimilar in a number of ways. Describe how they differ in terms of

   a. their protein characteristics
   b. their presence and abundance in cells
   c. their interactions with DNA
   d. their role in DNA packaging

Answer:
   a. The five histones (H1, H2A, H2B, H3, and H4) are small, basic, positively charged proteins that are rich in arginine and lysine. Histones H2A, H2B, H3, and H4 are among the most highly conserved of all proteins. In contrast to the small number of histones, there are many types of nonhistones. They are variable in size and are usually negatively charged, acidic proteins.
   b. The five histones are present in all cells of all eukaryotic organisms. Relative to the total amount of DNA, they are present in fairly constant amounts. In contrast, nonhistones differ in number and type between cell types in an organism, at different times in the same cell type, and from organism to organism. As a class, nonhistones may be between 50 and 100 percent of the mass of DNA.
   c. Histones are positively charged and mostly interact with the negatively charged sugar-phosphate backbone of DNA. Nonhistones are typically negatively charged

and can interact with the positively charged histones or with DNA. The HMG class of nonhistones binds to the minor groove of the DNA and causes DNA bending.

d. The histones H2A, H2B, H3, and H4 are fundamental to packaging DNA into nucleosome core particles and the formation of the 10-nm nucleofilament. Histone H1 is involved in compaction of this nucleofilament to produce a 30-nm chromatin fiber. Nonhistones are important in higher-order packaging of DNA. They form the protein scaffold that anchors the looped domains of the 30-nm chromatin fiber and they control the degree of chromosome condensation.

10.36   Rearrangements at the end of 16p (the short arm of chromosome 16) underlie a variety of common human genetic disorders, including β-thalassemia (a defect in hemoglobin metabolism caused by mutations in the β-globin gene), mental retardation, and the adult form of polycystic kidney disease. Recently, the determination of approximately 285-kilobase (kb) pairs of DNA sequence at the end of human chromosome 16p has allowed very detailed analysis of the structure of this chromosome region. The first functional gene lies about 44 kb from the region of simple telomeric sequences and about 8 kb from the telomere-associated sequences. Analysis of sequences proximal (nearer the centromere) to the first gene reveals a sinusoidal variation in GC content, with GC-rich regions associated with gene-rich areas and AT-rich regions associated with *Alu*-dense areas. The β-globin gene lies about 130 kb from the telomere-associated sequences.

a. Discuss these findings in light of the current view of telomere structure and function as presented in the text.

b. What new information have the preceding data revealed about the distribution of SINEs in the terminus of 16p? (SINEs and LINEs are, respectively, short and long interspersed nuclear elements.)

Answer:

a. These findings support the view that telomeres are specialized chromosome structures with two distinct structural components, simple telomeric sequences and telomere-associated sequences. They show that functional genes do not reside in the telomeric region, consistent with the view that telomeres are heterochromatic and function in DNA replication and stability. They document the structure of telomeric and near-telomeric regions. For example, they show the considerable distance over which the telomere-associated sequences are found, about 36 kb, and give a sense of the number, size, and density of genes in the region near this telomere.

b. At least in this region, *Alu* sequences are found more often in AT-rich areas. These areas are not as gene rich as adjacent GC-rich areas. Thus, this class of moderately repetitive sequences, as well as the genes in this area, appear to have a nonrandom distribution.

11

DNA REPLICATION

CHAPTER OUTLINE

THINKING ANALYTICALLY

Because of its fundamental importance in genetics and in treating diseases such as cancer, there has been a considerable interest in understanding the mechanism of DNA replication. At this point, we understand the mechanism of DNA replication in considerable detail. For example, we know that DNA replicates semiconservatively, and we know the nature of the complex set of proteins and enzymes that act in a sequential, coordinated fashion to replicate DNA. The best way to understand and retain the details of this complex process is to place them in context. Start by developing a general understanding of the steps of DNA replication. Ask yourself: What must happen first? How is this accomplished? What happens next? How is that accomplished? Then focus on understanding the activities of each of the enzymes and proteins used in each step. To help develop a contextual understanding and establish a mental image of each of the steps, refer to the text figures that summarize the key events of DNA replication.

The following related exercise will help you assess your understanding and retention of the material. After reading each section, close the text and sketch out the aspect of DNA replication that has just been presented. Then relate the figure you have sketched to the previous steps of DNA replication or DNA recombination. This will force you to confront unclear concepts and help you identify details you might have missed.

REVIEW OF KEY TERMS, SYMBOLS, AND CONCEPTS

In your own words, write a brief, precise definition of each term in the groups below. Check your definitions using the text. Then develop a concept map using the terms in each list.

1	2	3
semiconservative DNA replication	semidiscontinuous DNA replication	semidiscontinuous DNA replication
conservative DNA replication	origin of replication	supercoiling
dispersive DNA replication	replication bubble	DNA gyrase
Meselson-Stahl experiment	initiator protein	rolling circle model
CsCl equilibrium gradient	DNA helicase, primase	λ phage
^{15}N, ^{14}N	primosome	temperate phage
buoyant density	replisome	lytic pathway
harlequin chromosomes	RNA primer	lysogenic pathway
BUdR	template	"sticky" ends
base analog	SSB protein	cos site
DNA polymerase I, II, III	replication fork	concatamer
dNTP, dNMP, Mg^{2+}	Okazaki fragments	ter
Kornberg enzyme	bidirectional replication	leading strand
template strand	DNA polymerase I, III	lagging strand
non-template strand	$5' \rightarrow 3'$ synthesis	
primer	proofreading	
	$3' \rightarrow 5'$, $5' \rightarrow 3'$ exonuclease	
	leading, lagging strand	
	DNA ligase	
	polA1, polA1ex1 mutants	
	temperature-sensitive mutant	

4	5
eukaryote	DNA recombination
cell cycle	Holliday model
cyclins	Holliday intermediate

4	5
Cdk	branch migration
origin recognition complex	patched duplex
replicator selection	spliced duplex
prereplicative complexes	gene conversion
DNA polymerase α, δ, ϵ	mismatch repair
DNA replication	1:3, 3:1 segregation
DNA repair	
proofreading ability	
replicon	
ARS	
telomere	
telomerase	
reverse transcription	
reverse transcriptase, *TERT*	
TLC1, EST1, TEL1, TEL2 genes	
H3-H4 tetramer	
H2A-H2B dimer	
histone chaperone	

QUESTIONS FOR PRACTICE

Multiple Choice Questions

1. Consider the Meselson-Stahl experiment where *E. coli* were grown in ^{15}N medium. Which model of DNA replication is eliminated by analyzing DNA after exactly one round of replication in ^{14}N medium?
   a. semiconservative
   b. dispersive
   c. conservative

2. The enzymes most directly concerned with catalyzing DNA synthesis are DNA
   a. ligases
   b. exonucleases
   c. polymerases
   d. primases

3. Which of the following is not essential for the in vitro synthesis of DNA?
   a. magnesium ions
   b. DNA polymerase I
   c. DNA primase
   d. a DNA fragment

4. Which enzyme untwists the parent strands of DNA during replication?
   a. DNA primase
   b. DNA helicase
   c. DNA gyrase
   d. DNA ligase

5. Which of the following do not provide evidence for semiconservative DNA replication?
   a. the Meselson-Stahl experiment
   b. harlequin chromosomes
   c. the existence of ARS elements
   d. Okazaki fragments

6. Which of the following *E. coli* enzymes have proofreading activity?
   a. DNA polymerase I
   b. DNA ligase
   c. DNA primase
   d. DNA polymerase III

7. DNA replication in certain viruses, such as the circular φX174, is achieved
   a. using a rolling circle model and semidiscontinuous DNA replication
   b. using a rolling circle model and continuous DNA replication
   c. without using an origin of replication
   d. using DNA fragmentation

8. During chromosome replication, new histones are translated during
   a. G_1 and S, and nucleosomes are re-formed using new and old histones
   b. G_1 and S, and new nucleosomes are formed with only these new histones
   c. S, and nucleosomes are re-formed using new and old histones
   d. S, and new nucleosomes are formed with only these new histones

9. Which of the following molecular events does not occur during reciprocal genetic recombination?
   a. A heteroduplex region forms between two DNA strands of nonsister chromatids.
   b. Branch migration occurs as a Holliday intermediate rotates.
   c. Endonuclease cleavage of the Holliday intermediate produces a patched duplex and a parental chromosome.
   d. There are two double-stranded breaks in nonsister chromatids that are rejoined to produce recombinant chromosomes.
   e. Mismatch repair of sites in heteroduplex DNA results in conversion of one allele to another.

10. What is the function of the enzyme telomerase in mammals?
    a. to replicate telomeres in all cell types
    b. to replicate and expand simple telomeric repeats in germ cells and tumor cells
    c. to expand the length of simple telomeric repeats in brain cells
    d. to replicate telomere-associated sequences and simple telomeric repeats in all cells

Answers: 1c, 2c, 3c, 4b, 5c, 6b,c, 7a, 8c, 9d, 10b

Thought Questions

1. Describe the mechanism of semidiscontinuous DNA replication.

2. In his investigation of the requirements for in vitro synthesis of DNA, Kornberg found that an absolute minimum of four components was necessary. What were these components, and why in particular were each and all necessary? Also, what limitations did his in vitro method have as compared to in vivo synthesis, which involves other components, such as helicase, primase, and ligase?

3. Describe how eukaryotic DNA is assembled into nucleosomes after DNA replication. Where do the new nucleosomes come from?

4. What phenotype would you expect to see in a mutant in the *E. coli polA* gene that showed heat-sensitive 3′→5′ exonuclease activity, but normal 5′→3′ exonuclease activity at all temperatures?

5. In humans, how are simple telomeric repeats replicated in a fast-dividing population of cells such as epithelial cells? Are all of them replicated at each cell division?

6. The text describes evidence that telomere length is under genetic control and that in mammals, only immortal cells (e.g., tumor cells, germ cells) have telomerase activity. Why might this be important? In particular, why might it be important for some cells *not* to have telomerase activity?

7. Explain how the Holliday model for reciprocal genetic recombination can lead to both parental and recombinant products.

Thought Questions for Media

After reviewing the media on the *iGenetics* CD-ROM, try to answer these questions.

1. In the animation depicting the experiment by Meselson and Stahl that showed that DNA replication was semiconservative, how did they initially label DNA with ^{15}N?

2. In the animation depicting DNA synthesis using DNA polymerase, when proofreading occurs, what type of molecule is released?

3. In the animation depicting the semidiscontinuous model of DNA replication, in which direction is DNA synthesis on the lagging strand and which direction is this relative to the movement of the replication fork?

1. How can you experimentally determine if a replication fork proceeds bidirectionally or unidirectionally?

2. How can you experimentally determine whether DNA primase randomly starts synthesizing an RNA primer or whether it primes DNA synthesis only at specific sites?

3. In the text, the function of *E. coli* DNA polymerase II was not described. Based on the iActivity, what genetic analysis could you perform to demonstrate that it is involved in DNA repair?

SOLUTIONS TO TEXT PROBLEMS

11.1   Describe the Meselson-Stahl experiment, and explain how it showed that DNA replication is semiconservative.

Answer: Bacterial DNA was uniformly labeled with heavy nitrogen by growing *E. coli* for many generations in media containing "heavy" nitrogen, ^{15}N. Such ^{15}N DNA has a greater buoyant density than ^{14}N DNA and can be differentiated from ^{14}N DNA by its banding position in a CsCl density gradient. Semiconservative DNA replication was demonstrated by placing *E. coli* with ^{15}N DNA into ^{14}N media, and following the density of the DNA that was present in cells after growth proceeded through each of several successive generations. If DNA replication were conservative, after one cell division and one round of DNA replication one would expect to see two distinct, equally dense bands, one corresponding to ^{14}N DNA and one corresponding to ^{15}N DNA. This was not seen. If DNA replication were either semiconservative or dispersive, after one cell division and one round of DNA replication one would expect to see one band consisting of DNA with a density halfway between ^{14}N DNA and ^{15}N DNA. This was seen. To further distinguish between semiconservative and dispersive replication, the consequences of another round of cell division and DNA replication were examined. In semiconservative replication, one would expect two bands, one of density halfway between ^{14}N DNA and ^{15}N DNA, and one having ^{14}N DNA density. In dispersive replication, only DNA with both ^{14}N DNA and ^{15}N DNA would be seen. The density of the two bands seen (one at the level of ^{14}N DNA and one at the level of half ^{14}N DNA and half ^{15}N DNA) supported the semiconservative model. See text Figure 11.2.

11.2   In the Meselson-Stahl experiment, ^{15}N-labeled cells were shifted to ^{14}N medium at what we can designate as generation 0.
   a. For the semiconservative model of replication, what proportion of ^{15}N–^{15}N, ^{15}N–^{14}N, and ^{14}N–^{14}N would you expect to find after one, two, three, four, six, and eight replication cycles?
   b. Answer (a) in terms of the conservative model of DNA replication.

Answer: Key: ^{15}N–^{15}N DNA = HH; ^{15}N–^{14}N DNA = HL; ^{14}N–^{14}N DNA = LL.
   a. Generation 1: all HL; 2: ½ HL, ½ LL; 3: ¼ HL, ¾ LL; 4: ⅛ HL, ⅞ LL; 6: 1/32 HL, 31/32 LL; 8: 1/128 HL, 127/128 LL.
   b. Generation 1: ½ HH, ½ LL; 2: ¼ HH, ¾ LL; 3: ⅛ HH, ⅞ LL; 4: 1/16 HH, 15/16 LL; 6: 1/64 HH, 63/64 LL; 8: 1/256 HH, 255/256 LL.

11.3   A spaceship lands on Earth and with it a sample of extraterrestrial bacteria. You are assigned the task of determining the mechanism of DNA replication in this organism.
         You grow the bacteria in unlabeled medium for several generations, and then grow it in the presence of ^{15}N for exactly one generation. You extract the DNA and subject it to CsCl centrifugation. The banding pattern you find is as follows:

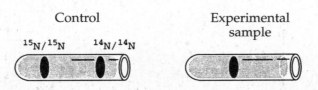

Control   Experimental
                                   sample
$^{15}N/^{15}N$ $^{14}N/^{14}N$

It appears to you that this is evidence that DNA replicates in the semiconservative manner, but you are wrong. Why? What other experiment could you perform (using the same sample and technique of CsCl centrifugation) that would further distinguish between semiconservative and dispersive modes of replication?

Answer: The CsCl centrifugation result eliminates the possibility of the conservative model of replication, but is still consistent with either semiconservative or dispersive models of DNA replication. To distinguish between these two possibilities using the same sample and the technique of CsCl centrifugation, one could denature the DNA and then subject the single-stranded sample to CsCl centrifugation. This could be done in practice by using an alkaline CsCl gradient, as the two DNA strands will denature at high pH. The expected results are shown below.

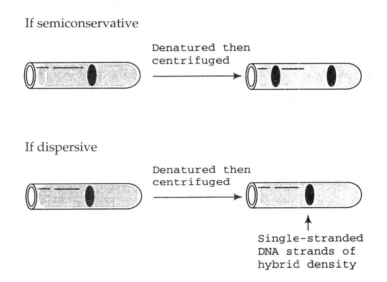

11.4   The elegant Meselson-Stahl experiment was among the first experiments to contribute to what is now a highly detailed understanding of DNA replication. Consider this experiment again in light of current molecular models by answering the following questions:

a. Does the fact that DNA replication is semiconservative mean that it must be semidiscontinuous?

b. Does the fact that DNA replication is semidiscontinuous ensure that it is also semiconservative?

c. Do any properties of known DNA polymerases ensure that DNA is synthesized semiconservatively?

Answer:

a. That DNA replication is semiconservative does not mean that it also must be semidiscontinuous. For example, if DNA polymerase were able to synthesize DNA in both the $3' \rightarrow 5'$ and $5' \rightarrow 3'$ directions, DNA replication could proceed continuously on both DNA strands while still being semiconservative. Alternatively, if each of the two old strands were completely (or even substantially) unwound, and replication were initiated from the 3′ end (or 3′ region) of each, it could proceed continuously (or mostly continuously) in a $5' \rightarrow 3'$ direction along each strand.

b. That DNA replication is semidiscontinuous does ensure that it is semiconservative. In the semidiscontinuous model, each old, separated strand serves as a template for a new strand. This is the essence of the semiconservative model.

c. That DNA polymerase synthesizes just one new strand from each "old" single-stranded template and can synthesize in only one direction (5'→3') ensures that replication is semiconservative.

11.5 List the components necessary to make DNA in vitro, using the enzyme system isolated by Kornberg.

Answer: DNA can be synthesized in vitro using the Kornberg enzyme (DNA polymerase I), all four dNTPs (dATP, dGTP, dCTP, dTTP), magnesium ions (Mg^{2+}), and a fragment of DNA that will serve as a template.

11.6 How do we know that the Kornberg enzyme is not the main enzyme involved in DNA synthesis for chromosome duplication in the growth of *E. coli?*

Answer: The primary evidence that the Kornberg enzyme is not the main enzyme for DNA synthesis in vivo stems from an analysis of the growth and biochemical phenotypes of two mutants affecting DNA polymerase I: *polA1* and *polAex1*. The mutant *polA1* lacks 99 percent of polymerase activity, but is nonetheless able to grow, replicate its DNA, and divide. The conditional mutant *polAex1* retains most of the polymerizing activity at the restrictive temperature of 42°C, but is still unable to replicate its chromosomes and divide (it has lost the enzyme's 5'→3' exonuclease activity). In this way, the analysis of the *polA1* and *polAex1* mutants indicated that there must be other DNA-polymerizing enzymes in the cell.

11.7 Kornberg isolated DNA polymerase I from *E. coli*. DNA polymerase I has an essential function in DNA replication. What are the functions of the enzyme in DNA replication?

Answer: The 5'→3' exonuclease activity of DNA polymerase I is essential for DNA replication as a lack of this activity in the conditional mutant *polAex1* (at the restrictive temperature of 42°C) results in the failure of DNA replication. DNA polymerase I functions to remove the RNA primer synthesized during the initiation of DNA replication and replace this RNA with DNA. When DNA polymerase III, the main synthetic enzyme for DNA polymerization, reaches an RNA primer, it dissociates from the DNA. DNA polymerase I functions to continue synthesis of the DNA in a 5'→3' direction. It simultaneously removes the RNA primer using its 5'→3' exonuclease activity and replaces the RNA with DNA nucleotides.

11.8 Suppose you have a DNA molecule with the base sequence T A T C A going from the 5' to the 3' end of one of the polynucleotide chains. The building blocks of the DNA are drawn as in the following figure.

Use this shorthand system to diagram the completed double-stranded DNA molecule, as proposed by Watson and Crick.

Answer: The deoxyribonucleotides have a triphosphate group at the 5' end of the molecule and a hydroxyl group at the 3' end of the molecule. This information establishes the polarity of each of the deoxyribonucleotides (dATP, dGTP, dTTP, and dCTP). Since the phosphodiester bond will form between the 5' and 3' ends of adjacent nucleotides, the 5'→3' bonding of the T A T C A oligonucleotide can be represented as:

The complementary strand will base-pair with antiparallel orientation:

11.9   Base analogs are compounds that resemble the natural bases found in DNA and RNA, but are not normally found in those macromolecules. Base analogs can replace their normal counterparts in DNA during in vitro DNA synthesis. Researchers studied four base analogs for their effects on in vitro DNA synthesis using *E. coli* DNA polymerase. The results were as follows, with the amounts of DNA synthesized expressed as percentages of the DNA synthesized from normal bases only.

	Normal Bases Substituted by the Analog			
Analog	A	T	C	G
A	0	0	0	25
B	0	54	0	0
C	0	0	100	0
D	0	97	0	0

Which bases are analogs of adenine? Of thymine? Of cytosine? Of guanine?

Answer: None are analogs of adenine, B and D are analogs of thymine, C is an analog of cytosine, and A is an analog of guanine.

11.10 Concerning DNA replication:

a. Describe (draw) models of continuous, semidiscontinuous, and discontinuous DNA replication.

b. What was the contribution of Reiji and Tuneko Okazaki and colleagues with regard to these replication models?

Answer: Since DNA replication is known to be semidiscontinuous, it is helpful to first consider the details of this model. As the other models are constructed, reference to the semidiscontinuous model can identify the problems associated with them.

a. A *semidiscontinuous model* is shown in text Figures 11.5 and 11.6. Its main features are that:

i.   DNA replication proceeds only in the 5'→3' direction.

ii. Because DNA replication proceeds only in one direction, it is continuous on the 3'→5' template strand (the leading strand), while it is discontinuous on the 5'→3' template strand (the lagging strand). This is why it is *semi*discontinuous.

iii. Replication initiates at a replication origin, where a complex of proteins and enzymes forms at a replication fork. Initially, helicase untwists the DNA and SSBs (single-strand binding proteins) bind the single-stranded DNA to protect it from nuclease attack and keep it unwound.

iv. Since DNA polymerase can only polymerize DNA if a 3'-OH is provided, an RNA primer must be synthesized by a primase-enzyme complex (the primosome). The primosome binds to the single-stranded DNA and synthesizes a short RNA primer in the 5'-to-3' direction.

v.   Once an RNA primer is synthesized, DNA polymerase III (a complex of proteins) adds nucleotides onto the 3'-OH in the 5'→3' direction, using the unwound DNA as a template. The newly synthesized DNA is thus synthesized onto an RNA primer. This is referred to as an Okazaki fragment. On the lagging strand, synthesis moves "away" from the fork, and synthesis must be reinitiated frequently. On the leading strand, synthesis proceeds "into" the fork. The proteins involved in replication are associated into a replisome, so that the DNA polymerase III of the lagging strand is complexed with DNA polymerase III on the leading strand. The replisome that forms at a replication fork moves as a unit along the DNA, enabling new DNA to be synthesized efficiently on both the leading and lagging strands.

vi. On the lagging strand, DNA polymerase III will (after about 1,000 to 2,000 bp) encounter the 5' end of an RNA primer of the previously synthesized Okazaki fragment. At this point, it dissociates and is replaced by DNA polymerase I, which uses its 5'→3' exonuclease activity to excise the RNA primer and replace it with DNA.

vii. Once the RNA primer is replaced with DNA, the gap between adjacent DNA fragments is sealed by DNA ligase.

A *continuous model* would need to consider what we know about the semiconservative nature of DNA replication, as well as the enzymes known to be involved in DNA replication. In particular, the model would have to

• produce new double helices that each have one old and one new strand
• use DNA polymerases that synthesize DNA only in the 5'→3' direction
• use DNA polymerases requiring a 3'→5' template
• use DNA polymerases that need a primer to initiate replication

One problem to be solved by any model of continuous-only DNA replication is to ensure that the DNA is completely replicated. For example, if DNA replication starts in the middle of a linear chromosome, and proceeds continuously in only a $5'{\rightarrow}3'$ direction using a $3'{\rightarrow}5'$ template strand, half of that template strand will remain unreplicated. This problem might be overcome in different ways, depending on whether the chromosome is circular or linear. Each solution, however, presents additional problems.

Suppose DNA replication occurs in a circular chromosome. As in the semidiscontinuous model, a protein or set of proteins would recognize an origin-of-replication site within the DNA molecule. To create a site for initiation of DNA synthesis, an endonuclease would nick one strand. Helicase would then bind the free $5'$ end and start unwinding it. DNA primase would not be needed, as the free $3'$-OH would serve as the end of a "primer." The complementary ($3'{\rightarrow}5'$) strand would be used as a template for new DNA synthesis. A protein such as SSB would bind to the unwound, nontemplate strand (the strand with an exposed $5'$ end) to keep it separate from the template strand. In some unknown manner, this strand would need to be kept separate and perhaps unwound throughout the replication of the template strand. This raises a first problem, as this might be a daunting task given the size of a chromosome compared to the size of a cell. Since DNA replication is semiconservative, this strand (which lacks a priming site) would also need to be replicated. The mechanism by which this would occur is unknown and raises a second problem. In a replication fork proceeding down the length of the template strand, DNA polymerase III would add nucleotides based on the sequence of the $3'{\rightarrow}5'$ template strand, continuously until the site of the origin was reached. Some mechanism would be needed to mark this site so that DNA polymerase III would stop synthesis. A ligation reaction would be needed to join the two ends of the newly synthesized DNA. There are multiple approaches to solving the two problems posed by this model, but in reality, neither problem occurs due to the semidiscontinuous nature of DNA replication.

For linear eukaryotic chromosomes, it would be difficult to initiate DNA synthesis at the very ends of chromosomes. But this would not always be required, as in germline cells of eukaryotes, the simple telomeric repeats at the very ends of chromosomes are replicated by telomerase. However, in somatic cells, a distinct mechanism would be needed to identify origin-of-replication sites very near to, but not at, the ends of the chromosomes. After an endonuclease made a nick near the end of a linear chromosome, a DNA polymerase III–like enzyme could add nucleotides $5'{\rightarrow}3'$ based on the sequence of the $3'{\rightarrow}5'$ template strand. A replication fork would proceed down the length of the template strand, replicating just one of the strands. Presumably, the other strand would be bound by an SSB-like protein as it was unwound. (Recall the first problem above.) At the other end of a linear chromosome, a similar process would occur.

In both of these models of continuous DNA replication, no DNA primase was required, and no RNA primer was used. Certainly, other models of continuous DNA replication might be devised that use DNA primase and RNA primers. However, there is then the question of how to remove the primers, and so a DNA polymerase I–like enzyme must be used. One significant problem with these models is that they require one of the two strands of the parental DNA molecule to remain separate from the template strand until it is replicated. Given the sizes of chromosomes in prokaryotes and eukaryotes relative to their cell sizes, this is a considerable problem. It is a problem that is solved by having a replication fork synthesize DNA semidiscontinuously, so that there are only very short stretches of unwound, SSB-bound DNA.

A *discontinuous model* of DNA replication would be nearly identical to the semidiscontinous model, except that DNA replication would be discontinuous not only on the lagging strand, but also on the leading strand. Presumably, DNA polymerase III would "fall off" the leading-strand template periodically and then come back "on board" to continue synthesizing DNA. No new priming would be required, as there would be a free 3'-OH available for DNA polymerase III onto which to add deoxyribonucleotides.

b. Reiji Okazaki provided the primary evidence that DNA synthesis must be discontinuous on at least one strand. He demonstrated the existence of Okazaki fragments, short segments of DNA that have short RNA primers at their 5' ends. Their existence indicated that DNA synthesis must be periodically re-primed on the lagging strand and therefore that it must be discontinuous on that strand. (See also the answer to problem 11.16.)

11.11   The following events, steps, or reactions occur during *E. coli* DNA replication:

Column A	Column B
____ a. Unwinds the double helix	A. Polymerase I
____ b. Prevents reassociation of complementary bases	B. Polymerase III
____ c. Is an RNA polymerase	C. Helicase
____ d. Is a DNA polymerase	D. Primase
____ e. Is the "repair" enzyme	E. Ligase
____ f. Is the major elongation enzyme	F. SSB protein
____ g. Is a 5'→3' polymerase	G. Gyrase
____ h. Is a 3'→5' polymerase	H. None of these
____ i. Has 5'→3' exonuclease function	
____ j. Has 3'→5' exonuclease function	
____ k. Bonds free 3'-OH end of a polynucleotide to a free 5'monophosphate end of a polynucleotide	
____ l. Bonds 3'-OH end of a polynucleotide to a free 5' nucleotide triphosphate	
____ m. Separates daughter molecules and causes supercoiling	

For each entry in column A, select the appropriate entry in column B. Each entry in A may have more than one answer, and each entry in B can be used more than once.

Answer: a. C; b. F; c. D; d. A, B; e. A; f. B; g. A, B, D; h. H; i. A; j. A, B; k. E; l. A, B, (D, after the first two bases of the RNA primer are positioned); m. G

11.12   How long would it take *E. coli* to replicate its entire genome (4.2×10^6 bp), assuming a replication rate of 1,000 nucleotides per second at each fork with no pauses?

Answer: In *E. coli,* replication proceeds bidirectionally from the replication origin (*oriC*). Thus, each replication fork proceeds halfway down the length of the circular *E. coli* chromosome, at which point the replication forks meet and replication is complete. Thus, it will take

$$\frac{4.2 \times 10^6 \text{ bp}}{2 \text{ replication forks}} \times \frac{1 \text{ second}}{1,000 \text{ bp}} \times \frac{1 \text{ minute}}{60 \text{ seconds}} = 35 \text{ minutes}$$

11.13 A diploid organism has 4.5×10^8 bp in its DNA. This DNA is replicated in 3 minutes. Assuming all replication forks move at a rate of 10^4 bp per minute, how many replicons (replication units) are present in this organism's genome?

Answer: Since a replication fork moves at a rate of 10^4 bp per minute and each replicon has two replication forks moving in opposite directions, in one replicon, replication would be occurring at a rate of 2×10^4 bp/minute. Since all of the organism's DNA is replicated in 3 minutes, the number of replicons in the diploid genome is

$$\frac{4.5 \times 10^8 \text{ bp}}{3 \text{ minutes}} \times \frac{1 \text{ replicon}}{2 \times 10^4 \text{ bp/minute}} = 7,500$$

11.14 Describe the molecular action of the enzyme DNA ligase. What properties would you expect an *E. coli* cell to have if it had a temperature-sensitive mutation in the gene for DNA ligase?

Answer: DNA ligase catalyzes the formation of a phosphodiester bond between the 3'-OH and the 5' monophosphate groups on either side of a single-strand DNA gap, sealing the gap. See Figure 11.7. Temperature-sensitive ligase mutants would be unable to seal such gaps at the restrictive (high) temperature, leading to fragmented lagging strands and presumably cell death. If a biochemical analysis were performed on DNA synthesized after *E. coli* were shifted to a restrictive temperature, there would be an accumulation of DNA fragments the size of Okazaki fragments. This would provide additional evidence that DNA replication must be discontinuous on one strand.

11.15 Chromosome replication in *E. coli* commences from a constant point, called the origin of replication. It is known that DNA replication is bidirectional. Devise a biochemical experiment to prove that the *E. coli* chromosome replicates bidirectionally. (Hint: Assume that the amount of gene product is directly proportional to the number of genes.)

Answer: Assume the amount of a gene's product is directly proportional to the number of copies of the gene present in the *E. coli* cell. Assay the enzymatic activity of genes at various positions in the *E. coli* chromosome during the replication period. Then some genes (those immediately adjacent to the origin) will double their activity very shortly after replication begins. Relate the map position of genes having doubled activity to the amount of time that has transpired since replication was initiated. If replication is bidirectional, there should be a doubling of the gene products both clockwise and counterclockwise from the origin.

11.16 Reiji Okazaki concluded that both strands could not replicate continuously. What evidence led him to this conclusion?

Answer: Reiji Okazaki and his colleagues added a pulse of radioactive DNA precursor, ^{3}H-thymidine, to cultures of *E. coli* for a small fraction of its generation time (0.5 percent). To prevent the further incorporation of additional radioactive precursor into the DNA after this period, they chased this pulse with a large amount of nonradioactive thymidine. They then sampled the bacterial DNA at successive intervals as DNA replication continued. When DNA was sampled soon after the labeling period, most of the radioactivity was found in DNA with very low molecular weights, in DNA fragments between 100 and 1,000 nucleotides long. When DNA was sampled at longer intervals, the radioactivity was found in DNA with much higher molecular weight, in much longer DNA fragments. The short DNA fragments seen in the early samplings indicate that DNA replication normally involves the synthesis of short DNA segments—the Okazaki fragments—that are then linked together. This provided evidence that DNA replication occurs in a discontinuous fashion on the lagging strand.

11.17 A space probe returns from Jupiter and brings with it a new microorganism for study. It has double-stranded DNA as its genetic material. However, studies of replication of the alien DNA reveal that, although the process is semiconservative, DNA synthesis is continuous on both the leading-strand and the lagging-strand templates. What conclusions can you draw from that result?

Answer: Clearly, DNA replication in the Jovian bug does not occur as it does in *E. coli*. Assuming that the double-stranded DNA is antiparallel as it is in *E. coli*, the Jovian DNA polymerases must be able to synthesize DNA in the $5' \rightarrow 3'$ direction (on the leading strand) as well as in the $3' \rightarrow 5'$ direction (on the lagging strand). This is unlike any DNA polymerase on Earth.

11.18 Phage such as λ and T4 are packaged from concatamers.
   a. What are concatamers, and what type of DNA replication is responsible for producing concatamers?
   b. In what ways does this type of DNA replication differ from that used by *E. coli*?

Answer:
   a. Concatamers are multiple copies of a chromosome linked end to end, produced by rolling circle replication.
   b. Rolling circle replication initiates differently from the bidirectional replication used by *E. coli*. In rolling circle replication, a nick is made in one strand of a double-stranded circle, and the 5' end is displaced. This creates a replication fork, leaving a single-stranded stretch of DNA that serves as a template for the addition of dNTPs to the free 3' end by DNA polymerase III. Unlike initiation in *E. coli*, initiator protein is not used. Once a replication fork and a replisome is formed, the intact circular DNA acts as a leading-strand template. New DNA synthesis occurs continuously as the 5' cut end is displaced from the circular molecule. The displaced strand serves as the lagging-strand template on which DNA synthesis occurs discontinuously. As leading-strand synthesis using the parental DNA circle continues, a concatamer is formed. The concatamer is cleaved into unit-length chromosomes prior to packaging in phage particles. This is also unlike DNA replication in *E. coli*, where no concatamer is formed.

11.19   Although λ is replicated into a concatamer, linear unit-length molecules are packaged into phage heads.

    a. What enzymatic activity is required to produce linear unit-length molecules, how does it produce molecules that contain a single complete λ genome, and what gene encodes the enzyme involved?

    b. What types of ends are produced when this enzyme acts on DNA, and how are these important in the λ life cycle?

Answer:

    a. The DNA endonuclease encoded by the *ter* gene recognizes sequences at *cos* sites appearing just once within a λ genome. It makes a staggered cut at these sites to produce the unit-length linear DNA molecules that are packaged.

    b. The *ter* enzyme produces complementary ("sticky") 12-base-long, single-stranded ends. After λ infects *E. coli,* these ends pair and gaps in the phosphodiester backbone are sealed by DNA ligase to produce a closed circular molecule. This molecule recombines into the *E. coli* chromosome if the lysogenic pathway is followed, or replicates using rolling circle replication if a lytic pathway is followed.

11.20   M13 is an *E. coli* bacteriophage whose capsid holds a closed circular DNA molecule with 2,221 T, 1,296 C, 1,315 G, and 1,575 A nucleotides. M13 lacks a gene for DNA polymerase and so must use bacterial DNA polymerases for replication. Unlike λ or T4, this phage does not form concatamers during replication and packaging.

    a. Suppose the M13 chromosome were replicated in a manner similar to the way the *E. coli* chromosome is replicated, using semi-discontinuous replication from a double-stranded circular DNA template. How would the semi-discontinuous DNA replication mechanism discussed in the text need to be modified?

    b. Suppose the M13 chromosome were replicated in a manner similar to the way the λ chromosome is replicated using rolling circle replication. How would the rolling circle replication mechanism discussed in the text need to be modified?

Answer:

    a. Since M13 has a closed circular genome with $(A) \neq (T)$ and $(G) \neq (T)$, it must have a single-stranded DNA genome. Bidirectional replication would require the initial synthesis of a complementary strand. To produce many phage, many rounds of bidirectional replication would be necessary. However, upon completion of replication and prior to packaging, the nongenomic strand of the resulting double-stranded molecules would need to be selectively degraded.

    b. To produce single-stranded molecules with the same sequence and base composition as the packaged M13 genome, rolling circle replication must use a complementary template. Therefore, DNA polymerase must initially synthesize the genome's complementary strand to make a double-stranded molecule. Then, a nuclease could nick the genomic strand to create a displacement fork. Continuous rolling circle replication using the intact complementary strand as the leading-strand template and without discontinuous replication on the displaced genomic strand will generate single-stranded M13 genomes. To prevent concatamer formation, the newly replicated DNA must cleaved by an endonuclease after exactly one genome has been replicated. To form a closed circle, the molecule's ends would need to be ligated to each other.

11.21   Compare and contrast eukaryotic and prokaryotic DNA polymerases.

Answer: Multiple DNA polymerases have been identified in all cells: There are 5 in prokaryotes and 15 or more in eukaryotes. All DNA polymerases synthesize DNA from a primed strand in the $5' \rightarrow 3'$ direction using a template. In both eukaryotes and prokaryotes, certain DNA polymerases are used for replication, while others are used for repair. Prokaryotes and eukaryotes differ in how many polymerases they use, and how they use them, in each of these processes.

In *E. coli*, DNA polymerase I and III function in DNA replication. Both have $3' \rightarrow 5'$ exonuclease activity that is used in proofreading. DNA polymerase III is the main synthetic enzyme and can exist as a core enzyme with three polypeptides or as a holoenzyme with an additional six different polypeptides. DNA polymerase I consists of one polypeptide. Unlike DNA polymerase III, it has the $5' \rightarrow 3'$ exonuclease activity needed to excise RNA from the 5' end of Okazaki fragments. DNA polymerases II, IV, and V function in DNA repair.

In eukaryotes, nuclear DNA replication requires three DNA polymerases: Pol α/primase, Pol δ, and Pol ε. After primase initiates new strands in replication by making about 10 nucleotides of an RNA primer, Pol α extends them by adding about 30 nucleotides of DNA. The RNA/DNA primers are extended by Pol δ and Pol ε on the leading and lagging strand. However, it is not clear which enzyme synthesizes which strand. Other DNA polymerases function in DNA repair and mitochondrial DNA replication.

11.22   What mechanism do eukaryotic cells employ to keep their chromosomes from replicating more than once per cell cycle?

Answer: Chromosomes are prevented from replicating more than once by ensuring that no origin of replication is used a second time in a cell cycle. This is accomplished by initiating DNA replication in two temporally separate steps. The first step occurs in G_1 and involves replicator selection. Prereplicative complexes form on each replicator, but replication initiation does not occur at this time. Only after passage of the cell from G_1 phase to S phase can the second step, replication initiation, occur. Replication initiation occurs only once per origin because after a cell enters S phase, new prereplicative complexes can no longer form.

Replication initiation is limited to S phase by cyclin-dependent kinases. Their activity is needed to activate the prereplicative complexes to initiate replication, but new prereplicative complexes cannot form when their activity is present. Active cyclin-dependent kinases are not present in G_1, so the prereplicative complexes can form in G_1. They cannot initiate replication until S, when active cyclin-dependent kinases are present. Chromosomes are prevented from replicating more than once per cell cycle because once a cell is in S phase, active cyclin-dependent kinases prevent new prereplicative complexes from forming.

11.23   Autoradiography is a technique that allows radioactive areas of chromosomes to be observed under the microscope. The slide is covered with a photographic emulsion, which is exposed by radioactive decay. In regions of exposure, the emulsion forms silver grains on being developed. The tiny silver grains can be seen on top of the (much larger) chromosomes. Devise a method to find out which regions in the human karyotype replicate during the last 30 minutes of the S phase. (Assume a cell cycle in which the cell spends 10 hours in G_1, 9 hours in S, 4 hours in G_2, and 1 hour in M.)

Answer: Assuming cells spend 4 hours in G_2, there are 4.5 hours from the last 30 minutes of S to metaphase in M. Late-replicating chromosomal regions can be identified by adding ^3H thymidine to the medium, waiting 4.5 hours, and then preparing a slide of metaphase chromosomes. Chromosomal regions displaying silver grains are late replicating, as cells that were at earlier stages of S when the ^3H was added will be unable to reach metaphase in 4.5 hours.

11.24   In typical human fibroblasts in culture, the G_1 period of the cell cycle lasts about 10 hours, S lasts about 9 hours, G_2 takes 4 hours, and M takes 1 hour. Suppose you added radioactive (^3H) thymidine to the medium, left it there for 5 minutes, and then washed it out and put it with an ordinary medium.

   a.   What percentage of cells would you expect to become labeled by incorporating the ^3H-thymidine into their DNA?

   b.   How long would you have to wait after removing the ^3H-medium before you would see labeled metaphase chromosomes?

   c.   Would one or both chromatids be labeled?

   d.   How long would you have to wait if you wanted to see metaphase chromosomes containing ^3H in the regions of the chromosomes that replicated at the beginning of the S period?

   Answer:

   a.   Assuming that the fibroblast culture is completely asynchronous, and that cells are distributed in all stages of the cell cycle, 10/24 of the cells would be in G_1, 9/24 in S, 4/24 in G_2, and 1/24 in M. Only cells in S are replicating their DNA, so only 9/24 of the cells are capable of incorporating ^3H-thymidine into their DNA. (9/24)(100%) = 37.5%.

   b.   If a cell in the very last stages of S phase incorporated ^3H-thymidine into its DNA, it would have to proceed through G_2 (4 hours) before entering M. Not until M are metaphase chromosomes seen. Therefore, wait a little over 4 hours to see labeled metaphase chromosomes.

   c.   DNA replication is semiconservative, meaning that each of the two double strands is used as a template for the synthesis of a new, complementary strand. When ^3H-thymidine is incorporated into the newly synthesized DNA, it will be incorporated into each new complementary strand, and so be in each chromatid that is synthesized.

   d.   Cells that took up ^3H-thymidine into their DNA at the beginning of S phase would need to complete the S phase, and then go through G_2 and mitotic prophase before label could be seen in metaphase chromosomes. This would take a little over 13 hours.

      The experiment that is outlined in this problem gives one an elegant means to determine which chromosomal region replicates early in S phase and which chromosomal region replicates late in S phase. Proceed by systematically collecting cells at various time points after the ^3H-thymidine pulse. Spread their chromosomes on slides; cover the chromosomes with a photographic emulsion that is able to detect decay of ^3H particles. After a period of time, develop the emulsion to detect sites on chromosomes that were replicating at a given stage of S phase. This kind of experimentation supported a view that there are early- and late-replicating regions of eukaryotic chromosomes. The late-replicating regions are characteristically heterochromatic and include centromeric regions.

11.25   Suppose you performed the experiment in problem 11.24, but left the radioactive medium on the cells for 16 hours instead of 5 minutes. How would your answers change?

Answer:

a. Sixteen hours is longer than any single stage of the cell cycle. In particular, a cell that just entered G_2 at the beginning of the labeling period would, at the end of a 16-hour labeling period, be in early S phase. Therefore, it would be labeled. Indeed, every cell would be labeled.

b. Labeled metaphase chromosomes would already be present by the time the ^{3}H-thymidine was removed, as some of the cells in G_1 at the beginning of the labeling period would go through S and G_2, and be in M after 16 hours.

c. Both chromatids would be labeled, as in a 5-minute pulse of ^{3}H-thymidine.

d. Consider what would happen if metaphase chromosomes were examined in cells collected periodically after the radioactive medium was washed out. Start with cells sampled immediately after the radioactive medium is washed out. Since these cells were left in the radioactive medium for 16 hours, and there is only a little over 13 hours between the beginning of the S period and metaphase of mitosis, the previous S phase of these cells was spent entirely in the presence of ^{3}H-thymidine. Any metaphase chromosomes seen at this point in time would be labeled in their entirety, so these chromosomes would not be useful to identify regions replicated at a particular time during the S period. Suppose instead that cells were sampled after about 6 hours in nonradioactive medium. These cells would have been in the first part of G_1 when the radioactive medium was added and spent about the first 8 hours of their S period in the radioactive medium. All of their chromosomal regions would be labeled except for the regions that replicated late in the S period. By sampling cells over increasing lengths of time, and comparing the results of different samplings, chromosomal regions that replicated at successively earlier stages could be identified. Finally, consider the cell sample taken after 13 hours in nonradioactive medium. These cells would have just entered the S period when the nonradioactive medium had been washed out. Since they have only spent the beginning of their S period in radioactive medium, chromosomal regions that replicated early in the S period would be labeled. However, since the culture was left in radioactive medium for 16 hours, the parents of these cells spent about the last hour of their S period in the presence of ^{3}H-thymidine. Thus, these cells would have chromosomes that were labeled both at the beginning and at the end of the S period. Only by comparison of this sample to samples obtained at earlier time points might these regions be distinguished.

11.26   In Figure 11.3 in the text semiconservative DNA replication is visualized in eukaryotic cells using the harlequin chromosome–staining technique.

a. Explain what the harlequin chromosome–staining technique is and how it provides evidence for semiconservative DNA replication in eukaryotes.

b. Propose a hypothesis to explain why, in the figure, some chromatids appear to contain segments of both DNA containing T and DNA containing BUdR, while others appear to consist entirely of DNA with T or DNA with BUdR.

Answer:

a. Harlequin chromosomes are prepared by allowing tissue culture cells to undergo two rounds of DNA replication in the presence of BUdR. After BUdR (a base analog that replaces T) becomes incorporated in the DNA, metaphase chromosomes stained with Giemsa stain and a fluorescent dye have one darkly stained and one lightly stained chromatid. Since BUdR-DNA stains less intensely than T-DNA, the presence of two differently stained chromatids indicates that one chromatid has two BUdR-strands and the other has one BUdR- and one T-labeled strand.

This supports the semiconservative model of DNA replication. After one round of replication, each DNA molecule has one (new) BUdR- and one (old) T-labeled strand. After two rounds of replication, each of these molecules produces one DNA molecule with two BudR-labeled strands (the light chromatid) and one DNA molecule with one BUdR- and one T-labeled strand (the dark chromatid).

    b. Chromosomes with chromatids containing segments of T-labeled DNA and BUdR-labeled DNA have had a sister-chromatid exchange (mitotic crossing-over).

11.27   When the eukaryotic chromosome duplicates, the nucleosome structures must duplicate. Discuss how synthesis of histones is related to the cell cycle, and discuss how new nucleosomes are assembled at replication forks.

Answer: New histones are synthesized to complex with newly synthesized DNA. The five histone genes are transcribed starting near the end of G_1, just prior to S. The mRNAs that are produced are translated throughout S, producing new histones that will be assembled into nucleosomes as the chromosomes are duplicated. Nucleosomes assemble and disassemble from DNA as the replication fork passes a replicating DNA region. New nucleosomes are assembled from components of both old and new histones with the aid of histone chaperones.

Each parental histone core of a nucleosome separates into an H3-H4 tetramer and two copies of an H2A-H2B dimer. The parental H2A-H2B dimers are released to a pool of H2A-H2B dimers that include newly synthesized and assembled H2A-H2B dimers. The parental H3-H4 tetramer is transferred directly to one of the two replicated DNA helices past the fork. The other DNA helix is bound by a new H3-H4 tetramer. The DNA-bound H3-H4 tetramers initiate nucleosome assembly drawing from the pool of H2A-H2B dimers. Thus, a new nucleosome will have either a parental or new H3-H4 tetramer and a pair of H2A-H2B dimers that may be parental-parental, parental-new, or new-new.

11.28   A mutant *Tetrahymena* has an altered repeated sequence in its telomeric DNA. What change in the telomerase enzyme would produce this phenotype?

Answer: Telomerase synthesizes the simple-sequence telomeric repeats at the ends of chromosomes. The enzyme is made up of both protein and RNA, and the RNA component has a base sequence that is complementary to the telomere repeat unit. The RNA component is used as a template for the telomere repeat, so that if the RNA component were to be altered, the telomere repeat would be as well. Thus, the mutant in this question is likely to have an altered RNA component.

11.29   What is the evidence that telomere length is regulated in cells, and what are the consequences of the misregulation of telomere length?

Answer: Evidence that telomere length is regulated in cells has come from the analysis of mutations of the yeast *TEL1* and *TEL2* genes, which cause cells to maintain their telomeres at a new, shorter-than-wild-type length. These analyses demonstrate that the telomere length is under genetic control. Insights into the consequences of misregulation of telomere length have also come from the analysis of yeast mutants. Analysis of mutations that affect telomerase function (e.g., *TLC1*) and telomere length (*EST1*) have shown that telomerase activity is necessary for long-term cell viability.

These inferences are also supported by observation that mammalian somatic cells lack telomerase, but cells that are "immortal," such as germ cells and cancer cells, do not. Cells that lack telomerase activity will have progressively shorter telomeres as the cells proceed through repeated cycles of mitosis because of the failure to replicate those ends. Cells without telomerase are able to divide only a finite number of times before they become inviable because of the eventual loss of their telomeric sequences.

11.30   What is gene conversion? How does the Holliday model for genetic recombination allow for gene conversion?

Answer: Gene conversion occurs when one allele directs the conversion of a partner allele to its own form. In the Holliday model, strand exchange in the region of a gene with two different alleles generates a heteroduplex with a mismatch. Excision and repair of the mismatch by DNA synthesis can convert one of the alleles to its partner allele, resulting in gene conversion. See text Figure 11.18.

11.31   Crosses were made between strains, each of which carried one of three different alleles of the same gene, *a*, in yeast. For each cross, some unusual tetrads resulted at low frequencies. Explain the origin of each of these tetrads:

Cross:	$a1$ $a2^+$	$a1$ $a3^+$	$a2$ $a3^+$
	$\times$	$\times$	$\times$
	$a1^+$ $a2$	$a1^+$ $a3$	$a2^+$ $a3$
Tetrads:	$a1^+$ $a2$	$a1^+$ $a3$	$a2^+$ $a3$
	$a1^+$ $a2^+$	$a1^+$ $a3$	$a2^+$ $a3^+$
	$a1$ $a2$	$a1^+$ $a3^+$	$a2$ $a3^+$
	$a1$ $a2^+$	$a1$ $a3^+$	$a2$ $a3^+$

Answer: The crosses involve alleles at just one gene, and in each case, tetrads containing doubly mutant and/or completely normal spores are recovered due to events associated with intragenic recombination. In the $a1\,a2^+ \times a1^+\,a2$ cross, each allele shows a 2:2 segregation pattern, and the middle two spores are recombinants that result from cleaving a Holliday intermediate between the $a1$ and $a2$ sites to produce spliced duplexes. For the $a1\,a3^+ \times a1^+\,a3$ and the $a2\,a3^+ \times a2^+\,a3$ crosses, there is evidence that the segregation of one of the alleles in the tetrad has resulted from gene conversion caused by mismatch repair of heteroduplex DNA. The $a1\,a3^+ \times a1^+\,a3$ cross shows 2:2 segregation of $a3^+:a3$ and 3:1 segregation of $a1^+:a1$ resulting from gene conversion of an $a1$ allele to $a1^+$. In the $a2\,a3^+ \times a2^+\,a3$ cross, the $a2$ allele segregates in a Mendelian fashion while the $a3$ allele segregates 3:1 $a3^+:a3$ as a result of gene conversion of one $a3$ allele to $a3^+$.

11.32 From a cross of *y1 y2⁺* × *y1⁺ y2*, where *y1* and *y2* are both alleles of the same gene in yeast, the following tetrad type occurs at very low frequencies:

$$
\begin{array}{ll}
y1^+ & y2 \\
y1 & y2 \\
y1 & y2 \\
y1 & y2^+
\end{array}
$$

Explain the origin of this tetrad at the molecular level.

Answer: The tetrad is not a PD, NPD, or T tetrad. The tetrad shows evidence of gene conversion of both alleles: The wild-type *y1⁺* allele has undergone conversion to *y1*, and the wild-type *y2⁺* allele has undergone conversion to *y2*. When a heteroduplex is formed that contains a mismatch needing repair, the mismatch can be repaired to either the wild-type or the mutant allele. If strand exchange occurred and subsequent branch migration through the *y* gene generated mismatches at both the *y1* and *y2* sites, the strands could have been repaired to cause the indicated gene conversion events.

11.33 In *Neurospora* the *a*, *b*, and *c* loci are situated in the same arm of a particular chromosome. The location of *a* is near the centromere; *b* is near the middle, and *c* is near the telomere of the arm. Among the asci resulting from a cross of *ABC* × *abc*, the following ascus was found (the 8 spores are indicated in the order in which they were arranged in the ascus): *ABC, ABC, ABc, ABc, aBC, aBC, abc, abc*. How might this ascus have arisen?

Answer: The *A/a* locus shows a 2:2 (MI) segregation pattern, the *B/b* locus shows a 3:1 segregation pattern, while the *C/c* locus shows a 1:1:1:1 (MII) segregation pattern. From these patterns and the relative locations of the genes and their centromere, we can infer that a crossover occurred between *A/a* and *C/c* and gene conversion occurred at *B/b*. Therefore, this tetrad could have arisen from (1) the formation of a Holliday intermediate between the *A/a* and *C/c* loci, (2) resolution of the Holliday intermediate through a spliced duplex to give recombinants between the *A/a* and *C/c* loci, and (3) conversion of a *b* allele to *B* by mismatch repair of a heteroduplex at the *B/b* locus. See text Figures 11.17 and 11.18.

11.34 Mutants at the autosomal *rosy* (*ry*) locus in *Drosophila* have rosy eyes instead of the normal deep-red color. Two mutations, *ry²⁰⁶* and *ry²⁰⁹*, are point mutations at the *ry* gene. The *kar* and *ace* loci are each about 0.2 map units from *ry* with the order *kar–ry–ace*. Mutants at *kar* have karmoisin (bright red) eyes, while mutants at *ace* lack the enzyme acetylcholinesterase and are recessive lethal. A rosy-eyed *kar ry²⁰⁶ ace/kar⁺ ry²⁰⁹ ace⁺* female was crossed to a rosy-eyed *kar⁺ ry²⁰⁹ ace⁺/kar⁺ ry²⁰⁹ ace⁺* male. The vast majority of progeny had rosy eyes, but normal-eyed males and females were produced at a very low frequency. Testcrossing the normal-eyed male progeny (recall that recombination does not occur in *Drosophila* males) revealed that they received one of four types of chromosomes from their mother: *kar⁺ ry⁺ ace, kar ry⁺ ace⁺, kar ry⁺ ace,* or *kar⁺ ry⁺ ace⁺*. Explain the origin of each of these chromosomes at the molecular level.

Answer: Since both *ry²⁰⁶* and *ry²⁰⁹* are point mutations, rare reversion to *ry⁺* could produce *kar ry⁺ ace* and *kar⁺ ry⁺ ace⁺* chromosomes. The lack of flanking marker exchange in these chromosomes is also consistent with an ry^+ allele arising from gene conversion by mismatch repair. The ry^{206} allele on the kar ry^{206} ace chromosome was converted to ry^+ using the normal sites present in the ry locus on the kar^+ ry^{209} ace^+ chromosome. The ry^{209} allele on the kar^+ ry^{209} ace^+ chromosome was converted to ry^+ using the normal sites present in the ry locus on the kar ry^{206} ace chromosome. See text Figure 11.18. The chromosomes where the flanking kar and ace marker loci are exchanged (kar^+ ry^+ ace, kar ry^+ ace^+) result from intragenic recombination: a Holliday intermediate forming between the ry^{206} and ry^{209} sites was cleaved to produce spliced duplexes. See text Figure 11.17.

12

GENE FUNCTION

THINKING ANALYTICALLY

The problems in this chapter require you to make multiple connections. You need to:

1. Relate the genetic properties of a mutation to its visible phenotype.
2. Relate the visible phenotype to a biochemical deficit (a "biochemical phenotype").
3. Relate the biochemical deficit of a mutation back to its genetic properties.

As you make these connections, layers of complexity will unfold in a problem. Clues at one level (genetic properties of the mutations, mutant phenotypes, biochemical abnormalities found in the mutant) can be used to infer what might be happening at another level. Approach the problems systematically, attempting to connect these levels wherever possible.

REVIEW OF KEY TERMS, SYMBOLS, AND CONCEPTS

In your own words, write a brief definition of each term in the groups below. Check your definitions using the text. Then develop a concept map using the terms in each list.

1	2
inborn error of metabolism	enzyme
metabolic pathway	nonenzymatic protein
biochemical pathway	cystic fibrosis (CF)
nutritional mutant	transmembrane conductance regulator
auxotroph	active transport
prototroph	sickle-cell anemia, sickle-cell trait
ascus, asci, ascospore, conidia	hemoglobin
minimal medium	α, β polypeptide
mating type	electrophoresis
one-gene–one-enzyme hypothesis	amino acid substitution
one-gene–one-polypeptide hypothesis	hydrophilic
amino acid	multiple alleles
polypeptide	autosomal recessive mutation
protein	genetic counseling
N, C terminus	pedigree analysis
pleiotropic	carrier detection
alkaptonuria (AKU)	fetal analysis
phenylketonuria (PKU)	amniocentesis
albinism	chorionic villus sampling
Lesch-Nyhan syndrome	
Tay-Sachs disease	
recessive trait	
OMIM	

To relate biochemical pathways to mutant phenotypes, diagram as much of the biochemical pathway as you can, and then try to understand how specific phenotypic consequences arise when a particular step is blocked. Go slowly and be thorough. It is essential to remember that mutations in different steps of a biochemical pathway do not always cause identical mutant phenotypes. In addition, mutations affecting a particular step can have multiple consequences. They may lead to the absence of the final product and cause one phenotype. They may also lead to the accumulation of intermediate metabolites of the pathway and cause a different phenotype. Furthermore, mutations at different steps of a pathway may be able to be "rescued" if different biochemical intermediates are provided to the mutant.

QUESTIONS FOR PRACTICE

Multiple Choice Questions

1. An inborn error of metabolism is
   a. any biochemical abnormality that results from taking a drug

b. any heritable biochemical abnormality

c. any recessive mutation

d. any dominant mutation

2. Why was the one-gene–one-enzyme hypothesis recast as the one-gene–one-polypeptide hypothesis?

a. Genes can encode proteins that are not enzymes.

b. Some enzymes are not polypeptides.

c. Some enzymes have more than one polypeptide subunit.

d. Some polypeptides are not enzymes.

3. Suppose an individual was diagnosed with PKU as a child and was successfully treated. If that individual and a normal, noncarrier (pku^+/pku^+) partner have children, what is the probability that they will have offspring that are carriers for PKU?

a. 0.00

b. 0.25

c. 0.50

d. 1.00

4. In individuals affected with AKU,

a. a block in the pathway leads to the accumulation of homogentisic acid

b. if homogentisic acid is provided, the pathway can be completed

c. a block in the pathway leads to the accumulation of phenylalanine

d. if phenylalanine is provided, the pathway can be completed

5. Two *Neurospora* auxotrophs are unable to grow on minimal medium but are able to grow on minimal medium supplemented with arginine. When each is tested, only one can grow on minimal medium supplemented with ornithine, a biochemical precursor to arginine. Which statement(s) below are supported by these findings?

a. The auxotrophs are blocked in different biochemical steps.

b. At least one auxotroph is blocked in a step before ornithine is made.

c. Both of the auxotrophs are blocked in a step before ornithine is made.

d. Both of the auxotrophs accumulate ornithine.

e. a and b

6. In complex metabolic pathways such as the phenylalanine-tyrosine metabolic pathway,

a. blocks at different steps always result in the same phenotype

b. a block at one step results solely in the accumulation of the biochemical made just before that step

c. blocks at different steps can lead to very different phenotypes

d. a block at one step can lead to the accumulation of potentially toxic derivatives of biochemicals synthesized before that step

e. a and b.

f. c and d.

7. What is the difference between sickle-cell anemia and sickle-cell trait?

a. None; they refer to individuals equally affected with the same disease.

b. Individuals with sickle-cell anemia have severe disease and have two abnormal alleles; individuals with the sickle-cell trait have a less severe form of the disease, as they are heterozygotes with one disease allele and one normal allele.

c. Individuals with sickle-cell anemia have severe disease; The term *sickle-cell trait* refers to the disease allele present in their families, and not to symptomatic status.

d. Individuals with sickle-cell anemia have severe disease and are homozygous for a severe disease allele; individuals with sickle-cell trait have mild disease and are homozygous for a less severe disease allele.

8. Both Tay-Sachs disease and Lesch-Nyhan syndrome are examples of human diseases that are
   a. pleiotropic
   b. caused by a deficiency in one enzyme activity
   c. recessive
   d. a, b, and c

9. An example of a heritable disease that results from a biochemical defect in a nonenzymatic protein is
   a. PKU
   b. AKU
   c. sickle-cell anemia
   d. Tay-Sachs disease

10. Which of the following is *not* true about both amniocentesis and chorionic villus sampling?
    a. Both can be performed with an acceptably low risk as early as the eighth week of pregnancy.
    b. Each procedure has a risk of fetal loss.
    c. One is more likely than the other to provide an accurate diagnosis.
    d. Both are valid, useful procedures in genetic counseling.

Answers: 1b, 2c, 3d, 4a, 5e, 6f, 7b, 8d, 9c, 10a

Thought Questions

1. If you were a genetic counselor, how would you counsel the following couple who are contemplating having children? The prospective father is apparently normal, except for a nervous twitch of his nose, but his maternal uncle had PKU as a child and his brother, who plays in a heavy metal band, is hard of hearing. The prospective mother also appears normal, but her father has hemophilia and her brother, who plays in the same band as her brother-in-law, is also hard of hearing. You know that hemophilia is often X linked and that PKU and sometimes deafness are autosomal recessive.

2. Distinguish between the procedures of amniocentesis and chorionic villus sampling. How is each useful in genetic counseling? What reasons are there for or against choosing each procedure?

3. All states currently require testing for PKU in infants. Do you expect the frequency of PKU to decline or increase over time? Justify your answer.

4. In the United States, the frequency of PKU is about 1 in 16,000 newborns. The frequency of cystic fibrosis is about 1 in 4,000 newborns. While all states currently require testing for PKU in infants, not all states require testing for cystic fibrosis. Both are autosomal recessive disorders. What are the risks associated with screening for an inherited disease? What reasons might there be not to routinely test for cystic fibrosis (or PKU) in newborns? What reasons might there be to routinely test for PKU (or cystic fibrosis) in newborns?

5. Over 200 hemoglobin mutants have been detected. Explain why some hemoglobin mutations do not lead to as severe sickle-cell anemia phenotypes as others.

6. Some alleles of sickle-cell anemia, when heterozygous with a normal allele, confer some resistance to malarial infection. What consequences might this have on selection for this allele in different human populations?

7. Keep in mind your answers to questions 5 and 6 as you respond to the following question. Suppose you have the ability to make point mutations in a gene that encodes an enzyme, and are able to determine the percentage of normal enzyme activity that remains in the

mutant enzyme. What effects would you expect to see if you systematically mutate the gene for the enzyme, changing only one amino acid at a time?

8. Different diseases are more prevalent in different ethnic groups. For example, within the U.S. population, Tay-Sachs disease is more prevalent in individuals of Ashkenazi Jewish descent, while certain types of sickle-cell anemia are more prevalent in some individuals of Hispanic, Native American, and African American descent. Given this situation and a history of documented discrimination based on ethnicity in the United States, what concerns must be addressed if genetic testing is to be used for carrier testing and prenatal diagnosis with the aim of improving the quality of health care?

9. As you have seen in the study of PKU, Tay-Sachs disease, Lesch-Nyhan syndrome, and sickle-cell anemia, mutations that result in the blockage of a single enzymatic step or the function of a single protein can have pleiotropic phenotypes. Generate a hypothesis about a general principle from these data.

10. What functions do (should) genetic counseling services provide? What do they look for, how do they proceed, what do they recommend, and what, if any, advice do (should) they give?

11. In many states in the United States, legislation is being developed to address the use and privacy of information obtained from genetic testing. What specific protections can be legislated for genetic testing to be performed without fears of stigmatization and discrimination? Under what circumstances, if any, should genetic testing be allowed for diseases for which there is no treatment? No cure? For some diseases with very similar symptoms, genetic tests may be useful to distinguish between diseases and improve therapy. Should laws governing the use of genetic tests as diagnostic tests (in symptomatic individuals) differ from laws governing the use of genetic tests as tests for predicting risk (in asymptomatic individuals)?

12. What is meant by "obtaining informed consent" prior to undergoing a genetic test? What information is needed to give informed consent prior to undergoing a genetic test?

Thought Questions for Media

After reviewing the media on the *iGenetics* CD-ROM, try to answer these questions.

1. In the animation depicting the experiments by Beadle and Tatum, how can *Neurospora* be propagated asexually?
2. How did Beadle and Tatum obtain nutritional mutations?
3. Why did Beadle and Tatum use two rounds of screening for auxotrophs?

1. Why is the arginine pathway present in *Neurospora* also present in humans?
2. Suppose a block is made in a single enzymatic step of the arginine pathway in either humans or *Neurospora*. What is the consequence to the level of the precursor compound synthesized before the blocked enzymatic step?
3. How might citrullinemia, argininosuccinic aciduria, or OTC deficiency lead to ammonia buildup? How might you implement effective therapeutic treatments?

SOLUTIONS TO TEXT PROBLEMS

12.1   Most enzymes are proteins, but not all proteins are enzymes. What are the functions of enzymes, and why are they essential for living organisms to carry out their biological functions?

Answer: Enzymes are macromolecules that catalyze chemical reactions. These reactions can be quite simple, such as the addition of a hydroxyl (−OH) group onto a simple compound, or they can be exceedingly complex, such as the replication of an entire chromosome. Enzymes are essential to biological systems because they provide catalytic power and usually increase reaction rates (the rate of production of the product of the chemical reaction) by at least a millionfold. Nearly all of the chemical transformations made in the body proceed at very low rates in the absence of enzymes. Enzymes enable cells to efficiently process nutrients, synthesize essential biochemicals such as amino acids, and synthesize and degrade proteins, nucleic acids, lipids, and other cellular macromolecules. Therefore, enzymes are necessary to catalyze reactions that produce needed compounds and perform essential processes.

12.2   What was the significance of Archibald Garrod's work, and why do you expect that it was not appreciated by his contemporaries?

Answer: Garrod's work provided the first evidence of a specific relationship between genes and enzymes. By studying families with alkaptonuria, Garrod and Bateson inferred that enzyme deficiencies are genetically controlled traits. Garrod then demonstrated that the position of a block in a metabolic pathway can be determined by the accumulation of the chemical compound that precedes the blocked step. One likely reason that his contemporaries did not appreciate the significance of his work was that—in 1902 when Garrod studied alkaptonuria—there was not a wide understanding of Mendelian principles (the rediscovery of Mendel's work occurred just a few years earlier). Consequently, the concept that genes control traits was not established, and there was no understanding that genes make protein products.

12.3   Phenylketonuria (PKU) is an inheritable metabolic disease of humans; its symptoms include mental deficiency. This phenotypic effect results from

    a. the accumulation of phenylketones in the blood
    b. the absence of phenylalanine hydroxylase
    c. a deficiency of phenylketones in the blood
    d. a deficiency of phenylketones in the diet

Answer: a

12.4   If a person were homozygous for both PKU and alkaptonuria (AKU), would you expect him or her to exhibit the symptoms of PKU, AKU, or both? Refer to the following pathway:

Phenylalanine

↓ (blocked in PKU)

tyrosine → DOPA → melanin

↓

p-Hydroxyphenylpyruvic acid

↓

Homogentisic acid

↓ (blocked in AKU)

Maleylacetoacetic acid

Answer: A double homozygote should have PKU, but not AKU. The PKU block should prevent most homogentisic acid from being formed, so it could not accumulate to high levels and cause AKU.

12.5 Refer to the pathway shown in Question 12.4. What effect, if any, would you expect PKU or AKU to have on pigment formation? Explain your answer.

Answer: The block in PKU leads to decreased tyrosine levels and so should lead to a decrease in melanin formation and pigmentation. However, some tyrosine will be obtained from food, and so the block in the pathway can be partially circumvented in this way. It has been reported that PKU patients are sometimes lighter in pigmentation than their normal relatives.

Since the block in AKU patients lies after (downstream from) the formation of melanin, pigmentation should not be affected in AKU patients. However, the levels of the products of an enzymatic reaction can affect the efficiency with which that reaction proceeds. If the products of an enzymatic reaction accumulate, the reaction may be inhibited. This in turn can lead to the accumulation of the substrates of the reaction. In this case, the accumulation of homogentisic acid in AKU individuals may ultimately lead to elevation of tyrosine levels. This tyrosine might be available for conversion to melanin. Hence, if AKU has any effect, it may be to increase pigmentation levels.

Using the same logic as in problem 12.4, an individual with both PKU and AKU will show symptoms of PKU, but not AKU. Therefore, it is likely that there will be decreased tyrosine levels and a decrease in melanin formation and pigmentation.

12.6 Define the term *autosomal recessive mutation,* and give some examples of diseases that are caused by autosomal recessive mutations. Explain how two parents who do not display any symptoms of a given disease (albinism, or any of the diseases you have named) can have two or even three children who have the disease. How can these same parents have no children with the disease?

Answer: Autosomes are chromosomes that are found in two copies in both males and females. That is, an autosome is any chromosome except the X and Y chromosome. Since individuals have two of each type of autosome, they have two copies of each gene on an autosome. The forms of the gene, or the alleles at the gene, can be the same or different. They can have either two normal alleles (homozygous for the normal allele), one normal allele and one mutant allele (heterozygous for the normal and mutant alleles), or two mutant alleles (homozygous for the mutant allele). A recessive mutation is one that exhibits a phenotype only when it is homozygous. Therefore, an autosomal recessive mutation is a mutation that occurs on any chromosome except the X or Y and that causes a phenotype only when homozygous. Heterozygotes exhibit a normal phenotype.

Of the diseases discussed in this chapter, many are autosomal recessive. For example, phenylketonuria, albinism, Tay-Sachs disease, and cystic fibrosis are autosomal recessive diseases. Heterozygotes for the disease allele are normal, but homozygotes with the disease allele are affected. For phenylketonuria and albinism, homozygotes are affected because they lack a required enzymatic function. In these cases, heterozygotes have a normal phenotype because their single normal allele provides sufficient enzyme function.

Parents contribute one of their two autosomes to their gametes, so that each offspring of a couple receives an autosome from each parent. If in a particular conception each of two heterozygous parents contributes a chromosome with the normal allele, the offspring will be homozygous for the normal allele and be normal. If in a particular conception one of the two heterozygous parents contributes a chromosome with the normal allele and the other parent contributes a chromosome with the mutant allele, the offspring will be heterozygous but be normal. If in a particular conception each parent contributes the chromosome with the mutant allele, the offspring will be homozygous for the mutant allele and develop the disease. Therefore, heterozygous parents can have both normal and affected children. Since each conception is independent, two heterozygous parents can have all normal, all affected, or any mix of normal and affected children.

12.7   Consider sickle-cell anemia as an example of a devastating disease that is the result of an autosomal recessive genetic mutation on a specific chromosome. Explain what a molecular or genetic disease is. Compare and contrast this disease with a disease caused by an invading microorganism such as a bacterium or virus.

Answer: A genetic disease such as sickle-cell anemia is caused by a change in DNA that alters levels or forms of one or more gene products. In turn, the altered forms or levels of gene products result in changes in cellular functions, which in turn cause a disease state. The examples given in this chapter demonstrate that genetic diseases can be associated with mutations in single genes that affect their protein products. For example, sickle-cell anemia is caused by mutations in the gene for β-globin. Mutations lead to amino acid substitutions that cause the β-globin polypeptide to fold incorrectly. This in turn leads to sickled red blood cells and anemia. The environment can have a significant effect on disease severity and many genetic diseases are treatable. For example, PKU can be treated by altering diet. Unlike diseases caused by an invading microorganism or other external agent that are subject to the defenses of the human immune system and that generally have short-lived clinical symptoms and treatments, genetic diseases are caused by heritable changes in DNA that are associated with chronically altered levels or forms of one or more gene products.

12.8   A breeder of Irish setters has a particularly valuable show dog that he knows is descended from the famous bitch Rheona Didona, who carried a recessive gene for atrophy of the retina. Before he puts the dog to stud, he must ensure that it is not a carrier for this allele. How should he proceed?

Answer: One method is to testcross the male to a retinal atrophic female. Retinal atrophic pups (expected half of the time) would indicate that the male is heterozygous. If the retinal atrophy is associated with a known biochemical phenotype (e.g., an altered enzyme activity), a preferred method is to take a tissue biopsy or blood sample from the male and assess whether normal levels of enzyme activity are present in the male. Recessive mutations in genes for many enzymes, when heterozygous with a normal allele, lead to reduced levels of enzyme activity even though no visible phenotype is observed.

12.9   What problems might we encounter if we accept the one-gene–one-enzyme concept as completely accurate, and what have we discovered about this concept? What work led to that discovery?

Answer: Beadle and Tatum discovered that one gene specified one enzyme when they showed that nutritional mutations (auxotrophs) in *Neurospora* were associated with defects in specific steps of a biochemical pathway. Mutations in different steps accumulated the biochemical intermediate made before the enzymatic block, and could be rescued, or grown, if there were addition of a biochemical intermediate made after the enzymatic block. This genetic analysis indicated that the normal gene specified an enzyme in the pathway that converted one biochemical intermediate to another.

The one-gene–one-enzyme concept is inaccurate in situations where an enzyme is composed of more than one polypeptide subunits, and different polypeptide subunits are encoded by separate genes. In such cases, multiple genes encode one enzyme. If two genes encode different polypeptide subunits that form one enzyme, then mutations in each gene can result in the same phenotype. That is, mutations in either of the two genes will result in the same enzyme deficiency and the accumulation of the same precursor product. This discovery led to the more accurate hypothesis is that one gene encodes one polypeptide.

12.10 a^+, b^+, c^+, and d^+ are independently assorting Mendelian genes controlling the production of a black pigment. The alternate alleles that give abnormal functioning of these genes are a, b, c, and d. A black individual of genotype $a^+/a^+\ b^+/b^+\ c^+/c^+\ d^+/d^+$ is crossed with a colorless individual of genotype $a/a\ b/b\ c/c\ d/d$ to produce a black F_1. Then $F_1 \times F_1$ crosses are done. Assume that a^+, b^+, c^+, and d^+ act in a pathway as follows:

$$\text{colorless} \xrightarrow{a^+} \text{colorless} \xrightarrow{b^+} \text{colorless} \xrightarrow{c^+} \text{brown} \xrightarrow{d^+} \text{black}$$

a. What proportion of the F_2 progeny is colorless?
b. What proportion of the F_2 progeny is brown?

Answer: The F_1 cross is $a^+/a\ b^+/b\ c^+/c\ d^+/d \times a^+/a\ b^+/b\ c^+/c\ d^+/d$.

a. A colorless F_2 individual would result if an individual has an *a/a*, *b/b*, and/or *c/c* genotype. This would consist of many possible genotypes. Rather than identify all of these combinations, use the fact that the proportion of colorless individuals = 1 − the proportion of pigmented individuals. The proportion of pigmented individuals ($a^+/-$ $b^+/-$ $c^+/-$) is $\frac{3}{4} \times \frac{3}{4} \times \frac{3}{4} = \frac{27}{64}$. The chance of not obtaining this genotype is $1 - \frac{27}{64} = \frac{37}{64}$.

b. A brown individual is $a^+/-$ $b^+/-$ $c^+/-$ d/d. The proportion of brown individuals is $\frac{3}{4} \times \frac{3}{4} \times \frac{3}{4} \times \frac{1}{4} = \frac{27}{256}$.

12.11   Using the genetic information given in Problem 12.10, now assume that a^+, b^+, and c^+ act in a pathway as follows:

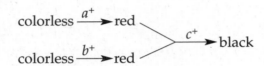

Black can be produced only if both red pigments are present; that is, c^+ converts the two red pigments together into a black pigment.

a. What proportion of the F$_2$ progeny is colorless?
b. What proportion of the F$_2$ progeny is red?
c. What proportion of the F$_2$ progeny is black?

Answer:

a. Colorless individuals are a/a b/b $-/-$. The chance of obtaining such an individual from a cross of a^+/a b^+/b $c^+/c \times a^+/a$ b^+/b c^+/c is $\frac{1}{4} \times \frac{1}{4} = \frac{1}{16}$.

b. There are two ways to determine the proportion of red individuals. First, notice that red individuals are obtained either when only one of the a^+ or b^+ functions is present (no matter what genotype is at c), or when both a^+ and b^+ functions are present but c^+ is not. Thus, these phenotypes are obtained from the following genotypes: $a^+/-$ b/b $-/-$, a/a $b^+/-$ $-/-$, or $a^+/-$ $b^+/-$ c/c. The probability of obtaining one of these genotypes is

$$(\tfrac{3}{4} \times \tfrac{1}{4} \times 1) + (\tfrac{1}{4} \times \tfrac{3}{4} \times 1) + (\tfrac{3}{4} \times \tfrac{3}{4} \times \tfrac{1}{4}) = \tfrac{33}{64}$$

A second way is to use the information from (a) and (c). From (c), one has that $\frac{27}{64}$ individuals are black, so the remaining $\frac{37}{64}$ are either red or colorless. From (a), $\frac{1}{16} = \frac{4}{64}$ are colorless. Thus, there are $\frac{(37-4)}{64} = \frac{33}{64}$ that are red.

c. Black individuals are formed only if both red pigments are available and are then converted to black pigment. Thus, an individual must be $a^+/-$ $b^+/-$ $c^+/-$. The chance of obtaining such an individual is $\frac{3}{4} \times \frac{3}{4} \times \frac{3}{4} = \frac{27}{64}$.

12.12   Three genes on different chromosomes are responsible for three enzymes that catalyze the same reaction in corn:

$$\text{colorless compound} \xrightarrow{a^+,\, b^+,\, c^+} \text{red compound}$$

The normal functioning of any one of these genes is sufficient to convert the colorless compound to the red compound. The abnormal functioning of these genes is designated by a, b, and c, respectively.

a. A red a^+/a^+ b^+/b^+ c^+/c^+ is crossed with a colorless a/a b/b c/c to give a red F_1, a^+/a b^+/b c^+/c. The F_1 is selfed. What proportion of the F_2 progeny is colorless?

b. It turns out that another step is involved in the pathway—one that is controlled by gene d^+, which assorts independently of a^+, b^+, and c^+:

$$\text{colorless compound 1} \xrightarrow{d^+} \text{colorless compound 2} \xrightarrow{a^+, b^+, c^+} \text{red compound}$$

The inability to convert colorless 1 to colorless 2 is designated d. A red a^+/a^+ b^+/b^+ c^+/c^+ d^+/d^+ is crossed with a colorless a/a b/b c/c d/d. The F_1 corn are all red. The red F_1 corn are now selfed. What proportion of the F_2 corn is colorless?

Answer:

a. Since any of the normal alleles a^+, b^+, or c^+ is sufficient to catalyze the reaction leading to color, in order for color to fail to develop, all three normal alleles must be missing. That is, the colorless F_2 must be a/a b/b c/c. The chance of obtaining such an individual is $\frac{1}{4} \times \frac{1}{4} \times \frac{1}{4} = \frac{1}{64}$.

b. Now, colorless F_2 are obtained if *either* one or both steps of the pathway are blocked. That is, colorless F_2 are obtained in either of the following genotypes: d/d $-/-$ $-/-$ $-/-$ (the first or both steps blocked) or $d^+/-$ a/a b/b c/c (second step blocked). The chance of obtaining such individuals is

$$\left(\frac{1}{4} \times 1 \times 1 \times 1\right) + \left(\frac{3}{4} \times \frac{1}{4} \times \frac{1}{4} \times \frac{1}{4}\right) = (64 + 3)/256 = \frac{67}{256}$$

12.13 In *Drosophila*, the recessive allele bw causes a brown eye, and the (unlinked) recessive allele st causes a scarlet eye. Flies homozygous for both recessives have white eyes. The genotypes and corresponding phenotypes, then, are as follows:

$bw^+/-$	$st^+/-$	red eye
bw/bw	$st^+/-$	brown
$bw^+/-$	st/st	scarlet
bw/bw	st/st	white

Outline a hypothetical biochemical pathway that would produce this type of gene interaction. Demonstrate why each genotype shows its specific phenotype.

Answer: Since the bw/bw $st^+/-$ individuals have brown pigmentation but only have st^+ function the st^+ gene must make brown pigment. Similarly, the scarlet pigmentation of $bw^+/-$ st/st individuals indicates that the bw^+ gene must make scarlet pigment. The combination of the brown and scarlet pigments in $bw^+/-$ $st^+/-$ individuals leads to red eyes. The absence of either pigment in bw/bw st/st individuals leads to white (colorless) eyes.

The production of two different colored pigments, with each affected by a separate gene, suggests that there are two distinct biochemical pathways. Each of these pathways might have a number of steps. One pathway produces a brown pigment and the other produces a scarlet pigment. This can be illustrated in the following diagram:

Two pathways, each with an unknown number of steps:

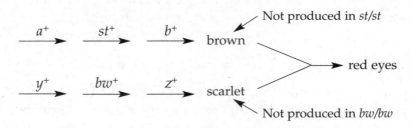

12.14 In J. R. R. Tolkien's *The Lord of the Rings*, the Black Riders of Mordor ride steeds with eyes of fire. As a geneticist, you are very interested in the inheritance of the fire-red eye color. You discover that the eyes contain two types of pigments—brown and red—that are usually bound to core granules in the eye. In wild-type steeds, precursors are converted by these granules to the afore said pigments, but in steeds homozygous for the recessive X-linked gene *w* (white eye), the granules remain unconverted and a white eye results. The metabolic pathways for the synthesis of the two pigments are shown in Figure 12.A:

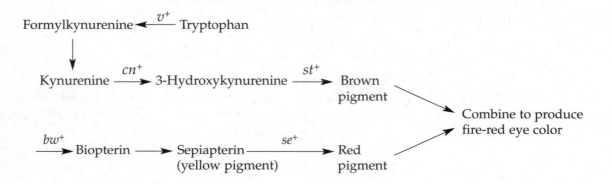

Each step of the pathway is controlled by a gene: Mutation *v* results in vermilion eyes, *cn* results in cinnabar eyes, *st* results in scarlet eyes, *bw* results in brown eyes, and *se* results in black eyes. All these mutations are recessive to their wild-type alleles, and all are unlinked. For the following genotypes, show the proportions of steed eye phenotypes that would be obtained in the F_1 of the given matings:

a. $w/w\ bw^+/bw^+\ st/st \times w^+/Y\ bw/bw\ st^+/st^+$

b. $w^+/w^+\ se/se\ bw/bw \times w/Y\ se^+/se^+\ bw^+/bw^+$

c. $w^+/w^+\ v^+/v^+\ bw/bw \times w/Y\ v/v\ bw/bw$

d. $w^+/w^+\ bw^+/bw\ st^+/st \times w/Y\ bw/bw\ st/st$

Answer:

a. ½ $w/w^+\ bw/bw^+\ st/st^+$, fire-red-eyed daughters; ½ $w/Y\ bw/bw^+\ st/st^+$, white-eyed sons

b. $w/w^+\ se/se^+\ bw/bw^+$ and $w^+/Y\ se/se^+\ bw/bw^+$, all fire-red eyes

c. $w/w^+\ v/v^+\ bw/bw$ and $w^+/Y\ v/v^+\ bw/bw$, all brown eyes

d. ¼ w^+/w or w^+/Y, *bw/bw⁺ st/st⁺*, fire-red eyes; ¼ w^+/w or w^+/Y, *bw/bw st/st⁺*, brown eyes; ¼ w^+/w or w^+/Y, *bw/bw⁺ st/st*, scarlet eyes; ¼ w^+/w or w^+/Y, *bw/bw st/st*, (the color of 3-hydroxykynurenine plus the color of the precursor to biopterin, or colorless = white)

12.15   Upon infection of *E. coli* with bacteriophage T4, a series of biochemical pathways result in the formation of mature progeny phages. The phages are released after lysis of the bacterial host cells. Suppose that the following pathway exists:

$$A \xrightarrow{\text{enzyme}} B \xrightarrow{\text{enzyme}} \text{mature phage}$$

Suppose also that we have two temperature-sensitive mutants that involve the two enzymes catalyzing these sequential steps. One of the mutations is cold sensitive (*cs*), in that no mature phages are produced at 17°C. The other is heat sensitive (*hs*), in that no mature phages are produced at 42°C. Normal progeny phages are produced when phages carrying either of the mutations infect bacteria at 30°C. However, let us assume that we do not know the sequence of the two mutations. Two models are therefore possible:

$$(1) \quad A \xrightarrow{hs} B \xrightarrow{cs} \text{phage}$$

$$(2) \quad A \xrightarrow{cs} B \xrightarrow{hs} \text{phage}$$

Outline how you would experimentally determine which model is the correct model without artificially lysing phage-infected bacteria.

Answer: Wild type T4 will produce progeny phages at all three temperatures. Consider what will happen under each model if *E. coli* is infected with a doubly mutant phage (one step is cold sensitive, one step is heat sensitive), and the growth temperature is shifted between 17° and 42° during phage growth.

    Suppose model (1) is correct and cells infected with the double mutant are first incubated at 17° and then shifted to 42°. Progeny phages will be produced and the cells will lyse, as each step of the pathway can be completed in the correct order. In model (1), the first step, *A* to *B*, is controlled by a gene whose product is heat-sensitive but not cold-sensitive. At 17°, the enzyme works, and A will be converted to B. While phage are at 17°, the second, cold-sensitive step of the pathway prevents the production of mature phage. However, when the temperature is shifted to 42°, the accumulated B product can be used to make mature phage, so that lysis will occur.

    Under model (1), a temperature shift performed in the reverse direction does not allow for growth. When *E. coli* cells are infected with a doubly mutant phage and placed at 42°, the heat-sensitive first step precludes the accumulation of B. When the culture is shifted to 17°, B can accumulate, but now, the second step cannot occur, so no progeny phage can be produced. Therefore, if model (1) is correct, lysis will be seen only in a temperature shift from 17° to 42°.

    If model (2) is correct, growth will be seen only in a temperature shift from 42° to 17°. Hence, the correct model can be deduced by performing a temperature shift experiment in each direction and observing which direction allows progeny phage to be produced.

12.16 Four mutant strains of *E. coli* (*a*, *b*, *c*, and *d*) all require substance X to grow. Four plates were prepared, as shown in the following figure:

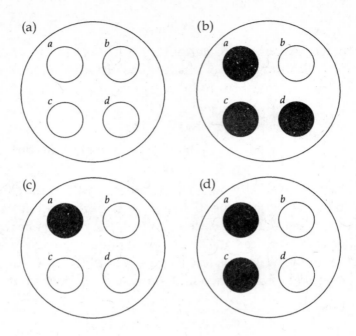

In each case, the medium was minimal, with just a trace amount of substance X, to allow a small amount of growth of the mutant cells. On plate (a) cells of mutant strain *a* were spread over the entire surface of the agar and grew to form a thin lawn (continuous bacterial growth over the plate). On plate (b) the lawn is composed of mutant *b* cells, and so on. On each plate, cells of the four mutant types were inoculated over the lawn, as indicated by the circles. Dark circles indicate luxuriant growth. This experiment tests whether the bacterial strain spread on the plate can feed the four strains inoculated on the plate, allowing them to grow. What do these results show about the relationship of the four mutants to the metabolic pathway leading to substance X?

Answer: A strain blocked at a later step in the pathway accumulates a metabolic interme-diate that can "feed" a strain blocked at an earlier step. It secretes the metabolic inter-mediate into the medium, thereby providing a nutrient to bypass the earlier block of another strain. Consequently, a strain that feeds all others (but itself) is blocked in the last step of the pathway, while a strain that feeds no others is blocked in the first step of the pathway. Mutant *a* is blocked in the earliest step in the pathway because it cannot feed any of the others. Mutant *c* is next because it can supply the substance *a* needs but cannot feed *b* or *d*. Mutant *d* is next, and mutant *b* is last in the pathway because it can feed all the others. The pathway is *a*→*c*→*d*→*b*.

12.17 Two mutant strains of *Neurospora* lack the ability to make compound Z. When crossed, the strains usually yield asci of two types: (1) those with spores that are all mutant and (2) those with four wild-type and four mutant spores. The two types occur in a 1:1 ratio.

a. Let *c* represent one mutant, and let *d* represent the other. What are the genotypes of the two mutant strains?

b. Are *c* and *d* linked?

c. Wild-type strains can make compound Z from the constituents of the minimal medium. Mutant *c* can make Z if supplied with X but not if supplied with Y, while mutant *d* can make Z from either X or Y. Construct the simplest linear pathway of the synthesis of Z from the precursors X and Y, and show where the pathway is blocked by mutations *c* and *d*.

Answer:

a. $c/c\ d^+/d^+$ and $c^+/c^+\ d/d$

b. The diploid $c/c^+\ d/d^+$ produces equal numbers of PD tetrads (4 $c\ d^+$ and 4 $c^+\ d$ spores, all mutant) and NPD tetrads (4 $c^+\ d^+$ and 4 $c\ d$ spores, 4 wild type and 4 mutant), so *c* and *d* are not linked.

c. The pathway is Y→X→Z. Mutant *c* is blocked in the synthesis of X from Y. Mutant *d* is blocked before the synthesis of Y.

12.18 The following growth responses (where + = growth and 0 = no growth) of mutants 1–4 were seen on the related biosynthetic intermediates A, B, C, D, and E:

	Growth On				
Mutant	**A**	**B**	**C**	**D**	**E**
1	+	0	0	0	0
2	0	0	0	+	0
3	0	0	+	0	0
4	0	0	0	+	+

Assume that all intermediates are able to enter the cell, that each mutant carries only one mutation, and that all mutants affect steps after B in the pathway. Which of the following schemes best fits the data with regard to the biosynthetic pathway?

(a) 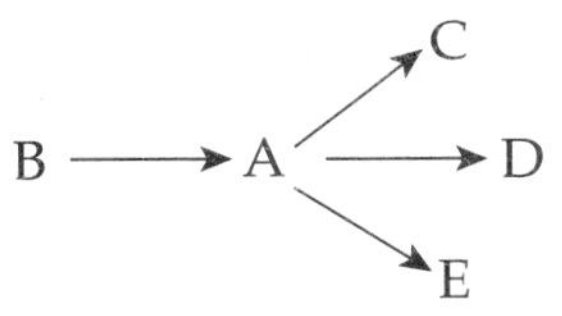

(b) 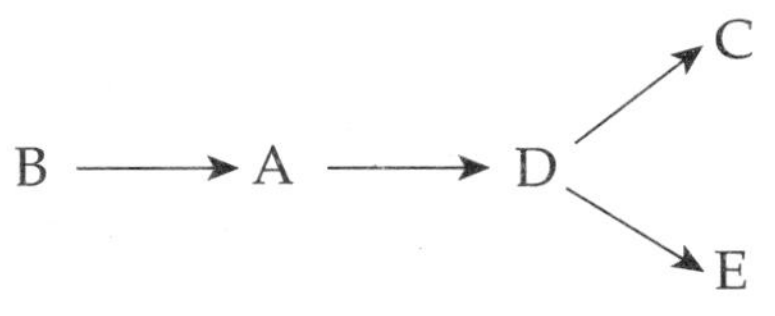

(c) 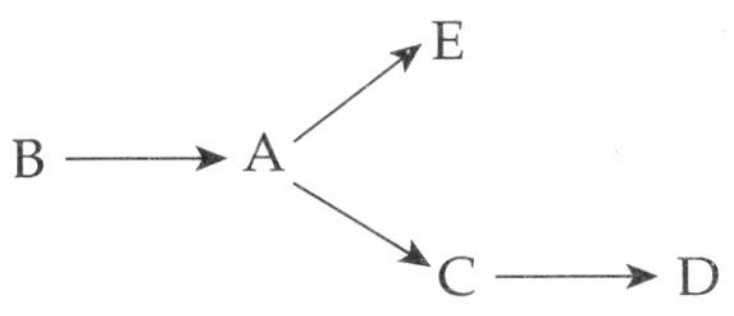

(d) 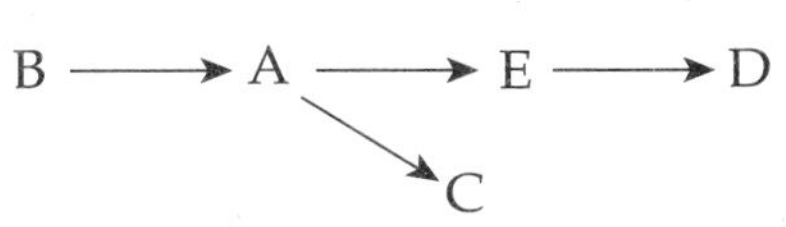

Answer: One approach to this problem is to try to sequentially fit the data to each pathway, as if each were correct. Check where each mutant *could* be blocked (remember, each mutant carries only one mutation), whether the mutant would be able to grow if supplemented with the *single* nutrient that is listed, and whether the mutant would not be able to grow if supplemented with the "no growth" intermediate. It will not be possible to fit the data for mutant 4 to pathway (a) the data for mutants 1 and 4 to pathway (b) or data for mutants 3 and 4 to pathway (c). The data for all mutants can be fit only to pathway (d). Thus, (d) must be the correct pathway.

A second approach to this problem is to realize that in any linear segment of a biochemical pathway (a segment without a branch), a block early in the segment can be circumvented by *any* metabolites that normally appear later in the same segment. Consequently, if two (or more) intermediates can support growth of a mutant, they normally are made after the blocked step in the same linear segment of a pathway. From the data given, compounds D and E both circumvent the single block in mutant 4. This means that compounds D and E lie after the block in mutant 4 on a linear segment of the metabolic pathway. The only pathway where D and E lie in an unbranched linear segment is pathway (d). Mutant 4 could be blocked between A and E in this pathway. Mutant 4 cannot be fit to a *single* block in any of the other pathways that are shown, so the correct pathway is (d).

12.19 A *Neurospora* mutant has been isolated in the laboratory in which you are working. This mutant cannot make an amino acid that we will call Y. Wild-type *Neurospora* cells make Y from a cellular product X through a biochemical pathway involving three intermediates called c, d, and e. How would you demonstrate that your mutant contains a defective gene for the enzyme that catalyzes the d→e reaction?

Answer: If the enzyme that catalyzes the d→e reaction is missing, the mutant strain should accumulate d and be able to grow on minimal medium to which e is added. In addition, it should not be able to grow on minimal medium or on minimal medium to which X, c, or d is added but should grow if Y is added. Therefore, plate the strain on these media and test which allow for growth of the mutant strain and which intermediate is accumulated if the strain is plated on minimal medium.

12.20 Upon learning that the diseases listed in the following table are caused by a missing enzyme activity, a medical student proposes the therapies shown in the rightmost column:

Disease	Missing Enzyme Activity	Proposed Therapy
Tay-Sachs disease	N-acetylhexosaminidase A, which catalyzes the formation of ganglioside G_{M3} from gangliosideG_{M2}	administer ganglioside G_{M3} (by feeding or injection)
Lesch-Nyhan syndrome	HGPRT, an enzyme used in purine metabolism	administer purines
Phenylketonuria	phenylalanine hydroxylase, which catalyzes the formation of tyrosine from phenylalanine	administer tyrosine

a. Explain why each of the proposed therapies will be ineffective in treating the associated disease. For which diseases would symptoms worsen if the proposed therapy were followed?
b. Vitamin D–dependent rickets results in muscle and bone loss and is caused by a deficiency of 25-hydroxycholecalciferol 1 hydroxylase, an enzyme that catalyzes the formation of 1,25-dihydroxycholecalciferol (vitamin D) from 25-hydroxycholecalciferol. Unlike any of the situations in (a), patients can be effectively treated by daily administration of the product of

the enzymatic reaction, 1,25-dihydroxycholecalciferol (vitamin D). If you assayed for levels of serum 25-hydroxycholecalciferol in patients, what would you expect to find? Why is treatment with the product of the enzymatic reaction effective here, but not in the situations described in (a)?

Answer:

a. In each of these diseases, the lack of an enzymatic step leads to the toxic accumulation of a precursor or its byproduct. The proposed treatments are ineffective because they do not prevent the accumulation of the toxic precursor. Administering purines will worsen Lesch-Nyhan syndrome, since the disease results from purine accumulation.

b. The loss of 25-hydroxycholecalciferol 1 hydroxylase should lead to increased serum levels of 25-hydroxycholecalciferol, the precursor it acts upon. Since administration of the end product of the reaction, 1,25-dihydroxycholecalciferol (vitamin D), is an effective treatment, this disease is unlike those in part a. It appears that this disease is caused by the loss of the reaction's end product and not the accumulation of its precursor.

12.21 Four strains of *Neurospora*, all of which require arginine but have unknown genetic constitution, have the following nutrition and accumulation characteristics:

		Growth On			
Strain	Minimal Medium	Ornithine	Citrulline	Arginine	Accumulates
1	–	–	+	+	Ornithine
2	–	–	–	+	Citrulline
3	–	–	–	+	Citrulline
4	–	–	–	+	Ornithine

Pairwise complementation tests of the four strains gave the following results (+ = growth on minimal medium, and 0 = no growth on minimal medium):

	4	3	2	1
1	0	+	+	0
2	0	0	0	
3	0	0		
4	0			

Crosses among mutants yielded prototrophs in the following percentages:

1×2: 25 percent
1×3: 25 percent
1×4: none detected among 1 million progeny (ascospores)
2×3: 0.002 percent
2×4: 0.001 percent
3×4: none detected among 1 million progeny (ascospores)

Analyze the data and answer the following questions.

a. How many distinct mutational sites are represented among these four strains?

b. In this collection of strains, how many types of polypeptide chains (normally found in the wild type) are affected by mutations?

c. Write the genotypes of the four strains, using a consistent and informative set of symbols.

d. Determine the map distances between all pairs of linked mutations.

e. Determine the percentage of prototrophs that would be expected among ascospores of the following types: (1) strain 1 × wild type; (2) strain 2 × wild type; (3) strain 3 × wild type; (4) strain 4 × wild type

Answer: Use the nutrition and accumulation data to infer where each mutant is blocked. 1 accumulates ornithine, and can grow only if citrulline or arginine is added. Hence, 1 is blocked in the conversion of ornithine to citrulline. 2 and 3 are blocked in the conversion of citrulline to arginine. Mutant 4 is more complex. It can grow only if supplemented with arginine, but accumulates ornithine. 4 must be a double mutant and is blocked in the conversion of ornithine to citrulline as well as from citrulline to arginine. This gives the following pathway:

$$\text{precursor} \longrightarrow \text{ornithine} \xrightarrow{1,\,4} \text{citrulline} \xrightarrow{2,\,3,\,4} \text{arginine}$$
$$\text{Enzyme:} \quad A \qquad\qquad\qquad B \qquad\qquad\qquad C$$

Use the complementation test data to determine whether two mutants affect the same function. Diploids constructed from two mutants that are able to grow on minimal media indicate that the mutants complement each other: One provides a function missing in the other. Failure to grow indicates no complementation: Both are missing the same function. 2 and 3 complement 1, so 2 and 3 affect a different step than 1. 2 and 3 fail to complement each other, so they affect the same step. 4 fails to complement 1, 2, or 3 and affects both steps. This is consistent with the deduced pathway: 1 and 4 affect enzyme B, and 2, 3, and 4 affect enzyme C.

Start analyzing the recombination data by assigning symbols to the genes and writing out the genotypes and crosses. Let b^+ encode enzyme B and c^+ encode enyzme C. Then $1 = b^1\,c^+$, $2 = b^+\,c^2$, $3 = b^+\,c^3$, and $4 = b^4\,c^4$. The prototrophic ascospres produced in the crosses 1 × 2, 1 × 3, and 2 × 3 are $b^+\,c^+$ and are half of the recombinant progeny (e.g., in 1 × 2, parental types are $b^1\,c^+$ and $b^+\,c^2$, recombinant types are $b^+\,c^+$ and $b^1\,c^2$). The results can be tabulated as:

Cross	Genotypes	$b^+\,c^+$ Spores
1 × 2	$b^1\,c^+ \times b^+\,c^2$	25.0%
1 × 3	$b^1\,c^+ \times b^+\,c^3$	25.0%
1 × 4	$b^1\,c^+ \times b^4\,c^4$	$< 1 \times 10^{-6}$
2 × 3	$b^+\,c^2 \times b^+\,c^3$	0.002%
2 × 4	$b^+\,c^2 \times b^4\,c^4$	0.001%
3 × 4	$b^+\,c^3 \times b^4\,c^4$	$< 1 \times 10^{-6}$

Use the recombination frequencies to infer whether the *b* and *c* genes are linked and whether two alleles at one gene affect the same site or are separable by recombination and affect different sites. 1×2 and 1×3 give 25 percent prototrophs (show 50 percent recombination), so *b* and *c* are unlinked. 2×3 produces very few prototrophs, as they can only arise from rare, intragenic recombination. The 0.002 percent prototrophs (= half the recombinants) indicate that c^2 and c^3 are 0.004 mu apart. 4 fails to recombine with 1 or 3, indicating that b^4 and b^1 affect the same site, and that c^4 and c^3 affect the same site. The 0.001 percent prototrophs from 2×4 arise when a c^+ allele is obtained after intragenic recombination between c^2 and c^4 *and* the b^+ allele (in mutant 2) assorts with it into the ascospore. When c^+ assorts with b^4 (half the time), a prototrophic ascospore is not obtained. This is why only 0.001 percent prototrophs are recovered from 2×4, instead of the 0.002 percent prototrophs recovered from 2×3. This results in the following map:

a. Three distinct mutational sites exist based on the recombination analysis: one in the *b* gene and two in the *c* gene.

b. Two polypeptide chains are affected based on the complementation analysis: one encoded by the *b* gene and one encoded by the *c* gene.

c. The genotypes are as given in the table above.

d. There are 0.004 mu between the two mutant sites in the *c* gene identified by c^2 and c^3. The *b* and *c* genes are unlinked.

e. The *b* and *c* genes assort independently of one another. Thus, strain 1, 2, or 3, when mated with the wild type, will give 50 percent prototrophs and 50 percent auxotrophs. As strain 4 is a double mutant, it will give only 25 percent prototrophs. Four equally frequent genotypes would be expected from the cross $b^4 c^4 \times + +$. Only one of the four will be $+ +$.

12.22 Alleles of a wild-type gene can be thought of as giving a normal phenotype because they confer a particular, normal amount of gene function. Mutant alleles can alter the level of function in a variety of ways. So-called loss-of-function alleles can be thought of as eliminating or decreasing gene function. Among the loss-of-function alleles are amorphic alleles (which eliminate gene function) and hypomorphic alleles (which decrease gene function). A hypermorphic allele is an overproducer that makes more of the wild-type product. So-called gain-of-function alleles have novel functions. Among the gain-of-function alleles are antimorphic alleles (which are antagonistic to the function of a wild-type allele) and neomorphic alleles (which provide a new, substantially altered function to the gene).

Consider the following hypothetical situation: In the production of purple pigment in sweet peas (see Chapter 4, p. 93), the wild-type (+) product of the p^+ gene is an enzyme P that converts white pigment to purple pigment. This enzyme activity can be measured in tissue extracts and mixtures of tissue extracts. The table below describes the results obtained when enzyme activity is measured in extracts and mixtures of extracts from strains with different p^+ genotypes (all are c^+/c^+):

Genotype	Percent of +/+ Activity	Percent of +/+ Activity When Mixed 50 : 50 with +/+ Extract	(A) Homozygote Phenotype	(B) Heterozygote Phenotype	(C) Hemizygote Phenotype	(D) Allele Classification
p^+/p^+	100	100	_____	_____	_____	_____
p^1/p^1	20	60	_____	_____	_____	_____
p^2/p^2	0	50	_____	_____	_____	_____
p^3/p^3	300	200	_____	_____	_____	_____
p^4/p^4	0	5	_____	_____	_____	_____
p^5/p^5	0^a	50^b	_____	_____	_____	_____

[a]Produces red, not purple, pigment
[b]Produces red and purple pigments

Complete the columns of the table, filling in (A) the phenotype expected in a homozygote; (B) the phenotype expected in a heterozygote (allele/+); (C) the phenotype expected in a hemizygote (an individual that is heterozygous for the allele and a deletion for the locus); (D) the classification of the allele using the definitions set forth at the beginning of the problem.

Answer: A single p^+ allele provides 50 percent of the enzyme activity seen in a p^+/p^+ homozygote. Since p^+ is dominant (i.e., $P-C/C$ plants are purple), this appears to be enough activity to provide for a wild-type phenotype. If a plant with less than 50 percent of normal activity does not synthesize enough purple pigment for a wild-type phenotype (e.g., 25 percent of normal activity gives a light purple flower) and a plant with more than 100 percent of normal activity produces noticeably darker purple pigmentation, the following phenotypes should be seen.

Genotype	Percent of +/+ Activity	Percent of +/+ Activity When Mixed 50 : 50 with +/+ Extract	(A) Homozygote Phenotype	(B) Heterozygote Phenotype	(C) Hemizygote Phenotype	(D) Allele Classification
p^+/p^+	100	100	purple	purple	purple	wild type
p^1/p^1	20	60	light purple	purple	very light purple	hypomorph
p^2/p^2	0	50	white	purple	white	amorph
p^3/p^3	300	200	very dark purple	dark purple	dark purple	hypermorph
p^4/p^4	0	5	white	very light purple	white	antimorph
p^5/p^5	0^a	50^b	red	reddish purple	red	neomorph

[a]Produces red, not purple, pigment
[b]Produces red and purple pigments

12.23   Glutathione (GSH) is important for a number of biological functions, including the prevention of oxidative damage in red blood cells, the synthesis of deoxyribonucleotides from ribonucleotides, the transport of some amino acids into cells, and the maintenance of protein conformation. Mutations that have lowered levels of glutathione synthetase (GSS), a key enzyme in the synthesis of glutathione, result in one of two clinically distinguishable disorders. The severe form is characterized by massive urinary excretion of 5-oxoproline (a chemical derived from a synthetic precursor to glutathione), metabolic acidosis (an inability to regulate physiological pH appropriately), anemia, and central nervous system damage. The mild form is characterized solely by anemia. The characterization of GSS activity and the GSS protein in two affected patients, each with normal parents, is given in the following table:

Patient	Disease Form	GSS Activity in Fibroblasts (percent of normal)	Effect of Mutation on GSS Protein
1	severe	9%	Arginine at position 267 replaced by tryptophan
2	mild	50%	Aspartate at position 219 replaced by glycine

a. What pattern of inheritance do you expect these disorders to exhibit?

b. Explain the relationship of the form of the disease to the level of GSS activity.

c. How can two different amino acid substitutions lead to dramatically different phenotypes?

d. Why is 5-oxoproline produced in significant amounts only in the severe form of the disorder?

e. Is there evidence that the mutations causing the severe and mild forms of the disease are allelic (in the same gene)?

f. How might you design a test to aid in prenatal diagnosis of this disease?

Answer:

a. Since normal parents have affected offspring, the disease appears to be recessive. However, since patients with 50 percent of GSS activity have a mild form of the disease, individuals may show mild symptoms if they are heterozygous (mutant/+) for a mutation that eliminates GSS activity. In a population, individuals having the disease may not all show identical symptoms, and some may have more severe disease than others. The severity of the disease in an individual will depend on the nature of the person's GSS mutation and, possibly, whether the person is heterozygous or homozygous for the disease allele. The alleles discussed here appear to be recessive.

b. Patient 1, with 9 percent of normal GSS activity, has a more severe form of the disease, whereas patient 2, with 50 percent of normal GSS activity, has a less severe form of the disease. Thus, increased disease severity is associated with less GSS enzyme activity.

c. The two different amino acid substitutions may disrupt different regions of the enzyme's structure (consider the effect of different amino acid substitutions on hemoglobin's function, discussed in the text). As amino acids vary in their polarity and charge, different amino acid substitutions within the same structural region could have different chemical effects on protein structure. This, too, could lead to different levels of enzymatic function. (For a discussion of the chemical differences between amino acids, see Chapter 14.)

d. By analogy with the disease PKU discussed in the text, 5-oxoproline is produced only when a precursor to glutathione accumulates in large amounts due to a block in a biosynthetic pathway. When GSS levels are 9 percent of normal, this occurs. When GSS levels are 50 percent of normal, there is sufficient GSS enzyme activity to partially complete the pathway and prevent high levels of 5-oxoproline.

e. The mutations are allelic (in the same gene), since both the severe and the mild forms of the disease are associated with alterations in the same polypeptide that is a component of the GSS enzyme. (Note that while the data in this problem suggest that the GSS enzyme is composed of a single polypeptide, they do not exclude the possibility that GSS has multiple polypeptide subunits encoded by different genes.)

f. If GSS is normally found in fetal fibroblasts, one could, in principle, measure GSS activity in fibroblasts obtained via amniocentesis. The GSS enzyme level in cells from at-risk fetuses could be compared with that in normal control samples to predict disease due to inadequate GSS levels. Some variation in GSS level might be seen, depending on the allele(s) present. Since more than one mutation is present in the population, it is important to devise a functional test that assesses GSS activity, rather than a test that identifies a single mutant allele.

12.24 You have been introduced to the functions and levels of proteins and their organization. List as many protein functions as you can and give an example of each.

Answer: Proteins serve many different functions. Some examples are given in the table on the following page.

12.25 We know that the function of any protein is tied to its structure. Give an example of how a disruption of a proteins structure by mutation can lead to a distinctive phenotypic effect.

Answer: Many examples are given in the text. Here are two:

Hb-S mutations in the β-globin gene change the sixth amino acid from the N terminus of the β-globin polypeptide. This amino acid is changed from a negatively charged, acidic amino acid (glutamic acid) to an uncharged, neutral amino acid (valine). This single amino acid change causes the β-globin polypeptide to fold in an abnormal way, which in turn causes sickling of red blood cells and sickle-cell anemia.

The most common mutation causing cystic fibrosis (CF) is a mutation in the CFTR (cystic fibrosis transmembrane regulator) gene called ΔF508. This mutation deletes three consecutive base pairs of DNA and results in the deletion of a single amino acid in the CFTR protein. This amino acid is deleted in a region of the CFTR protein important for binding ATP. Since this region contributes to the function of the protein—to transport chloride ions across cell membranes—this single amino acid deletion affects the function of the protein and causes CF.

12.26 The human β-globin gene provides an excellent example of how the sequence of nucleotides in a gene is eventually expressed as a functional protein. Explain how mutations in the β-globin gene can cause an altered phenotype. How can two different mutations in the same gene cause very different disease phenotypes?

Answer: Many single base-pair mutations in the β-globin gene cause amino acid substitutions at different sites in the β-polypeptide chain. They can have different effects, depending on the type of amino acid substitution and its position. Some of them, such as the Hb-S mutation described in the answer to problem 12.25, alter the charge of an amino acid and thereby affect the folding of the protein. This in turn either affects the function of the hemoglobin molecule and/or the shape of the red-blood cell, causing anemia. Not all such mutations are equally severe. In the Hb-C mutation, the sixth amino acid from

Protein Function	Examples and Their Roles
Enzyme	DNA polymerase I—synthesizes DNA, proofreads, excises an RNA primer
	DNA ligase—seals a nick in a DNA sugar-phosphate backbone
	Phenylalanine hydroxylase—converts phenylalanine to tyrosine
	Topoisomerase—introduces/removes negative supercoils in DNA
Structural protein	Tubulin—polymerizes to form microtubules, used in chromosome segregation at anaphase
	Myosin—a muscle protein used in muscle contraction
	Actin—polymerizes to form actin filaments, which provide structural integrity to the cell
Viral protein	Proteins making up the coat of T2, TMV
Receptor	Acetylcholine receptor—binds the neurotransmitter acetylcholine, found on the surface of muscle cells
Transporter	Hemoglobin—transports oxygen
Ion channel	CFTR—transports chloride ions
Regulatory molecule	Transcription factors—determine where mRNA transcripts are produced (see Chapter 11)
Bind DNA	Single-stranded DNA-binding protein—keeps single-stranded DNA strands intact and from renaturing within a replication bubble during DNA synthesis
Antibody	Immunoglobulins—immune function
Blood type	A, B, O glycoproteins—specify blood type

the N-terminus of the β-polypeptide chain is changed from glutamic acid to lysine. Both of these amino acids are hydrophilic. Consequently, the conformation of the hemoglobin molecule is not as altered as in the Hb-S mutation, and the Hb-C mutation causes only mild anemia.

12.27 Consider the human hemoglobin variants shown in text Figure 12.11. What would you expect the phenotype to be in people heterozygous for the following two hemoglobin mutations?

a. Hb Norfolk and Hb-S

b. Hb-C and Hb-S

Answer:

a. From text Figure 12.11, Hb-Norfolk affects the α-chain whereas Hb-S affects the β-chain of hemoglobin. Since each chain is encoded by a separate gene, there remains one normal allele at the genes for each of α- and β-chains in a double heterozygote. Thus, some normal hemoglobin molecules form and double heterozygotes do not have severe anemia. However, unlike double heterozygotes for two different, completely recessive mutations that lie in one biochemical pathway, these heterozygotes exhibit an abnormal phenotype. This is because some mutations in the α- and β-chains of hemoglobin show partial dominance. In particular, Hb-S/+ heterozygotes show symptoms of anemia if there is a sharp drop in oxygen tension, so these double heterozygotes exhibit mild anemia.

b. Both Hb-C and Hb-S affect the sixth amino acid of the β-chain. The Hb-C mutation alters the normal glutamate to lysine, while the Hb-S mutation alters it to valine. Since both mutations affect the β-chain, no normal hemoglobin molecules are present. According to the text, only one type of β-chain is found in any one hemoglobin molecule. Therefore, an Hb-C/Hb-S heterozygote has two types of hemoglobin: those with Hb-C β-chains and those with Hb-S β-chains. The individuals would exhibit anemia, perhaps intermediate in phenotype between homozygotes for Hb-C and Hb-S.

12.28   Devise a rapid screen to detect new mutations in hemoglobin, and critically evaluate which types of mutations your screen can and cannot detect.

Answer: One rapid screening method would be to isolate and lyse red blood cells, and then use gel electrophoresis to separate hemoglobin proteins based on their electric charge as shown in text Figure 12.8. This screen would detect a mutation causing an amino acid substitution that also alters the electric charge of hemoglobin. It would not detect a mutation that does not alter the charge of hemoglobin and so would not detect all mutations that alter the function of hemoglobin.

12.29   What can prospective parents do to reduce the risk of bearing offspring who have genetically based enzyme deficiencies?

Answer: At one level, the answer to this question is straightforward. Couples who are prospective parents can undergo genetic testing for mutations that result in enzyme deficiencies. If both members of a couple are heterozygous for mutations in the same gene, then they may elect to undergo genetic counseling to assess the risk of having an affected child and to discuss the options available to them concerning prenatal screening.

In practice, however, this question raises very complex issues. First, a decision to undergo genetic testing is associated with certain risks. While the risks associated with drawing blood for the testing procedure itself can be minimal, there are additional risks with very real, significant consequences. These include a risk of emotional discomfort upon learning that one is (or is not) a carrier, risks to personal and familial relationships, and potential risks to privacy, insurability, employment, discrimination, and stigmatization. Second, it is difficult to know which mutations to test for. Sometimes this issue can be rationally addressed if there is a clear family history of a particular inherited disease. However, this may not be the case. For example, for autosomal recessive diseases, family members may show no signs of a disease because they are heterozygous, and some individuals will not know their familial history. In addition, because the genes we have are shared within our ethnic group, the risks we have for different inherited diseases are related to our ethnicity. Given this, genetic screening and testing can raise issues related to cultural and societal values, historical (and perhaps current) levels of discrimination, and the availability and quality of health care services. Third, this question raises personal, philosophical, and religious issues. Some individuals may not be willing to undergo genetic testing out of deeply held personal beliefs. While some couples may be willing to undergo genetic testing and prenatal testing so that they can use this information in planning, they may not be willing to electively terminate a fetus who has a genetically based enzyme deficiency. With current and anticipated advances in genomics (see Chapters 17 and 18), these and other policy, societal, and ethical issues related to genetic screening and genetic testing are becoming more significant.

12.30   Some methods used to gather fetal material for prenatal diagnosis are invasive and therefore pose a small, but very real, risk to the fetus.

   a. What specific risks and problems are associated with chorionic villus sampling and amniocentesis?

   b. How are these risks balanced with the benefits of each procedure?

   c. Fetal cells are reportedly present in maternal circulation after about 8 weeks of pregnancy. However, the number of cells is very low, perhaps no more than one in several million maternal cells. To date, it has not been possible to isolate fetal cells from maternal blood in sufficient quantities for routine genetic analysis. If the problems associated with isolating fetal cells from maternal blood were overcome, and sufficiently sensitive methods were developed to perform genetic tests on a small number of cells, what would be the benefits of performing such tests on these fetal cells?

   Answer:

   a. Amniocentesis is typically done after 12 weeks of pregnancy (after the first trimester), while chorionic villus sampling can be done between 8 and 12 weeks of pregnancy. In either sampling method, there is a risk of fetal loss. While chorionic villus sampling can allow parents to learn if the fetus has a genetic defect earlier than amniocentesis, there is more risk of fetal loss and more chance of an inaccurate diagnosis due to the contaminating presence of maternal cells.

   b. These sampling methods offer parents an opportunity to learn about the genetic constitution and health of the fetus early in a pregnancy. If the fetus is diseased, early prenatal diagnosis allows parents to make an informed decision about the pregnancy. They can seek genetic counseling and consider what options might be available to them: to treat the diseased fetus in utero, to plan for the care of a diseased child, or to terminate the pregnancy.

   c. Such a method would allow for early, *noninvasive* prenatal diagnosis. This would substantially reduce the risk to the fetus and the discomfort to the mother. Since a blood sample could be obtained by many health care professionals and shipped off-site for processing, it would also eliminate the need for a highly skilled physician and sophisticated on-site support services to perform the sampling procedure. Consequently, it would allow for more accessible prenatal testing at lower cost.

12.31   Many autosomal recessive mutations that cause disease in newborns can be diagnosed and treated. However, only a few inherited diseases are routinely tested for in newborns. Explore the basis for which tests are performed by answering the following questions concerning testing for PKU, which is required for newborns throughout the United States, and testing for CF, which is only done if a newborn or infant shows symptoms consistent with a diagnosis of CF.

a. What are the relative frequencies of PKU and CF in newborns, and how, if at all, are these frequencies related to mandated testing?

b. What is the basis of the Guthrie test used for detecting PKU, and what features of the test make it useful for screening large numbers of newborns efficiently?

c. Multiple diagnostic tests have been developed for CF. Some are DNA based while others indirectly assess CFTR protein function. An example of the latter is a test that measures salt levels in sweat. In CF patients, salt levels are elevated due to diminished CFTR protein function. Although the ΔF508 mutation discussed in the text is common in patients with a severe form of CF, other CF mutations are associated with less severe phenotypes. Tests assessing CFTR protein function may not reliably distinguish normal newborns from newborns with mild forms of CF. What challenges do the types of available tests and the range of disease phenotypes present in a population pose for implementing diagnostic testing?

d. Discuss the importance of testing for PKU and CF at birth relative to the time that therapeutic intervention is required. Under what circumstances is testing newborns for CF warranted?

Answer:

a. In Caucasians, PKU occurs in about 1 in 12,000 births while CF occurs in about 1 in 2,000 births. In African-Americans and Asians, the CF frequency is 1 in 17,000 and 1 in 90,000, respectively. Given their relative frequencies in Caucasians, the choice of which diseases have mandated testing is not based on disease frequency alone.

b. The Guthrie test is a simple clinical screen for phenylalanine in the blood. A drop of blood is placed on a filter paper disc and the disc placed on solid culture medium containing *B. subtilis* and β-2-thienylalanine. The β-2-thienylalanine normally inhibits the growth of *B. subtilis,* but the presence of phenylalanine prevents this inhibition. Therefore, the amount of growth of *B. subtilis* is a measure of the amount of phenylalanine in the blood. The test provides an easy, relatively inexpensive means to reliably quantify blood phenylalanine levels, making it an effective preliminary screen for PKU in newborn infants.

c. Mandated diagnostic testing requires a highly accurate test—one that has very low false-positive and false-negative rates—as misdiagnosis of a genetic disease in a genetically normal individual has significant potential for emotional distress in the family of the misdiagnosed child and misdiagnosis of an affected individual as normal may delay necessary therapeutic treatment. A set of mutations with a range of different disease phenotypes may make it difficult to employ a single easy-to-use test. For example, different mutations may make it impossible to use just one DNA-based test and non-DNA-based tests that are effective at diagnosing severe disease phenotypes may not be equally effective at diagnosing mild disease forms, as they may give results that overlap with those from normal individuals.

d. Testing for PKU in newborns is essential for early intervention to prevent the toxic accumulation of phenylketones and the resulting neurological damage in early infancy. Unless it is documented that intervention in newborns is critical for CF disease management, testing for CF in newborns is less critical. Testing is warranted to confirm a diagnosis when severe CF symptoms are apparent in a newborn.

12.32 Reflecting on your answers to Problem 12.31, state why newborns are *not* routinely tested for recessive mutations that cause uncurable diseases such as Lesch-Nyhan syndrome and Tay-Sachs disease?

Answer: There is not a cure or a therapy for these diseases, and individuals with these diseases do not begin to exhibit symptoms until one or more years after birth. It is difficult to justify routine testing in this situation since the information gained from testing will not impact on the disease course or outcome.

12.33　Mr. and Mrs. Chávez have a son that is afflicted with Lesch-Nyhan syndrome and are now expecting a second child. Mr. and Mrs. Lieberman have a son with Tay-Sachs disease and are also expecting a second child. Both couples are concerned that their second child might develop the disease seen in their son and so discuss their situation with a genetic counselor. After taking their family histories, she describes a set of tests that can provide information that might help the couples.

　a. What different types of tests can be done to aid in carrier detection and fetal analysis, and what are their advantages and disadvantages?

　b. How would you determine whether the disease seen in each couple's son results from a new mutation or has been transmitted from one or both of the parents?

　c. Fetal analyses of each pregnancy identify a normal XX karyotype, and DNA-based testing on fetal cells reveals the presence of both normal and mutant alleles of the relevant gene. Have these fetal analyses helped to resolve each couples concerns? Why or why not?

Answer:

　a. Tests can be DNA based and determine the genotype of a parent or fetus or be biochemically based and determine some aspect of the individual's physiology. For example, the Guthrie test determines the relative amount of phenylalanine in a drop of blood to assess whether an individual has PKU; enzyme assays can determine whether a person has a complete or partial enzyme deficiency; gel electrophoresis can determine whether a person has an altered α- or β-globin that might be associated with anemia. DNA-based tests assess the presence or absence of a specific mutation, and are normally employed only when there is already suspicion that an individual may carry that mutation (e.g., the couple has already had an affected offspring). Biochemical tests typically focus on assessing gene function, so they are often used in screens. However, they may not provide detailed information about which gene or biochemical step is affected and require that the biochemical activity be present in the tested cell population, such as cells obtained from an amniocentesis.

　b. Lesch-Nyhan syndrome is caused by an X-linked mutation and the affected son's X came from his mother, so use a DNA-based test to evaluate whether Mrs. Chávez is a carrier for the same mutation as her son. Tay-Sachs disease is caused by an autosomal recessive mutation and each parent contributed one of their autosomes to the affected son, so use a DNA-based test to evaluate whether each parent is heterozygous for an allele present in the affected son. If any of the tested parents do not carry the mutation present in their affected son, that son has a new mutation.

　c. Each conception has produced a female carrier. Mr. and Mrs. Lieberman should be relieved, as their daughter is heterozygous for a recessive disease and will not develop Tay-Sachs disease. However, Mr. and Mrs. Chávezes' concerns have not been resolved as the female carrier of an X-linked recessive disorder may be symptomatic due to random X-chromosome inactivation. If the X chromosome bearing the normal allele is inactivated in most cells, the mutant allele will be expressed and their daughter will develop Lesch-Nyhan syndrome. If the X chromosome bearing the mutant allele is inactivated in most cells, the normal allele will be expressed and their daughter will be normal. In this case, if the mutation is not new, she would be like her carrier mother. The fetal cells tested from amniocentesis or chorionic villus sampling may not be representative of the X chromosome inactivation pattern of all fetal cells, so it is not possible to infer from these tests alone whether Mr. and Mrs. Chávezes' daughter will develop Lesch-Nyhan syndrome.

12.34   In evaluating my teacher, my sincere opinion is that:

a. He or she is a swell person whom I would be glad to have as a brother-in-law or sister-in-law.

b. He or she is an excellent example of how tough it is when you do not have either genetics or environment going for you.

c. He or she may have okay DNA to start with, but somehow all the important genes got turned off.

d. He or she ought to be preserved in tissue culture for the benefit of other generations.

Answer: Your choice!

13

GENE EXPRESSION: TRANSCRIPTION

CHAPTER OUTLINE

THINKING ANALYTICALLY

As transcription and RNA processing are central to gene expression, they have been investigated intensely. Our substantial understanding of these processes is reflected in the wealth of detailed information presented in this chapter. As was the case for understanding DNA replication, the best way to understand and retain this information is to organize the details and place them in context. As you read the chapter, continually refer to the text figures that summarize the key events of transcription and RNA processing. Try explaining the figures to yourself out loud to enhance your understanding. To clarify and solidify your understanding of the material after you have read the chapter once, go through it again and categorize its detailed information.

To start categorizing the information, develop a list of terms and identify their chemical nature. Is the term a region of DNA, of RNA, or part of a protein? After this, relate the term to its function in the overall process of gene expression. If it pertains to DNA, how does it relate to the structure of a gene or a transcription unit? If it pertains to RNA, how does it relate to the RNA being synthesized or being processed? If it pertains to a part of a protein, how is it involved in transcription or RNA processing?

To help you relate a term to its function, construct sketches of the processes of transcription and transcript processing. Diagram how the structure of a gene at the DNA level relates to the structure of its primary transcript at the RNA level. Then diagram how the structure of the primary transcript relates to the structure of the processed, mature transcript.

REVIEW OF KEY TERMS, SYMBOLS, AND CONCEPTS

In your own words, write a brief, precise definition of each term in the groups on this and the next page. Check your definitions using the text. Then develop a concept map using the terms in each list.

1	2	3	4
central dogma	prokaryotic RNA	RNA polymerase II	RNA polymerase II
replication	polymerase	protein-coding genes	transcription unit
transcription	transcription	mRNA, snRNA	leader, trailer,
translation	initiation	regulatory elements	coding sequences
primary transcript	promoter	general transcription	5′, 3′ UTR
precursor RNA	closed, open	factors	precursor mRNA
mRNA	promoter complex	regulatory factors	pre-, mature mRNA
tRNA	holoenzyme	core promoter	posttranscriptional
rRNA	core enzyme	promoter proximal	modification
snRNA	sigma factor	elements	monocistronic
RNA polymerase	σ^{70}, σ^{32}, σ^{54}, σ^{23}	Goldberg-Hogness,	5′ cap
RNA polymerases	heat shock, stress	TATA, CAAT,	5′ 7-methyl guanosine
I, II, III	consensus sequence	GC boxes	5′-5′, 2′-5′ bonds
prokaryotic genes	−10, −35 regions	enhancer element	capping enzyme
eukaryotic genes	Pribnow box	activator	methylation
structural gene	elongation	adapter	3′ polyadenylation
protein-coding gene	proofreading activity	upstream activator	poly(A) site, tail
nucleolus	transcription bubble	sequence (UAS)	poly(A) polymerase
nucleoplasm	rho-dependent	TBP, TFIIB, D, E, F,	RNA endonuclease
nucleus	termination	H, TAF	CPSF, CstF, CFI, CFII,
gene regulatory	type I, II terminators	initial committed	PAB II
elements	hairpin loop	complex	introns, exons
5′→3′, 3′→5′	twofold symmetry	minimal	intervening,
coding strand	coupled transcription	transcription	expressed sequence
template strand	and translation	initiation complex	hnRNA, gRNA
nontemplate strand	transcription unit	complete	mRNA splicing
NTPs	leader, coding,	transcription	branch-point sequence
	trailer sequences	initiation	3′, 5′ splice sites
	5′, 3′ UTR	complex	snRNA, snRNP
	polycistronic	preinitiation	spliceosome
	upstream, downstream	complex	U1, U2, U4, U5, U6
			RNA lariat structure
			RNA editing

5	6	7
E. coli RNA polymerase	RNA polymerase III	spliceosome
RNA polymerase I	5S rRNA genes	snRNPs ("snurps")
30S, 50S subunits	tRNA genes	self-splicing

5	6	7
70S ribosome	tDNA	group I introns
23S, 16S, 5S rRNA	internal control region (ICR)	protein-independent reaction
40S, 60S subunits	pre-tRNA molecule	secondary structure
80S ribosome	codon	lariat molecule
28S, 18S, 5.8S, 5S rRNA	anticodon	*Tetrahymena* pre-mRNA
ribosomal protein	posttranscriptional	28S spacer sequence removal
rDNA	modification	28S intron removal
rRNA transcription unit (*rrn*)	TFIIIA, B, C	RNA enzyme
precursor rRNA	transcription initiation	ribozyme
RNase III	factor	RNA ligase
rDNA repeat unit	cloverleaf	
spacer sequence	stem-loop structure	
nontranscribed spacer (NTS)	RNA ligase	
internal transcribed spacer (ITS)	RNA endonuclease	
external transcribed spacer (ETS)		

Next, distinguish between processes and characteristics in prokaryotes and eukaryotes. While there are some similarities between prokaryotes and eukaryotes, there are many, many differences. The fact that three RNA polymerases are active in eukaryotes results in three sets of data. For each RNA polymerase, there are distinct requirements for initiation and termination. Different DNA sequences are used, and different protein factors are involved in each case. It may help to prepare a list of analogous sites and factors, pairing prokaryotic elements with those doing similar jobs in eukaryotes.

QUESTIONS FOR PRACTICE

Multiple Choice Questions

1. Each row of the table below lists an item composed of DNA, RNA, and/or protein. Indicate an item's chemical composition by placing an X in the appropriate column(s).

Item	DNA	RNA	Protein
sigma factor			
promoter			
terminator			
transcription factor			
regulatory factor			
Pribnow box			
transcription initiation complex			
spacer sequences			
NTS			

(continued)

Item	DNA	RNA	Protein
snRNP			
spliceosome			
rho			
promoter element			
RNA polymerase core enzyme			
ribozyme			
consensus sequence			
U2			
enhancer			

2. Now indicate whether each item is found only in prokaryotes, only in eukaryotes, or in both prokaryotes and eukaryotes by placing an X in the appropriate column.

Item	Only in Prokaryotes	Only in Eukaryotes	Found in Both
sigma factor			
promoter			
terminator			
transcription factor			
regulatory factor			
Pribnow box			
transcription initiation complex			
spacer sequences			
NTS			
snRNP			
spliceosome			
rho			
promoter element			
RNA polymerase core enzyme			
ribozyme			
consensus sequence			
U2			
enhancer			

3. Which one statement is correct?
   a. During transcription, the template strand is read in a 3'→5' direction.
   b. During transcription, the template strand is read in a 5'→3' direction.
   c. During transcription, the nontemplate strand is read in a 3'→5' direction.
   d. During transcription, the nontemplate strand is read in a 5'→3' direction.

4. Which one statement is correct?
   a. During transcription, an RNA is synthesized in the 5'→3' direction.
   b. During transcription, an RNA is synthesized in the 3'→5' direction.
   c. During transcription, DNA is synthesized in the 5'→3' direction.
   d. During transcription, DNA is synthesized in the 3'→5' direction.

5. Which one of the following is incorrect?
   a. DNA polymerases can proofread but RNA polymerases cannot.
   b. Some RNA polymerases, but no DNA polymerases, are sensitive to α-amanitin.
   c. Only RNA polymerases can initiate the formation of a polynucleotide.
   d. RNA polymerases synthesize in a 3'→5' direction while DNA polymerases synthesize in a 5'→3' direction.

6. Which one of the following is essential for RNA polymerase to bind an *E. coli* promoter?
   a. sigma factor
   b. snRNP
   c. TFIID
   d. U1

7. Which one of the following is essential for RNA polymerase II to bind a eukaryotic promoter?
   a. sigma factor
   b. *rho*
   c. TFIID
   d. U1

8. Which one of the following is incorrect?
   a. Prokaryotic and eukaryotic mRNAs have 5' untranslated leader sequences.
   b. Only prokaryotic mRNAs are polyadenylated at their 3' ends.
   c. In prokaryotes, transcription and translation are coupled.
   d. Consensus sequences can be found in both DNA and RNA.
   e. In eukaryotes, RNA splicing occurs in the nucleus.

Answers:

1. A promoter, terminator, Pribnow box, NTS, promoter element, and enhancer are composed of DNA. U2 and a ribozyme are composed of RNA. A sigma factor, transcription factor, regulatory factor, *rho*, and the RNA polymerase core enzyme are composed of protein. Spacer and consensus sequences can be found in both RNA and DNA. A transcription initiation complex contains proteins that interact with DNA. Spliceosomes and snRNPs contain both RNA and proteins.

2. Sigma factor, the Pribnow box, *rho*, and the RNA polymerase core enzyme are only found in prokaryotes (*E. coli*). An NTS, snRNP, spliceosome, ribozyme, enhancer, and U2 are only found in eukaryotes. A promoter, terminator, transcription factor, regulatory factors, transcription initiation complex, spacer sequence, promoter element, and consensus sequence can be found in both eukaryotes and prokaryotes.

3a, 4a, 5d, 6a, 7c, 8b

Thought Questions

1. Compare and contrast each of the following groups of terms with respect to location, function, and host organism (prokaryote or eukaryote).
   a. Pribnow box and TATA element
   b. promoter elements and enhancer elements
   c. the promoters bound by RNA polymerase I, II, III
   d. rRNA genes in prokaryotes and eukaryotes

2. Do prokaryotes or eukaryotes have more RNA polymerase (per unit total protein)? Why should the two groups of organisms differ?

3. How does *rho*-dependent termination differ from *rho*-independent termination? What is the role of the hairpin loop in each of these events? Is the hairpin loop found in DNA or RNA?

4. What takes the place of sigma factor in eukaryotes?

5. Recall that there was no simple relationship between the C value and the structural and organizational complexity of an organism. Do you expect a relationship to exist between the number of rDNA repeat units and the structural or organizational complexity of the organism? The amount of hnRNA? The amount of cytoplasmic mRNA? Why or why not?

6. The promoters for RNA polymerase II have been highly conserved and thus differ little among eukaryotes. However, the promoters for RNA polymerase I show significant differences among eukaryotes. Why might this be the case? (Hint: Consider the number of genes read by each polymerase.)

7. Consider what will happen to an mRNA once it is processed, and then address the following questions: How precise does RNA splicing need to be? How precise does transcription initiation need to be? How precise does polyadenylation need to be?

8. What posttranscriptional modifications are routinely made to a eukaryotic mRNA? Where do they occur in the cell?

Thought Questions for Media

After reviewing the media on the *iGenetics* CD-ROM, try to answer these questions.

1. At which end of the template strand does transcription begin?
2. When trascription is complete, which strand of DNA is the final RNA molecule identical to if the thymine in DNA is replaced with uracil in RNA?
3. What purpose does the sequence AAUAAA serve in a pre-mRNA's 3′ UTR?

1. Suppose a mutation in the β-globin gene introduced a new 3′ splice site within its first intron.
   a. In what different ways could the mutation affect the reading frame of the mature mRNA?
   b. If the mature mutant mRNA were translated in the same reading frame after the mutant site, what would its effect be on the β-globin protein?
   c. If the mutation caused the mRNA to be translated in a different reading frame, what would be its effect on the β-globin protein? (Remember that when mRNAs are "read" out of their correct reading frames, stop codons are often introduced.)

2. Suppose a point mutation occurred in the first intron of the β-globin gene but did not change either the branch point, the 5′ splice site or the 3′ splice site sequence. Could such a mutation ever affect the level of mRNA transcripts available for translation? How might your answer change if the mutation were a deletion or insertion of 100 base pairs of DNA? (Recall that the first intron is 130 base pairs long.)

3. In the iActivity, the first multiple-choice question describes two forms of beta-thalassemia: beta-plus and beta-zero. Beta-plus thalassemia occurs when a reduced amount of β-globin protein is produced, while beta-zero thalassemia occurs when no β-globin protein is produced.
   a. Do you expect the phenotypes of these forms of thalassemia to be identical in heterozygotes having a disease allele and a normal allele? Why or why not?
   b. Will they be identical in homozygotes for a disease allele? Why or why not?

SOLUTIONS TO TEXT PROBLEMS

13.1   Compare DNA and RNA with regard to structure, function, location, and activity. Also, how do these molecules differ with regard to the polymerases used to synthesize them?

Answer: While both DNA and RNA are composed of linear polymers of nucleotides, their bases and sugars differ. DNA contains deoxyribose and thymine, while RNA contains ribose and uracil. Their structures also differ. DNA is frequently double stranded, while RNA is usually single stranded. Single-stranded RNAs are capable of forming stable, functional, and complex stem-loop structures, such as those seen in tRNAs. Double-stranded DNA is wound in a double helix and packaged by proteins into chromosomes, either as a nucleoid body in prokaryotes or within the eukaryotic nucleus. After being transcribed from DNA, RNA can be exported into the cytoplasm. If it is mRNA, it can be bound by ribosomes and translated. Eukaryotic RNAs are highly processed before being transported out of the nucleus. DNA functions as a storage molecule, while RNA functions either as a messenger (mRNA carries information to the ribosome), or in the processes of translation (rRNA functions as part of the ribosome, tRNA brings amino acids to the ribosome), or in eukaryotic RNA processing (snRNA functions within the spliceosome).

Both DNA polymerases and RNA polymerases catalyze the synthesis of nucleic acids in the 5′→3′ direction. Both use a DNA template and synthesize a nucleic acid polynucleotide that is complementary to the template. However, DNA polymerases require a 3′-OH to add onto, while RNA polymerases do not. That is, RNA polymerases can initiate chains without primers, while DNA polymerases cannot. Furthermore, RNA polymerases usually require specific base-pair sequences as signals to initiate transcription.

13.2   All base pairs in the genome are replicated during the DNA synthesis phase of the cell cycle, but only some of the base pairs are transcribed into RNA. How is it determined which base pairs of the genome are transcribed into RNA?

Answer: The DNA sequences that are transcribed into RNA are determined using two general principles. First, signals in the DNA base sequences identify the specific region to be transcribed. Only regions bounded by transcription initiation and transcription termination signals are transcribed. In regions bounded by these signals, only one strand is ordinarily transcribed, so that transcripts are formed in a 5′→3′ direction using a single DNA template strand. Second, transcription within a defined region occurs only if additional transcription-inducing molecules are present. Some of these molecules recognize the signals within the DNA that define the transcription unit and are used in the transcription process.

13.3   Discuss the similarities and differences between the *E. coli* RNA polymerase and eukaryotic RNA polymerases.

Answer: Both eukaryotic and *E. coli* RNA polymerases transcribe RNA in a 5′→3′ direction using a 3′→5′ DNA template strand. There are many differences between the enzymes, however. In *E. coli,* a single RNA polymerase core enzyme is used to transcribe genes. In eukaryotes, there are three types of RNA polymerase molecules: RNA polymerase I, II, and III. RNA polymerase I synthesizes 28S, 18S, and 5.8S rRNA and is found in the nucleolus. RNA polymerase II synthesizes hnRNA, mRNA, and some snRNAs and is nuclear. RNA polymerase III synthesizes tRNA, 5S rRNA, and some snRNAs and is also nuclear.

Each RNA polymerase uses a unique mechanism to identify those promoters at which it initiates transcription. In prokaryotes such as *E. coli,* a sigma factor provides specificity to the sites bound by the four-polypeptide core enzyme, so that it binds to promoter sequences. The holoenzyme loosely binds a sequence lying about 35 bp before transcription initiation (the −35 region), changes configuration, and then tightly binds a region lying about 10 bp before transcription initiation (the −10 region) and melts about 17 bp of DNA around this region. The two-step binding to the promoter orients the polymerase on the DNA and facilitates transcription initiation in the 5′→3′ direction. After about eight or nine bases are formed in a new transcript, sigma factor dissociates from the holoenzyme, and the core enzyme completes the transcription process. Although the principles by which eukaryotic RNA polymerases bind their promoters are similar in that they use a set of ancillary protein factors—transcription factors—the details are quite different. In eukaryotes, each of the three types of RNA polymerases recognizes a different set of promoters by using a polymerase-specific set of transcription factors, and the mechanisms of interaction are different.

13.4   What are the most significant differences between the organization and expression of prokaryotic genes and eukaryotic genes?

Answer: Prokaryotic and eukaryotic genes differ in their gene structure, in their RNA processing, and in how transcription is coupled to translation.

Structure: Prokaryotic genes are defined by an upstream promoter, an RNA-coding sequence, and a downstream terminator. While these three features also exist in eukaryotic genes, eukaryotic promoters are more complex, and nearby or distant enhancer and silencer elements can strongly affect the level of transcription of eukaryotic genes. The RNA-coding sequences of eukaryotes can be interrupted with introns. Finally, prokaryotic mRNAs are often polycistronic, containing the amino acid–coding information for more than one gene. In contrast, eukaryotic mRNAs are generally monocistronic, containing the amino acid–coding information from just one gene. A notable exception to this general principle is found in *C. elegans* where polycistronic mRNAs are found at certain loci.

Processing: Prokaryotic genes lack introns, while eukaryotic genes typically have one or more introns, and, therefore, a transcribed region can be larger than the size of a mature mRNA. The excision of introns from primary mRNAs is only one aspect of the processing of eukaryotic RNAs. In addition, eukaryotic mRNAs are modified at both their 5′ and 3′ ends: They are capped and polyadenylated.

Coupling of transcription and translation: Since prokaryotes lack a nucleus, transcription is directly coupled to translation. In eukaryotes, mRNAs must be processed and then transported out of a nucleus before they are translated by ribosomes in the cytoplasm.

These three differences provide cell types in a multicellular eukaryote with three means to regulate gene expression. Gene expression can be regulated at each of these three levels: transcription, mRNA processing, and translation.

13.5   Discuss the molecular events involved in the termination of RNA transcription in prokaryotes. In what ways is this process fundamentally different in eukaryotes?

Answer: Termination of transcription in *E. coli* is signaled by controlling elements (sequences within the DNA) called terminators. Two classes of terminators exist, *rho*-independent (Type I) and *rho*-dependent (Type II). Both Type I and Type II termination events lead to the cessation of RNA synthesis and the release of both the RNA chain and the RNA polymerase from the DNA.

Type I terminators utilize sequences with twofold symmetry lying about 16–20 bp upstream of the termination point to signal the termination site. See text Figure 13.5. The hairpin loop that forms when a twofold symmetric sequence is transcribed, plus a string of Us, lead to termination, perhaps by destabilizing the RNA-DNA hybrid in the terminator region. Type II terminators lack the structure of Type I terminators, and instead use an ATP-activated *rho* protein that binds to recognition sequences in the transcribed termination region. The binding of *rho* leads to the hydrolysis of the ATP, and the release of the transcript and the RNA polymerase from the DNA template.

In eukaryotes, transcription termination differs depending on the RNA polymerase under consideration. RNA polymerase I transcribes repeated 18S, 5.8S, and 28S rDNA clusters as single transcription units. Transcription of each unit is terminated at a specific site. The termination site lies in the nontranscribed spacer (NTS) between adjacent rDNA transcription units.

RNA polymerase II transcribes genes for mRNAs and some snRNAs. Transcription termination for these genes is fundamentally different from prokaryotic transcription termination, as these genes lack specific transcription termination sequences at their 3′ ends. mRNA transcription can continue for hundreds or thousands of nucleotides downstream of the protein-coding sequence until it is past a poly(A) site in the RNA. The poly(A) site is recognized and cleaved by a complex set of proteins. It is positioned 10 to 30 nucleotides after an AAUAAA sequence and is followed by a GU-rich or U-rich sequence. Once the RNA is cleaved, poly(A) polymerase adds A nucleotides onto the 3′-OH of the RNA to produce a poly(A) tail.

RNA polymerase III transcribes genes for 5S rRNA, tRNA and some snRNAs. Termination events for RNA polymerase III were not discussed in this chapter.

13.6 More than 100 promoters in prokaryotes have been sequenced. One element of these promoters is sometimes called the Pribnow box, named after the investigator who compared several *E. coli* and phage promoters and discovered a region they held in common. Discuss the nature of this sequence. (Where is it located and why is it important?) Another consensus sequence appears a short distance from the Pribnow box. Diagram the positions of the two prokaryotic promoter elements relative to the start of transcription for a typical *E. coli* promoter.

Answer: The Pribnow box is the −10 element of the prokaryotic promoter. It is located at −10 bp relative to the starting point of transcription and has the consensus sequence 5′-TATAAT-3′ (in the coding strand). A second consensus sequence is located about 35 bp before transcription initiation (the −35 region). Each of these sequences is important for the binding of RNA polymerase and initiation of RNA transcription. At the start of transcription initiation, the holoenzyme loosely binds the −35 region, then changes configuration and tightly binds the −10 region. It melts about 17 bp of DNA around this region. The two-step binding to the promoter orients the polymerase on the DNA and facilitates transcription initiation in the 5′→3′ direction, as shown in Figure 13.4 and the following cartoon:

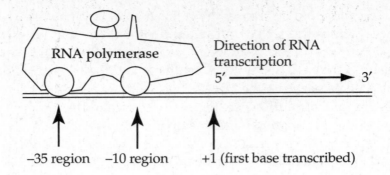

13.7   An *E. coli* transcript with the first two nucleotides 5'-AG-3' is initiated from the following segment of double-stranded DNA:

5'-TAGTGTATTGACATGATAGAAGCACTCTTACTATAATCTCAATAGCTACG-3'
3'-ATCACATAACTGTACTATCTTCGTGAGAATGATATTAGAGTTATCGATGC-5'

a. Where is the transcription start site?

b. What are the approximate locations of the regions that bind the RNA polymerase homoenzyme?

c. Does transcription elongation proceed towards the right or left?

d. Which DNA strand is the template strand?

e. Which DNA strand is the RNA-coding strand?

Answer:

a, b. There are multiple 5'-AG-3' sequences in each strand, and transcription may proceed in either direction. Determine the correct initiation site by locating the −10 and −35 consensus sequences recognized by RNA polymerase and σ^{70}. Good −35 (TTGACA) and −10 (TATAAT) consensus sequences are found on the top strand, starting at the 8th and 32nd bases from the 5' end, respectively, indicating that the initiation site is the 5'-AG-3' starting at the 44th base from the 5' end of that strand.

c. Transcription proceeds from left to right in this example.

d. the bottom (3'→5') strand

e. the top (5'→3') strand

13.8   Figure 13.A shows the sequences, given 5'→3', that lie upstream a subset of *E. coli* genes transcribed by RNA polymerase and σ^{70}. Carefully examine the sequences in the −10 and −35 regions, and then answer the following questions:

a. The −10 and −35 regions have the consensus sequences 5'-TATAAT-3' and 5'-TTGACA-3', respectively. How many of the genes that are listed have sequences that perfectly match the −10 consensus? How many have perfect matches to the −35 consensus?

b. Based on your examination of these sequences, what does the term *consensus sequence* mean?

c. What is the function of these consensus sequences in transcription initiation?

d. More generally, what might you infer about a DNA sequence if it is part of a consensus sequence?

e. None of these promoters have perfect consensus sequences, but some have better matches than others. Speculate what consequence this might have on the efficiency of transcription initiation.

Gene	−35 Region	−10 Region	Initiation Region
lac	ACCCAGGCTTTACACTTTATGGCTTCCGGCTCGTATGTTGTGTGGAATTGTGAGCGG		
lac1	CCATCGAATGGCGCAAAACCTTTCGCGGTATGGCATGATAGCGCCCGGAAGAGAGTC		
galP2	ATTTATTCCATGTCACACTTTTCGCATCTTTGTTATGCTATGGTTATTTCATACCAT		
araB,A	GGATCCTACCTGACGCTTTTTATCGCAACTCTCTACTGTTTCTCCATACCCGTTTTT		
araC	GCCGTGATTATAGACACTTTTGTTACGCGTTTTTGTCATGGCTTTGGTCCCGCTTTG		
trp	AAATGAGCTGTTGACAATTAATCATCGAACTAGTTAACTAGTACGCAAGTTCACGTA		
bioA	TTCCAAAACGTGTTTTTTGTTGTTAATTCGGTGTAGACTTGTAAACCTAAATCTTTT		
bioB	CATAATCGACTTGTAAACCAAATTGAAAAGATTTAGGTTTACAAGTCTACACCGAAT		
tRNA^Tyr	CAACGTAACACTTTACAGCGGCGCGTCATTTGATATGATGCGCCCCGCTTCCCGATA		
rrnD1	CAAAAAAATACTTGTGCAAAAAATTGGGATCCCTATAATGCGCCTCCGTTGAGACGA		
rrnE1	CAATTTTTCTATTGCGGCCTGCGGAGAACTCCCTATAATGCGCCTCCATCGACACGG		
RRNa2	AAAATAAATGCTTGACTCTGTAGCGGGAAGGCGTATTATGCACACCCCGCGCCGCTG		

Answer: In the following figure, possible −10 and −35 regions of each sequence are boxed and consensus sequences are in boldface type.

Gene	−35 Region	−10 Region	Initiation Region
lac	ACCCAGGCTT **TACACT** TTATGGCTTCCGGCTCG **TATGTT** GTGTGGAATTGTGAGCGG		
lac1	CCATCGAATG **GCGCAA** AACCTTTCGCGGTATGG **CATGAT** AGCGCCCGGAAGAGAGTC		
galP2	ATTTATTCCAT **GTCACA** CTTTTCGCATCTTTGT **TATGCT** ATGGTTATTTCATACCAT		
araB,A	GGATCCTAC **CTGACG** CTTTTTATCGCAACTCTC **TACTGT** TTCTCCATACCCGTTTTT		
araC	GCCGTGATTA **TAGACA** CTTTTGTTACGCGTTTT **TGTCAT** GGCTTTGGTCCCGCTTTG		
trp	AAATGAGCTG **TTGACA** ATTAATCATCGAACTAG **TTAACT** AGTACGCAAGTTCACGTA		
bioA	TTCCAAAAC **GTGTTT** TTTGTTGTTAATTCGGTG **TAGACT** TGTAAACCTAAATCTTTT		
bioB	CATAATCGAC **TTGTAA** ACCAAATTGAAAAGATT **TAGGTT** TACAAGTCTACACCGAAT		
tRNA^Tyr	CAACGTAACAC **TTTACA** GCGGCGCGTCATTTGA **TATGAT** GCGCCCCGCTTCCCGATA		
rrnD1	CAAAAAAATAC **TTGTGC** AAAAAATTGGGATCCC **TATAAT** GCGCCTCCGTTGAGACGA		
rrnE1	CAATTTTTCTA **TTGCGG** CCTGCGGAGAACTCCC **TATAAT** GCGCCTCCATCGACACGG		
RRNa2	AAAATAAATGC **TTGACT** CTGTAGCGGGAAGGCG **TATTAT** GCACACCCCGCGCCGCTG		

a. Two sequences (*rrnD1, rrnE1*) have perfect matches with the −10 consensus sequence, one sequence (*trp*) has a perfect match with the −35 consensus sequence.

b. Examining these sequences illustrates the fact that the identification of a consensus sequence in a promoter or another DNA sequence does not mean that the same sequence will always be present in a specific region. A consensus sequence is identified by determining the most common base at a particular position in a specified region in a large number of aligned sequences. For example, in the set of sequences shown here, the most frequently found bases in each position of the −10 region are 5'-TATGAT-3'. The most frequent bases in all of the positions except for the fourth position match −10 region consensus sequence (5'-TATAAT-3').

c. These are the sites that the RNA polymerase holoenzyme initially recognizes and binds to during transcription initiation.

d. It is likely to have some functional importance, perhaps because it, like these consensus sequences, is recognized by a protein.

e. Sequences closer to the consensus sequence may be more easily recognized by the RNA polymerase holoenzyme. This could result in more efficient transcription initiation so that the gene is more frequently transcribed.

13.9   The single RNA polymerase of *E. coli* transcribes all of its genes, even though these genes do not all have identical promoters.

a. What different types of promoters are found in the genes of *E. coli*?

b. How is the single RNA polymerase of *E. coli* able to initiate transcription even though it uses different types of promoters?

c. Why might it be to *E. coli*'s advantage to have genes with different types of promoters?

Answer:

a. *E. coli* promoters vary with the type of sigma factor that is used to recognize them. More than four types of promoters exist, each having different recognition sequences. Most promoters have -35 and -10 sequences that are recognized by σ^{70}. Other promoters have consensus sequences that are recognized by different sigma factors, which are used to transcribe genes needed under altered environmental conditions such as heat shock and stress (σ^{32}), limiting nitrogen (σ^{54}), or when cells are infected by phage T4 (σ^{23}).

b. Although there is one core RNA polymerase enzyme, different RNA polymerase holoenzymes are formed from different sigma factors. Promoter recognition is determined by the sigma factor.

c. Utilizing different sigma factors allows for a quick response to altered environmental conditions (for example, heat shock, low N_2, phage infection) by the coordinated production of a set of newly required gene products.

13.10   Three different RNA polymerases are found in all eukaryotic cells, and each is responsible for synthesizing a different class of RNA molecules. How do the characteristics of the eukaryotic RNA polymerases differ in terms of their cellular location and products?

Answer: RNA polymerase I transcribes the major rRNA genes that code for 18S, 5.8S, and 28S rRNAs and is found exclusively in the nucleolus; RNA polymerase II transcribes the protein-coding genes to produce mRNA molecules and some snRNAs and is found in the nucleoplasm; RNA polymerase III transcribes the 5S rRNA genes, the tRNA genes, and some small nuclear RNAs and is found in the nucleoplasm.

In the cell, the 18S, 5.8S, 28S, and 5S rRNAs are structural and functional components of the ribosome, which functions during translation. The mRNAs are translated to produce proteins. The tRNAs bring amino acids to the ribosome to donate to the growing polypeptide chain during protein synthesis. Small nuclear RNAs function in nuclear processes such as RNA splicing and processing.

13.11   *E. coli* RNA polymerase can transcribe all the genes of *E. coli*, but the three eukaryotic RNA polymerases transcribe only specific, nonoverlapping subsets of eukaryotic genes. What mechanisms are used to restrict the transcription of each of the three eukaryotic polymerases to a particular subset of eukaryotic genes?

Answer: The initiation of transcription at different genes by RNA polymerases I, II, and III is restricted through the use of three distinct sets of transcription factors that recognize three distinct classes of promoters. Since the genes transcribed by each polymerase have uniquely organized controlling regions and the transcription factors that interact with each polymerase recognize only the regulatory elements in one type of controlling region, transcription by each polymerase is specific and nonoverlapping. See Figure 13.7 for initiation by RNA polymerase II.

13.12 Figure 13.3 shows the structure of a prokaryotic gene, showing its promoter, RNA-coding sequence, and terminator region. Modify the figure to show the general structures of eukaryotic genes transcribed by

a. RNA polymerase II

b. RNA polymerase III

Answer:

a.

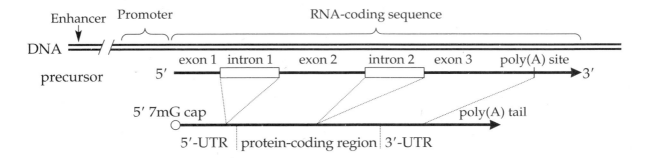

b.

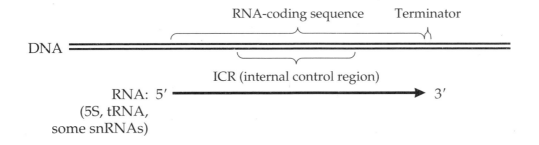

13.13 A piece of mouse DNA was sequenced as follows (a space is inserted after every tenth base for ease in counting; (. . .) means a lot of unspecified bases):

```
AGAGGGCGGT   CCGTATCGGC   CAATCTGCTC   ACAGGGCGGA
TTCACACGTT   GTTATATAAA   TGACTGGGCG   TACCCCAGGG
TTCGAGTATT   CTATCGTATG   GTGCACCTGA   CT(...)
GCTCACAAGT   ACCACTAAGC(...).
```

What can you see in this sequence to indicate that it might be all or part of a transcription unit?

Answer: By convention, DNA sequence is given 5'→3'. This is the same polarity as the antitemplate strand, the strand that is complementary to the 3'→5' template strand used for transcription. By using this convention, analyzing a DNA sequence is made easier. If a region of the sequence is transcribed, the RNA will have an identical sequence, except that U will replace T.

   Transcription units should be flanked at their 5' ends by promoter elements. Therefore, approach this problem by surveying the sequence for the GC, CAAT, and TATA box consensus sequences. The figure below summarizes their location and the approximate site of transcription initiation. An mRNA that could be produced (a conceptual mRNA) is given in lowercase letters. This mRNA has two signals (underlined) that could be used by the ribosome to recognize the start and termination of translation. Initiation of a polypeptide chain is signaled by a 5'-AUG-3' codon (encoding a methionine). Chain termination is signaled by a UAA stop codon. Features of translation are described in Chapter 14.

<pre>
              −80   −65   −50
          GC element   CAAT box   GC element
  5'   AGAGGGCGGT CCGTATCGGC CAATCTGCTC AGAGGGCGGA

                            −30
                          TATA box
       TTCACACGTT GTTATATAAA TGACTGGGCG TACCCCAGGG

  +1 (approx., conceptual)
                   transcription   potential
                   initiation   translation start
           TTCGAGTATT CTATCGTATG GTGCACCTGA CT(...)
  mRNA   5'uauu cuaucguaug gugcaccuga cu(...)

  potential translation stop
       GCTCACAAGT ACCACTAAGC (...)
       gcucacaagu accacuaagc (...)
</pre>

13.14 How do the structures of mRNA, rRNA, and tRNA differ? Hypothesize a reason for the difference.

Answer: Although mRNA, rRNA, and tRNA are single stranded, they can differ substantially in the secondary and tertiary structures achieved by intramolecular base-pairing interactions. These interactions, in the case of rRNA and tRNA, are essential for them to fulfill their different functions.

   mRNA typically has a 5' untranslated leader sequence, a protein-coding sequence, and a 3' untranslated trailer sequence (see Figure 13.8). In eukaryotes, mRNAs are processed to have 5' caps and 3' poly(A) tails, and to remove intronic sequences (see Figures 13.9b, 13.10, 13.11, and 13.12).

   rRNAs are of three sizes in prokaryotes (5S, 16S, 23S) and four sizes in eukaryotes (5S, 5.8S, 18S, and 28S). They are single stranded, and three rRNAs are produced by processing a single rRNA precursor transcript (an exception is eukaryotic 5S rRNA) (see Figures 13.18 and 13.19). They fold in a complex manner and have both structural and functional roles in the ribosome.

tRNA molecules are small, about 4S, and consist of a single chain of 75 to 90 nucleotides. There is extensive base pairing between different parts of the tRNA molecule, allowing it to form a cloverleaf secondary-structure having three or four loops. The tRNA molecule functions to carry amino acids to the ribosome. An amino acid is attached to its 3' end in the cytoplasm for delivery to the ribosome. After an anticodon in loop II binds to the mRNA, the ribosome transfers the amino acid to a growing polypeptide chain. There is extensive posttranscriptional modification of tRNA bases, so that tRNAs contain modified bases such as inosine, ribothymidine, pseudouridine, dihydrouridine, methylguanosine, dimethylguanosine, and methylinosine. See Figure 13.20.

13.15   Many eukaryotic mRNAs, but not prokaryotic mRNAs, contain introns. Describe how these sequences are removed during the production of mature mRNA.

Answer: Introns are removed from a pre-mRNA molecule by the action of a spliceosome. Consensus sequences at the intron-exon boundaries and within the intron itself are recognized by spliceosome components as the spliceosome assembles around the intronic sequence. First, U1 snRNP binds to the 5' splice site by base-pairing interactions between the U1 snRNA and the 5' splice site sequence. Then, U2 snRNP binds to a branch-point region inside of the intron. Following the association of a preassembled U4/U6/U5 particle with the bound U1 and U2 snRNPs, U4 snRNP dissociates from the complex, allowing the formation of an active spliceosome. Splicing proceeds via cleavage of the 5' exon-intron junction, attaching the 5' end of the intron sequence to a branch-point sequence via an unusual 2'-5' bond, and then cleavage at the 3' exon-intron junction. When the two exons are ligated together, a lariat structure composed of intron sequences is released. See Figures 13.12 and 13.13.

13.16   How is the mechanism of group I intron removal different from the mechanism used to remove the introns in most eukaryotic mRNAs? Speculate as to why these different mechanisms for intron removal might have evolved and how each might be advantageous to a eukaryotic cell.

Answer: Group I introns are self-splicing. Unlike the removal of introns from most eukaryotic mRNAs, removal of group I introns does not require a spliceosome containing snRNAs and protein-splicing factors. In group I introns, the intron folds in a protein-independent manner to catalyze the splicing reaction.

While both splicing mechanisms result in the joining of RNA sequences flanking an intron, the more widespread use of spliceosomes to remove introns may reflect advantages associated with this mechanism. For example, it may allow greater flexibility in the sequence composition of an intron, as specific secondary structures do not have to form, while still allowing for the precise removal of intronic sequences (using consensus sequences at the 5' and 3' splice sites and the branch point). Spliceosome processing of introns may also allow for greater diversity in intron size. The group I intron in the *Tetrahymena* pre-rRNA is 413 bp. In contrast, introns processed by spliceosomes can be from tens to thousands of base pairs. As is discussed in Chapter 20, intron removal using spliceosomes can also be regulated using protein factors. This allows for an additional level of control over gene expression. Thus, splicing mechanisms using spliceosomes allow for greater diversity in intron composition and gene expression than self-splicing mechanisms.

13.17   Distinguish between the following terms: *leader sequence, trailer sequence, coding sequence, intron, spacer sequence, nontranscribed spacer sequence, external transcribed spacer sequence,* and *internal transcribed sequence.* Give examples of actual molecules in your answer.

Answer: Leader and *trailer sequences* are used to denote untranslated sequences in mRNA not containing protein-coding information that lie at the 5′ end (for leader) or 3′ end (for trailer). See text Figure 13.12.

    Coding sequences are sequences in mRNA that are translated by the ribosome. These sequences encode information (in the form of a triplet code—see Chapters 1 and 14) that can be read by the ribosome to direct the synthesis of a polypeptide. *Introns* are sequences present in eukaryotic precursor RNA molecules that are spliced out in the mature, functional molecule. Introns are present in pre-mRNAs as well as in pre-rRNAs and pre-tRNAs.

    Nontranscribed spacer sequences are sequences, such as those in the repeated rDNA genes, that lie between adjacent, repeating transcription units. *External transcribed spacer sequences* are found in pre-rRNA transcripts in eukaryotes (from transcription of 18S, 28S, and 5.8S pre-rRNAs). They are located immediately upstream of the 5′ end of the 18S sequence and downstream of the 3′ end of the 28S sequence. The *internal transcribed spacer sequences* are located on either side of the 5.8S sequence. See text Figures 13.18 and 13.19.

13.18   Discuss the posttranscriptional modifications and processing events that take place on the primary transcripts of eukaryotic rRNA and protein-coding genes.

Answer: In eukaryotes, 18S, 28S, and 5.8S rRNA genes are transcribed from the rDNA into a single pre-rRNA molecule, which is then processed by removing the internal and external spacer sequences, leading to the production of the mature rRNAs. (See Figure 13.19 in text.) The eukaryotic 5S rRNA is transcribed separately to produce mature rRNA molecules that need no further processing. Eukaryotic 5S rRNAs are imported into the nucleolus, where they are assembled with the other mature rRNAs and the ribosomal protein subunits to produce functional ribosomal subunits. Some *Tetrahymena* pre-rRNAs have self-splicing introns (group I) in the 28S rRNA. Remember, the removal of the intron in the 28S rRNA via self-splicing is a process that is separate from the cleavage of spacer sequences from the mature rRNAs. (See Figure 13.14 in text.)

    Protein-coding eukaryotic transcripts synthesized by RNA polymerase II are extensively processed. A 5′-5′ bonded m^{7}Gppp cap is added to the 5′ end of the nascent transcript when the chain is 20 to 30 nucleotides long. A poly(A) tail of variable length can be added to the 3′ end of the transcript, and spliceosomes remove intronic sequences. The site of poly(A) addition, as well as splice site selection, can be regulated, so that more than one alternatively processed mature mRNA is sometimes produced from a single precursor mRNA transcript. Some mRNAs are posttranscriptionally edited: Nucleotides are inserted, deleted, or converted from one base to another.

13.19   Small RNA molecules such as snRNAs and gRNAs play essential roles in eukaryotic transcript processing.

a. Where are these molecules found in the cell, and what roles do they have in transcript processing?

b. How is the abundance of snRNAs related to their role in transcript processing?

Answer:

a. snRNAs are found in the nucleus and are used in RNA splicing. They form snRNPs by complexing with proteins. snRNPs assemble around intronic sequences (starting with U1 snRNP binding to the 5′ splice site and U2 snRNP binding to the branch-point sequence) to form spliceosomes that remove the intronic sequences and splice together exonic sequences. See text Figure 13.13.

   gRNAs are found in the nucleus and mitochondria and are involved in RNA editing, the posttranscriptional insertion or deletion of nucleotides or the modification of one base to another. In trypanosomes where RNA editing results in the removal and insertion of U nucleotides, the gRNA pairs with the mRNA transcript and is thought to be responsible for cleaving the transcript, templating the missing U nucleotides, and ligating the transcript back together again.

b. snRNAs are highly abundant with at least 10^5 copies per cell. Their abundance reflects the large number of transcripts with introns that must be processed before the transcripts are exported from the nucleus into the cytoplasm for translation.

13.20   Describe the organization of the ribosomal DNA repeating unit of a higher eukaryotic cell.

Answer: rDNA in a eukaryotic cell is organized into tandem arrays of repeating units, each of which have an 18S, 5.8S, and 28S rRNA gene. In between each repeating unit is a nontranscribed spacer. The three genes within each unit are transcribed by RNA polymerase I as a 45S precursor rRNA molecule in the nucleolus. 5′ and 3′ flanking spacer sequences (external transcribed spacer sequences) as well as internal spacer sequences are removed, leading to the production of 18S, 5.8S, and 28S rRNAs. See text Figure 13.19.

13.21   Which of the mutations that follow would be likely to be recessive lethals in humans? Explain your reasoning.
   a. deletion of the U1 genes
   b. a single base-substitution mutation in the U1 gene that prevented U1 snRNP from binding to the 5′-GU-3′ sequence found at the 5′ splice junctions of introns
   c. deletion within intron 2 of β-globin
   d. deletion of four bases at the end of intron 2 and three bases at the beginning of exon 3 in β-globin

Answer: A recessive lethal is a mutation that causes death when it is homozygous—that is, when only mutant alleles are present. Heterozygotes for such mutations can be viable. Recessive lethal mutations result in death because some essential function is lacking. Neither copy of the gene functions, so the organism dies.

   a. Deletion of the U1 genes will be recessive lethal, as U1 snRNA is essential for the identification of the 5′ splice site in RNA splicing. Incorrect splicing would lead to nonfunctional gene products for many genes, a nonviable situation.
   b. This mutation would prevent U1 from base-pairing with 5′ splice sites and thus, by the same reasoning as in (a), would be recessive lethal.

c. If a deletion within intron 2 did not affect a region important for its splicing (for example, the branch point or the regions near the 5' or 3' splice sites), it would have no effect on the mature mRNA produced. Consequently, such a mutation would lack a phenotype if it were homozygous. However, if the splicing of intron 2 were affected and the mRNA altered, such a mutation, if homozygous, could result in the production of only nonfunctional hemoglobin, leading to severe anemia and death.

d. The deletion described would affect the 3' splice site of intron 2, leading to, at best, aberrant splicing of that intron. If the mutation were homozygous, only a nonfunctional protein would be produced, resulting in severe anemia and death.

13.22 The following figure shows the transcribed region of a typical eukaryotic protein-coding gene.

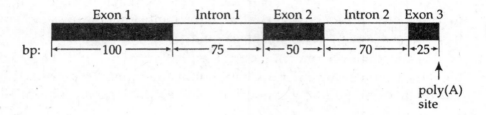

What is the size (in bases) of the fully processed, mature mRNA? Assume a poly(A) tail of 200 As in your calculations.

Answer: 1 (5' m^{7}G cap) + 100 (exon 1) + 50 (exon 2) + 25 (exon 3) + 200 (poly(A) tail) = 376 bases.

13.23 Most human obesity does not follow Mendelian inheritance patterns, because body fat content is determined by a number of interacting genes and environmental variables. Insights into how specific genes function to regulate body fat content have come from studies of mutant, obese mice. In one mutant strain, *tubby (tub)*, obesity is inherited as a recessive trait. A comparison of the DNA sequence of the *tub$^+$* and *tub* alleles has revealed a single base pair change: Within the transcribed region, a 5' G-C base pair has been mutated to a T-A base pair. The mutation causes an alteration of the initial 5' base of the first intron. Therefore, in the homozygous *tub/tub* mutant, a longer transcript is found. Propose a molecularly based explanation for how a single base change causes a nonfunctional gene product to be produced, why a longer transcript is found in *tub/tub* mutants, and why the *tub* mutant is recessive.

Answer: The first two bases of an intron are typically 5'-GU-3' and are essential for base-pairing with the U1 snRNA during spliceosome assembly. A GC-to-TA mutation at the initial base pair of the first intron impairs base pairing with the U1 snRNA, so that the 5' splice site of the first intron is not identified. This causes the retention of the first intron in the *tub* mRNA and a longer mRNA transcript in *tub/tub* mutants. When the mutant *tub* mRNA is translated, retention of the first intron could result in the introduction of amino acids not present in the *tub$^+$* protein or, if the intron contained a chain termination (stop) codon, premature translation termination and the production of a truncated protein. In either case, a nonfunctional gene product is produced.

The *tub* mutation is recessive because the single *tub*⁺ allele in a *tub/tub*⁺ heterozygote produces mRNAs that are processed normally, and when these are translated, enough normal (*tub*⁺) product is produced to obtain a wild-type phenotype. Only the *tub* allele produces abnormal transcripts. When both copies of the gene are mutated in *tub/tub* homozygotes, no functional product is made and a mutant, obese phenotype results.

13.24   Which of the deletions that follow could occur in a single mutational event in a human? Explain.

    a. deletion of 10 copies of the 5S ribosomal RNA genes only

    b. deletion of 10 copies of the 18S rRNA genes only

    c. simultaneous deletion of 10 copies of the 18S, 5.8S, and 28S rRNA genes only

    d. simultaneous deletion of 10 copies each of the 18S, 5.8S, 28S, and 5S rRNA genes

Answer:

    a. This could occur. Since the 5S genes are clustered, 10 copies could be removed in one deletion.

    b. This could not occur. The 18S genes are part of a cluster of 18S, 5.8S, and 28S genes. The 18S genes cannot be deleted without also deleting the others.

    c. This could occur. The genes are clustered.

    d. This could not occur. The 18S, 5.8S, and 28S genes are clustered at a locus separate from the 5S genes.

13.25   During DNA replication in a mammalian cell, a mistake occurs: Ten wrong nucleotides are inserted into a 28S rRNA gene. The mistake is not corrected. What will probably be its effect on the cell?

Answer: The cell will most likely be just fine. The remaining 1,000 copies of the 28S rRNA genes are normal, so this error will have little phenotypic consequence.

13.26   Choose the correct answers, noting that each blank may have more than one correct answer and that each answer (1 through 4) could be used more than once.

Answers:

1. eukaryotic mRNAs

2. prokaryotic mRNAs

3. transfer RNAs

4. ribosomal RNAs

    a. _____ have a cloverleaf structure

    b. _____ are synthesized by RNA polymerases

    c. _____ display one anticodon each

    d. _____ are the template of genetic information during protein synthesis

    e. _____ contain exons and introns

    f. _____ are of four types in eukaryotes and only three types in *E. coli*

    g. _____ are capped on their 5′ end and polyadenylated on their 3′ end

Answer:

a. 3

b. 1, 2, 3, 4

c. 3

d. 1, 2

e. 1 (note that some tRNAs and rRNAs have introns)

f. 4

g. 1

14

GENE EXPRESSION: TRANSLATION

CHAPTER OUTLINE

THINKING ANALYTICALLY

Much of the material presented in this chapter is detailed and descriptive. The initiation, elongation, and termination of translation are all complex events. Each involves ribosomal subunits, RNAs, and stage-specific factors that differ in prokaryotes and eukaryotes. As in the past two chapters, placing the information in context will help you learn and retain it.

There are several levels in which to organize this chapter's information. First, consider the level of spatial organization. Associate the different types of RNA (rRNA, mRNA, tRNA) and factors involved in translation with the different components of the translation machinery in the cell. Then construct a temporal framework to follow these components through the stages of translation. For prokaryotes, and then for eukaryotes, what RNAs and factors are components of the first initiation complex? What RNAs and factors are present during elongation? What new factors are required for termination? Then come back to the level of spatial organization. For eukaryotes, once proteins are synthesized, what factors are used to shuttle them to the appropriate location?

REVIEW OF KEY TERMS, SYMBOLS, AND CONCEPTS

In your own words, write a brief, precise definition of each term in the groups below. Check your definitions using the text. Then, develop a concept map using the terms in each list.

1	2	3	4
protein	genetic code	translation	endoplasmic
polypeptide	triplet code	aminoacyl-tRNA	reticulum
amino acid	codon	synthetase	membrane-bound
α-carbon	frameshift mutation	codon recognition	ribosome
R group	revertant	specificity of codon	signal hypothesis
peptide bond	reversion	recognition	signal sequence
acidic, basic, neutral	random copolymers	initiation	signal recognition
polar, neutral	ribosome-binding	elongation	particle (SRP)
nonpolar amino	assay	termination	docking protein
acid	cell-free, protein-	aminoacyl-tRNA	cotranslational
C, N terminus	synthesizing	charged tRNA	transport
primary structure	system	transformylase	signal peptidase
secondary structure	reading frame	fMet, tRNA.fMet	Golgi apparatus
tertiary structure	comma-free code	tRNA.Met	cisternal space
quaternary structure	nonoverlapping code	30S, 70S initiation	glycoprotein
conformation	universal, degenerate	complex	
α-helix	code	ribosome-binding site	
β-pleated sheet	start, stop codon	Shine-Dalgarno	
chaperone	sense, nonsense codon	sequence	
	wobble hypothesis	IF1, IF2, IF3, eIFs	
		cap binding protein	
		PABP	
		Kozak sequence	
		scanning model	
		peptidyl transferase	
		P, A, E sites	
		EF-Tu, EF-Ts, EF-G, GTP	
		translocation	
		polyribosome	
		polysome	
		termination factor	
		RF1, RF2, RF3, eRF1, eRF3	
		UGA, UAG, UAA codon	

Keep in mind the polarity of DNA and RNA when analyzing DNA sequences, inferring transcripts that are produced, and identifying translation reading frames. Transcripts are always synthesized in the 5′→3′ direction using the 3′→5′ DNA template strand, and polypeptides are always synthesized starting at their N-terminus using codons that start relatively nearer to the 5′ end of the mRNA. The rhyme "5 to 3, N to C" may help you remember the orientations of mRNA and polypeptides. Pay close attention to maintain the correct mRNA, tRNA anticodon, and DNA polarity.

One of the highlights of this chapter is the presentation of a set of elegant experiments used to decipher the genetic code. These technically demanding experiments addressed one of the fundamental questions in genetics: How is information in DNA used to code for proteins? Consider them carefully, and consider how interpreting them often uses probabilistic thinking. The product rule, which states that the frequency of two independent events occurring together is the product of their separate probabilities, allows one to deduce the appropriate codons that specify particular amino acids. Consider the following related problem:

   a. A synthetic mRNA is generated by randomly polymerizing adenine and guanine ribonucleotides. What codons would be formed, and what would be their relative frequency, if the synthesis began with a limitless mixture of 3 adenine to 1 guanine?

   b. When the mRNAs generated in (a) are added to a cell-free translation system, polypeptides are formed that, on average, have about 55 percent lysine, about 19 percent arginine, about 19 percent glutamic acid, and about 7 percent glycine. What can you infer about codon usage from these data (without consulting a table of the genetic code)?

Answer: First, determine what possible codons (three-nucleotide combinations) could be generated. Then determine the frequency of that codon by multiplying together the probability of obtaining each base in one codon. Since there is a 3 A : 1 G ratio, the chance of finding an A is ¾, and the chance of finding a G is ¼.

$P(\text{AAA}) = (\frac{3}{4})^3 = 0.421875 \sim 42\%$
$p(\text{AAG}) = (\frac{3}{4})^2 (\frac{1}{4}) = 0.140625 \sim 14\%$
$p(\text{AGA}) = (\frac{3}{4})^2(\frac{1}{4}) = 0.140625 \sim 14\%$
$p(\text{AGG}) = (\frac{3}{4})(\frac{1}{4})^2 = 0.046875 \sim 5\%$
$p(\text{GAA}) = (\frac{1}{4})(\frac{3}{4})^2 = 0.140625 \sim 14\%$
$p(\text{GAG}) = (\frac{1}{4})^2(\frac{3}{4}) = 0.046875 \sim 5\%$
$p(\text{GGA}) = (\frac{1}{4})^2(\frac{3}{4}) = 0.046875 \sim 5\%$
$p(\text{GGG}) = (\frac{1}{4})^3 = 0.015625 \sim 2\%$

Since AAA accounts for about 42 percent of the total codons that are translated, and only lysine constitutes at least 42 percent of the amino acid content of the polypeptides, AAA must encode lysine. By similar reasoning, GGG must encode glycine. Since there is only 7 percent glycine, and no codon alone accounts for 7 percent of the total codon usage, glycine must be encoded by at least two codons. GGG must be one of these, since if any other two codons encoded glycine, the amount of glycine would exceed 7 percent. The remaining glycine must be encoded by one of the GAG, GGA, or AGG codons, since they are represented 5 percent of the time. The remaining lysine must be encoded by either AAG, AGA, or GAA, as no combination of other codons would give 55 percent lysine. This leaves arginine to be encoded by one of the AAG, AGA, GAA codons and one of the AGG, GAG, GGA codons. Similarly, glutamic acid must be encoded by one of the AAG, AGA, GAA codons and one of the AGG, GAG, GGA codons. To make additional inferences, more experimentation (e.g., with different copolymers) must be done.

QUESTIONS FOR PRACTICE

Multiple Choice Questions

1. The term *secondary structure* refers to a polypeptide's
   a. sequence of amino acids
   b. structure that results from local interactions between the residues of different amino acids
   c. folding that results from local and distant interactions between the residues of different amino acids
   d. interactions with a second polypeptide chain

2. The eukaryotic equivalent of the Shine-Dalgarno sequence is the
   a. 60S large ribosomal subunit
   b. 40S small ribosomal subunit
   c. 5' cap of the mRNA
   d. poly(A) tail

3. In which stage of translation is GTP not used?
   a. initiation
   b. elongation
   c. translocation
   d. termination

4. Which of the following are the same in both prokaryotes and eukaryotes?
   a. elongation factors
   b. stop codons
   c. the use of fMet.tRNA
   d. AUG initiation codons
   e. both b and d

5. Specificity in translation is obtained by all of the following except
   a. using specific tRNA synthetases for each amino acid
   b. using an anticodon
   c. using a degenerate triplet code having codons for each amino acid
   d. using RNA polymerase

6. Where can coding sequences not be found?
   a. introns
   b. exons
   c. mRNA
   d. hnRNA

7. In an unprocessed, newly synthesized eukaryotic protein,
   a. the N terminus is encoded by sequences in the 5' region of the mRNA
   b. the C terminus is encoded by sequences in the 5' region of the mRNA
   c. the first amino acid is always f-methionine
   d. the last amino acid is always encoded by the codon AAA

8. During translation in *E. coli,*
   a. an intact 70S ribosome binds the mRNA during initiation
   b. Shine-Dalgarno sequences present in the mRNA align it with the 23S rRNA
   c. both initiation and elongation require energy in the form of GTP
   d. termination occurs when the ribosome recognizes a chain termination codon with the aid of EF-Tu

9. The RNAs that are used in eukaryotic translation include all of the following except:
   a. 5S rRNA

    b. 28S rRNA
    c. mRNA
    d. tRNA
    e. snRNA

10. In eukaryotes, proteins destined to remain in the cell are distributed by all of the following except
    a. packaging into secretory vesicles
    b. complexing with an SRP using an N terminal signal sequence
    c. using posttranslational transport

Answers: 1b, 2c, 3d, 4e, 5d, 6a, 7a, 8c, 9e, 10a

Thought Questions

1. The following sequence of bases in *E. coli* DNA is part of a gene. Mark the region that, if transcribed, will form the first codon to be translated. Mark the region that, if transcribed, will form a Shine-Dalgarno sequence. Is the strand shown the template or non-template strand? (Hint: Find the initiation codon. Be careful about polarity!)

    5'-CCGCATTCCTCCGGCGGGACCTACT-3'

2. Why is the code for methionine not degenerate? (Hint: What special role does methionine play in translation?)

3. What are the components of a cell-free, protein-synthesizing system? What is the function of each?

4. Describe three complementary methods by which the genetic code was deciphered. Is any single method sufficient to completely decipher the code? What advantages and disadvantages does each method have?

5. The genetic code is nearly universal. To what extent does this indicate that the code was "settled on" very early in the evolution of cells?

6. Consider which bases can wobble to base-pair with each other. What structural similarities exist between bases at the 3' end of a codon that can wobble to base-pair with the base at the 5' end of the anticodon?

7. What percentage of 70S ribosomes do you expect to be associated with mRNA in prokaryotes? What percentage of 80S ribosomes do you expect to be associated with mRNA in eukaryotes?

8. Discuss the various ways in which proteins within a cell are targeted for secretion. How can a genetic approach be used to dissect the secretory pathway?

9. How do prokaryotes and eukaryotes differ in the mechanisms they use to identify the first codon in a reading frame? In what ways are the mechanisms similar?

10. How are posttranscriptional modifications to mRNA important for translation?

Thought Questions for Media

After reviewing the media on the *iGenetics* CD-ROM, try to answer these questions.

1. How does the ribosome IF-GTP complex identify where it should bind to the mRNA?

2. What are the functions of EF-Tu and EF-G in translation elongation?

3. When in translation is GTP hydrolysis required?

4. Review the A, E, and P sites on the ribosome. Which site never has a charged tRNA in it? Which site is only bound by a charged tRNA or a RF? Which site is almost always occupied by a tRNA with an attached polypeptide?

5. Peptidyl transferase provides two functions, one during translation elongation and one during translation termination. Are they the same enzymatic function?

1. Based on viewing the iActivity, what different features of a coding region mutation would cause it to be associated with a mutant phenotype?

2. What type of coding region mutations have no effect whatsoever on protein function?

3. What type of coding region mutations are likely to have little, if any, effect on protein function?

SOLUTIONS TO TEXT PROBLEMS

14.1   The form of genetic information used directly in protein synthesis is (choose the correct answer):

a. DNA

b. mRNA

c. rRNA

d. tRNA

Answer: b

14.2   Most genes encode proteins. What exactly is a protein, structurally speaking? List some of the functions of proteins.

Answer: A protein is composed of one or more polypeptides, each of which is composed of a folded, linear chain of amino acids. Proteins function in many different ways, including as enzymes to catalyze biochemical reactions, as structural molecules (e.g., actin, myosin, tubulin), as receptors (e.g., on the cell surface, for small signaling molecules such as neurotransmitters), as transporters (e.g., a protein that transports sugars across a membrane), as ion channels (e.g., the protein defective in cystic fibrosis, CFTR, transports Cl^- ions), and as regulatory molecules (e.g., transcription factors).

14.3   In each of the following cases stating how a certain protein is treated, indicate what level(s) of protein structure would change as the result of the treatment:

a. Hemoglobin is stored in a hot incubator at 80°C.

b. Egg white (albumin) is boiled.

c. RNase (a single polypeptide enzyme) is heated to 100°C.

d. Meat in your stomach is digested (gastric juices contain proteolytic enzymes).

e. In the β-polypeptide chain of hemoglobin, the amino acid valine replaces glutamic acid at the number-six position.

Answer:

a. The hemoglobin will dissociate into its four component subunits, because the heat will destabilize the ionic bonds that stabilize the quaternary structure of the protein. An individual subunit's tertiary structure may also be altered, because the thermal energy of the heat may destabilize the folding of the polypeptide.

b. The protein will denature. Its tertiary structure is destabilized by heating, so it does not retain a pattern of folding that allows it to be soluble.

c. The protein will denature when its tertiary structure is destabilized by heating. Unlike albumin, RNase will renature if cooled slowly and will reestablish its normal, functional tertiary structure.

d. It is likely that the meat proteins will be denatured when their tertiary and quaternary structures are destabilized by the acid conditions of the stomach. Then the primary structure of the polypeptides will be destroyed as they are degraded into their amino-acid components by proteolytic enzymes in the digestive tract.

e. Valine is a neutral, nonpolar amino acid, unlike the acidic glutamic acid. (See Figure 14.2 in text.) A change in the chemical properties of the sixth amino acid may alter the function of the hemoglobin molecule by affecting multiple levels of protein structure. Since it is an amino-acid substitution, it changes the primary structure of the β-polypeptide. This change could affect local interactions between amino acids lying near it, and in doing so, alter the secondary structure of the β-polypeptide. It could also affect the folding patterns of the protein and alter the tertiary structure of the β-polypeptide. Finally, the sixth amino-acid residue is known to be important for inter-actions between the subunits of hemoglobin molecules (see Figures 12.9–12.11) because some mutations which alter that amino acid result in sickle-cell anemia. Thus, this change could alter the quaternary structure of hemoglobin.

14.4   Bovine spongiform encephalopathy (BSE) ("mad cow") disease and the human version, Creutzfeldt-Jakob disease (CJD), are characterized by the deposition of amyloid—insoluble, nonfunctional protein deposits—in the brain. In these diseases, amyloid deposits contain an abnormally folded version of the prion protein. Whereas the normal prion protein has lots of alpha-helical regions and is soluble, the abnormally folded version has α-helical regions converted into β-pleated sheets and is insoluble. Curiously, small amounts of the abnormally folded version can trigger the conversion of an α-helix to a β-pleated sheet in the normal protein, making the abnormally folded version infectious.

a. Some cases of CJD may have arisen from ingesting beef having tiny amounts of the abnormally folded protein. What would you expect to find if you examined the primary structure of the prion protein in the affected tissues? What levels of protein structural organization are affected in this form of prion disease?

b. Answer the questions posed in part (a) for cases of CJD in which susceptibility to CJD is inherited due to a rare mutation in the gene for the prion protein.

Answer:

a. The primary structure, or amino-acid sequence, of the prion protein would be unchanged, as the disease is caused not by a mutation, but rather by misfolding of the prion protein. One misfolded protein can convert a normally folded protein to the mis-folded state, so the misfolded proteins are infectious. The secondary structure is affected, as α-helical regions are misfolded into β-pleated sheets. This is likely to lead to an altered tertiary structure resulting in the formation of amyloid.

b. If a genetic mutation led to an amino-acid substitution, it would affect the primary structure of the prion protein. A particular amino-acid substitution in the prion protein could make it more susceptible to being misfolded and lead, as in (a), to changes in its secondary and tertiary structures.

14.5   The structure and function of the rRNA and protein components of ribosomes have been investigated by separating those components from intact ribosomes and then using reconstitution experiments to determine which of these components are required for specific ribosomal activities.

a. Contrast the components of prokaryotic ribosomes with those of eukaryotic ribosomes.

b. What is the function of ribosomes, what steps are used by ribosomes to carry out this function, and which components of ribosomes are active in each step?

Answer: Prokaryotic and eukaryotic ribosomes each have two subunits with parallel functions and components. However, as presented in Chapter 13, the detailed constituents differ. The 70S prokaryotic ribosome has a 30S small subunit containing a 16S rRNA and 20 different proteins, and a 50S large subunit, containing a 23S rRNA and 5S rRNA and 34 different proteins. The larger 80S eukaryotic ribosome has a 40S small subunit, containing an 18S rRNA and about 35 proteins, and a 60S large subunit, containing 28S, 5.8S, and 5S rRNAs and about 50 ribosomal proteins. In both eukaryotic and prokaryotic ribosomes, the proteins serve as structural units that help to organize the rRNA into key ribozyme elements.

Ribosomes function in translation, which proceeds through three phases: initiation, elongation, and termination. During the initiation phase of translation, the small subunit, with the aid of initiation factors, is bound to the mRNA and the initiating aminoacyl-tRNA (tRNA.fMet in prokaryotes, an initiator Met-tRNA in eukaryotes). The large subunit binds, forming an initiation complex with the initiator tRNA in the P site of the ribosome. During elongation, a second aminoacyl-tRNA is bound to the A site in the ribosome, the peptidyl transferase activity of the ribosome catalyzes the formation of the peptide bond, a now-uncharged tRNA is released (from the E site of the large subunit), and the ribosome translocates one codon down the mRNA. Translation elongation continues, so that codons are matched to amino acids in a growing polypeptide chain, until a stop (nonsense) codon is reached. Stop codons are recognized by release factors, which initiate three processes: the release of the polypeptide from the tRNA in the P site, the release of the tRNA from the ribosome, and the dissociation of the two ribosomal subunits and the release factor from the mRNA.

14.6   What is the evidence that the rRNA component of the ribosome serves more than a structural role?

Answer: Multiple lines of evidence support the view that the rRNA component of the ribosome serves more than a structural role. First, the 3′ end of the 16S rRNA is important for identifying where the small ribosomal subunit should bind the mRNA. It has a sequence that is complementary to the Shine-Dalgarno sequence, the ribosome-binding site (RBS) in the mRNA. Mutational analyses demonstrated that the 3′ end of the 16S rRNA must base-pair with this mRNA sequence for correct initiation of translation. Second, the 23S rRNA is required for peptidyl transferase activity. Evidence that the peptidyl transferase consists entirely of RNA comes from studies of the atomic structure of the large ribosomal subunit and is supported by experiments that show that peptidyl transferase activity remains following the depletion of the 50S subunit proteins, but not after the digestion of rRNA with ribonuclease T1.

14.7   The term *genetic code* refers to the set of three-base code words (codons) in mRNA that stand for the 20 amino acids in proteins. What are the characteristics of the code?

Answer:
1. Triplet: Three nucleotides specify either the insertion of an amino acid into a polypeptide chain or a chain termination event.
2. Continuous: An mRNA-coding region is read in a continuous fashion without skipping nucleotides.

3. Nonoverlapping: A coding region is read in a nonoverlapping manner. Triplets are read sequentially.

4. Nearly universal: Nearly all organisms use the same code, with exceptions found in mammalian mitochondria and the nuclear genomes of some protozoa.

5. Degenerate: More than one codon typically codes for a single amino acid.

6. Signals start and stop: The code has translation stop (chain termination) and translation start signals. AUG, which encodes methionine within a coding region, is usually used as a start codon. One of three nonsense (*not sensing* an amino acid) codons are used as chain termination codons.

7. Wobble: The 5′ base of some anticodons wobbles, so that some tRNAs can recognize multiple codons (coding for the same amino acid).

14.8   If the codon were four bases long, how many codons would exist in a genetic code?

Answer: At each position of a four-base-long codon, there will be one of four possible bases. So, there are $4 \times 4 \times 4 \times 4 = 4^4 = 256$ possible codons. This is far more than is needed to code for 20 amino acids. On the other hand, if there were only two bases in a codon, there would be only $4 \times 4 = 4^2 = 16$ possible codons, too few to uniquely specify 20 amino acids. Codons three bases long provide for $4 \times 4 \times 4 = 4^3 = 64$ possible codons.

14.9   Base-pairing wobble occurs in the interaction between the anticodon of the tRNAs and the codons. On a theoretical level, determine the minimum number of tRNAs needed to read the 61 sense codons.

Answer: Figure 14.7 and Table 14.1 aid in answering this question. The answer is given in the following table:

Amino Acid	tRNAs Needed	Rationale
Ile	1	3 codons can use 1 tRNA (wobble)
Phe	1	2 codons can use 1 tRNA (wobble)
Tyr	1	"
His	1	"
Gln	1	"
Asn	1	"
Lys	1	"
Asp	1	"
Glu	1	"
Cys	1	"
Trp	1	1 codon
Met	2	Single codon, but need one tRNA for initiation and one tRNA for elongation

Amino Acid	tRNAs Needed	Rationale
Val	2	4 codons: 2 can use 1 tRNA (wobble)
Pro	2	"
Thr	2	"
Ala	2	"
Gly	2	"
Leu	3	6 codons: 2 can use 1 tRNA (wobble)
Arg	3	"
Ser	3	"
Total	32	61 codons

14.10   Antibiotics have been highly useful in elucidating the steps of protein synthesis. If you have an artificial messenger RNA with the sequence AUGUUUUUUUUUUUUU . . . , it will produce the following polypeptide in a cell-free protein-synthesizing system: fMet-Phe-Phe-Phe. . . . Suppose that, in your search for new antibiotics, you find one called putyermycin, which blocks protein synthesis. When you try it with your artificial mRNA in a cell-free system, the product is fMet-Phe. What step in protein synthesis does putyermycin affect? Why?

Answer: Since a dipeptide is formed, translation initiation is not affected, nor is the first step of elongation—the binding of a charged tRNA in the A site and the formation of a peptide bond. However, since *only* a dipeptide is formed, it appears that translocation is inhibited.

14.11   Describe the reactions involved in the aminoacylation (charging) of a tRNA molecule.

Answer: Aminoacylation of tRNAs occurs enzymatically using aminoacyl-tRNA synthetases. To provide for specificity in charging a tRNA, a separate aminoacyl-tRNA synthetase exists for each amino acid. This prevents an inappropriate amino acid from becoming attached to a tRNA. If an inappropriate amino acid were attached to a tRNA, the wrong amino acid would be inserted into a growing polypeptide chain, even with correct codon-anticodon base-pairing between the mRNA and tRNA.

Aminoacyl-tRNA synthetases catalyze two sequential reactions to attach an amino acid to its appropriate tRNA. The first reaction results in the formation of an aminoacyl-AMP complex from the amino acid and ATP. The second reaction results in a covalent bond between the carboxyl group of the amino acid and the last ribose at the 3' end of the tRNA. See text Figure 14.9.

14.12   If the initiating codon of an mRNA were altered by mutation, what might be the effect on the transcript?

Answer: In both prokaryotes and eukaryotes, the initiating codon of a transcript is AUG, recognized by a special charged tRNA (tRNA.fMet in prokaryotes, initiator Met-tRNA in eukaryotes). This first codon is recognized in part because it is in a particular context: In prokaryotes, it is near the Shine-Dalgarno sequence, while in eukaryotes, it is embedded in a Kozak sequence. It is possible that if it were mutated, the 30S (in prokaryotes) or 40S (in eukaryotes) initiation complex would not form correctly, since the initiator tRNA would not find the correct initiator codon. The consequence of this would be that the mRNA would not be translated. On the other hand, if another AUG codon was very close by (and in eukaryotic mRNAs, embedded in a Kozak-type sequence), it might be used as an initiation codon, but the mRNA might be translated at a lower efficiency.

14.13   What differences are found in the initiation of protein synthesis between prokaryotes and eukaryotes? What differences are found in the termination of protein synthesis between prokaryotes and eukaryotes?

Answer: In both prokaryotes and eukaryotes, protein synthesis is initiated using special initiator tRNAs. In prokaryotes, methionine is attached to tRNA.fMet by methionine-tRNA synthetase, and transformylase catalyzes the addition of a formyl group to the methionine. The resulting fMet-tRNA.fMet is used at the AUG initiation codon. In eukaryotes, the N-terminal methionine is not formylated, but a special initiator methionine-tRNA is still used.

In both prokaryotes and eukaryotes, an initiation complex containing the small ribosomal subunit, initiator factors, GTP, and the mRNA forms sequentially. In prokaryotes, after IF1, IF2, IF3, and GTP bind to the 30S ribosomal subunit, fMet-tRNA.fMet and the mRNA attach to form the 30S initiation complex. The mRNA is aligned using an internal ribosome-binding site (the Shine-Dalgarno sequence) lying near the AUG initiation codon that base-pairs with sequences at the 3' end of the 16S rRNA. After the 50S ribosomal subunit binds, GTP is hydrolyzed and IF1 and IF2 are released, leaving a 70S initiation complex. In eukaryotes, after eIF4A binds to the 5' cap of the mRNA, a complex of the 40S ribosomal subunit, the initiator Met-tRNA.Met, GTP, and additional eIFs bind. Only after the 40S ribosomal subunit scans and finds the initiator codon does the 60S ribosomal subunit bind. This displaces the eIFs and produces the 80S initiation complex.

In both prokaryotes and eukaryotes, chain termination is signaled by one of three stop codons (UAG, UAA, UGA). These stop codons do not code for an amino acid, but are recognized with the help of termination factors or release factors. Of the three RFs in *E. coli*, RF1 recognizes UAA and UAG, RF2 recognizes UAA and UGA, and RF3 stimulates chain termination. Eukaryotes have a single release factor, eRF, that recognizes all three stop codons. The release factors trigger release of the polypeptide from the tRNA in the P site of the ribosome, release of the tRNA from the ribosome, and dissociation of the two ribosomal subunits from the mRNA. The initiating amino acid is usually cleaved from the completed peptide.

14.14   In Chapter 13, we saw that eukaryotic mRNAs are posttranscriptionally modified at their 5' and 3' ends. What role does each of these modifications play in translation?

Answer: A eukaryotic mRNA is modified to contain a 5' 7-methyl-G cap and a 3' poly(A) tail. The 5' cap is required early in translation initiation—it binds to the eIF-4F complex just prior to the binding of a complex of the 40S ribosomal subunit, the initiator Met-tRNA, and other eIF proteins. Transcription initiation is stimulated by the looping of the poly(A) tail close to the 5' end. This occurs when the poly(A) binding protein (PAB) binds to eIF-4G, which is part of the eIF-4F complex.

14.15 Translation usually initiates at an `AUG` codon near the 5′ end of an mRNA, but mRNAs often have multiple `AUG` triplets near their 5′ ends. How is the correct initiation `AUG` codon identified in prokaryotes? How is it correctly identified in eukaryotes?

Answer: In both prokaryotes and eukaryotes, the initiator `AUG` codon is present in a sequence context that helps define it as the initiator codon. In prokaryotes, the correct `AUG` is found downstream of a purine-rich ribosome-binding site (RBS). The RBS is complementary to a pyrimidine-rich region at the 3′ end of the 16S rRNA. The formation of complementary base pairs between the mRNA and the 16S rRNA allows the ribosome to locate the correct `AUG` within the mRNA for the initiation of protein synthesis. In eukaryotes, the correct `AUG` codon is identified by scanning for an `AUG` codon embedded in a Kozak sequence. After the 5′ cap is bound by eIF-4F, a complex of the 40S ribosomal subunit with the initiator Met-tRNA, several eIF proteins, and `GTP` binds the mRNA. Together with other eIFs, the complex scans along the mRNA for an initiator `AUG` codon embedded in a Kozak sequence.

14.16 Energy is required during multiple steps during translation. What are the sources of this energy, which steps require energy for their completion, and what is energy used for during these steps?

Answer: Energy obtained from the hydrolysis of ATP or GTP is used to charge tRNAs and to facilitate macromolecular interactions and rearrangements during initiation, elongation and translocation as shown in the following table:

Step	Energy Source	Use of Energy
Charging tRNA	hydrolysis of ATP to AMP	Aminoacylation: formation of an aminoacyl-AMP by an aminoacyl-tRNA synthetase
Initiation	hydrolysis of GTP to GDP	Formation of the 70S initiation complex: attachment of 50S subunit and release of IF1 and IF2
Elongation	hydrolysis of GTP to GDP	Binding of aminoacyl-tRNA to the ribosome and release of EF-Tu-GDP
Translocation	hydrolysis of GTP to GDP	Translocation and displacement of the uncharged tRNA away from the P site. After EF-G-GTP binds to the ribosome, GTP hydrolysis may alter the structure of EF-G to facilitate the translocation event.

14.17 Random copolymers were used in some of the experiments that revealed the characteristics of the genetic code. For each of the following ribonucleotide mixtures, give the expected codons and their frequencies, and give the expected proportions of the amino acids that would be found in a polypeptide directed by the copolymer in a cell-free protein-synthesizing system.

a. 4 A : 6 C

b. 4 G : 1 C

c. 1 A : 3 U : 1 C

d. 1 A : 1 U : 1 G : 1 C

Answer: Determine the expected amino acids in each case by calculating the expected frequency of each kind of triplet codon that might be formed and inferring from these what types and frequencies of amino acids would be used during translation.

a. 4 A : 6 C gives 2^3 = 8 codons, specifically AAA, AAC, ACC, ACA, CCC, ACA, CAC, and CAA. Since there is 40% A and 60% C,

$$P(\text{AAA}) = 0.4 \times 0.4 \times 0.4 = 0.064, \text{ or } 6.4\% \text{ Lys}$$
$$P(\text{AAC}) = 0.4 \times 0.4 \times 0.6 = 0.096, \text{ or } 9.6\% \text{ Asn}$$
$$P(\text{ACC}) = 0.4 \times 0.6 \times 0.6 = 0.144, \text{ or } 14.4\% \text{ Thr}$$
$$P(\text{ACA}) = 0.4 \times 0.6 \times 0.4 = 0.096, \text{ or } 9.6\% \text{ Thr (24\% Thr total)}$$
$$P(\text{CCC}) = 0.6 \times 0.6 \times 0.6 = 0.216, \text{ or } 21.6\% \text{ Pro}$$
$$P(\text{CCA}) = 0.6 \times 0.6 \times 0.4 = 0.144, \text{ or } 14.4\% \text{ Pro (36\% Pro total)}$$
$$P(\text{CAC}) = 0.6 \times 0.4 \times 0.6 = 0.144, \text{ or } 14.4\% \text{ His}$$
$$P(\text{CAA}) = 0.6 \times 0.4 \times 0.4 = 0.096, \text{ or } 9.6\% \text{ Gln}$$

b. 4 G : 1 C gives 2^3 = 8 codons, specifically GGG, GGC, GCG, GCC, CGG, CGC, CCC, and CCG. Since there is 80% G and 20% C,

$$P(\text{GGG}) = 0.8 \times 0.8 \times 0.8 = 0.512, \text{ or } 51.2\% \text{ Gly}$$
$$P(\text{GGC}) = 0.8 \times 0.8 \times 0.2 = 0.128, \text{ or } 12.8\% \text{ Gly (64\% Gly total)}$$
$$P(\text{GCG}) = 0.8 \times 0.2 \times 0.8 = 0.128, \text{ or } 12.8\% \text{ Ala}$$
$$P(\text{GCC}) = 0.8 \times 0.2 \times 0.2 = 0.032, \text{ or } 3.2\% \text{ Ala (16\% Ala total)}$$
$$P(\text{CGG}) = 0.2 \times 0.8 \times 0.8 = 0.128, \text{ or } 12.8\% \text{ Arg}$$
$$P(\text{CGC}) = 0.2 \times 0.8 \times 0.2 = 0.032, \text{ or } 3.2\% \text{ Arg (16\% Arg total)}$$
$$P(\text{CCC}) = 0.2 \times 0.2 \times 0.2 = 0.008, \text{ or } 0.8\% \text{ Pro}$$
$$P(\text{CCG}) = 0.2 \times 0.2 \times 0.8 = 0.032, \text{ or } 3.2\% \text{ Pro (4\% Pro total)}$$

c. 1 A : 3 U : 1 C gives 3^3 = 27 different possible codons. Of these, one will be UAA, a chain-terminating codon. Since there is 20% A, 60% U, and 20% C, the probability of finding this codon is $0.6 \times 0.2 \times 0.2 = 0.024$, or 2.4%. All of the remaining 26 (97.6%) codons will be sense codons. Proceed in the same manner as in (a) and (b) to determine their frequency, and determine the kinds of amino acids expected. To take the frequency of nonsense codons into account, divide the frequency of obtaining a particular amino acid considering all 27 possible codons by the frequency of obtaining a sense codon. This gives

$$(0.8/0.976)\% = 0.82\% \text{ Lys}$$
$$(3.2/0.976)\% = 3.28\% \text{ Asn}$$

$(12.0/0.976)\% = 12.3\%$ Ile
$(9.6/0.976)\% = 9.84\%$ Tyr
$(19.2/0.976)\% = 19.67\%$ Leu
$(28.8/0.976)\% = 29.5\%$ Phe
$(4.0/0.976)\% = 4.1\%$ Thr
$(0.8/0.976)\% = 0.82\%$ Gln
$(3.2/0.976)\% = 3.28\%$ His
$(4.0/0.976)\% = 4.1\%$ Pro
$(12.0/0.976)\% = 12.3\%$ Ser

It is likely that the chains produced would be relatively short due to the chain-terminating codon.

d. 1 A:1 U:1 G:1 C will produce $4^3 = 64$ different codons, all possible in the genetic code. The probability of each codon is $\frac{1}{64}$, so there will be a $\frac{3}{64}$ chance of a codon being chain terminating. With those exceptions, the relative proportion of amino-acid incorporation is dependent directly on the codon degeneracy for each amino acid. Inspecting the table of the genetic code in Figure 14.7 and taking the frequency of nonsense codons into account yields the following table:

Amino Acid	Number of # Codons	Frequency
Trp	1	$\frac{1}{61} = 1.64\%$
Met	1	1.64%
Phe	2	$2/61 = 3.28\%$
Try	2	3.28%
His	2	3.28%
Gln	2	3.28%
Asn	2	3.28%
Lys	2	3.28%
Asp	2	3.28%
Glu	2	3.28%
Cys	2	3.28%
Ile	3	$3/61 = 4.92\%$
Val	4	$4/61 = 6.56\%$
Pro	4	6.56%
Thr	4	6.56%
Ala	4	6.56%
Gly	4	6.56%
Leu	6	$6/61 = 9.84\%$
Arg	6	9.84%
Ser	6	9.84%

14.18   What would the minimum word (codon) size need to be if, instead of four, the number of different bases in the mRNA were

a. two

b. three

c. five

Answer: The minimum word size must be able to uniquely designate 20 amino acids, so the number of combinations must be at least 20. The following table gives the number of combinations as a function of word size:

	Word Size	Number of Combinations
a.	5	$2^5 = 32$
b.	3	$3^3 = 27$
c.	2	$5^2 = 25$

14.19   Suppose that, at stage A in the evolution of the genetic code, only the first two nucleotides in the coding triplets led to unique differences and that any nucleotide could occupy the third position. Then, suppose there was a stage B in which differences in meaning arose, depending on whether a purine (A or G) or pyrimidine (C or U) was present at the third position. Without reference to the number of amino acids or multiplicity of tRNA molecules, how many triplets of different meaning can be constructed out of the code at stage A? at stage B?

Answer: Stage A: $4^2 = 16$ different meaningful triplets
Stage B: $4^2 \times 2 = 32$ different meaningful triplets

14.20   A gene encodes a polypeptide 30 amino acids long containing an alternating sequence of phenylalanine and tyrosine. What are the sequences of nucleotides corresponding to this sequence in each the following:

a. the DNA strand that is read to produce the mRNA, assuming Phe = UUU and Tyr = UAU in mRNA

b. the DNA strand that is not read

c. tRNAs

Answer:

a. 3'-TAC AAA ATA AAA ATA AAA ATA AAA ATA...-5' (The first fMet or Met is removed following translation of the mRNA.)

b. 5'-ATG TTT TAT TTT TAT TTT TAT TTT TAT...-3'

c. 3'-AAA-5' is the anticodon for Phe, and 3'-AUA-5' is the anticodon for Tyr

14.21   A segment of a polypeptide chain is Arg-Gly-Ser-Phe-Val-Asp-Arg. It is encoded by the following segment of DNA:

Which strand is the template strand? Label each strand with its correct polarity (5′ and 3′).

Answer: The template strand is the one read to produce the mRNA. It is complementary to the mRNA, and has an opposite polarity. The nontemplate strand has the same 5′→3′ polarity as the mRNA, and if U is replaced by T, the same sequence. Use the polypeptide segment and a table of the genetic code to determine the possibilities that exist for the first three codons of the mRNA (read 5′→3′) and the sequence of the nontemplate DNA strand. Let N = any nucleotide, Y = a pyrimidine (C or U), and R = a purine (G or A).

	amino acids:	Arg	-	Gly	-	Ser
potential codons:	5′	AGR		GGN		AGY 3′
		or				or
		CGN				UCN
nontemplate strand sequence:	5′	AGR		GGN		AGY 3′
		or				or
		CGN				TCN

By comparing the nontemplate strand sequence to the one given, the top strand is the nontemplate strand, with the 5′ end on the right side. The template strand is the bottom strand, and transcription occurs from right to left. The C terminus and the N terminus of the polypeptide segment can also be determined from this information.

```
----3′-GGCTAGCTGCTTCCTTGGGGA-5′----
        |||||||||||||||||||||
----5′-CCGATCGACGAAGGAACCCCT-3′----
```

Transcribed into: 3′ GGC UAG CUG CUU CCU UGG GGA 5′
Translated into: C-Arg-Asp-Val-Phe-Ser-Gly-Arg-N

14.22 Two populations of RNAs are made by the random combination of nucleotides. In population A the RNAs contain only A and G nucleotides (3 A : 1 G), whereas in population B the RNAs contain only A and U nucleotides (3 A : 1 U). In what ways other than amino acid content will the proteins produced by translating the population A RNAs differ from those produced by translating the population B RNAs?

Answer: In population A, the codons that can be produced encode Lys (AAA, AAG), Arg (AGG, AGA), Glu (GAG, GAA), and Gly (GGA, GGG). All of these are sense codons, so long polypeptide chains containing these amino acids will be synthesized. In population B, the codons that can be produced are the sense codons for Lys (AAA), Asn (AAU), Ile (AUA, AUU), Tyr (UAU), Leu (UUA), Phe (UUU), and a nonsense codon (UAA). The frequency of the nonsense codon will be ($\frac{1}{4} \times \frac{3}{4} \times \frac{3}{4}$) = $\frac{9}{64}$ = 0.14, or 14 percent. Thus, the polypeptides formed in population B will, on average, be shorter than those formed in population A. If a nonsense codon appears about 14 percent of the time, there will be, on average, 1/0.14 = 7.14 codons from one nonsense codon to the next. On average, six sense codons will lie in between a nonsense codon, so polypeptides will be synthesized that are about six amino acids long.

14.23   In *E. coli*, a particular tRNA normally has the anticodon 5'-GGG-3', but because of a mutation in the tRNA gene, the tRNA has the anticodon 5'-GGA-3'.

   a. What codon would the normal tRNA recognize?

   b. What codon would the mutant tRNA recognize?

   Answer:

   a. The normal tRNA has the anticodon 5'-GGG-3', which binds to the codons 5'-CCC-3' and (because of wobble) 5'-CCU-3'. Both of these codons encode Pro (proline).

   b. The mutant tRNA has the anticodon 5'-GGA-3', which recognizes the codons 5'-UCC-3' and (because of wobble) 5'-UCU-3'. Since the amino acid that is attached to the tRNA is unaffected by a mutant anticodon, the mutant tRNA will continue to carry proline to the ribosome.

   There are two consequences of the anticodon mutation. First, since 5'-UCC-3' and 5'-UCU-3' encode Ser (serine), the mutant tRNA will compete with the normal tRNA.Ser in the cell for binding to UCC and UCU codons during the translation of an mRNA. In a percentage of UCC codons, this leads to the insertion of proline instead of serine into a polypeptide chain. Second, unless the cell normally has an additional tRNA.Pro able to bind to the codons 5'-CCC-3' and 5'-CCU-3' (e.g., a tRNA with the anticodon 5'-IGG-3'), these codons would no longer be able to be read as sense codons. Consequently, when the ribosome encounters these codons in an mRNA, it will stall and chain termination will occur. This would lead to truncated proteins.

14.24   A protein found in *E. coli* normally has the N-terminal amino-acid sequence Met-Val-Ser-Ser-Pro-Met-Gly-Ala-Ala-Met-Ser. ... A mutation alters the anticodon of a tRNA from 5'-GAU-3' to 5'-CAU-3'. What would be the N-terminal amino-acid sequence of this protein in the mutant cell? Explain your reasoning.

   Answer: The anticodon 5'-GAU-3' recognizes the codon 5'-AUC-3', which encodes Ile. The mutant tRNA anticodon 5'-CAU-3' would recognize the codon 5'-AUG-3', which normally encodes Met. The mutant tRNA would therefore compete with tRNA.Met for the recognition of the 5'-AUG-3' codon, and if successful, insert Ile into a protein where Met should be. Since a special tRNA.Met is used for initiation, only AUG codons other than the initiation AUG will be affected. Thus, this protein will have four different N terminal sequences, depending on which tRNA occupies the A site in the ribosome when the codon AUG is present there:

   Met-Val-Ser-Ser-Pro-**Ile**-Gly-Ala-Ala-**Ile**-Ser
   Met-Val-Ser-Ser-Pro-**Met**-Gly-Ala-Ala-**Ile**-Ser
   Met-Val-Ser-Ser-Pro-**Ile**-Gly-Ala-Ala-**Met**-Ser
   Met-Val-Ser-Ser-Pro-**Met**-Gly-Ala-Ala-**Met**-Ser

14.25   The gene encoding an *E. coli* tRNA containing the anticodon 5'-GUA-3' mutates so that the anticodon now is 5'-UUA-3'. What will be the effect of this mutation? Explain your reasoning.

   Answer: The normal tRNA recognizes the codon 5'-UAC-3', and so must have carried the amino acid tyrosine. The altered anticodon will recognize the codon 5'-UAA-3', a chain termination codon. Consequently, a tyrosine will be inserted when the nonsense codon

UAA in an mRNA is positioned in the A site of the ribosome. This will result in read-through of the mRNA some of the time (when the termination factor does not compete for binding to the chain termination codon), and the addition of amino acids onto the C terminus of the protein. mRNAs having UAG and UGA chain termination codons will not be affected.

14.26 The following diagram shows the normal sequence of the coding region of an mRNA, along with six mutant versions of the same mRNA:

Normal:	AUGUUCUCUAAUUAC(...)AUGGGGUGGGUGUAG
Mutant *a*:	AUGUUCUCUAAUUAG(...)AUGGGGUGGGUGUAG
Mutant *b*:	AGGUUCUCUAAUUAC(...)AUGGGGUGGGUGUAG
Mutant *c*:	AUGUUCUCGAAUUAC(...)AUGGGGUGGGUGUAG
Mutant *d*:	AUGUUCUCUAAAUAC(...)AUGGGGUGGGUGUAG
Mutant *e*:	AUGUUCUCUAAUUC(...)AUGGGGUGGGUGUAG
Mutant *f*:	AUGUUCUCUAAUUAC(...)AUGGGGUGGGUGUGG

Indicate what protein would be formed in each case, where (. . .) denotes a multiple of three unspecified bases.

Answer: First rewrite the sequences so that the codons can be readily seen, noting the mutations (underlined):

Normal:	AUG UUC UCU AAU UAC (...) AUG GGG UGG GUG UAG
Mutant *a*:	AUG UUC UCU AAU UA<u>G</u> (...) AUG GGG UGG GUG UAG
Mutant *b*:	A<u>G</u>G UUC UCU AAU UAC (...) AUG GGG UGG GUG UAG
Mutant *c*:	AUG UUC UC<u>G</u> AAU UAC (...) AUG GGG UGG GUG UAG
Mutant *d*:	AUG UUC UCU AA<u>A</u> UAC (...) AUG GGG UGG GUG UAG
Mutant *e*:	AUG UUC UCU AAU <u>UC. ..)A UGG GGU GGG UGU AG.</u>
Mutant *f*:	AUG UUC UCU AAU UAC (...) AUG GGG UGG GUG U<u>GG</u>

Mutants *a, b, c, d,* and *f* are point mutations in which one base has been substituted for another. Mutant *e* is a deletion of a single base that results in a shift in the reading frame of the mRNA (a frameshift mutation). Translating each sequence using the genetic code, determine the proteins that would be formed if these sequences were translated:

normal:	AUG UUC UCU AAU UAC ... AUG GGG UGG GUG UAG
	Met Phe Ser Asn Tyr ... Met Gly Trp Val Stop

a:	AUG UUC UCU AAU UA<u>G</u> ... AUG GGG UGG GUG UAG
	Met Phe Ser Asn <u>Stop</u>

Mutant *a* is a nonsense mutation and results in premature chain termination.

b:	A<u>G</u>G UUC UCU AAU UAC ... AUG GGG UGG GUG UAG
	Met Gly Trp Val Stop

Mutant *b* mutates the initiation codon, so that a polypeptide would be formed (if formed at all) using a downstream initiation codon. It results in a polypeptide missing amino acids at its N terminus.

c: AUG UUC UC<u>G</u> AAU UAC ... AUG GGG UGG GUG UAG
        Met Phe <u>Ser</u> Asn Tyr ... Met Gly Trp Val Stop

Mutant *c* changes a base in the 3′ end of the codon. This does not alter the amino acid that is inserted. It will be "silent" and have no phenotypic effect.

d: AUG UUC UCU AA<u>A</u> UAC ... AUG GGG UGG GUG UAG
        Met Phe Ser <u>Lys</u> Tyr ... Met Gly Trp Val Stop

Mutant *d* changes a base in the 3′ end of the codon. This does alter the amino acid that is inserted. It is a missense mutation, resulting in the insertion of a Lys instead of an Asn.

e: AUG UUC UCU AAU UC. ..A UGG GGU GGG UGU AG.
        Met Phe Ser Asn Ser ... Trp Gly Gly Cys ?

Mutant *e* is a single base-pair deletion and results in a frameshift mutation that alters the reading frame of the protein. All amino acids inserted following Asn are likely to be incorrect. It is conceivable that a stop codon could be read in the region that is indicated by the . . ., leading to premature chain termination.

f: AUG UUC UCU AAU UAC ... AUG GGG UGG GUG U<u>GG</u>
        Met Phe Ser Asn Tyr ... Met Gly Trp Val Trp..

Mutant *f* changes a base in the chain-terminating UAG codon so that the amino acid Trp will now be inserted. It will result in the placement of additional amino acids onto the C terminus of the protein.

14.27   The following diagram shows the normal sequence of a particular protein, along with several mutant versions of it:

> Normal: Met-Gly-Glu-Thr-Lys-Val-Val-...-Pro
> Mutant 1: Met-Gly
> Mutant 2: Met-Gly-Glu-Asp
> Mutant 3: Met-Gly-Arg-Leu-Lys
> Mutant 4: Met-Arg-Glu-Thr-Lys-Val-Val-...-Pro

For each mutant, explain what mutation occurred in the coding sequence of the gene, where (. . .) denotes a multiple of three unspecified bases.

Answer: One approach to this problem is to infer the possible coding sequence(s) that could be used for the normal protein (using N = any nucleotide, R = purine, Y = pyrimidine) and then examine this sequence to deduce what possible mutations could have resulted in the mutant proteins.

Based on the normal amino acid sequence, the mRNA sequence is:

```
amino acid sequence:    Met-Gly-Glu-Thr-Lys-Val-Val-...-Pro
   potential mRNA: 5'-AUG GGN GAR ACN AAR GUN GUN ... CCN-3'
```

In mutant 1, premature chain termination has occurred. This could have occurred if, in the DNA transcribed into the third [GAR (Glu)] codon, a GC base pair was changed to a TA base pair. This would lead to a UAR (stop) codon.

```
normal sequence:    Met-Gly-Glu-Thr-Lys-Val-Val-...-Pro
normal   mRNA: 5'-AUG GGN GAR ACN AAR GUN GUN ... CCN-3'
mutant   mRNA: 5'-AUG GGN UAR ACN AAR GUN GUN ... CCN-3'
mutant sequence:    Met-Gly-Stop
```

It could also have occurred if a TA base-pair insertion mutation occurred in the DNA so that a U was transcribed in between the normal second and third codons, resulting in a UGA (stop) third codon.

```
amino acid sequence:    Met-Gly-Glu-Thr-Lys-Val-Val-...-Pro
potential mRNA:     5'-AUG GGN GAR ACN AAR GUN GUN ... CCN-3'
mutant mRNA:        5'-AUG GGN UGA RAC NAA RGU NGU N.. .CC-3'
mutant sequence:        Met-Gly-Stop
```

In mutant 2, premature chain termination has occurred after a wrong amino acid has been inserted. To explain both of these results as a consequence of a single mutational event, try either insertion or deletion mutations that would alter the reading frame. One possible explanation is that a GC base-pair insertional mutation in the DNA resulted in a G being inserted after the third codon. If the N of the fourth codon were a U, then such a frameshifting insertion would change the Thr to Asp and also introduce a chain termination codon into the fifth codon position. If the R in the GAR coding for Glu (the third amino acid) is a G, this mutation could also be caused by a GC or AT base-pair insertional mutation after the first two nucleotides coding for Glu. As before, assume the N of the fourth codon is a U.

```
amino acid sequence:    Met-Gly-Glu-Thr-Lys-Val-Val-...-Pro
potential mRNA:     5'-AUG GGN GAR ACN AAR GUN GUN ... CCN-3'
mutant mRNA:        5'-AUG GGN GAR GAC UAA RGU NGU N.. .CC-3'
mutant sequence:        Met-Gly-Glu-Asp-Stop
```

In mutant 3, the situation is similar to that with mutant 2. This time, however, several wrong amino acids are inserted before chain termination. To explain all of these consequences as the result of a single mutational event, check for the consequences of deletions or insertions in the region of the second and third codons. One possible explanation is that a deletion mutation in the DNA resulted in the N of the second codon being deleted. If, as in mutant 2, the N of the fourth codon is a U, the R of the third codon is a G, the R of the fifth codon is an A, and the N of the sixth codon is an A, the mutant sequence would be obtained.

```
normal sequence:   Met-Gly-Glu-Thr-Lys-Val-Val ... Pro
normal mRNA:    5'-AUG GGN GAR ACN AAR GUN GUN ... CCN-3'
normal mRNA:    5'-AUG GGN GAG ACU AAA GUA GUN ... CCN-3'
mutant mRNA:    5'-AUG GGG AGA CUA AAG UAG UN. ..C CN-3'
mutant sequence:   Met-Gly-Arg-Leu-Lys-Stop
```

In mutant 4, the normal second amino acid (Gly) has been replaced with Arg. Arg is encoded by AGR or CGN, while Gly is encoded by GGN. If a GC base pair were substituted for a CG base pair in the DNA so that the first G of the second codon were replaced by a C, Arg would be inserted as the second amino acid.

```
normal sequence:   Met-Gly-Glu-Thr-Lys-Val-Val ... Pro
normal mRNA:    5'-AUG GGN GAR ACN AAR GUN GUN ... CCN-3'
mutant mRNA:    5'-AUG CGN GAR ACN AAR GUN GUN ... CCN-3'
mutant sequence:   Met-Arg-Glu-Thr-Lys-Val-Val ... Pro
```

14.28   The N terminus of a protein has the sequence Met-His-Arg-Arg-Lys-Val-His-Gly-Gly. A molecular biologist wants to synthesize a DNA chain that could encode this portion of the protein. How many possible DNA sequences can encode this polypeptide?

Answer: Because of the redundancy in the genetic code, a number of different DNA strands could serve as coding (nontemplate) strands for this sequence. If N = any of the four nucleotides, Y = either pyrimidine nucleotide (C, T), and R = either purine nucleotide (A, G), then the amino acid sequence could be encoded by any of the following strands:

```
Met-His-    Arg   -   Arg    -Lys-Val-His-Gly-Gly
ATG CAY (AGR or CGN) (AGR or CGN) AAR GTN CAY GGN GGN
```

Calculate the number of different sequences by multiplying the possibilities (using the product and sum rules):

```
ATG CAY (AGR or CGN) (AGR or CGN) AAR GTN CAY GGN GGN
 1 ×  2 ×  (2 + 4)   ×  (2 + 4)  × 2 × 4 × 2 × 4 × 4 = 18,432
```

14.29   In the recessive condition in humans known as sickle-cell anemia, the β-globin polypeptide of hemoglobin is found to be abnormal. The only difference between it and the normal β-globin is that the sixth amino acid from the N terminal end is valine, whereas the normal β-globin has glutamic acid at this position. Explain how this amino acid substitution occurred in terms of differences in the DNA and the mRNA.

Answer: Both GAA and GAG code for glutamic acid, while GUU, GUC, GUA, and GUG code for valine. The simplest explanation is that there was an AT-to-TA change in the DNA, at the 17th base pair in the coding region of the gene. In this event, the 6th codon, instead of being GAA or GAG, would be GUA or GUG and encode valine.

14.30   Cystic fibrosis is an autosomal recessive disease in which the cystic fibrosis transmembrane conductance regulator (CFTR) protein is abnormal. The transcribed portion of the cystic fibrosis gene spans about 250,000 base pairs of DNA. The CFTR protein, with 1,480 amino acids, is translated from an mRNA of about 6,500 bases. The most common mutation in this gene results in a protein that is missing a phenylalanine at position 508 (ΔF508).

   a. Why is the RNA coding sequence of this gene so much larger than the mRNA from which the CFTR protein is translated?

   b. About what percentage of the mRNA together makes up 5′ untranslated leader, and 3′ untranslated trailer, sequences?

   c. At the DNA level, what alteration would you expect to find in the ΔF508 mutation?

   d. What consequences might you expect if the DNA alteration you describe in (c) occurred at random in the protein-coding region of the cystic fibrosis gene?

Answer:

   a. If the primary mRNA for this gene is 250 kb, it must be substantially processed by RNA splicing (removing introns) and polyadenylation to a smaller mature mRNA.

   b. A 1,480 amino acid protein requires $1,480 \times 3 = 4,440$ bases of protein-coding sequence. This leaves $6,500 - 4,440 = 2,060$ bases of 5′ untranslated leader and 3′ untranslated trailer sequence in the mature mRNA—about 32 percent.

   c. The ΔF508 mutation could be caused by a DNA deletion for the three base pairs encoding the mRNA codon for phenylalanine. This codon is UUY (Y = U or C), and the DNA sequence of the nontemplate strand is 5′-TTY-3′. The segment of DNA containing these bases would be deleted in the appropriate region of the gene.

   d. If positioned at random and solely within a gene's coding region (that is, not in 3′ or 5′ untranslated sequences or in intronic sequences), a deletion of three base pairs results either in an mRNA missing a single codon or an mRNA missing bases from two adjacent codons. If three of the six bases from two adjacent codons were deleted, the remaining three bases would form a single codon. In this case, an incorrect amino acid might be inserted into the polypeptide at the site of the left codon, and the amino acid encoded by the right codon would be deleted. If the 3′ base of the left codon were deleted, it would be replaced by the 3′ base of the right codon. Since the code is degenerate and wobble occurs in the 3′ base, this type of deletion might not alter the amino acid specified by the left codon. The adjacent amino acid would still be deleted, however.

14.31   Antibiotics have been useful in determining whether cellular events depend on transcription or translation. For example, actinomycin D is used to block transcription, and cycloheximide (in eukaryotes) is used to block translation. In some cases, though, surprising results are obtained after antibiotics are administered. Adding actinomycin D, for example, may result in an increase, not a decrease, in the activity of a particular enzyme. Discuss how this result might come about.

Answer: Some genes can inhibit the activity of others. An increase in an enzyme's activity will be seen if actinomycin D blocks the transcription of a gene that codes for an inhibitor of the enzyme's activity.

15

DNA MUTATION, DNA REPAIR, AND TRANSPOSABLE ELEMENTS

THINKING ANALYTICALLY

Mutation is a fundamental process in genetics and evolution. This chapter presents the conceptual framework geneticists use to consider the origin of mutations. Then it presents a detailed discussion of the types and causes of mutation, strategies to identify mutants, the nature of mutagens and how mutagens can be identified, and the biological mechanisms used to repair mutations. The chapter closes with a discussion of transposons, ubiquitous elements in prokaryotes and eukaryotes that are capable of moving within a genome, and in so doing, causing mutation. The material presented here is both highly conceptual and quite detailed. Focus first on the concepts, and, as in previous chapters, learn to use the terms and definitions to explain the concepts. To follow some of the complicated processes described in this chapter, it is essential to clearly understand the precise meaning of each term.

REVIEW OF KEY TERMS, SYMBOLS, AND CONCEPTS

In your own words, write a brief, precise definition of each term in the groups below. Check your definitions using the text. Then develop a concept map using the terms in each list.

1	2	3
mutation	mutagen	DNA repair
adaptation	carcinogen	mismatch repair, proofreading
fluctuation test	spontaneous mutation	direct correction, direct reversal
random mutation	mutation rate	mutator mutation, gene
base-pair mutation	mutation frequency	photoreactivation
base-pair substitution	tautomer, tautomeric shift	light repair
somatic vs. germ-line mutation	depurination, deamination	photolyase
gene vs. chromosomal mutation	mutational hot spot	excision (dark) repair, NER
point mutation	induced mutation	base excision repair
transition, transversion mutation	ionizing, nonionizing radiation	glycosylase
missense mutation	thymine dimer	O^6-methyl-guanine methyltransferase
nonsense mutation	SOS response	translesion DNA synthesis
neutral mutation	base analog	SOS response
silent mutation	5BU, AZT, MMS	methyl-directed mismatch repair
frameshift mutation	base-modifying agent	xeroderma pigmentosum
forward mutation	nitrous acid, hydroxylamine	ataxia-telangiectasia
reverse mutation (reversion)	intercalating agent	Fanconi anemia
true vs. partial reversion	acridine, proflavin	Bloom syndrome
suppressor mutation, gene	site-specific in vitro mutagenesis	Cockayne syndrome
second-site mutation	Ames test	hereditary nonpolyposis colon cancer
intragenic, intergenic suppressor		
nonsense suppressor, suppression		

4	5	6
mutant screen	transposable element	transposon vs. retrotransposon
visible mutation	transposition	mutator gene
replica plating	transposition event	replicative vs. conservative transposition
nutritional mutation	insertion sequence (IS) element	homologous recombination
auxotrophic mutation	IS module	
conditional mutation		

4	5	6
temperature-sensitive mutant	terminal inverted repeat	deletion
resistance mutation	transposase	inversion
	target site	translocation
	target site duplication	*Ac, Ds* controlling elements
	transposon (Tn)	null mutation
	nonhomologous	autonomous element
	recombination	nonautonomous element
	composite transposon	stable allele
	noncomposite transposon	unstable (mutable) allele
	Tn*10,* Tn*3*	donor site
	replicative transposition	*Ty* element
	conservative (nonreplicative),	long terminal (direct) repeat
	cut-and-paste transposition	delta repeat
	simple insertion	target site duplication
	cointegrate	retrovirus
	cointegration model	reverse transcriptase
	transposase	*P* element
	resolvase	LINEs, SINEs
	β-lactamase	Alu, L1 elements
	F factor	
	plasmid	
	episome	

As you consider the molecular basis of mutation, it will help to review the chemical structure of the bases to clearly visualize the action of mutagens. Use diagrams as you consider the effects of specific mutations and repair processes on the structure of DNA. Use diagrams to follow how a mutant DNA sequence results in altered codon usage in an mRNA. Proceed methodically (and fastidiously, so as not to overlook a subtle point such as changes in polarity) through any analysis involving base-pair changes. Use terms carefully, remembering that mutations affect the bases in DNA and then, indirectly, the bases in transcribed RNA and the amino acids in proteins.

Pay especially close attention to the sections on suppressor mutations and the screening procedures used to isolate new mutations. After you organize the presented information, use diagrams and flowcharts to visualize the relationships between the terms and processes in these areas.

The material on transposons is broadly descriptive, so it will be helpful to categorize this information. Distinguish between the various categories of transposable elements by considering transposable elements from four perspectives: (1) structure, (2) mode of transposition, (3) the effect of a transposed element on the structure of a gene or chromosome, and (4) the effect of a transposed element on the function of a gene. As you read this section of the chapter, refer frequently to the text figures to visualize more clearly the different transposon structures and modes of transposition.

QUESTIONS FOR PRACTICE

Multiple Choice Questions

Match the best choice below to each of the descriptions in questions 1–8.

- a. missense mutation
- b. transversion mutation
- c. neutral mutation
- d. forward mutation
- e. nonsense mutation
- f. transition mutation
- g. frameshift mutation
- h. suppressor mutation

1. _____ A purine-pyrimidine base pair that is mutated to a different purine-pyrimidine base pair

2. _____ A point mutation in the DNA that changes a codon in the mRNA and causes an amino acid substitution that does not alter the function of the translated protein

3. _____ A DNA change in which a GC base pair is replaced by a TA base pair

4. _____ A mutation that results in the addition or deletion of a base pair within the coding region of a gene

5. _____ Any point mutation that is expressed as a change from the wild type to the mutant phenotype

6. _____ A base pair change in the DNA that results in the change of an mRNA codon to either UAG, UAA, or UGA

7. _____ A mutation in the DNA that changes a codon in the mRNA and causes an amino acid substitution that may or may not produce a change in the function of a translated protein

8. _____ Any new mutation that restores some or all of the wild-type phenotype to a previously isolated mutation

9. The Ames test is used to screen for
   a. potential mutagens and carcinogens
   b. the presence of enol forms of thymine and guanine
   c. frameshift mutations
   d. spontaneous mutations

10. In the Ames test, rat liver extracts are used
    a. to chemically alter and detoxify potential mutagens
    b. to chemically alter and toxify potential mutagens
    c. to determine if an environmental chemical that itself is not mutagenic may become mutagenic when processed in the liver
    d. all of the above

11. A population of cells is subjected to UV irradiation and then placed in the dark. There will be significant
    a. production of base tautomers
    b. breakage of phosphodiester bonds
    c. formation of pyrimidine dimers
    d. photolyase activity

12. Single bacterial cells sensitive to the infection of a phage are inoculated into a large number of separate culture dishes and allowed to grow in parallel. After many generations, a sample is taken from each culture, inoculated with the phage, and separately plated. Which result is expected?
    a. All plates will have similar numbers of bacterial colonies.
    b. No bacterial colonies will be seen on any of the plates.

    c. Only one plate in 10^6 will have bacterial colonies.

    d. Some plates will have no bacterial colonies, some will have a few, and some will have many.

13. How are transposition and recombination fundamentally different?
    a. Recombination requires DNA homology but transposition does not.
    b. Transposition requires DNA homology but recombination does not.
    c. Only during transposition can a piece of DNA be moved from a virus to a cell.
    d. Only during recombination can a deletion occur.

14. Which of the following can result from the insertion of a transposon in bacteria?
    a. gene inactivation
    b. an increase or decrease in transcriptional activity of a gene
    c. deletions and insertions
    d. all of the above

15. Which of the following can be consequences of transposition in eukaryotes?
    a. no effect on a nearby gene
    b. production of a mutation by insertional mutagenesis
    c. increased or decreased transcription from a nearby promoter
    d. all of the above

16. In what ways are bacterial IS and Tn elements alike?
    a. Both have inverted repeat sequences at their ends.
    b. Both integrate into target sites and cause a target site duplication.
    c. Both contain a transposase gene.
    d. Both contain antibiotic resistance genes.
    e. Both always use replicative transposition.
    f. a, b, and c
    g. all of the above

17. Reverse transcriptase is used by which two of the following transposable elements? [Mark two letters.]
    a. the *Ty* element in yeast
    b. an IS element in bacteria
    c. the Tn*10* element in bacteria
    d. an L1 element in humans

18. In a bacterial IS element, how does transposase know the boundaries of the DNA to be transposed?
    a. There are special recognition sites at the ends of the IS element.
    b. It recognizes the target site of insertion.
    c. It recognizes the IR sequences.
    d. It recognizes the delta sequences.

19. In corn, the insertion of an autonomous element within a gene results in the production of an allele that mutates at high frequency. Why?
    a. Although an autonomous element cannot excise, it is highly mutable.
    b. An autonomous element has transposase and can excise, thereby causing a mutation.
    c. Autonomous alleles are unstable if nonautonomous elements are also in the genome.
    d. Autonomous alleles generally insert into promoter regions.

20. *Ty* elements in yeast are similar to retroviruses because (choose the correct answer)
    a. both *Ty* elements and retroviruses utilize an RNA transposition intermediate
    b. both *Ty* elements and retroviruses utilize reverse transcriptase

   c. introns are usually found in retroviruses and *Ty* elements

   d. both (a) and (b)

21. Which of the following is *not* true about the L1 family of LINE elements in mammals?

   a. They are retrotransposons.

   b. They comprise about 5 percent of the genome.

   c. All elements in one individual are full-length elements.

   d. They are capable of causing disease by insertional mutagenesis during an individual's lifetime.

Answers: 1f, 2c, 3b, 4g, 5d, 6e, 7a, 8h, 9a, 10d, 11c, 12d, 13a, 14d, 15d, 16f, 17a & d, 18c, 19b, 20d, 21c

Thought Questions

1. What is the evidence that mutations occur spontaneously and at a low frequency, regardless of selection for or against them?

2. The proofreading ability of DNA polymerases is extremely good, but not quite perfect. What error rates are associated with DNA replication, from whence do they arise, and how might this be important in evolutionary terms?

3. Distinguish between a reversion mutation, an intragenic suppressor mutation, and an intergenic suppressor mutation.

4. Distinguish between a missense and a nonsense mutation and between a missense, a silent, and a neutral mutation.

5. How would you classify the following mutations: (a) a single base change in a promoter region that affects the transcription of a gene; (b) a single base change in a promoter region that has no effect; (c) a DNA sequence difference between two strains that has no apparent phenotypic effect?

6. What is the basis for the Ames test? What are its specific merits? Are there any kinds of mutagens that it might not detect (e.g., those that specifically cause point mutations, frameshifts, small deletions, or chromosomal rearrangements)?

7. We have obtained significant insight into how DNA is repaired from studies in bacteria and yeast, as well as from the molecular genetic analyses of disease genes from humans with defective DNA repair. (a) Speculate as to the phenotype(s) of bacteria and yeast mutations that are defective in DNA repair. (b) How might comparison of these three systems be beneficial?

8. In *Drosophila*, the first mutations isolated (e.g., the first white-eyed mutant) were spontaneous. Currently, most new mutations are isolated by using mutagens. Why are mutagens currently employed, and what benefit do mutagens have for the study of genetics and development? How is site-specific in vitro mutagenesis used to study the function of cloned genes?

9. What spontaneous mutation rates are found in different organisms? Are these rates relatively similar or quite different? Why do you think this is so?

10. Dyes are routinely used to stain nuclei in bacteriological and histological preparations as well as to visualize DNA and RNA molecules separated by size using agarose gel electrophoresis. Some of these dyes bind to the backbone of nucleic acids (e.g., the major groove of double-stranded DNA), while others intercalate between bases. How would you specifically test whether a particular dye is mutagenic? How would you use the results of the test to design general precautions for laboratory workers using these dyes?

11. If a substance is mutagenic, is it necessarily carcinogenic? Why or why not? If a substance is carcinogenic, is it necessarily mutagenic? Why or why not?

12. Contrast conservative and replicative transposition. How can conservative transposition of the *Ac* element result in gene duplication? Can a particular IS element behave in both ways?

13. Consider how the distinction between direct and inverted repeats is central to the recognition of insertion sequences. The insertion sequence itself contains inverted terminal repeats. Thus, inverted repeats are present as part of the transposable element both before and after the insertion. On the other hand, the act of insertion results in a staggered cut in the host DNA. When the gaps are filled in after insertion, direct repeats will be generated flanking the inverted terminals of the transposable element. Indeed, insertion sequences have been identified by the presence of a pair of inverted repeats flanked by direct repeats. Using this logic, find the insertion sequence in the following single-stranded length of DNA:

CGTAGCCATTTGCGATATGCATCCGAATATCGCAAGCCATGCCA

14. In what situations can a nonautonomous element behave like an autonomous element? Give specific examples in corn and *Drosophila*.

15. Following the mobilization of *Ty* elements in yeast, two mutants were recovered that affected the activity of enzyme X. One of these mutants lacked enzyme activity and was a null mutation. The other mutant had three times as much enzyme activity. Both mutants had a *Ty* element inserted into the X gene. Generate a hypothesis to explain how this is possible.

16. What differences and similarities are there between transposable elements that utilize an RNA intermediate and those that do not?

17. What kinds of repetitive elements are SINEs and LINEs? Support for a hypothesis that a particular repetitive element is a transposon can come from analyzing its DNA sequence for open reading frames that encode transposase-like products. In humans, Alu sequences do not encode any of the enzymes that are presumably needed for retrotransposition. What aspects of their structure suggest that they might be capable of retrotransposition? What evidence is there that Alu sequences are retrotransposons capable of moving via an RNA intermediate, much like the yeast *Ty* element?

18. In humans, insertion of transposable elements has occasionally been associated with disease. An L1 element inserted into the gene for blood-clotting factor VIII resulted in hemophilia in two unrelated individuals. An Alu element insertion disrupted pre-mRNA processing at the gene for neurofibromatosis. In each case, the insertion was not in either of the affected individuals' parents and so appeared to be a result of an insertional event during the affected individuals' lifetime. Given that each of these classes of elements represents a substantial fraction of moderately repetitive DNA, how likely is it that most spontaneous mutations are due to insertional mutagenesis by L1 elements? If it is not very likely, why not?

Thought Questions for Media

After reviewing the media on the *iGenetics* CD-ROM, answer these questions.

1. What are the consequences of a nonsense mutation?
2. What are the consequences if a nonsense mutation is suppressed?
3. Suppose the mutation in the *Nonsense Mutation and Nonsense Suppressor Mutation* animation was an AT-to-GC transition instead of an AT-to-TA transversion.
   a. What would be the result of this mutation on the gene's protein product?
   b. Explain how this mutation could be suppressed using a mechanism similar to that illustrated in the animation.

4. The *Mutagenic Effects of 5BU* animation shows how 5BU can introduce mutations that are either AT-to-GC transition mutations or GC-to-AT transition mutations. For each type of mutation:
   a. How many rounds of replication are involved in the formation of the mutation?
   b. In which round(s) of replication must 5BU be in its normal form?
   c. In which round(s) of replication must 5BU be in its rare form?

5. If 5BU is incorporated into DNA, but never shifts to its rare form during DNA replication, can it still cause a transition mutation?

6. Based in the information in the animation, is there a way for 5BU to cause a transversion mutation?

7. In principle, a mutagenic compound can cause either forward (new) mutations or reverse mutations.
   a. Does the Ames test assess the frequency of reverse mutations (say, *his* to *his*$^+$) or the frequency of forward mutations (say, *his*$^+$ to *his*)? How?
   b. Why is it advantageous for the Ames test to assess this type of mutation frequency?
   c. Does this strategy influence which types of mutagenic compounds can be detected using the Ames test?

8. According to the *Ames Test Protocol* animation, why is the S9 liver extract added to the bacteria during the Ames test?

9. Suppose two different transposable elements are mobilized and insert themselves into two different genes. In gene A, transposon IS1 inserts itself into the translated region. In gene B, transposon IS2 inserts itself at a site three base pairs downstream from the TATA box, but before the start of the transcribed region.
   a. What similarities do you expect to find between the sequences of the transposable element inserted at each gene?
   b. In addition to the sequence of the transposable elements, what sequence alterations do you expect to find in the region near the site of insertion?
   c. What features do the sequence alterations in (b) have, and how do they arise?
   d. How might the transposon insertions influence the functions of genes A and B? Are they similar or different? Why?

1. In the iActivity *A Toxic Town*, you observed that vinyl chloride causes a missense transversion mutation in the Ames test. Although the allowable EPA concentration of vinyl chloride is 2 μg/L, it is present at 140 μg/L in the groundwater of Russellville.
   a. Can you infer that it is the vinyl chloride in the drinking water that causes missense transversion mutations in the residents of Russellville? Why or why not?
   b. Suppose the concentration of vinyl chloride in a neighboring village is 2 μg/L, the EPA allowable concentration. Is the water safe to drink? What concerns, if any, would you have?
   c. Would your answers to (a) and (b) change if you knew that most of the adult residents of Russellville and the neighboring village smoked three packs of cigarettes a day?

2. In the iActivity *A Toxic Town*, 1,2-dichloroethene was not mutagenic in the Ames test, but was found in the groundwater at about six times the allowable EPA concentration. Since it is not mutagenic, why does the EPA bother to regulate how much 1,2-dichloroethene can be found in groundwater? What other factors might be used to determine the allowable groundwater concentration of a chemical?

3. Suppose that sequence analysis like that performed in the iActivity shows that a chemical causes a silent transition. Since it does not alter the amino-acid sequence of a polypeptide, should you still be concerned about the chemical as a potential mutagen?

4. In the *The Genetics Shuffle* iActivity, you were able to determine that Tn*10* transposition is conservative. Does this allow you to infer whether Tn*10* transposition proceeds through a DNA or RNA intermediate?

5. Suppose you performed an experiment similar to the one illustrated in the *The Genetics Shuffle* iActivity to identify the mode of transposition of an uncharacterized transposon. You observe that none of the colonies exhibit any sectoring and that they are either blue or white. After picking several blue and several white colonies, you use PCR to amplify the region surrounding the *donor* DNA in each colony. You sequence the PCR products and compare the sequences to those of the molecules used to generate the heteroduplexes. What do you find, and how do you explain your findings?

SOLUTIONS TO TEXT PROBLEMS

15.1   Mutations are (choose the correct answer)
   a. caused by genetic recombination
   b. heritable changes in genetic information
   c. caused by faulty transcription of the genetic code
   d. usually, but not always, beneficial to the development of the individuals in which they occur

   Answer: b

15.2   Answer true or false: Mutations occur more frequently if there is a need for them.

   Answer: False. Mutations occur spontaneously at a more or less constant frequency, regardless of selective pressure. Once they occur, however, they can be selected for or against, depending on the advantage or disadvantage they confer. It is important not to confuse the frequency of mutations with selection.

15.3   Which of the following is *not* a class of mutation?
   a. frameshift
   b. missense
   c. transition
   d. transversion
   e. none of the above (meaning all are classes of mutation)

   Answer: e

15.4   Ultraviolet light usually causes mutations by a mechanism involving (choose the correct answer)
   a. one-strand breakage in DNA
   b. light-induced change of thymine to alkylated guanine
   c. induction of thymine dimers and their persistence or imperfect repair
   d. inversion of DNA segments
   e. deletion of DNA segments
   f. all of the above

   Answer: c. The key to this answer is the word *usually*. The other choices might apply rarely, but not usually.

15.5   The amino-acid sequence shown in the following table was obtained from the central region of a particular polypeptide chain in the wild-type and in several mutant bacterial strains:

Codon

	1	2	3	4	5	6	7	8	9
Wild type ...	Phe	Leu	Pro	Thr	Val	Thr	Thr	Arg	Trp
Mutant 1: ...	Phe	Leu	His	His	Gly	Asp	Asp	Thr	Val
Mutant 2: ...	Phe	Leu	Pro	Thr	Met	Thr	Thr	Arg	Trp
Mutant 3: ...	Phe	Leu	Pro	Thr	Val	Thr	Thr	Arg	
Mutant 4: ...	Phe	Pro	Pro	Arg					
Mutant 5 ...	Phe	Leu	Pro	Ser	Val	Thr	Thr	Arg	Trp

For each mutant, say what change in DNA level has occurred, whether the change is a base-pair substitution mutation (transversion or transition, missense or nonsense) or a frameshift mutation, and in which codon the mutation occurred. (Refer to the codon dictionary in Figure 14.7, p. 362.)

Answer: Start methodically and write out the potential codons for the normal protein. Let N represent any nucleotide, R a purine (A or G), and Y a pyrimidine (C or U). Then the codons can be written as:

	1	2	3	4	5	6	7	8	9
Normal:	Phe	Leu	Pro	Thr	Val	Thr	Thr	Arg	Trp
Codon:	UUY	UUR	CCN	ACN	GUN	ACN	ACN	CGN	UGG
		or						or	
		CUN						AGR	

Mutant 1: The alteration of all of the amino acids after codon 2 suggests that this mutation is a frameshift, either a deletion or an addition of a base pair in the DNA region that codes for the mRNA near codons 2 and 3. His is CAY, which can be generated from CCN by the insertion of a single A between the two Cs. If this is the case, one has:

	1	2	3	4	5	6	7	8	9
Mutant 1:	Phe	Leu	His	His	Gly	Asp	Asp	Thr	Val
Frameshift:	UUY	UUR	CAC	NAC	NGU	NAC	NAC	NCG	NUG
		or						or	or
		CUN						NGR	RUG

Now identify the unknown bases, given the amino acid sequence:
    To code for His, the new fourth codon must be CAC.
    To code for Gly, the new fifth codon must be GGU.
    To code for Asp, the new sixth codon must be GAC.
    To code for Asp, the new seventh codon must be GAC.
    To code for Thr, the new eighth codon must be ACG.
    To code for Val, the new ninth codon must be GUG.

Thus, one has:

	1	2	3	4	5	6	7	8	9
Mutant 1:	Phe	Leu	His	His	Gly	Asp	Asp	Thr	Val
Codon:	UUY	UUR	CAC	CAC	GGU	GAC	GAC	ACG	GUG
		or							
		CUN							

This analysis indicates that the codons used in the normal protein are:

	1	2	3	4	5	6	7	8	9
Normal:	Phe	Leu	Pro	Thr	Val	Thr	Thr	Arg	Trp
Codon:	UUY	UUR	CCC	ACG	GUG	ACG	ACA	CGG	UGG
		or							
		CUN							

Mutant 2: Codon 5 in mutant 2 encodes Met, instead of Val. This single change could be caused by a point mutation, where the G in codon GUG is changed to an A, leading to an AUG codon. This would occur by a CG-to-TA transition in the DNA.

Mutant 3: Mutant 3 has a normal amino-acid sequence, but is prematurely terminated, indicating that a nonsense mutation occurred. The nonsense mutation occurred in the ninth codon. If an A base is substituted (for one of the Gs) or inserted (before or after the first G), one could have either UAG or UGA. Thus, either a frameshift or base substitution (CG-to-TA transition) occurred.

Mutant 4: Mutant 4 shows premature termination at codon 5 (indicating a nonsense codon there) and has missense mutations at codons 2 and 4. Compare the possible sequences of the mutant with the normal sequence to see if one mutational event can account for all of these phenomena.

	1	2	3	4	5	6	7	8	9
Normal:	Phe	Leu	Pro	Thr	Val	Thr	Thr	Arg	Trp
Normal	UUY	UUR	CCC	ACG	GUG	ACG	ACA	CGG	UGG
Codon:		or							
		CUN							
Mutant 4:	Phe	Pro	Pro	Arg	Stop				
Possible	UUY	CCN	CCN	AGR	UAA				
Mutant		(CCC)	(CCA)	or	UAG				
Codons:				CGN	UGA				

Consider codon 2 carefully. For Pro to be encoded by CCN in the mutant protein and be obtained by a single change from the second codon, Leu (in the normal protein) must be encoded by CUN. If the U were deleted (by an AT deletion in the DNA) and the N were a C, a CCC Pro codon and a frameshift would result. Codon 3 would become CCA and still code for Pro, codon 4 would become CGG and code for Arg, and codon 5 would become UGA, a nonsense codon.

Mutant 5: Mutant 5 shows an alteration of only the fourth amino acid (Thr to Ser), suggesting a point mutation. If the fourth codon were changed from ACG to UCG (by a TA-to-AT transversion in the DNA), this missense mutation would occur.

15.6   In mutant strain X of *E. coli,* a leucine tRNA that recognizes the codon 5'-CUG-3' in normal cells has been altered so that it now recognizes the codon 5'-GUG-3'. A missense mutation, which affects amino-acid 10 of a particular protein is suppressed in mutant X cells.

   a. What are the anticodons of the two Leu tRNAs, and what mutational event has occurred in mutant X cells?

   b. What amino acid would normally be present at position 10 of the protein (without the missense mutation)?

   c. What amino acid would be put in at position 10 if the missense mutation is not suppressed (i.e., in normal cells)?

   d. What amino acid is inserted at position 10 if the missense mutation is suppressed (i.e., in mutant X cells)?

Answer:

   a. If the normal codon is 5'-CUG-3', the anticodon of the normal tRNA is 5'-CAG-3'. If a mutant tRNA recognizes 5'-GUG-3', it must have an anticodon that is 5'-CAC-3'. The mutational event was a CG-to-GC transversion.

   b. Since a leucine-bearing (mutant) tRNA can suppress the mutation, presumably leucine is normally present at position 10.

   c. The mutant tRNA recognizes the codon 5'-GUG-3', which codes for Val. In normal cells, a Val-tRNA.Valine would recognize the codon and insert valine.

   d. Leu

15.7   A researcher using a model eukaryotic experimental system has identified a temperature-sensitive mutation, *rpIIA^{ts}*, in a gene that encodes a protein subunit of RNA polymerase II. This mutation results from a missense mutation. Mutants have a recessive lethal phenotype at the higher, restrictive temperature but grow at the lower, permissive (normal) temperature. To identify genes whose products interact with the subunit of RNA polymerase II, the researcher designs a screen to isolate mutations that will act as dominant suppressors of the temperature-sensitive recessive lethal mutation.

   a. Explain how a new mutation in an interacting protein could suppress the lethality of the temperature-sensitive original mutation.

   b. In addition to mutations in interacting proteins, what other type of suppressor mutations might be found?

   c. Outline how the researcher might select for the new suppressor mutations.

   d. Do you expect the frequency of suppressor mutations to be similar to, much greater than, or much less than the frequency of new mutations at a typical eukaryotic gene?

   e. How might this approach be used generally to identify genes whose products interact to control transcription?

Answer:

a. The temperature sensitivity of the *rpIIA^{ts}* mutant could be due to a single amino-acid change in the protein subunit of RNA polymerase II that causes it to be nonfunctional at the restrictive temperature. If its inability to function is due to a change in its secondary or tertiary structure that prevents it from interacting with another protein during transcription, then a mutation in this second interacting protein might be compensatory. Such a mutation would effectively suppress the initial mutation, as it would allow for transcription even at the restrictive temperature.

b. A new mutation in a second protein would be an intergenic suppressor mutation. The original mutation could also be suppressed by reverting the missense mutation or by an intragenic suppressor mutation. In an intragenic suppressor mutation, a particular second site within the protein would need to be mutated so as to compensate for the initial mutation.

c. One approach is to treat *rpIIA^{ts}/rpIIA^{ts}* individuals with a mutagen, and mate them to *rpIIA^{ts}/rpIIA^{ts}* individuals at the permissive temperature. A second-site suppressor could be selected for by removing the parents and shifting the progeny to the restrictive temperature. Since all of the offspring are *rpIIA^{ts}/rpIIA^{ts}*, only offspring carrying a new mutation capable of suppressing the recessive lethality of the *rpIIA^{ts}* mutation will survive.

d. Second-site suppressors will be quite rare, and will appear at a much lower frequency than mutations in a typical eukaryotic gene. To suppress a particular defect, a very specific new mutation must occur. Hence, the vast majority of mutations that are induced by a mutagen will lack the specific compensatory ability of a suppressor mutation.

e. Since intergenic suppressors may result from mutations in interacting proteins, this approach could be used to identify genes for proteins that interact during transcription.

15.8   The mutant *lacZ-1* was induced by treating *E. coli* cells with acridine, whereas *lacZ-2* was induced with 5BU. What kinds of mutants are these likely to be? Explain. How could you confirm your predictions by studying the structure of the β-galactosidase in these cells?

Answer: Acridine is an intercalating agent and so can be expected to induce frameshift mutations. 5BU is incorporated into DNA in place of T. During DNA replication, it is likely to be read as C by DNA polymerase because of a keto-to-enol shift. This results in point mutations, usually TA-to-CG transitions. Considering these expectations, we find that *lacZ-1* would probably result in a completely altered amino-acid sequence after some point, although it might be truncated (due to the introduction—out of frame—of a nonsense codon). In either case, it would be very likely to have a different molecular weight and charge and so migrate differently than the wild type would during gel electrophoresis (see Figure 12.8, p. 318). *lacZ-2* is likely to contain a single amino-acid difference, due to a missense mutation, although it, too, could contain a nonsense codon. A missense mutation might lead to the protein's having a different charge, while a nonsense codon would lead to a truncated protein that would have a lower molecular weight. Both would migrate differently during gel electrophoresis.

15.9 a. The sequence of nucleotides in an mRNA is

5'-AUGACCCAUUGGUCUCGUUAG-3'

Assuming that ribosomes could translate this mRNA, how many amino acids long would you expect the resulting polypeptide chain to be?

b. Hydroxylamine is a mutagen that results in the replacement of an AT base pair for a GC base pair in the DNA; that is, it induces a transition mutation. When hydroxylamine was applied to the organism that made the mRNA molecule shown in (a), a strain was isolated in which a mutation occurred at the 11th position of the DNA that coded for the mRNA. How many amino acids long would you expect the polypeptide made by this mutant to be? Why?

Answer:

a. The codons read as

5'-AUG-ACC-CAU-UGG-UCU-CGU-UAG-3'

The last codon is a nonsense (chain termination) codon, while the others are sense codons. The chain would be six amino acids long.

b. The new sequence would be

5'-AUG-ACC-CAU-UAG-...

Since UAG is a nonsense (chain termination) codon, the new chain would be only three amino acids long.

15.10 In a series of 94,075 babies born in a particular hospital in Copenhagen, 10 were achondroplastic dwarfs (an autosomal dominant condition). Two of these 10 had an achondroplastic parent. The other 8 achondroplastic babies each had two normal parents. What is the apparent mutation rate at the achondroplasia locus?

Answer: There were eight new mutations in 94,073 normal couples. Since the phenotype is dominant, the phenotype is seen when just one of the parental genes is mutated. There were 2 × 94,073 copies of the gene that could have undergone mutation. Therefore, the apparent mutation rate at this locus is

$$8/(2 \times 94{,}073) = 8/188{,}146 = 4 \times 10^{-5} \text{ mutations per locus per generation}$$

15.11 Three of the codons in the genetic code are chain-terminating codons for which no naturally occurring tRNAs exist. Just like any other codons in the DNA, though, these codons can change as a result of base-pair changes in the DNA. Confining yourself to single base-pair changes at a time, determine which amino acids could be inserted in a polypeptide by mutation of these chain-terminating codons: (a) UAG, (b) UAA, and (c) UGA. (The genetic code is listed in Figure 14.7, p. 362.)

Answer:

Nucleotide Altered	Codon					
	UAG	**Code**	**UAA**	**Code**	**UGA**	**Code**
First	AAG	Lys	AAA	Lys	AGA	Arg
	CAG	Gln	CAA	Gln	CGA	Arg
	GAG	Glu	GAA	Glu	GGA	Gly
Second	UUG	Leu	UUA	Leu	UUA	Leu
	UCG	Ser	UCA	Ser	UCA	Ser
	UGG	Trp	UGA	Stop	UAA	Stop
Third	UAC	Tyr	UAC	Tyr	UGC	Cys
	UAU	Tyr	UAU	Tyr	UGU	Cys
	UAA	Stop	UAG	Stop	UGG	Trp

15.12   The amino-acid substitutions in the following figure occur in the α and β chains of human hemoglobin.

Ala (1)　　　　　　　Val (3) ——————— Met (4)

Glu (2)

Pro (14) — Gln (13)　　　　　Lys (6) — Thr (7) — Ser (8)

Gly (5)

Asp (10)

Tyr (12) — His (11)　　　　　Asn (9)

Those amino acids connected by lines are related by single nucleotide changes. Propose the most likely codon or codons for each of the numbered amino acids. (Refer to the genetic code in Figure 14.7, p. 362.)

Answer:

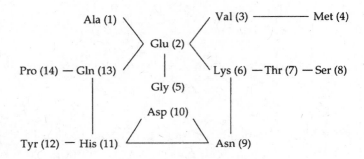

15.13 Charles Yanofsky studied the tryptophan synthetase of *E. coli* in an attempt to identify the base sequence specifying this protein. The wild type gave a protein with a glycine in position 38. Yanofsky isolated two *trp* mutants, *A23* and *A46*. Mutant *A23* had Arg instead of Gly at position 38, and mutant *A46* had Glu at position 38. Mutant *A23* was plated on minimal medium, and four spontaneous revertants to prototrophy were obtained. The tryptophan synthetase from each of the four revertants was isolated, and the amino acids at position 38 were identified. Revertant 1 had Ile, revertant 2 had Thr, revertant 3 had Ser, and revertant 4 had Gly. In a similar fashion, three revertants from *A46* were recovered, and the tryptophan synthetase from each was isolated and studied. At position 38, revertant 1 had Gly, revertant 2 had Ala, and revertant 3 had Val. A summary of these data is given in the following figure.

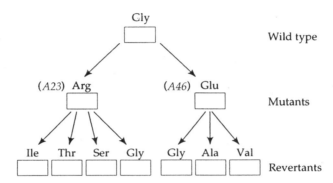

Using the genetic code in Figure 14.7, p. 362, deduce the codons for the wild type, for the mutants *A23* and *A46*, and for the revertants, and place each designation in the space provided in the figure.

Answer:

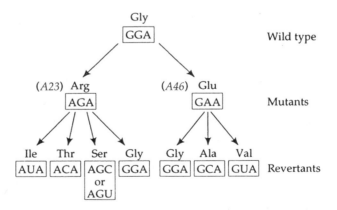

15.14 Consider an enzyme chewase from a theoretical microorganism. In the wild-type cell, the chewase has the following sequence of amino acids at positions 39 to 47 (reading from the amino end) in the polypeptide chain:

-Met-Phe-Ala-Asn-His-Lys-Ser-Val-Gly-
　39　 40　 41　 42　 43　 44　 45　 46　 47

A mutant organism that lacks chewase activity was obtained. The mutant was induced by a mutagen known to cause single base-pair insertions or deletions. Instead of making the complete chewase chain, the mutant made a short polypeptide chain only 45 amino acids long. The first 38 amino acids were in the same sequence as the first 38 of the normal chewase, but the last seven amino acids were as follows:

-Met-Leu-Leu-Thr-Ile-Arg-Val-
　39　　40　　41　　42　　43　　44　　45

A partial revertant of the mutant was induced by treating it with the same mutagen. The revertant that made a partly active chewase has the following sequence of amino acids at positions 39 to 47 in its amino acid chain:

-Met-Leu-Leu-Thr-Ile-Arg-Gly-Val-Gly
　39　　40　　41　　42　　43　　44　　45　　46　　47

Using the genetic code given in Figure 14.7, p. 362, deduce the nucleotide sequences for the mRNA molecules that specify this region of the protein in each of the three strains.

Answer: Using the genetic code, and letting N represent any nucleotide, R a purine, and Y a pyrimidine, the wild-type RNA can be denoted as:

-Met-Phe-Ala-Asn-His-Lys-Ser-Val-Gly-
　39　　40　　41　　42　　43　　44　　45　　46　　47
　　　　　　　　　　　　　　　　　　　　UCN
AUG UUY GCN AAY CAY AAR or GUN GGN
　　　　　　　　　　　　　　　　　　　　AGY

Since the mutants were obtained using a mutagen that causes single base-pair insertions or deletions, the mutants have frameshift mutations. Therefore, a base pair should be missing or added in codon 40. Determine which insertion or deletion gives a frameshift with the appropriate amino-acid sequence by comparing the mutant amino-acid sequence with that of the wild type. You will find that if the Y of codon 40 is deleted, an appropriate sequence can be obtained, providing the unknown nucleotides are as specified below:

-Met-Leu-Leu-Thr-Ile-Arg-Val-(stop)
　39　　40　　41　　42　　43　　44　　45
AUG UUG CNA AYC AYA ARA GYG UNG
　　　　　N=U Y=C Y=U R=G Y=U N=A

This means that the original sequence was

-Met-Phe-Ala-Asn-His-Lys-Ser-Val-Gly-
　39　　40　　41　　42　　43　　44　　45　　46　　47
AUG UUY GCU AAC CAU AAG AGU GUA GGN

In the revertant, the reading frame is restored at codon 46, the Val (GUG) at codon 45 is altered to a Gly (GGN), and the chain termination codon at position 46 (UAG) is altered to a Val (GUN). This would have occurred if a G was inserted before or after the first G in codon 45. One would have

```
-Met-Leu-Leu-Thr-Ile-Arg-Gly-Val-Gly-
  39   40   41   42   43   44   45   46   47
 AUG  UUG  CUA  ACC  AUA  AGA  GGU  GUA  GGN
```

15.15   DNA polymerases from different organisms differ in the fidelity of their nucleotide insertion; however, even the best DNA polymerases make mistakes, usually mismatches. If such mismatches are not corrected, they can become fixed as mutations after the next round of replication.

   a.   How does DNA polymerase attempt to correct mismatches during DNA replication?

   b.   What mechanism is used to repair such mismatches if they escape detection by DNA polymerase?

   c.   How is the mismatched base in the newly synthesized strand distinguished from the correct base in the template strand?

Answer:

   a.   Many, but not all, DNA polymerases have a proofreading capacity that causes them to stall at mismatches during DNA replication. When present, the $3' \rightarrow 5'$ exonuclease activity associated with the polymerase will excise mismatched bases, so that they can be replaced. See text discussion in Chapter 11, p. 286–288, and Chapter 15, p. 394.

   b.   Shortly after DNA replication, mismatches can still be repaired by using a set of enzymes that recognize and excise mismatches in the newly synthesized DNA strand. In *E. coli,* mismatch repair is initiated when single base-pair mismatches or small base-pair additions or deletions in newly replicated DNA are bound by the MutS protein. The newly replicated strand is distinguished from the parental strand by the presence of an unmethylated A nucleotide in a 5'-GATC-3' sequence close to the mismatch. The MutL and MutH proteins bring the unmethylated GATC close to the mismatch and form a complex with MutS. MutH nicks the new, unmethylated DNA strand and a section of this DNA strand, including the mismatch, is excised by an exonuclease. DNA polymerase III and ligase repair the gap, producing the correct base pair. See Figure 15.17.

   c.   In *E. coli,* the parental strand can be distinguished from the newly replicated strand by methylation of the A in the sequence 5'-GATC-3'. As a newly replicated DNA strand is not methylated until a short time after its synthesis, hemimethylation at this frequently appearing site allows for the identification of which strand is newly synthesized. Since only the strand with the unmethylated A is nicked by the MutH protein and excised by an exonuclease during mismatch repair, only the newly synthesized strand and not the parental strand is resynthesized by DNA polymerase III.

15.16   Two mechanisms in *E. coli* were described for the repair of thymine dimer formation after exposure to ultraviolet light: photoreactivation and excision (dark) repair. Compare these mechanisms, indicating how each achieves repair.

Answer: Photoreactivation requires the enzyme photolyase, which, when activated by a photon of light with a wavelength between 320 and 370 nm, splits the dimers apart.

Dark repair does not require light, but requires several different enzymes. First, the uvrABC endonuclease makes two single-stranded nicks, on the 5′ side and the 3′ side of the dimer. Then an exonuclease excises the 12-nucleotide-long segment of one strand between the nicks, including the dimer. Next, DNA polymerase I fills in the single-stranded region in the 5′→3′ direction. Finally, the gap is sealed by DNA ligase.

15.17   DNA damage by mutagens has very serious consequences for DNA replication. Without specific base pairing, the replication enzymes cannot specify a complementary strand, and gaps are left after the passing of a replication fork.

   a. What response has *E. coli* developed to large amounts of DNA damage by mutagens? How is this response coordinately controlled?

   b. Why is this response itself a mutagenic system?

   c. What effects would loss-of-function mutations in *recA* or *lexA* have on *E. coli*'s response?

   Answer:

   a. *E. coli* has developed the SOS response to respond to large amounts of DNA damage. When such damage occurs, the RecA protein becomes activated and stimulates LexA to cleave itself. Since the LexA protein functions as a repressor for about 17 genes, whose products are involved in DNA damage repair (it binds to a 20-nucleotide regulatory sequence called the SOS box), this results in the coordinate transcription of these genes. After the products of these genes "correct" the DNA damage, RecA is inactivated, and newly synthesized LexA represses the expression of the genes.

   b. The response is mutagenic because in some instances (e.g., in the repair of C^C dimers), the SOS repair system becomes stalled because of a mismatch. The delay is long enough for C to be deaminated to U. This results in a CG-to-TA transition.

   c. If there are loss-of-function mutations in both *recA* and *lexA*, or a loss-of-function mutation only in *lexA*, there would be no functional LexA protein to repress transcription of the 17 genes whose protein products are involved in the SOS response; this would result in constitutive activation of the SOS response. If the loss-of-function mutation is only in *recA*, however, heavy DNA damage would not result in activating the RecA protein, so RecA could not stimulate the LexA protein to cleave itself to induce the SOS response. Instead, the LexA protein would continue to repress the DNA repair genes in the SOS system. *E. coli* with a loss-of-function mutation in *recA* but with a normal *lexA* gene would, therefore, be deficient in SOS repair of heavy DNA damage and consequently would be sensitive to UV irradiation and to X-rays. *E. coli* with a *recA* loss-of-function mutation were observed to have this phenotype by Ann Dee Margulies and Alvin Clark, who discovered the *recA* gene in 1964.

15.18   After a culture of *E. coli* cells was treated with the chemical 5-bromouracil, it was noted that the frequency of mutants was much higher than normal. Mutant colonies were then isolated, grown, and treated with nitrous acid; some of the mutant strains reverted to wild type.

   a. In terms of the Watson-Crick model, diagram a series of steps by which 5BU may have produced the mutants.

   b. Assuming the revertants were not caused by suppressor mutations, indicate the steps by which nitrous acid may have produced the back mutations.

Answer:

a. In its normal state, 5-bromouracil is a T analog, base-pairing with A. In its rare state, it resembles C and can base-pair with G. It will induce an AT-to-GC transition as follows:

$$
\begin{array}{ccccccc}
\mid\mid & & \mid\mid & & \mid\mid & & \mid\mid \\
A\text{-}T & \rightarrow & A\text{-}5BU & \rightarrow & G\text{-}5BU & \rightarrow & G\text{-}C \\
\mid\mid & & \mid\mid & & \mid\mid & & \mid\mid
\end{array}
$$

b. Nitrous acid can deaminate C to U, resulting in a CG-to-TA transition.

15.19   A single, hypothetical strand of DNA is composed of the following base sequence, where A indicates adenine, T indicates thymine, G indicates guanine, C denotes cytosine, U denotes uracil, BU is 5-bromouracil, 2AP is 2-amino-purine, BU-enol is a tautomer of 5BU, 2AP-imino is a rare tautomer of 2AP, HX is hypoxanthine, X is xanthine, and 5' and 3' are the numbers of the free, OH-containing carbons on the deoxyribose part of the terminal nucleotides:

```
5'-T-HX-U-A-G-BU-enol-2AP-C-BU-X-2AP-imino-3'
```

a. Opposite the bases of the hypothetical strand, and using the shorthand of the base sequence, indicate the sequence of bases on a complementary strand of DNA.
b. Indicate the direction of replication of the new strand by drawing an arrow next to the new strand of DNA from (a).
c. When postmeiotic germ cells of a higher organism are exposed to a chemical mutagen before fertilization, the resulting offspring expressing an induced mutation are almost always mosaics for wild-type and mutant tissue. Give at least one reason that these mosaics, and not so-called complete or whole-body mutants, are found in the progeny of treated individuals.

Answer:

a, b.
```
5'-T - HX - U - A - G - BU-enol - 2AP - C - BU - X - 2AP-imino - 3'
3'-A-  C - A - T - C-    G   - T  -G- A - C-     C      - 5'
←
```

c. If postmeiotic germ cells are treated, a single mutated site in one strand of the double helix will not be repaired until mitotic division begins during embryogenesis. Given the semiconservative replication of DNA, the normal strand will be replicated into two normal strands. If the mutated site on the other strand is not repaired, when the strand is replicated it will give rise to a double helix with two mutant strands. In this scenario, the products of DNA replication at the first mitotic division consist of one normal and one mutant helix. This will produce two different cell types, resulting in a mosaic individual.

For some mutagens, such as BU or 2AP, there is a variation in this method of mosaic production. If the BU (or 2AP) remains in the DNA strand it was incorporated into, the first mitotic division may result in a normal double helix (the new strand is copied from the old normal strand), and a double helix with the BU (or 2AP) paired to a "wrong" base (due to the rare form of BU). In this case, a mutant site will be produced on one of the daughter double helices that result after the second mitotic division (at the four-cell stage). In principle, a mutant cell could be introduced at any mitotic division when BU is incorporated in its rare form during DNA synthesis. In each case, a mosaic individual would be produced, but the later the developmental stage at which the mutation was introduced, the fewer the mutation-bearing cells.

The following information applies to problems 15.20 through 15.24. A solution of single-stranded DNA is used as the template in a series of reaction mixtures and has the base sequence

where A = adenine, G = guanine, C = cytosine, T = thymine, H = hypoxanthine, and HNO_2 = nitrous acid. Use the shorthand system shown in the sequence, and draw the products expected from the reaction mixtures. Assume that a primer is available in each case.

15.20   The DNA template + DNA polymerase + dATP + dGTP + dCTP + dTTP + Mg^{2+}.

Answer:

15.21 The DNA template + DNA polymerase + dATP + dGMP + dCTP + dTTP + Mg^{2+}.

Answer: The absence of dGTP leads to a block in polymerization after the first two bases:

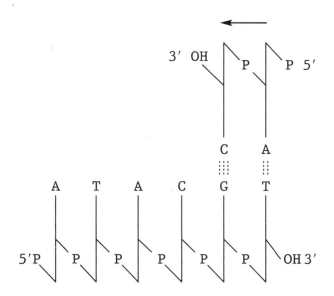

15.22 The DNA template + DNA polymerase + dATP + dHTP + dGMP + dTTP + Mg^{2+}.

Answer: While the dHTP can substitute for dGTP, there is no dCTP, so polymerization cannot continue past the first base:

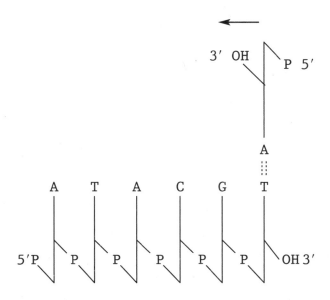

15.23 The DNA template is pretreated with HNO$_2$ + DNA polymerase + dATP + dGTP + dCTP + dTTP + Mg^{2+}.

Answer: Pretreatment of the template with HNO$_2$ deaminates G to X, C to U, and A to H. X will still pair with C, but U pairs with A, and H pairs with C, causing "mutations" in the newly synthesized strand:

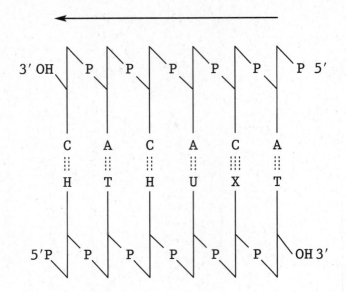

15.24   The DNA template + DNA polymerase + dATP + dGMP + dHTP + dCTP + dTTP + Mg^{2+}.

Answer: The dHTP will substitute for the absence of dGTP, and pair with C:

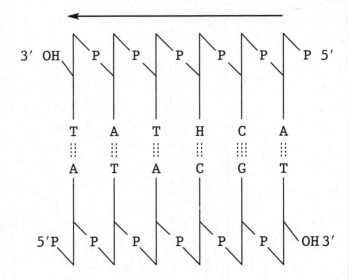

15.25   A strong experimental approach to determining the mode of action of mutagens is to examine the revertibility of the products of one mutagen by other mutagens. The following table presents data on the revertibility of *rII* mutations in phage T2 by various mutagens ("+" indicates majority of mutants reverted, "−" indicates almost no reversion; BU = 5-bromouracil, AP = 2-aminopurine, NA = nitrous acid, and HA = hydroxylamine):

Mutation	Proportion of Mutations Reverted by				Base-Pair Substitution Inferred
Induced by	**BU**	**AP**	**NA**	**HA**	
BU	+			–	
AP		–	+		
NA	+	+		+	
HA			+	–	GC→AT

Fill in the empty spaces.

Answer: Answer this problem using only the data provided. First consider what is known about HA. HA causes a GC-to-AT transition, and cannot revert mutations it induces. Therefore, HA cannot cause AT-to-GC transitions. HA also cannot revert the BU mutation. Consequently, BU must have caused a GC-to-AT transition.

Now consider the relationship between NA and HA. Since HA can revert NA mutations, the NA mutation was an AT-to-GC transition. This is consistent with the observation that NA can revert the HA mutation.

Now consider AP. AP cannot revert its own mutation, but it can revert the AT-to-GC transition caused by NA. Therefore the AP mutation was a GC-to-AT transition, and AP cannot induce AT-to-GC transitions. The AP mutation cannot be reverted by HA, because HA cannot induce AT-to-GC transitions. AP also could not revert the GC-to-AT transitions caused by BU or HA.

Because BU can revert its own GC-to-AT mutation, it can induce AT-to-GC transitions. Therefore, BU can revert the GC-to-AT mutations caused by AP and HA.

Because NA can cause AT-to-GC transitions, NA could revert the GC-to-AT transition caused by BU. It cannot be determined from these data whether NA can induce GC-to-AT transitions. If it can induce GC-to-AT transitions, it could revert its own mutations; if it cannot, it would not be able to revert its own mutations.

Mutation	Proportion of Mutations Reverted by				Base-Pair Substitution Inferred
Induced by	**BU**	**AP**	**NA**	**HA**	
BU	+	–	+	–	GC-to-AT
AP	+	–	+	–	GC-to-AT
NA	+	+	?	+	AT-to-GC
HA	+	–	+	–	GC-to-AT

15.26 a. Nitrous acid deaminates adenine to form hypoxanthine, which forms two hydrogen bonds with cytosine during DNA replication. After a wild-type strain of bacteria is treated with nitrous acid, a mutant is recovered that contains a protein with an amino-acid substitution: The mutant strain has a valine (Val) in a position where the wild-type strain has a methionine (Met). What is the simplest explanation for this observation?

   b. Hydroxylamine adds a hydroxyl (OH) group to cytosine, causing it to pair with adenine. Could mutant organisms like those in (a) be back mutated (returned to normal) using hydroxylamine? Explain.

Answer:

a. The codon for Met is 5'-AUG-3'. This would be encoded by 3'-TAC-5' in the template DNA strand, which pairs with 5'-ATG-3' in the nontemplate DNA strand. In this case, the A in the nontemplate strand was deaminated to form hypoxanthine and paired with C in the template strand. The new template was 3'-CAC-5' and the new codon was 5'-GUG-3', coding for Val.

b. In the mutant in (a), the template strand was 3'-CAC-5', and the nontemplate strand was 5'-GTG-3'. Since hydroxylamine acts on cytosine bases, it would act on the template strand. If it acted at the 3'-C, DNA replication would result in the nontemplate strand becoming 5'-ATG-3'. After an additional round of replication, one of the daughter cells would have a template strand that is 3'-TAC-5'. Thus, hydroxylamine could be used to obtain revertants.

15.27   A wild-type strain of bacteria produces a protein with the amino acid proline (Pro) at one site. Treatment of the strain with nitrous acid, which deaminates C to make it U, produces two different mutants. One mutant has a substitution of serine (Ser), and the other has a substitution of leucine (Leu) at the site.

   Treatment of the two mutants with nitrous acid now produces new mutant strains, each with phenylalanine (Phe) at the site. Treatment of these new Phe-carrying mutants with nitrous acid produces no change. The results are summarized in the following figure:

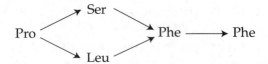

Using the appropriate codons, show how it is possible for nitrous acid to produce these changes and why further treatment has no influence. (Assume that only single nucleotide changes occur at each step.)

Answer: Nitrous acid deaminates C to make it U. U pairs with A, so treatment with nitrous acid leads to CG-to-TA transitions. Analyze how this treatment would affect the codons of this protein. Use N to represent any nucleotide and Y to represent a pyrimidine (U or C). Then the codons for Pro are CCN, the (relevant) codons for Ser are UCN, the codons for Leu are CUN, and the codons for Phe are UUY. (Nucleotides are written in the 5'→3' direction, unless specifically noted otherwise.)

   The codon CCN for Pro would be represented by CCN in the nontemplate DNA strand. Deamination of the 5'-C would lead to a nontemplate strand of UCN and a template strand of 3'-AGN-5'. This would produce a UCN codon encoding Ser. Deamination of the middle C would lead to a nontemplate strand of CUN and a template strand of 3'-GAN-5', producing a CUN codon encoding Leu.

   Further treatment of either mutant would result in deamination of the remaining C and a template strand of 3'-AAN-5'. This would result in a UUN codon. Since we are told that Phe is obtained, N must be C or U, and the template strand must be 3'-AAA-5' or 3'-AAG-5'.

   To explain why further treatment with nitrous acid has no effect, observe that nitrous acid acts via deamination and that T has no amine group. If the template strand were

$3'$-AAA-$5'$, the nontemplate strand would have been TTT. Since T cannot be deaminated, nitrous acid will have no effect on the nontemplate strand.

15.28   Three *ara* mutants of *E. coli* were induced by mutagen X. The ability of other mutagens to cause the reverse change (*ara* to *ara*$^+$) was tested, with the following results:

Frequency of *ara*$^+$ Cells Among Total Cells After Treatment

Mutant	None	BU	AP	HA	Frameshift
			Mutagen		
ara-1	1.5×10^{-8}	5×10^{-5}	1.3×10^{-4}	1.3×10^{-8}	1.6×10^{-8}
ara-2	2×10^{-7}	2×10^{-4}	6×10^{-5}	3×10^{-5}	1.6×10^{-7}
ara-3	6×10^{-7}	10^{-5}	9×10^{-6}	5×10^{-6}	6.5×10^{-7}

Assume that all *ara*$^+$ cells are true revertants. What base changes were probably involved in forming the three original mutations? What kinds of mutations are caused by mutagen X?

Answer: Use the revertant frequencies under the heading "None" to estimate the spontaneous reversion frequency.

   ara-1: BU and AP, but not HA or a frameshift, can revert *ara-1*. Both BU and AP cause CG-to-TA and TA-to-CG transitions, while HA causes only CG-to-TA transitions. If HA cannot revert *ara-1*, it must require a TA-to-CG transition to be reverted and must be caused by a CG-to-TA transition.

   ara-2: BU, AP, and HA, but not a frameshift, can revert *ara-2*. Since HA causes only CG-to-TA transitions, *ara-2* must have been caused by a TA-to-CG transition.

   ara-3: By the same logic as that for *ara-2*, *ara-3* must have been caused by a TA-to-CG transition.

   Provided that this is a representative sample, mutagen X appears to cause both TA-to-CG and CG-to-TA transitions. It does not appear to cause frameshift mutations.

15.29   As genes have been cloned for a number of human diseases caused by defects in DNA repair and replication, striking evolutionary parallels have been found between human and bacterial DNA repair systems. Discuss the features of DNA repair systems that appear to be shared in these two types of organism.

Answer: One of the most striking parallels between human and bacterial DNA repair systems has come from the analysis of mutations that affect mismatch repair (see problem 15.15). Mutations in any of the four human genes *hMSH2*, *hMLH1*, *hPMS1*, and *hPMS2* result in an increased accumulation of mutations in the genome and a hereditary predisposition to hereditary nonpolyposis colon cancer (HNPCC). Mutations in the *E. coli* genes *mutS* and *mutL* also result in an increased accumulation of mutations, and the normal alleles of these genes are essential for the initial stages of mismatch repair. The human gene *hMSH2* is homologous to *E. coli mutS*, and the other three human genes are homologous to *E. coli mutL*, indicating that aspects of mismatch repair are shared in these two organisms.

   Examples of additional human genetic defects that lack functions required for DNA repair include xeroderma pigmentosum, ataxia telangiectasia, Fanconi anemia, Bloom syndrome, and Cockayne syndrome (see Table 15.1, p. 398). While the specific DNA repair process affected by these mutations may not yet be sufficiently characterized for a direct comparison with a homologous process in *E. coli*, it has become clear that there are parallels in the types of repair processes that exist in the two organisms. For example, xeroderma pigmentosum and Fanconi anemia both affect repair of DNA damage induced by UV irradiation, whose mechanism of repair is well understood in *E. coli*.

15.30   Distinguish between prokaryotic insertion elements and transposons. How do composite transposons differ from noncomposite transposons?

Answer: Insertion elements (IS elements) are simpler in structure than transposons, and have a transposase gene flanked by perfect or nearly perfect inverted repeat sequences (IR sequences). The more complex transposable elements (Tn elements) exist in two forms, composite transposons and noncomposite transposons. Composite transposons have a central gene-bearing region flanked by IS elements. Transposition of these elements, and the genes they contain (e.g., genes for antibiotic resistance), occurs because of the function of the IS elements. Noncomposite transposons also contain genes, but do not terminate with IS elements. However, they do have inverted terminal repeated sequences that are required for transposition.

Both types of elements can integrate into target sites with which they have no homology, where they cause target site duplications. The transposition of IS elements and composite transposons requires use of the host cell's replicative enzymes and the transposase encoded in the element. Noncomposite transposons can transpose by two different mechanisms. Tn3-like noncomposite transposons utilize a replicative transposition mechanism employing a cointegrate—a fusion between a transposable element and the recipient DNA. Genes within Tn3 code for the enzymes transposase and resolvase that are needed for its transposition. Tn10-like composite transposons can move by a conservative (nonreplicative) transposition.

Upon insertion, all of the elements can cause mutations by a number of different mechanisms, including disruption of a gene's coding sequence or regulatory region.

15.31   What properties do bacterial and eukaryotic transposable elements have in common?

Answer: The structure and, at a general level, the function of eukaryotic and prokaryotic transposable elements are very similar. For example, both Tn and *Ac* elements have genes within them and have inverted repeats at their ends. Both prokaryotic and eukaryotic elements may affect gene function in a variety of different ways, depending on the element involved and how it integrates into or nearby a gene. The integration events of eukaryotic elements, like those of most prokaryotic transposable elements, involve nonhomologous recombination. Some eukaryotic elements, such as *Ty* elements in yeast and retrotransposons, move via an RNA intermediate, unlike the IS and Tn prokaryotic elements.

15.32   An IS element became inserted into the *lacZ* gene of *E. coli*. Later, a small deletion occurred that removed 40 base pairs near the left border of the IS element. The deletion removed 10 *lacZ* base pairs, including the left copy of the target site, and the 30 leftmost base pairs of the IS element. What will be the consequence of this deletion?

Answer: The left inverted repeat of the IS element has been removed, so that the two ends of this IS element are no longer homologous. The element will not be able to move out of this location and insert into another site.

15.33   Although the detailed mechanisms by which transposable elements transpose differ widely, some features underlying transposition are shared. Examine the shared and different features by answering the following questions.

    a. Use an example to illustrate different transposition mechanisms that require
       i. DNA replication of the element
      ii. no DNA replication of the element
      iii. an RNA intermediate

    b. What evidence is there that the inverted or direct terminal repeat sequences found in transposable elements are essential for transposition?

    c. Do all transposable elements generate a target-site duplication after insertion?

Answer:

    a. i. The transposon Tn3 is a transposable element whose transposition requires its DNA replication. Its transposition is illustrated in the cointegration model shown in Figure 15.22, p. 402. When the donor DNA containing the transposable element fuses with the recipient DNA, the transposable element becomes duplicated with one copy located at each junction between the donor and recipient DNA. After this cointegrate is resolved into two products, each has one copy of the transposable element.

       ii. The bacterial Tn10 element and the corn *Ac* element transpose without DNA replication using conservative transposition. After transposition, the element is lost from its original site of insertion.

      iii. *Ty* elements in yeast and *copia* elements in *Drosophila* transpose by an RNA intermediate. After synthesis of an RNA copy of the integrated DNA sequence, reverse transcription results in a new element that integrates at a new chromosomal location.

    b. Evidence that the inverted or direct terminal repeats found in transposable elements are essential for their transposition comes from the observation that mutations altering these sequences eliminate the ability of an element to transpose. These sequences are recognized by transposase during a transposition event.

    c. yes

15.34   In addition to single gene mutations caused by the insertion of transposable elements, the frequency of chromosomal aberrations such as deletions or inversions can be increased when transposable elements are present. How?

Answer: Deletions, inversions, and translocations can occur when homologous recombination occurs between two identical transposons inserted in different locations in the genome. To see how this occurs, draw out the chromosomes produced when a crossover occurs between two transposons inserted in different locations and orientations in the genome. If two transposons are inserted in the same orientation in one chromosome, pairing and homologous recombination between them results in a deletion of the segment lying between them. If two transposons are inserted in an opposite orientation in one chromosome, pairing and homologous recombination between them results in an inversion of the segment lying between them. If two transposons are inserted in different chromosomes, pairing and homologous recombination between them will result in a reciprocal translocation. Whether the translocation results in two single-centromere chromosomes or in a dicentric and acentric chromosome will depend on the orientation and position of the elements relative to their respective centromeres.

15.35　An understanding of the molecular structures of *Ac* and *Ds* elements, together with an understanding of the basis for their transposition, has allowed for a clearer, molecular-based interpretation of Barbara McClintock's observations. Ponder the significance of this interpretation on the acceptance of her work, and use your understanding of transposition of *Ac* and *Ds* elements in corn to propose an explanation for the following experiment.

Two different true-breeding strains of corn with colorless kernels A and B are crossed with each of two different true-breeding strains with purple kernels C and D. The F_1s from each cross are selfed, with the following results:

P:	A × C	A × D
F_1:	all purple	all purple
F_2:	3 purple : 1 colorless	3 purple : 1 colorless

P:	B × C	B × D
F_1:	all purple	all purple
F_2:	3 purple : 1 spotted*	3 purple : 1 colorless

*spotted = kernels with purple spots in a colorless background

Answer:

As is described in Figure 15.24, p. 406, the spotted-kernel phenotype arises from the transposition of *Ds* elements (incomplete transposons incapable of transposition themselves) out of the *C* gene during kernel development. This means that the mutation in colorless strain B is caused by an insertion of a *Ds* element in the *C* gene. It is able to transpose because the purple strain C has *Ac* elements—full-length transposons—that can activate *Ds* transposition.

Since spotted kernels are not seen in any of the other crosses, *Ds* elements are not transposed in them. There are two different explanations for this. First, since spotted kernels are not seen in the A × C cross, the colorless phenotype in strain A is not caused by the insertion of a *Ds* element. (It could result from just a point mutation in the *C* gene.) Second, since spotted kernels are not seen in the B × D cross, the D strain lacks *Ac* elements. For these reasons, the crosses A × C, A × D, and B × D all show only expected Mendelian patterns of inheritance.

15.36　A geneticist was studying glucose metabolism in yeast and deduced both the normal structure of the enzyme glucose-6-phosphatase (G6Pase) and the DNA sequence of its coding region. She was using a wild-type strain called A to study another enzyme for many generations when she noticed that a morphologically peculiar mutant had arisen from one of the strain A cultures. She grew the mutant up into a large stock and found that the defect in this mutant involved a markedly reduced G6Pase activity. She isolated the G6Pase protein from these mutant cells and found that it was present in normal amounts, but had an abnormal structure. The N-terminal 70 percent of the protein was normal. The C-terminal 30 percent was present but altered in sequence by a frameshift reflecting the insertion of 1 base pair, and the N-terminal 70 percent and the C-terminal 30 percent were separated by 111 new amino acids unrelated to normal G6Pase. These amino acids represented predominantly the AT-rich codons (Phe, Leu, Asn, Lys, Ile, Tyr). There were also two extra amino acids added at the C terminal end. Explain these results.

Answer: The extra 111 amino acids plus the one base-pair shift indicates that 334 base pairs were inserted into the G6Pase structural gene. Insertion of sequences is consistent with an initial *Ty* transposition into the G6Pase gene that was followed by recombination between its two deltas. Recombination between the two deltas would excise the *Ty* element but leave delta sequences behind in the G6Pase gene. If the delta element were 334 base pairs long, AT-rich, and positioned so that it would be translated and not generate a stop codon, it would yield the 111 new amino acids and one extra base pair, which would cause the frameshift. The two extra amino acids at the C terminal end of G6Pase were added presumably because the frameshift did not allow the normal termination codon to be read. In fact, delta elements are 334 base pairs long and are 70 percent AT, so have the characteristics of the inserted sequence.

15.37   Consider two theoretical yeast transposons, A and B. Each contains an intron, and each transposes to a new location in the yeast genome. Suppose you then examine the transposons for the presence of the intron. In the new locations, you find that A has no intron, but B does. What can you conclude about the mechanisms of transposon movement for A and B?

Answer: Since introns are spliced out only at the RNA level, a transposition event that results in the loss of an intron (such as that used by *Ty* elements) indicates that the transposition occurred via an RNA intermediate. Thus, A is likely to move via an RNA intermediate. The lack of intron removal during B transposition suggests that it uses a DNA→DNA transposition mechanism (either conservative or replicative, or some other mechanism).

15.38   After the discovery that *P* elements could be used to develop transformation vectors in *Drosophila melanogaster,* attempts were made to use them for the development of germ-line transformation in several different insect species. Charalambos Savakis and his colleagues successfully used a different transposable element found in *Drosophila*—the *Minos* element—to develop germ-line transformation in *Drosophila melanogaster* and in the medfly, *Ceratitus capitata,* a major agricultural pest present in Mediterranean climates.

   a.   What is the value of developing a transformation vector for an insect pest?

   b.   What basic information about the *Minos* element would need to be gathered before it could be used for germ-line transformation?

Answer:

   a.   The ability to modify a pest species genetically may provide a means to develop better strategies for pest control. For insects harboring parasites that affect human and animal populations (e.g., mosquitoes that are hosts to malaria), the ability to modify the insect host genetically may provide a means to influence the reproduction of a human or animal pathogen.

b. By analogy with the information used to develop *P* element–mediated transformation in *Drosophila*, it is essential to understand the mode of transposition of the element and to have identified the essential features of the element that allow it to transpose (the nature of the transposase and its regulation, the sequences the transposase recognizes to mediate transposition, etc.).

16

RECOMBINANT DNA TECHNOLOGY

CHAPTER OUTLINE

DNA CLONING
Restriction Enzymes
Cloning Vectors and DNA Cloning

RECOMBINANT DNA LIBRARIES
Genomic Libraries
Chromosome Libraries
cDNA Libraries

FINDING A SPECIFIC CLONE IN A DNA LIBRARY
Screening a cDNA Library
Screening a Genomic Library
Identifying Genes in Libraries by Complementation of Mutations
Using Heterologous Probes to Identify Specific DNA Sequences in Libraries
Using Oligonucleotide Probes to Identify Genes or cDNAs in Libraries

MOLECULAR ANALYSIS OF CLONED DNA
Restriction Mapping
Southern Blot Analysis of Sequences in the Genome
Northern Blot Analysis of RNA

DNA SEQUENCING

THE POLYMERASE CHAIN REACTION (PCR)
PCR Steps
Advantages and Limitations of PCR
Applications of PCR
RT-PCR and mRNA Quantification

REVIEW OF KEY TERMS, SYMBOLS, AND CONCEPTS

In your own words, write a brief, precise definition of each term in the groups on the following pages. Check your definitions using the text. Then develop a concept map using the terms in each list.

1	2	3
cloning	genomic library	library screening
genetic engineering	chromosome library	expression vector
recombinant DNA	cDNA, complementary DNA	autoradiogram
molecular cloning	cDNA library	autoradiography
cloning vector	partial digestion	replica plating
restriction enzyme	insert size	probe
restriction site	YAC, BAC, shuttle vectors	random-primer labeling
restriction endonuclease	*TEL, CEN, ARS* sequences	hexanucleotide random primers
twofold rotational symmetry	*TRP1, URA3* markers	Klenow fragment
palindrome	flow cytometry	radioactive labeling
sticky, staggered, blunt ends	oligo(dT) chains	nonradioactive labeling
5', 3' overhang	reverse transcriptase	digoxigenin-dUTP
anneal	RNase H	alkaline phosphatase
ligate	restriction site linker	chemiluminescent substrate
DNA ligase	adapter	colorimetric substrate
plasmid, BAC, YAC vector	subcloning	complementation
shuttle vector		heterologous probe
polylinker		oligonucleotide probe
multiple cloning site		guessmer
origin, *ori*		GenBank
selectable marker		
unique restriction site		
blue/white colony screening		
β-galactosidase		
omega fragment		
transformation		
electroporation		

4	5
restriction map	polymerase chain reaction (PCR)
physical map	amplification
agarose gel electrophoresis	amplimer
preparative agarose gel electrophoresis	denaturation
calibration curve	annealing
Southern blot analysis	extension
northern blot analysis	primer
dideoxy (Sanger) sequencing	*Taq* DNA polymerase
dideoxynucleotide	thermostable DNA polymerase
universal sequencing primer	Vent DNA polymerase
sequencing ladder	proofreading
open reading frame (ORF)	subcloning

4	5
computer database	DNA typing, fingerprinting, profiling
	ancient DNA
	RT-PCR
	real-time RT-PCR

THINKING ANALYTICALLY

The material in this chapter introduces core concepts in molecular genetics. It presents many techniques and ideas that are widely used throughout genetics. Many of these have been developed by coupling fundamental genetic principles with technological advances. For example, the development of the ability to clone DNA required the synthesis of information from bacterial genetics (e.g., how to transform cells, how to design and use selectable markers, how to perform bacterial crosses, how to design plasmids); DNA replication (the discovery, characterization, and purification of DNA polymerases and ligases); and the discovery, characterization, and purification of restriction enzymes. As you learn about molecular genetic methods, observe how they rely on a synthesis of information from different areas of genetics and biochemistry.

Learning this material requires you to understand the logic and fundamental genetic principles behind many different techniques. While it would be helpful to perform all of the techniques, this may not be an option. Indeed, the current repertoire of techniques is so great and so rapidly advancing that no one scientist has firsthand experience in all of them. Still, scientists learn to think effectively about them and apply them to their research questions. As you approach a new method, think through the steps of the method and consider the rationale for each step. Then consider how different methods can be used together or interchangeably.

This chapter discusses manipulations of DNA that cannot be directly sensed (with the eyes, ears, nose, or touch). Like material from the physical sciences—chemistry and physics—that considers molecular interactions, it requires visualization and abstract conceptualization. While the results of techniques presented in this chapter cannot be directly sensed, they are based on understanding and developing a mental image of the molecular structure, function, and activity of nucleic acids and proteins in test tubes, cells, and viruses under defined environmental conditions. When the techniques produce reasonable, expected results, they validate and sometimes expand our understanding of the structure and function of these molecules. For example, we have never *seen* the actual base sequences of a fragment of DNA, but we have developed a model of DNA structure from experimentation. *If* that model is correct, then a technique such as dideoxy sequencing of DNA should work. As you learn about an experimental method, identify the models being used. Consider how the results obtained with the method would be altered if the model were incorrect.

QUESTIONS FOR PRACTICE

Multiple Choice Questions

For questions 1–10, match each experimental goal with techniques useful to attain it. Multiple techniques may be used in achieving one goal. Some techniques will be used more than once.

EXPERIMENTAL GOAL	TECHNIQUE

1. Determine if a particular mRNA is expressed at a specific developmental stage.
2. Separate mRNA from a mixture of mRNA, rRNA, and tRNA prior to construction of cDNA.
3. Isolate a particular cDNA clone in a cDNA library using an antibody.
4. Determine DNA sequence similarity.
5. Determine the organization of introns in a gene.
6. Produce relatively large genomic DNA fragments for cloning in a library.
7. Identify plasmids with inserted DNA sequences.
8. Identify BACs with inserts homologous to a probe sequence.
9. Amplify a specific genomic DNA sequence.
10. Determine the size of a fragment of DNA inserted into a plasmid vector.

a. restriction mapping
b. partial digestion with a restriction enzyme that recognizes a 4-bp site
c. polymerase chain reaction
d. using an expression vector to produce a particular protein
e. northern blot analysis
f. Southern blot analysis
g. blue/white selection
h. using oligo(dT) to isolate RNA with poly(A) tails
i. lifts of bacterial colonies
j. agarose gel electrophoresis
k. dideoxy sequencing

11. Why are antibiotic resistance markers important components of plasmid cloning vectors?
    a. The plasmid must have resistance to accept DNA inserts.
    b. They allow identification of bacteria that have taken up a plasmid.
    c. They ensure the presence of the *ori* site.
    d. They ensure that the plasmid can be cut by a restriction enzyme.

12. When a PCR reaction is performed using genomic DNA as a template, a 1.5-kb product is amplified. When the same reaction is set up using cDNA as a template, a 0.8-kb product is amplified. A likely explanation for the different-sized products is that
    a. primers always bind to different sequences in different templates
    b. there is a mutation in the genomic DNA
    c. there is an intron in the gene
    d. the cDNA is degraded

13. Why is it advantageous to screen a genomic library constructed in a BAC vector instead of a plasmid vector?
    a. fewer inserts can be screened
    b. more inserts can be screened
    c. plasmid libraries contain only cDNA, not genomic DNA
    d. BAC libraries can be screened with an antibody

14. The restriction enzyme *Bam*HI cleaves a phosphodiester bond between two G - C base pairs at a six-base-pair site. The 5'→3' sequence of one of the sites is GGATCC. Which statement is true?
    a. *Bam*HI leaves a 5' overhang.
    b. *Bam*HI leaves a 3' overhang.
    c. *Bam*HI leaves a blunt end.
    d. DNA cleaved by *Bam*HI cannot be religated.

15. What advantages do BACs have over YACs?
    a. BACs can accommodate larger amounts of DNA than YACs.
    b. BACs are present in hundreds of copies per bacterium, making them easier to purify.
    c. BACs are selected using the *TRP1* and *URA3* markers, allowing an easier selection.
    d. BACs can be handled like regular plasmids, are more stable, and are easier to manipulate.

Answers: 1e, j; 2h; 3d, i; 4a, f, j, k; 5a, c, f, j, k; 6b, j; 7g (possibly also a, i, k); 8f, i, j; 9c; 10a, j, k; 11b, 12c, 13a, 14a, 15d

Thought Questions

1. Contrast PCR, RT-PCR, and real-time RT-PCR in terms of their templates, products, and applications.
2. For what purposes would you choose to use each of a plasmid, YAC or BAC library? When you could use any of them, why would you choose one over the others?
3. What kinds of organisms naturally make restriction enzymes? Of what use are they to the organism where they are naturally made?
4. Address the following questions to explore the kinds of results that could be obtained on a Southern blot made with genomic DNA.
    a. DNA is digested with a restriction enzyme that cuts at a defined 6-bp sequence having 2 G-C base pairs and 4 A-T base pairs. The genome has 40 percent G-C base pairs. What is the *average* fragment size?
    b. The cleaved DNA is separated by size using agarose gel electrophoresis. When the gel is stained, a smear of DNA is seen. Why is a smear and not one or a few bands of DNA seen?
    c. The DNA is transferred in a Southern blot to a membrane. When sequences transferred to the membrane are hybridized to a unique-sequence probe made from a fragment that is 4 kb in length, bands that are 0.5, 3, and 9 kb are seen. How do you interpret this result?
    d. The same blot is hybridized with a probe that has homology to moderately repetitive DNA. The probe has been made from a fragment that is 1.2 kb in size. The blot has about 40 bands, ranging in size from 3 to 17 kb. How do you interpret this result?
5. Construct a restriction map of a 10-kb DNA fragment using the following data:

ENZYMES USED	SIZES OF FRAGMENTS (in kb)
*Eco*RI	1, 4, 5
*Bam*HI	4, 6
*Hind*III	0.8, 1.5, 7.7
*Eco*RI and *Bam*HI	1, 2, 3, 4
*Eco*RI and *Hind*III	0.5, 0.8, 1, 3.2, 4.5
*Bam*HI and *Hind*III	0.8, 1.5, 2.5, 5.2
*Bam*HI, *Eco*RI, and *Hind*III	0.5, 0.8, 1, 2, 2.5, 3.2

6. Suppose you have cloned 11 kb of eukaryotic genomic DNA that includes a single gene. After analyzing the sequence of about 500 bp at each end of the clone, you believe you have evidence that these sequences contain the 5′ UTR and 3′ UTR sequences of the gene. What might this evidence be? How can you use the information you have in hand, together with Southern and northern blot analyses and PCR to (a) identify the length of the primary transcript of the gene, (b) identify the length of the mature mRNA of the gene, and (c) obtain a cDNA copy of the mature mRNA of the gene?

Thought Questions for Media

After reviewing the media on the *iGenetics* CD-ROM, try to answer these questions.

1. In the animation depicting restriction mapping, how is agarose gel electrophoresis able to separate fragments of cleaved DNA?
2. The rotating disk method is effective for orienting restriction fragment sites in a circular molecule relative to each other. Explain how it works.
3. Sometimes, when the rotating disk method is being used to locate restriction enzyme sites relative to each other, overlaying maps doesn't allow sites to be positioned. In some cases, an alternative mirror image map must be used. Why?
4. How could you modify the rotating disk method to map linear DNA fragments?
5. Automated DNA sequencers have made sequencing DNA both cheaper and easier. Why is it still important to know how to construct a restriction map?
6. The calibration curve shown in the animation is not linear over the entire length of the gel. However, it is approximately linear between about 1,000 and 10,000 base pairs. Why might it be linear over only part of the length of the gel?
7. Each of two enzymes, enzyme I and enzyme II, cuts a plasmid giving fragments of 2,506 and 5,160 bp. What is the best means to *quickly* and unambiguously determine how many different sites these enzymes recognize and their positions relative to each other?

1. What is the rationale behind deleting the *STA1, STA2,* and *STA2K* genes?
2. Why is it important to *entirely* delete each of the *STA1, STA2,* and *STA2K* genes?
3. Why are there differences between the beers made with the three recombinant yeast strains, since each has a deletion of a gene responsible for glucoamylase production?
4. Why is the *CUP1* gene needed in these experiments? How would you demonstrate that it had no effect on fermentation characteristics or beer quality?
5. Refer back to the frame that asks you to insert the *STA1* gene into the YEpD vector. Suppose there were two *Sna*BI sites flanking the *STA1* gene. Even though there is a unique *Sna*BI site in the YEpD vector, why wouldn't you use it instead of the *Bgl*II site?

SOLUTIONS TO TEXT PROBLEMS

16.1 The ability to clone and manipulate DNA fragments provides a set of tools for molecular biologists to investigate the structure and function of our genes and their protein products. What are the basic elements of research in recombinant DNA technology?

Answer: Recombinant DNA technology involves the isolation, cloning, manipulation, and expression of DNA sequences. Fundamental to it is the ability to cleave DNA using restriction enzymes, insert the cleaved DNA into cloning vectors (for example, plasmids) using an enzyme such as DNA ligase, and introduce (transform) these vectors into host cells, where they can be propagated and possibly express DNA sequences (that is, genes) of interest. DNA, once cloned, can be repurified and modified, altering parts of the DNA sequence. Since it is essential to be able to identify cloned sequences containing specific genes from libraries of cloned sequences, the basic elements of research also include the ability to screen genomic and cDNA libraries in different ways (expression screens, oligonucleotide probe screens, heterologous probe screens, etc.) that allow for the isolation of particular cloned sequences. The ability to sequence DNA and amplify specific regions using PCR should also be considered basic elements of research, as these widely used methods provide information about DNA molecules that is essential for research.

16.2 The ability of complementary nucleotides to base pair using hydrogen bonding, and the ability to selectively disrupt or retain accurate base pairing by treatment with chemicals (e.g., alkaline conditions) and/or heat is critical to many methods used to produce and analyze recombinant DNA. Give three examples of methods that rely on complementary base pairing, and explain what role complementary base pairing plays in each of these methods.

Answer: Examples of methods that utilize the hydrogen bonding in complementary base pairing include (1) the ligation of a DNA fragment with sticky ends into a site with complementary sticky ends in a cloning vector, (2) annealing of a labeled nucleic acid probe with a single-stranded DNA or RNA molecule attached to a membrane during screens of libraries and Southern and northern hybridizations, (3) annealing of an oligo(dT) primer to a poly(A) tail during the synthesis of cDNA from mRNA, and (4) annealing of a primer to a template during PCR. In each case, base pairing allows for nucleotides to interact in a sequence-specific manner essential for the procedure's success. For example, detection of membrane-bound single-stranded nucleic acid sequences in hybridization of probes to Southern blots requires complementary base pairing between a bound strand of DNA and complementary sequences in the probe.

16.3 Restriction endonucleases are naturally found in bacteria. What purposes do they serve?

Answer: Restriction enzymes serve to protect their hosts from infection by invading viruses and degrade any potentially infectious foreign DNA taken up by the cell (for example, by transformation). Since restriction enzymes digest DNA (restrict it) at specific sites, any foreign DNA will be cut up. To protect its own DNA from digestion by its restriction enzyme(s), a bacterium modifies (methylates) the sites recognized by its own restriction enzymes. This prevents cleavage at these sites.

16.4   A new restriction endonuclease is isolated from a bacterium. This enzyme cuts DNA into fragments that average 4,096 base pairs long. Like many other known restriction enzymes, the new one recognizes a sequence in DNA that has twofold rotational symmetry. From the information given, how many base pairs of DNA constitute the recognition sequence for the new enzyme?

Answer: The average length of the fragments produced indicates how often, on average, the restriction site appears. If the DNA is composed of equal amounts of A, T, C, and G, the chance of finding one specific base pair (A-T, T-A, G-C, or C-G) at a particular site is $\frac{1}{4}$. The chance of finding two specific base pairs at a site is $(\frac{1}{4})^2$. In general, the chance of finding n specific base pairs at a site is $(\frac{1}{4})^n$. Here, $1/4{,}096 = (\frac{1}{4})^6$, so the enzyme recognizes a six-base-pair site.

16.5   An endonuclease called *Avr*II ("a-v-r-two") cuts DNA whenever it finds the sequence

$$5'-CCTAGG-3'$$
$$3'-GGATCC-5'.$$

a. About how many cuts would *Avr*II make in the human genome, which contains about 3×10^9 base pairs of DNA, and in which 40 percent of the base pairs are G-C?

b. On average, how far apart (in terms of base pairs) will two *Avr*II sites be in the human genome?

c. In the cellular slime mold *Dictyostelium discoidium,* about 80 percent of the base pairs in regions between genes are A-T. On average, how far apart (in terms of base pairs) will two *Avr*II sites be in these regions?

Answer:

a. The enzyme recognizes a sequence that has two G-C base pairs, two C-G base pairs, one A-T base pair, and one T-A base pair in a particular order. Since 40 percent of the genome is composed of G-C base pairs, the chance of finding a G-C or C-G base pair is 0.20, and the chance of finding an A-T or a T-A base pair is 0.30. The chance of finding these six base pairs with this sequence is $(0.20)^4 \times (0.3)^2 = 0.000144$. A genome with 3×10^9 base pairs will have about 3×10^9 different groups of 6-bp sequences. Thus, the number of sites in the human genome is $(0.000144) \times (3 \times 10^9) = 432{,}000$.

b. 3×10^9 bp/432,000 sites $= 1/0.000144 = 6{,}944$ bp between sites

c. The chance of finding these six base pairs in a sequence having 80 percent A-T base pairs is $(0.10)^4 \times (0.4)^2 = 0.000016$, so two *Avr*II sites will be $1/0.000016 = 62{,}500$ bp apart.

16.6   About 40 percent of the base pairs in human DNA are G-C. On average, how far apart (in terms of base pairs) will the following sequences be?

a. two *Bam*HI sites   c. two *Not*I sites

b. two *Eco*RI sites   d. two *Hae*III sites

Answer: Since 40 percent of the base pairs in human DNA are G-C, the probability of finding a G-C or C-G base pair is 0.20 and the probability of finding a T-A or A-T base pair is 0.30. (This assumes that in any region of the genome, there will be an average of 40 percent G-C or C-G base pairs.) The probability of finding a particular restriction enzyme recognition sequence is given in the following table.

Enzyme	Recognition Sequence	Probability of Finding the Sequence	Average Distance Between Sites
*Bam*HI	5'-GGATCC-3' 3'-CCTAGG-5'	$(0.2)^4(0.3)^2 = 0.000144$	$1/0.000144 = 6{,}944$ bp
*Eco*RI	5'-GAATTC-3' 3'-CTTAAG-5'	$(0.2)^2(0.3)^4 = 0.000324$	$1/0.000324 = 3{,}086$ bp
*Not*I	5'-GCGGCCGC-3' 3'-CGCCGGCG-5'	$(0.2)^8 = 0.00000256$	$1/0.00000256 = 390{,}625$ bp
*Hae*III	5'-GGCC-3' 3'-CCGG-5'	$(0.2)^4 = 0.0016$	$1/0.0016 = 625$ bp

16.7 *E. coli*, like all bacterial cells, has its own restriction endonucleases that could interfere with the propagation of foreign DNA in plasmid vectors. For example, wild-type *E. coli* has a gene, *hsdR*, that encodes a restriction endonuclease that cleaves DNA that is not methylated at certain A residues. Why is it important to inactivate this enzyme by mutating the *hsdR* gene in strains of *E. coli* that will be used to propagate plasmids containing recombinant DNA?

Answer: If the enzyme is not inactivated, the restriction enzyme produced by the *hsdR* gene will cleave any DNA transformed into *E. coli* with the appropriate recognition sequence. This will make it impossible to clone DNA with the recognition sequence that is not methylated at the A in this sequence.

16.8 There are many varieties of cloning vectors that are used to propagate cloned DNA. One type of cloning vector used in *E. coli* is a plasmid vector. What features does a plasmid vector have that makes it useful for constructing and cloning recombinant DNA molecules?

Answer: Plasmids need three essential features to be utilized as cloning vectors:

1. A bacterial *ori*, or origin of replication sequence, to allow it to replicate in *E. coli*

2. A dominant selectable marker, such as antibiotic resistance, to allow selection of cells harboring the plasmid

3. At least one unique restriction enzyme cleavage site, so that DNA sequences cut with that enzyme can be spliced into the plasmid

Modern plasmid cloning vectors have been engineered to possess additional features that facilitate easier use as cloning vectors:

1. They are present in a high copy number, which facilitates purification of plasmid DNA.

2. They contain many unique restriction sites in a *polylinker* or *multiple cloning site*, to facilitate cloning fragments of DNA obtained after cleavage with a variety of different restriction enzymes.

3. The polylinker is inserted near the 5' end of the *lacZ* gene, which encodes β-galactosidase. Cells harboring a plasmid with an intact *lacZ* gene form blue colonies when grown on media with the β-galactosidase substrate X-gal. Cells harboring a plasmid whose *lacZ* gene has been interrupted by a cloned segment of DNA will not express functional β-galactosidase and will be white. Thus, one can tell if a bacterial colony harbors a plasmid with DNA insertion by its color.

4. They contain phage promoters flanking each side of the polylinker. These promoters are used to make in vitro RNA copies of the cloned DNA. These promoters can then be used for a variety of purposes, including making radioactively labeled RNA probes.

16.9 *E. coli* is a commonly used host for propagating DNA sequences cloned into plasmid vectors. Wild-type *E. coli* turns out to be an unsuitable host, however: Not only are the plasmid vectors "engineered," but so is the host bacterium. For example, nearly all strains of *E. coli* used for propagating recombinant DNA molecules carry mutations in the *recA* gene. The wild-type *recA* gene encodes a protein that is central to DNA recombination and DNA repair. Mutations in *recA* eliminate general recombination in *E. coli*, and render *E. coli* sensitive to UV light. How might a *recA* mutation make an *E. coli* cell a better host for propagating a plasmid carrying recombinant DNA? (Hint: What type of events involving recombinant plasmids and the *E. coli* chromosome will *recA* mutations prevent?) What additional advantage might there be to using *recA* mutants, considering that some of the *E. coli* cells harboring a recombinant plasmid could be accidentally released into the environment?

Answer: The *recA* mutation assists in preventing recombination between the host chromosome and the plasmid vector. This restricts propagation of the plasmid to the cytoplasm and maintains the integrity of cloned sequences. It also makes the host cell less viable if it is accidentally released into the environment, as it is less efficient at DNA repair and sensitive to UV light.

16.10 Much effort has been spent on developing cloning vectors that replicate in organisms other than *E. coli*.

a. Describe several different reasons one might want to clone DNA in an organism other than *E. coli*.

b. What is a shuttle vector, and why is it used?

c. Describe the salient features of a vector that could be used for cloning DNA in yeast.

Answer:

a. Vectors have been developed for transforming yeast as well as plant and animal cells. These are useful for studying cloned eukaryotic genes in a eukaryotic environment, commercial production of eukaryotic gene products (for example, drugs and antibodies), developing gene therapy, engineering crop plants, and developing transgenic animals.

b. Shuttle vectors are cloning vectors that can replicate in two or more host organisms. They are used to introduce DNA into organisms other than *E. coli*. A typical shuttle vector has the ability to replicate in *E. coli*, where it is usually easier and faster to do initial cloning and engineering steps. Once the appropriate recombinant DNA molecule is constructed in *E. coli*, it can be transferred into another organism without further subcloning.

c. A yeast shuttle vector should contain dominant selectable markers for both yeast (for example, *URA3*, which provides for uracil-independent growth when the vector is transformed into a *ura3* yeast) and *E. coli* (for example, ampicillin resistance); several unique restriction enzyme sites suitable for cloning foreign DNA; and, if it is not integrated into the yeast chromosome, sequences that allow it to replicate autonomously, as a plasmid, in both yeast cells and bacteria.

16.11   What is a cDNA library, and from what cellular material is it derived? How is a cDNA library used in cloning particular genes?

Answer: A complementary DNA, or cDNA, is a DNA copy of an mRNA. A cDNA library is a collection of clones containing the different cDNAs synthesized from the entire population of mRNA of a particular (usually eukaryotic) tissue or cell. It is constructed as follows: Initially, the mRNAs are isolated from the tissue or cell where they are expressed. The mRNA molecules are purified from the other RNAs present in cells by passing the total cellular RNA over an oligo-dT column, which binds the poly(A) tails of the mRNAs. After the other RNAs are washed through the column, the poly(A)$^+$ mRNAs are eluted from the column. Then, cDNAs are synthesized by annealing an oligo(dT) primer to the poly(A) tail of each mRNA and using the enzyme reverse transcriptase to synthesize a single-stranded DNA copy of each mRNA strand. This results in a collection of DNA-mRNA hybrids. RNase H is added to partially degrade the mRNA strands, leaving single-stranded complementary DNAs with short mRNA fragments attached. DNA polymerase I is added to synthesize a second DNA strand using the short mRNAs as primers. DNA ligase is used to ligate the DNA fragments of the second strand together. Finally, linkers or adapters are ligated onto the ends of the cDNAs (if linkers are used, the linkers are cleaved to generate sticky ends), and each member of the resulting collection of double-stranded DNAs is ligated into a restriction site in a cloning vector and propagated. Since there is a population of mRNA molecules in a particular tissue or cell, this procedure produces a population of clones, or a library, containing different cDNA inserts, that reflects the population of mRNA present in a particular tissue or cell. See text Figures 16.8, p. 430, and 16.9, p. 431.

If a cDNA library is constructed using mRNA isolated from a particular tissue, the cDNA inserts represent partial copies of genes transcribed in that tissue. Thus, each clone can be used to identify a gene expressed in that tissue. A cDNA library can be screened for cloned copies of particular mRNA transcripts either using the same methods that are used to screen genomic libraries, or by performing an expression screen. For example, just as for genomic libraries, a cDNA library could be screened by using a DNA probe containing part of the gene of interest, by using a heterologous probe, or by using an oligonucleotide probe. However, since a cDNA is a mature mRNA copy of a gene, if the cDNA library has been made in an expression vector, the cDNA can be expressed (that is, transcribed and the transcript translated). In this case, the protein encoded by a cDNA can be produced in a bacterial cell. If an antibody is available that binds to a protein expressed in a particular tissue, the antibody can be radioactively or nonradioactively labeled and used as a probe to identify clones expressing the protein. Such clones have cDNAs that encode the protein. This method of screening an expression vector library is described in detail in text Figure 16.10, p. 432.

16.12   Suppose you have cloned a eukaryotic cDNA and want to express the protein it encodes in *E coli*. What type of vector would you use, and what features must this vector have? How would this vector need to be modified to express the protein in a mammalian tissue culture cell?

Answer: Use an expression vector. Expression vectors have the signals necessary for DNA inserts to be transcribed and for these transcripts to be translated. In prokaryotes, the vector should have a prokaryotic promoter sequence upstream from the site where the cDNA is inserted, and possibly, a terminator sequence downstream of this site. In eukaryotes, a eukaryotic promoter would be needed, and a poly(A) site should be provided downstream from the site where the cDNA is inserted. If the cDNA lacked a start codon, a start AUG codon embedded in a Kozak consensus sequence would be needed upstream from the site where the cDNA is inserted so that the transcript can be efficiently translated. In the event that the cDNA lacked a start codon, care must be taken during the design of the cloning steps to ensure that the open reading frame (ORF) of the cDNA is in the same reading frame with the start codon provided by the vector.

16.13   Suppose you wanted to produce human insulin (a peptide hormone) by cloning. Assume that you could do this by inserting the human insulin gene into a bacterial host where, given the appropriate conditions, the human gene would be transcribed and then translated into human insulin. Which would be better to use as your source of the gene: human genomic insulin DNA or a cDNA copy of this gene? Explain your choice.

Answer: It would be preferable to use cDNA. Human genomic DNA contains introns, while cDNA synthesized from cytoplasmic poly(A)$^+$ mRNA does not. Prokaryotes do not process eukaryotic precursor mRNAs having intron sequences, so genomic clones will not give appropriate translation products. Since cDNA is a complementary copy of a functional mRNA molecule, the mRNA transcript will be functional, and when translated human (pro-)insulin will be synthesized.

16.14   You have inserted human insulin cDNA in the cloning vector pUC19 and transformed the clone into *E. coli*, but insulin was not expressed. Propose several hypotheses to explain why not.

Answer: If genomic DNA had been used, there could be concerns that an intron in the genomic DNA was not removed, since *E. coli* does not process RNAs as eukaryotic cells do. However, the cDNA is a copy of a mature mRNA, so this should not be a concern. Some other potential concerns are as follows.

First, in order for insulin to be expressed, it must be inserted in the correct reading frame, so that premature termination of translation does not occur and the correct polypeptide is produced. Therefore, check whether the insulin sequence is inserted in the correct reading frame.

Second, depending on the nature of the sequence inserted, a fusion protein with β-galactosidase, and not just human insulin, may have been produced. The polylinker in pUC19 is within the β-galactosidase gene. Sequences inserted into the polylinker, if inserted in the correct reading frame (the same one as used for β-galactosidase), will be translated into a β-galactosidase fusion protein. This protein would be greater in size than human insulin. This may not be acceptable, depending on the intended use of the recombinant protein. If a fusion protein was acceptable, it would be important to ensure that only the ORF (the open reading frame) of the insulin gene is properly inserted into the polylinker of the pUC19 vector.

Third, a complete copy of the human mRNA transcript for insulin may have been used, and not just the open reading frame. If transcribed, it would have features of eukaryotic transcripts but not features required for prokaryotic translation. Indeed, some of its 5′ UTR and 3′ UTR sequences may interfere with prokaryotic transcription and translation. For example, it will lack a Shine-Dalgarno sequence to specify where translation should initiate and identify the first AUG codon. In the pUC19 vector, a Shine-Dalgarno sequence is supplied after the promoter for the *lacZ* gene, since without an insert in the polylinker, β-galactosidase is produced. However, the cDNA may have 5′ UTR sequences which interfere with translation initiation in prokaryotes, or which contain stop codons, terminating translation of the β-galactosidase fusion protein.

Fourth, the cDNA may encode a protein that is posttranslationally processed in eukaryotic cells to become human insulin. The protein produced in *E. coli* may not be processed by *E. coli*. Depending on the type of posttranslational modification, it may be possible to modify (engineer) the cDNA to produce a protein similar to human insulin without requiring it to be posttranslationally modified by *E. coli*.

16.15   Three students are working as a team to construct a plasmid library from *Neurospora* genomic DNA. They want the library to have, on average, about 4-kb inserts. Each student proposes a different strategy for constructing the library, as follows:

> Mike: Cleave the DNA with a restriction enzyme that recognizes a 6-bp site, which appears about once every 4096 bp on average, and leaves sticky, overhanging ends. Ligate this DNA into the plasmid vector cut with the same enzyme, and transform the ligation products into bacterial cells.

> Marisol: Partially digest the DNA with a restriction enzyme that cuts DNA very frequently, say, once every 256 bp, and that also leaves sticky overhanging ends. Select DNA that is about 4 kb in size (e.g., purify fragments this size after the products of the digest are resolved by gel electrophoresis) and then ligate this DNA into a plasmid vector cleaved with a restriction enzyme that leaves the same sticky overhangs, and transform the ligation products into bacterial cells.

> Hesham: Irradiate the DNA with ionizing radiation, which will cause double-stranded breaks in the DNA. Determine how much irradiation should be used to generate, on average, 4-kb fragments and use this dose. Ligate linkers onto the ends of the irradiated DNA, digest the linkers with a restriction enzyme to leave sticky overhanging ends, ligate the DNA into a similarly digested plasmid vector, and then transform the ligation products into bacterial cells.

Which student's strategy will ensure that the inserts are representative of *all* of the genomic sequences? Why are the other students' strategies flawed?

Answer: Marisol's strategy will ensure that the inserts are representative of all of the genomic sequences. Partial digestion of genomic DNA will generate a population of overlapping fragments representative of the entire genome. When the library is screened, multiple overlapping clones from a region will be identified. Mike's strategy works in principle, but in practice has drawbacks. Analyzing a region requires each adjacent restriction fragment from that region to be cloned and recovered in a screen of a genomic library. Given the small size of the restriction fragments, the library will need to contain a very large number of clones and screening the library to find all of the adjacent clones in a region will be very laborious. In addition, large genes will be split into multiple pieces. Hesham's strategy is the least desirable. While using ionizing radiation to introduce double-strand breaks will result in the random fragmentation of DNA and produce a population of overlapping genomic DNA fragments, it will also introduce other types of DNA damage (see Chapter 15). Damage to the DNA may prevent its successful cloning, and sequences that can be cloned are unlikely to be identical to the genomic DNA, as bacterial DNA repair processes will lead to alterations in the DNA sequence.

16.16   Genomic libraries are important resources for isolating genes and for studying the functional organization of chromosomes. List the steps you would use to make a genomic library of yeast in a plasmid vector. In what fundamental way would you modify this procedure if you were making the library in a BAC vector?

Answer: A genomic library made in a plasmid vector is a collection of plasmids that have different yeast genomic DNA sequences into them. Like two volumes of a book series, two plasmids in the library will have identical vector sequences, but different yeast DNA inserts. Such a library is made as follows:

1. Isolate high-molecular-weight yeast genomic DNA by isolating nuclei, lysing them, and gently purifying their DNA.
2. Cleave the DNA into fragments that are 5–10 kb, an appropriate size for insertion into a plasmid vector. This can be done by cleaving the DNA with *Sau*3A for a limited time (i.e., performing a *partial* digest), and then selecting fragments of an appropriate size by either sucrose density centrifugation or agarose gel electrophoresis.
3. Digest a plasmid vector such as pUC19 with *Bam*HI. This will leave sticky ends that can pair with those left by *Sau*3A.
4. Mix the purified, *Sau*3A-digested yeast genomic DNA with the plasmid vector and DNA ligase.
5. Transform the recombinant DNA molecules into *E. coli.*
6. Recover colonies with plasmids by plating on media with ampicillin (pUC19 has a gene for resistance to this antibiotic), and with X-gal (to allow for blue-white selection to identify plasmids with inserts). Each colony will have a different yeast DNA insert, and all of the colonies comprise the yeast genomic library.

In a BAC vector, much larger DNA fragments—200 to 300 kb in size—would be used.

16.17   The human genome contains about 3×10^9 bp of DNA. How many 200-kb fragments would you have to clone into a BAC library to have a 90 percent probability of including a particular sequence?

Answer: From the text, $N = \ln(1 - p)/\ln(1 - f)$, where N is the necessary number of recombinant DNA molecules, p is the probability of including one particular sequence, and f is the fractional proportion of the genome in a single recombinant DNA molecule. Here, $p = 0.90, f = (2 \times 10^5)/(3 \times 10^9)$, so $N = 34,538$.

16.18   Some restriction enzymes leave sticky ends, while others leave blunt ends. It is more efficient to clone DNA fragments with sticky ends than DNA fragments with blunt ends. What is the best way to efficiently clone a set of DNA fragments having blunt ends?

Answer: Use a restriction site linker, a short segment of double-stranded DNA that contains a restriction site. The linker can be efficiently ligated onto blunt-ended DNA fragments. Digestion of the resulting DNA fragments with the restriction enzyme will then produce fragments with sticky ends. Their sticky ends allow for efficient ligation into plasmids digested with the same restriction enzyme. See text Figure 16.9, p. 431. To clone DNA fragments that have the restriction site found in the linker, use an adapter—a short, double-stranded piece of DNA with one sticky end and one blunt end.

16.19 A molecular genetics research laboratory is working to develop a mouse model for bovine spongiform encephalopathy (BSE) ("mad cow") disease, which is caused by misfolding of the prion protein. As part of their investigation, they want to investigate the structure of the gene for the prion protein in mice. They have a mouse genomic DNA library made in a BAC vector and a 2.1-kb-long cDNA for the gene. List the steps they should take to screen the BAC library with the cDNA probe.

Answer: Screen the BAC library in much the same manner as you would screen a plasmid genomic library (described in text Figure 16.11, p. 433) as follows:

1. Determine how many BACs you need to screen. Based on the answer to Question 16.17 and assuming mice and humans have similarly sized genomes, you would need to screen about 35,000 BACs to be 90 percent sure of obtaining a BAC clone with the gene.
2. Plate *E. coli* cells harboring the BACs onto bacterial plates with growth medium that selects for the presence of the BACs.
3. Either pick individual BAC-containing colonies, grow them up in microtiter plates, and then array them into grids on membranes, as is done for the cDNA clones in Figure 16.10, p. 432, or use a velveteen surface and replica plating to inoculate the bacterial colonies onto the surface of membranes.
4. After the BAC-containing bacterial colonies have grown, remove the membrane filters with the colonies from their culture dishes, lyse the bacteria that are growing on them, and allow the denatured DNA to bind the filter.
5. Make a probe by using the cDNA template and random-primer labeling as described in Box 16.1, p. 435, in the text.
6. Incubate the probe with the DNA filters in a heat-sealable plastic bag and allow the probe to hybridize to complementary BAC sequences that are bound to the filters.
7. Wash the filters free of unbound probe and then detect the location of the hybridized probe using autoradiography for a radioactively labeled probe or chemiluminescent detection for a nonradioactively labeled probe.
8. Pick the BAC colonies that have sequences complementary to the probe based on the locations of the hybridization signal.

16.20 Suppose a researcher wants to clone the genomic sequences that include a human gene for which a cDNA has already been obtained. She has available a variety of genomic libraries that can be screened with a probe made from the cDNA using the method described in Figure 16.11.

a. Assuming that each library has an equally good representation of the 3×10^9 base pairs in a haploid human genome, about how many clones should be screened if the researcher wants to be 95 percent sure of obtaining at least one hybridizing clone and
   i. the library is a plasmid library with inserts that are, on average, 7 kb?
   ii. the library is a YAC library with inserts that are, on average, 1 Mb?
b. What advantages and disadvantages are there to screening these different libraries?
c. What kinds of information might be gathered from the analysis of genomic DNA clones that could not be gathered from the analysis of cDNA clones?

Answer:

a. As in Question 16.16, $N = [\ln(1 - p)]/[\ln(1 - f)]$. Here, $p = 0.95$. For (i), $f = (7 \times 10^3)/(3 \times 10^9)$; for (ii), $f = (1 \times 10^6)/(3 \times 10^9)$. The number of clones required is (i) 1.28×10^6 plasmids and (ii) 8.99×10^3 YACs.

b. By screening libraries with larger average inserts, fewer clones must be screened. This advantage must be evaluated relative to the added difficulty of analyzing the larger inserts of positive clones. For example, restriction mapping 1 Mb is substantially more difficult than restriction mapping 7 kb. Since a single gene ranges in size from hundreds of base pairs to hundreds of kilobase pairs, an essential question to consider is how the cloned sequences will be analyzed and used once identified.

c. Genomic clones provide for the analysis of gene structure: intron/exon boundaries, transcriptional control regions, and polyadenylation sites. This analysis is important for evaluating how a gene's expression is controlled. Since mutations in regulatory regions can affect the expression of the gene, analysis of these regions is important for understanding the molecular basis of a mutation.

16.21   A scientist has carried out extensive studies on the mouse enzyme phosphofructokinase. He has purified the enzyme and studied its biochemical and physical properties. As part of these studies, he raised antibodies against the purified enzyme. What steps should he take to clone a cDNA for this enzyme?

Answer: The antibodies that recognize the purified enzyme should be used in a cDNA expression library screen as described in Figure 16.10, p. 432, in the text.

1. Prepare cDNAs using mRNA isolated from a mouse tissue where phosphofructokinase is abundant. See text Figure 16.8, p. 430.

2. Clone the cDNAs into a plasmid expression vector using linkers. See text Figure 16.9, p. 431.

3. Transform *E. coli* with the cDNA clones and plate the bacteria on a medium that selects for the presence of plasmids.

4. Transfer individual colonies to microtiter dishes, then grow and store the bacteria.

5. Print colonies to a membrane filter and grow them on the medium to express the protein products of the cloned cDNAs.

6. Remove the filters from their medium, lyse the cells in situ, and allow the protein products to be bound to the filters.

7. Label the antibodies with radioactivity and incubate the labeled antibodies with the membrane filters.

8. Wash off unbound antibody and use autoradiography to detect the location of bound antibody.

9. Pick the bacterial colonies that align with the radioactive signal, grow them up, and purify the cDNA-bearing plasmids from these colonies.

16.22   A researcher interested in the control of the cell cycle identifies three different yeast mutants whose rate of cell division is temperature sensitive. At low, permissive temperatures, the mutant strains grow normally and produce yeast colonies having a normal size. However, at elevated, restrictive temperatures, the mutant strains are unable to divide and produce no colonies. She has a yeast genomic library made in a plasmid shuttle vector, and wants to clone the genes affected by the mutants. What steps should she take to accomplish this objective?

Answer: She should clone the genes by complementation. First, cross the mutants into a genetic background to allow transformants to be selected using the *ura3* marker. Then, transform each mutant with a library containing wild-type sequences that has been made in a vector having the *ura3*⁺ gene. Plate the transformants at an elevated, restrictive temperature on media that also selects for *ura3*⁺. Colonies that grow have a plasmid that complements the cell division mutation—they are able to overcome the functional deficit of the mutation because the plasmid has provided a copy of the wild-type gene—and that also provides *ura3*⁺ function. Purify the plasmid from these colonies and characterize the cloned gene.

16.23   The amino-acid sequence of the actin protein is conserved among eukaryotes. Outline how you would use a genomic library of yeast prepared in a bacterial plasmid vector and a cloned cDNA for human actin to identify the yeast actin gene.

Answer: Since the actin amino-acid sequence is conserved, the DNA sequence will be somewhat conserved as well. Therefore, use the cloned cDNA for human actin as a heterologous probe to screen the yeast genomic library for the yeast actin gene. Proceed as described in text Figure 16.11, p. 433.

16.24   It's 3 a.m. Your best friend has awakened you with yet another grandiose scheme. He has spent the last two years purifying a tiny amount of a potent modulator of the immune response. He believes that this protein, by stimulating the immune system, could be the ultimate cure for the common cold. Tonight, he has finally been able to obtain the sequence of the first seven amino acids at the N terminus of the protein: Met-Phe-Tyr-Trp-Met-Ile-Gly-Tyr. He wants your help in cloning a cDNA for the gene so that he can express large amounts of the protein and undertake further testing of its properties. After you drag yourself out of bed and ponder the sequence for a while, what steps do you propose to take to obtain a cDNA for this gene?

Answer: Compare the amino acid sequence to the genetic code, and design a "guessmer"— a set of oligonucleotides that could code for this sequence. Here, the guessmer would have the sequence 5'-ATG TT(T or C) TA(T or C) TGG ATG AT (T, C or A) GG(A, G, T, or C) TA(T or C)-3', and be composed of 96 different oligonucleotides. Synthesize and then label these oligonucleotides, and use them as a probe to screen a cDNA library as described in Figure 16.11.

16.25   Explain how gel electrophoresis can be used to determine the size of a PCR product.

Answer: In gel electrophoresis, DNA fragments are separated by size. Since DNA is negatively charged due to its phosphates, it will migrate toward the positive pole in an electric field. Since, in an agarose gel, smaller DNA fragments will move more readily through the pores in the gel, smaller DNA fragments move through the gel more rapidly than larger DNA fragments. [Although larger DNA fragments have more phosphate groups and hence more *total* negative charge, they have the same amount of charge density (negative charge per unit mass) as smaller DNA fragments. Hence, the DNA is separated by size in this gel, not by charge.] To determine the size of a PCR fragment, it should be loaded into one well of an agarose gel, and a size standard, or marker, DNA sample should be loaded into another well of the gel. A current should be applied to induce the DNA samples to migrate through the gel. After electrophoresis, the gel should be stained with a dye that binds DNA, such as ethidium bromide. Ethidium bromide will complex with DNA and fluoresce under ultraviolet light. The fluorescent image should be photographed, and the distance each DNA band has migrated from the loading should be measured. Since the molecular sizes of the DNA fragments in the marker lane are known, a calibration curve can be drawn to relate the DNA size (in log kb) to the migration distance (in mm). Then the migration distance for the DNA band of the PCR product can be measured. By comparing it to the calibration curve, its size can be determined. See text Figure 16.14, p. 438, and Figure 16.15, p. 439.

16.26 Restriction endonucleases are used to construct restriction maps of linear or circular pieces of DNA. The DNA usually is produced in large amounts by recombinant DNA techniques. Generating restriction maps is like putting the pieces of a jigsaw puzzle together. Suppose we have a circular piece of double-stranded DNA that is 5,000 base pairs long. If this DNA is digested completely with restriction enzyme I, four DNA fragments are generated: Fragment *a* is 2,000 base pairs long, *b* is 1,400 base pairs long, *c* is 900 base pairs long, and *d* is 700 base pairs long. If, instead, the DNA is incubated with the enzyme for a short time, the result is incomplete digestion of the DNA: Not every restriction enzyme site in every DNA molecule will be cut by the enzyme, and all possible combinations of adjacent fragments can be produced. From an incomplete digestion experiment of this type, fragments of DNA were produced from the circular piece of DNA that contained the following combinations of the above fragments: *a-d-b, d-a-c, c-b-d, a-c, d-a, d-b,* and *b-c*. Finally, after digesting the original circular DNA to completion with restriction enzyme I, the DNA fragments are treated with restriction enzyme II under conditions conducive to complete digestion. The resulting fragments are 1,400, 1,200, 900, 800, 400, and 300 bp. Analyze all the data to locate the restriction enzyme sites as accurately as possible.

Answer: Sort out the results for the first enzyme by tabulating the results:

Complete Digestion		Partial Digestion	
Fragment	**Size (bp)**	**Fragment**	**Size (bp)**
a	2,000	*d-a-c*	3,600
b	1,400	*a-d-b*	3,100
c	900	*c-b-d*	3,000
d	700	*a-c*	2,900
Total plasmid	5,000	*d-a*	2,700
		b-c	2,300
		d-b	2,100

Consider the following: If an enzyme cuts a circular molecule once, it will produce one fragment. If an enzyme cuts a circular molecule twice, it will produce two fragments. (Diagram these situations to convince yourself of this.) Since four fragments are produced when enzyme I completely cleaves the plasmid, enzyme I must cut the plasmid at four sites. A partial digestion occurs when all four sites are not cut. For the partial digestion fragments that contain three fragments, two cuts were made in neighboring sites. Thus, the *d-a-c* fragment was released when two cuts were made at sites flanking fragment *b*, the *a-d-b* fragment was released when two cuts were made at sites flanking fragment *c*, and the *c-b-d* fragment was released when two cuts were made at sites flanking fragment *a*. For the partial digestion fragments containing only two fragments, cuts were made that flank both fragments. Thus, *a* is next to *c*, *d* is next to *a*, *b* is next to *c*, and *d* is next to *b*. This information can be used to order the fragments in the plasmid. Since *a* is next to *c* and *d*, the order must be *c-a-d*. Since *b* is next to *c* and *d*, the order must be *d-b-c*. Since *d* is next to *a* and *b*, the order must be *a-d-b*. Thus, the order of fragments in the plasmid is *c-a-d-b*.

When the plasmid is cleaved with both enzyme I and enzyme II, six fragments are produced. This indicates that the two enzymes together recognize six sites. Since enzyme I cleaves at four sites, enzyme II must cleave at two sites. Since the 1,400- and 900-bp fragments produced when enzyme I cleaves the plasmid remain intact in the double digestion, and the 2,000- and 700-bp fragments do not remain intact in the double digestion, enzyme II must cleave at sites within the 2,000- and 700-bp fragments (*a* and *d*). Given the 1,200- and 800-bp fragments produced with the double digestion, enzyme II must cleave *a* 800 bp from an enzyme I site. Given the 400- and 300-bp fragments produced with the double digestion, enzyme II must cleave *d* 300 bp from an enzyme I site. This gives the following map:

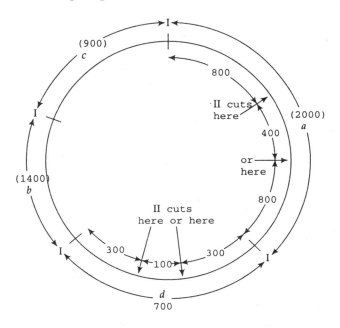

The ambiguities in enzyme II site positions could be resolved if one knew what size fragments were produced when the plasmid was digested to completion with enzyme II alone.

16.27   A piece of DNA 5,000 bp long is digested with restriction enzymes A and B, singly and together. The DNA fragments produced are separated by DNA electrophoresis and their sizes are calculated, with the following results:

	Digestion with	
A	**B**	**A + B**
2,100 bp	2,500 bp	1,900 bp
1,400 bp	1,300 bp	1,000 bp
1,000 bp	1,200 bp	800 bp
500 bp		600 bp
		500 bp
		200 bp

Each A fragment is extracted from the gel and digested with enzyme B, and each B fragment is extracted from the gel and digested with enzyme A. The sizes of the resulting DNA fragments are determined by gel electrophoresis, with the following results:

A Fragment	Fragments Produced by Digestion with B	B Fragment	Fragments Produced by Digestion with A
2,100 bp →	1,900, 200 bp	2,500 bp →	1,900, 600 bp
1,400 bp →	800, 600 bp	1,300 bp →	800, 500 bp
1,000 bp →	1,000 bp	1,200 bp →	1,000, 200 bp
500 bp →	500 bp		

Construct a restriction map of the 5,000-bp DNA fragment.

Answer: Construct a map stepwise, considering the relationship between the fragments produced by double digestion and the fragments produced by single-enzyme digestion. Start with the larger fragments. The 1,900-bp fragment produced by digestion with both A and B is a part of the 2,100-bp fragment produced by digestion with A, and the 2,500-bp fragment produced by digestion with B. Thus, the 2,500-bp and 2,100-bp fragments overlap by 1,900 bp, leaving a 200-bp A-B fragment on one side and a 600-bp A-B fragment on the other. One has:

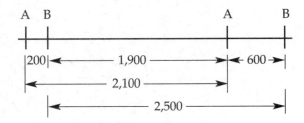

The map is extended in a stepwise fashion, until all fragments are incorporated into the map. The restriction map is:

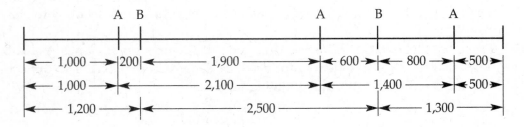

16.28   A colleague has sent you a 4.5-kb DNA fragment excised from a plasmid cloning vector with the enzymes *Pst*I and *Bgl*II (see text Table 16.1 for a description of these enzymes and the sites they recognize). Your colleague tells you that within the fragment there is an *Eco*RI site that lies 0.49 kb from the *Pst*I site.

   a. List the steps you would take to clone the *Pst*I-*Bgl*II DNA fragment into the plasmid vector pUC19 (described in text Figure 16.4).

   b. How would you verify that you have cloned the correct fragment and determine its orientation within the pUC19 cloning vector?

Answer:

a. The *Bgl*II enzyme leaves a 5'-GATC overhang, while the *Pst*I enzyme leaves a 3'-ACGT overhang. If the polylinker of the pUC19 vector could be cleaved to leave two identical overhangs, the 4.5-kb fragment could be directionally cloned into the vector. Examination of the restriction enzyme sites available in the polylinker of pUC19 reveals a *Pst*I site, but no *Bgl*II sites. There are two approaches to obtain the required 5'-GATC overhang. Examine the polylinker further and compare the sites in the polylinker to the enzymes described in text Table 16.1 to determine whether cleaving any of these sites leaves the same kind of overhang as *Bgl*II. This analysis identifies a *Bam*HI site that, if cut, would leave a 5'-GATC overhang, just like that of *Bgl*II. Thus, cleaving the vector with *Bgl*II would produce the appropriate sticky end. To directionally clone the insert, cleave the pUC19 vector with *Pst*I and *Bam*HI, allow the fragment to anneal to these sticky ends, and use DNA ligase to seal the gap in the phosphodiester backbones.

   Another approach to obtain appropriate sticky ends for directional cloning is to cut the polylinker with the enzyme *Sma*I (which leaves blunt ends), ligate a linker containing a *Bgl*II site onto the blunt end, and cleave the modified vector with *Bgl*II and *Pst*I.

b. Transform the ligated DNA into a host bacterial cell, and plate the cells on bacterial medium containing ampicillin and a substrate (X-gal) for β-galactosidase that turns blue when cleaved. This selects for bacterial colonies that harbor pUC19 plasmids and allows for a blue/white selection to identify colonies that have plasmids with inserts. Pick white colonies (which have an interrupted *lacZ* gene, and so do not cleave the substrate) and isolate plasmid DNA from them. Cleave the DNA with restriction enzymes and analyze the products using agarose gel electrophoresis to verify that the appropriate-sized fragments are recovered. Restriction with *Eco*RI should give two fragments, one that is 2.686 + 0.49 = 3.176 kb (vector plus the 0.49-kb *Eco*RI fragment of the insert) and one that is 4.5 − 0.49 = 4.01 kb (the insert minus the 0.49-kb fragment of the insert). A set of double digests (*Eco*RI + *Pst*I; *Eco*RI + *Bam*HI) will also be informative.

16.29 A 10-kb genomic DNA *Eco*RI fragment from a newly discovered insect is ligated into the *Eco*RI site of the pUC19 plasmid vector and transformed into *E. coli.* Plasmid DNA and genomic DNA from the insect are prepared and each DNA sample is digested completely with the restriction enzyme *Eco*RI. The two digests are loaded into separate wells of an agarose gel, and electrophoresis is used to separate the products by size.

a. What will be seen in the lanes of the gel after it is stained to visualize the size-separated DNAs?

b. What will be seen if the gel is transferred to a membrane to make a Southern blot, and the blot is probed with the 10-kb *Eco*RI fragment? (Assume the fragment does not contain any repetitive DNA sequence.)

Answer:

a. The lane with genomic DNA will have a smear, for there are many *Eco*RI sites in a genome and the distances between these sites will vary. The smear reflects the large number and many different sizes of *Eco*RI fragments. Since *Eco*RI recognizes a 6-bp site, the average size will be about 4,096 bp (assume the genome is 25 percent A, G, C, and T), and more intense staining will be seen around this size. The pUC19 plasmid has a single *Eco*RI restriction site into which the 10-kb insert has been cloned, so the lane with plasmid DNA will have two bands: the genomic DNA insert at 10 kb, and the plasmid DNA at 3 kb.

b. The probe will specifically detect the 10-kb *Eco*RI fragment, so signal will be seen in each lane at 10 kb.

16.30   During Southern blot analysis, DNA is separated by size using gel electrophoresis, and then transferred to a membrane filter. Before it is transferred, the gel is soaked in an alkaline solution to denature the double-stranded DNA, and then neutralized. Why is it important to denature the double-stranded DNA? (Hint: Consider how the membrane will be probed.)

Answer: The gel is soaked in an alkaline solution to denature the DNA to single-stranded form. It must be bound to the membrane in single-stranded form so that the probe can bind in a sequence-specific manner using complementary base pairing.

16.31   A researcher digests genomic DNA with the restriction enzyme *Eco*RI, separates it by size on an agarose gel, and transfers the DNA fragments in the gel to a membrane filter using the Southern blot procedure. What result would she expect to see if the source of the DNA and the probe for the blot is as described as follows?

a. The genomic DNA is from a normal human. The probe is a 2.0-kb DNA fragment obtained by excision with the enzyme *Eco*RI from a plasmid containing single-copy genomic DNA.

b. The genomic DNA is from a normal human. The probe is a 5.0-kb DNA fragment that is a copy of a LINE sequence (see Chapter 10, p. 273 and Chapter 15, p. 409) with an internal *Eco*RI site.

c. The genomic DNA is from a normal human. The probe is a 5.0-kb DNA fragment that is a copy of a LINE sequence that lacks an internal *Eco*RI site.

d. The genomic DNA is from a human heterozygous for a translocation (exchange of chromosome parts) between chromosomes 14 and 21. The probe is a 3.0-kb DNA fragment that is obtained by excision with the enzyme *Eco*RI from a plasmid containing single-copy genomic DNA from a normal chromosome 14. The translocation breakpoint on chromosome 14 lies within the 3.0-kb genomic DNA fragment.

e. The genomic DNA is from a normal female. The probe is a 5.0-kb DNA fragment containing part of the *testis determining factor* gene, a gene located on the Y chromosome.

Answer:

a. She should see a 2.0-kb band, as the 2.0-kb probe is a single-copy genomic DNA sequence.

b. LINE sequences are moderately repetitive DNA sequences, which may be distributed throughout the genome. Since the LINE sequence has an internal *Eco*RI site, each LINE element in the genomic DNA will be cut by *Eco*RI during preparation of the Southern blot. When the blot is incubated with the probe, both fragments will hybridize to the probe. The size of the fragments produced from each LINE element will vary according to where the element is inserted in the genome, and where the adjacent *Eco*RI sites are. Hence there will be many different-sized bands seen on the genomic Southern blot.

c. As in (b), there will be many different-sized bands on the genomic Southern blot. The sizes of the bands seen reflect the distances between *Eco*RI sites that flank a LINE element. All of the bands will be larger in size than the element, as the element is not cleaved by *Eco*RI. Counting the number of bands can give an estimate of the number of copies of the element in the genome.

d. Since the heterozygote has one normal chromosome 14, the probe will bind to the 3.0-kb *Eco*RI fragment derived from the normal chromosome 14. If the translocation is a reciprocal translocation, the remaining chromosome 14 is broken in two, and attached to different segments of chromosome 21. Since chromosome 14 has a break point in the 3.0-kb *Eco*RI fragment, the 3.0-kb fragment is now split into two parts, each attached to a different segment of chromosome 21. Consequently, the 3.0-kb probe spans the translation break point and will bind to two different fragments, one from each of the translocation chromosomes. The sizes of the fragments are determined by where the adjacent *Eco*RI sites are on the translocated chromosomes. Thus, the blot will have three bands, one of which is 3.0 kb.

e. Since the *TDF* gene is on the Y chromosome, no signal should be seen in a Southern blot prepared with DNA from a female having only X chromosomes.

16.32 The investigators described in Question 16.19 were successful in purifying a BAC-DNA clone containing the gene for the mouse prion protein. To narrow down which region of the BAC DNA contains the prion-protein gene, they purified the BAC DNA, digested it with the restriction enzyme *Not*I, and separated the products of the enzymatic digestion by size using gel electrophoresis. Then, they purified each of the relatively large *Not*I DNA fragments from the gel, digested each individually with the restriction enzyme *Bam*HI, and separated the products of each enzymatic digestion by size using gel electrophoresis. Finally, they transferred the size-separated DNA fragments from the agarose gel onto a membrane filter using the Southern blot technique, and allowed the DNA fragments on the filter to hybridize with a labeled cDNA probe. Figure 16.A shows the results that were obtained: The pattern of DNA bands seen after the BAC DNA is digested with *Not*I is shown in Panel A, the pattern of DNA bands seen after each *Not*I fragment is digested with *Bam*HI is shown in Panel B, and the pattern of hybridizing DNA fragments visible after probing the Southern blot is shown in Panel C.

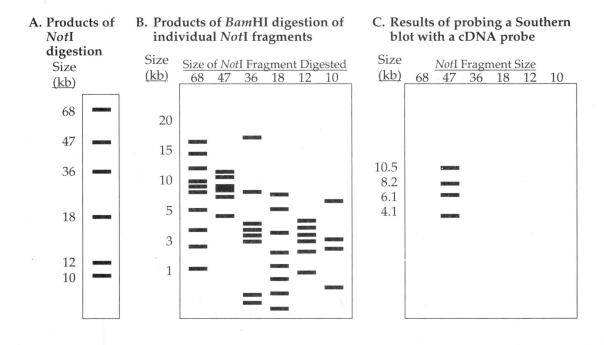

a. Note the scales (in kb) on the left of each figure. Why are relatively larger DNA fragments obtained with *Not*I than with *Bam*HI?

b. An alternative approach to identify the *Bam*HI fragments containing the prion-protein gene would be to digest the BAC DNA directly with *Bam*HI, separate the products by size using gel electrophoresis, make a Southern blot, and probe it with the labeled cDNA clone. Why might the researchers have added the additional step of first purifying individual large *Not*I fragments, and then separately digesting each with *Bam*HI before making the Southern blot?

c. Which *Not*I DNA fragment contains the gene for the mouse prion protein?

d. Which *Bam*HI fragments contain the gene for the mouse prion protein?

e. About what size is the RNA-coding region of the gene for the mouse prion protein? Why is it so much larger than the cDNA?

Answer:

a. *Not*I recognizes an 8-bp site, while *Bam*HI recognizes a 6-bp site. 8-bp sites appear about $\frac{1}{16}$ less frequently than 6-bp sites (See Question 16.6).

b. There are many *Bam*HI fragments in the BAC DNA insert, while there are fewer fragments in each *Not*I fragment. Digesting first with *Not*I allows regions of the BAC to be evaluated in an orderly, systematic manner and allows for the *Bam*HI fragments containing the gene to be more precisely identified and purified.

c. The 47-kb *Not*I fragment contains the gene, since it is the only *Not*I fragment that has sequences hybridizing to the cDNA.

d. The 10.5, 8.2, 6.1, and 4.1 *Bam*HI fragments contain the gene, since they hybridize to the cDNA probe.

e. The RNA-coding region is about 28.9 kb. It is larger than the cDNA since genomic DNA contains intronic sequences.

16.33 Sara is an undergraduate student who is doing an internship in the research laboratory described in Questions 16.19 and 16.32. Just before Sara started working in the lab, the restriction map in Figure 16.B was made of the 47-kb *Not*I restriction fragment containing the prion-protein gene (distances between restriction sites are in kb).

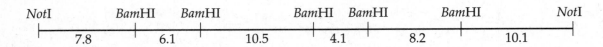

Since smaller DNA fragments cloned into plasmids are more easily analyzed than large DNA fragments cloned into BACs, Sara has been asked to "subclone" the 6.1-, 10.5-, 4.1-, and 8.2-kb *Bam*HI DNA fragments containing the prion-protein gene into the pUC19 plasmid vector. (See Figure 16.4 for a description of the pUC19 vector.) Her mentor gives her some intact pUC19 plasmid DNA, some of the purified 47-kb *Not*I fragment, and shows her where the lab's stocks of DNA ligase and *Bam*H1 are stored. Describe the steps Sara should take to complete her task. In your answer, address how she will identify plasmids that contain genomic DNA inserts, and how she will verify that she has identified clones containing each of the desired genomic *Bam*HI fragments.

Answer: Sara should digest the pUC19 and 47-kb *Not*I fragment DNAs with *Bam*HI. Then she should separate the digestion products of the 47-kb *Not*I fragment by size using gel electrophoresis and purify the 6.1-, 10.5-, 4.1-, and 8.2-kb genomic DNA fragments. Then she should set up four ligation reactions, one with each of the purified genomic DNA fragments and some of the *Bam*HI-digested pUC19 DNA. Then she should separately transform each of the ligations into *E. coli* and select for bacterial colonies with pUC19 having genomic DNA inserts. That is, she should plate the bacteria on media with ampicillin (to select for the presence of pUC19) and X-gal. Colonies with inserts will be white, while colonies without inserts will be blue. (X-gal will be metabolized to a blue compound by β-galactosidase if the *lacZ* gene in pUC19 has not been disrupted by a genomic DNA insert.) Finally, she should verify that the white colonies from each transformation have the desired genomic DNA inserts. She should grow up representative white colonies, purify plasmid DNA from them, digest the plasmid DNA with *Bam*HI, and separate it by size using gel electrophoresis. Colonies with the correct insert should show a 3-kb band (corresponding to the size of the pUC19 vector) and a band corresponding to the size of the appropriate genomic *Bam*HI fragment.

An alternative approach is to set up just a single ligation and transformation, mixing together all of the fragments produced by the *Bam*HI digestion of the *Not*I fragment with the *Bam*HI digestion of pUC19, and then sort out later which colonies have which genomic DNA inserts by plasmid purification and *Bam*HI restriction digestion.

16.34 Imagine that you have been able to clone the structural gene for an enzyme in a catecholamine biosynthetic pathway from the adrenal gland of rats. How could you use this cloned DNA as a probe to determine whether this same gene functions in the rat brain?

Answer: If the same gene functions in the brain, the gene should be transcribed into a precursor mRNA, processed to a mature mRNA, and then translated to produce the functional enzyme. Thus, transcripts for the gene should be found in the brain. To address this issue, label the cloned DNA, and use it to probe a northern blot having mRNA isolated from brain tissue. If the mRNA is rare, it may be prudent to use mRNA isolated from a specific region of the brain (such as the hypothalamus). An alternate, quite sensitive approach would be to sequence the cloned DNA, analyze the sequence to identify the coding region, and then design PCR primers that could be used to amplify cDNA made from mRNA isolated from various brain regions. To perform this RT-PCR (reverse transcriptase PCR), isolate mRNA, reverse transcribe it into cDNA, and then perform PCR. Obtaining an RT-PCR product in such an investigation would provide evidence that the gene is transcribed in the brain. In this alternative method, it would be important to be sure that no genomic DNA was present in the PCR amplification mixture, as the gene for the enzyme would be found in genomic DNA in both tissues.

16.35 A cDNA library is made with mRNA isolated from liver tissue. When a cloned cDNA from that library is digested with the enzymes *Eco*RI (E), *Hind*III (H), and *Bam*HI (B), the restriction map shown in the following figure (below), part (a) is obtained. When this cDNA is used to screen a cDNA library made with mRNA from brain tissue, three identical cDNAs with the restriction map shown in the following figure, part (b) are obtained. When either cDNA is used to synthesize a uniformly labeled ^{32}P-labeled probe and the probe is allowed to hybridize to a Southern blot prepared from genomic DNA digested singly with the enzymes *Eco*RI, *Hin*dIII, and *Bam*HI, an autoradiograph shows the pattern of bands in the following figure (below), part (c). When either cDNA is used to synthesize a uniformly labeled ^{32}P-labeled probe and used to probe a northern blot prepared with poly(A) RNA isolated from liver and brain tissues, the pattern of bands in part (d) of the figure is seen. Fully analyze these data and then answer the following questions.

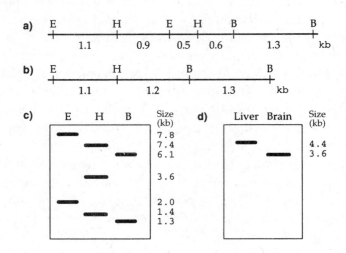

a. Do these cDNAs derive from the same gene?
b. Why are different-sized bands seen on the northern blot?
c. Why do the cDNAs have different restriction maps?
d. Why are some of the bands seen on the whole-genome Southern blot different sizes than some of the restriction fragments in the cDNAs?

Answer:

a. Since both cDNAs hybridize to the same bands on a genomic Southern blot, they are copies of mRNAs transcribed from the same sequences. Therefore, it is likely that they are from the same gene.

b. Different bands on the northern blot indicate that the primary mRNA for this gene may be processed differently in brain and liver tissue. For example, it is possible that the 0.8-kb size difference between the two bands reflects a 0.8-kb intron that is spliced out in brain tissue but is not spliced out in liver tissue.

c. The two cDNAs are copies of mRNAs found in two different tissues. The northern blot indicates that there are some differences between the mRNAs in the different tissues. Thus, it is not surprising that the restriction maps are not identical. Note that the ends of the restriction maps are identical (the same *Eco*RI-*Hin*dIII and *Bam*HI fragments), while the internal regions are not (the brain cDNA lacks the 0.5-kb *Eco*RI-*Hin*dIII fragment and some of each adjoining fragment).

d. The genomic Southern blot gives an indication of the gene organization at the DNA level, while the cDNA maps give an indication of the structure of the mRNA transcript(s). When the cDNA is used to probe genomic DNA sequences, it will hybridize to any sequences that are transcribed. Since restriction sites in the genome do not delineate where the transcribed regions are, the probe will hybridize to genomic DNA fragments that are only partly transcribed. That is, the probe will hybridize to transcribed sequences that are "connected to" nontranscribed sequences. Thus, the large (7.8-, 7.4-, 6.1-, and 3.6-kb) bands reflect the parts of the cDNA that hybridize to genomic DNA fragments that are only partly transcribed. Since they are the same fragment sizes that appear in the liver cDNA, the smaller fragments (2.0, 1.4, and 1.3 kb) represent fragments that are entirely transcribed. Based on these data, a possible gene organization is illustrated below:

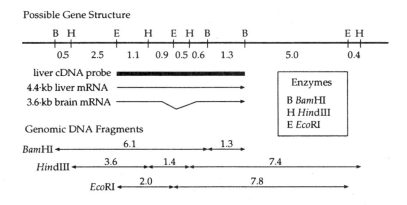

16.36 Draw the pattern of bands you would expect to see on a DNA sequencing gel if you annealed the primer 5'-CTAGG-3' to the following single-stranded DNA fragment and carried out a dideoxy sequencing experiment. Assume the dNTP precursors were all labeled.

3'-GATCCAAGTCTACGTATAGGCC-5'

Answer: The primer will anneal to the fragment, and be extended at its 3' end in four separate reactions, each with small amounts of a different dideoxynucleotide. In each reaction, some chains will be prematurely terminated when the dideoxynucleotide is incorporated. By using labeled dNTP precursors, all extension products will be labeled and be observed as distinct, labeled bands after separation on a denaturing polyacrylamide gel and signal detection.

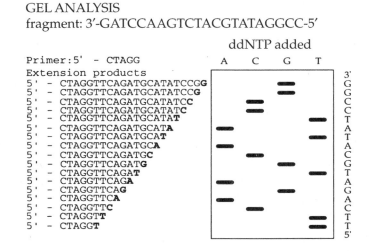

16.37   What information and materials are needed to amplify a segment of DNA using PCR?

Answer: In order to amplify a specific region, one needs to know the sequences flanking the target region so that primers able to amplify the target region can be designed. Once primers are synthesized, the polymerase chain reaction can be assembled. It contains a DNA template (genomic DNA, cDNA, or cloned DNA), the pair of primers that flank the DNA segment targeted for amplification, a heat-resistant DNA polymerase (*Taq*), the four dNTPs (dATP, dTTP, dGTP, and dCTP), and an appropriate buffer. See text Figure 16.21, p. 445.

16.38   In most PCR reactions, a DNA polymerase that can withstand short periods at very high (near boiling) temperatures is used. Why?

Answer: The polymerase chain reaction uses repeated cycles of denaturation, annealing, and extension. In the denaturation step, the reaction mixture is heated to 94°C to denature the DNA template. The annealing and extension steps are carried out at lower temperatures. By using a DNA polymerase able to withstand short periods at very high temperatures, the same reaction mixture can be subjected to multiple cycles of denaturation—there is no need to add additional DNA polymerase as the reaction proceeds through additional cycles. Usually, DNA polymerases purified from a thermophilic bacterium are used. For example, *Taq* DNA polymerase is purified from *Thermus aquaticus,* a bacterium that normally lives in hot springs. Since the bacterium normally lives in very hot environments, it has evolved to produce enzymes that are heat stable.

16.39   Both PCR and cloning allow for the production of many copies of a DNA sequence. What are the advantages of using PCR instead of cloning to amplify a DNA template?

Answer: PCR is a much more sensitive and rapid technique than cloning. Many millions of copies of a DNA segment can be produced from one DNA molecule in only a few hours using PCR. In contrast, cloning requires more DNA (ng to μg quantities) for restriction digestion and at least a week to proceed through all of the cloning steps.

16.40   PCR and RT-PCR can be used to quantify DNA and RNA levels. If you assume that each step of the PCR process is 100 percent efficient, how many copies of a template would be amplified after 30 cycles of a PCR reaction if the number of starting template molecules were
   a. 10
   b. 1,000
   c. 10,000

Answer: As shown in text Figure 16.21, p. 445, two unit-length double-stranded DNA amplimers are produced after the 3 cycle of PCR from one double-stranded DNA template molecule. If each step of the PCR process is 100 percent efficient, the number of amplimers geometrically increases in each subsequent cycle: In the 4th cycle there will be 4 amplimers, in the 5th cycle there will be 8 amplimers, and more generally, in the nth cycle there will be 2^{n-2} amplimers. In the 30th cycle, there will be $2^{28} = 2.68 \times 10^8$ molecules. A larger number of initial template molecules will lead to a proportional increase in amplimer production.

a. $10 \times 2^{28} = 2.68 \times 10^9$ molecules

b. $1,000 \times 2^{28} = 2.68 \times 10^{11}$ molecules

c. $10,000 \times 2^{28} = 2.68 \times 10^{12}$ molecules

Consider these answers with respect to the experimental observation that about 5 ng of DNA (about 2.3×10^9 copies of a 200-bp DNA fragment) is detected readily on an ethidium bromide stained agarose gel.

16.41   *Taq* DNA polymerase, which is commonly used for PCR, is a thermostable DNA polymerase that lacks proofreading activity. Other DNA polymerases, such as *Vent*, have proofreading activity.

a. What advantages are there to using a DNA polymerase for PCR that has proofreading activity?

b. Although some DNA polymerases are more accurate than others, all DNA polymerases used in PCR introduce errors at a low rate. Why are errors introduced in the first few cycles of a PCR amplification more problematic than errors introduced in the last few cycles of PCR amplification?

Answer: ·

a. Since *Taq* DNA polymerase lacks proofreading activity, base-pair mismatches that occur during replication go uncorrected. This means that some of the molecules produced in the PCR process will contain errors relative to the starting template. Enzymes with proofreading activity significantly reduce the introduction of errors.

b. If an error is introduced in the first few cycles of a PCR amplification, most of the derivative DNA molecules produced during subsequent cycles of PCR amplification will also contain the error. This happens since molecules produced in earlier cycles of PCR serve as templates for molecules synthesized in later cycles of PCR. Consequently, if an error is introduced in a later cycle in the PCR amplification process, fewer molecules will have the error.

16.42   Katrina purified a clone from a plasmid library made using genomic DNA and sequenced a 500-bp-long segment using the dideoxy sequencing method. Her twin sister Marina used PCR with *Taq* DNA polymerase to amplify the same 500-bp fragment from genomic DNA. Marina sequenced the fragment using the dideoxy sequencing method, and obtained the same sequence as Katrina did. She then cloned the fragment into a plasmid vector, and, following ligation and transformation into *E. coli*, sequenced several, independently isolated plasmids to verify that she cloned the correct sequence. Most of them have the same sequence as Katrina's clone, but Marina finds that about $\frac{1}{3}$ of them have a sequence that differs in one or two base pairs. None of the clones that differ from Katrina's clone are identical. Fearing she has done something wrong, Marina repeats her work, only to obtain the same results: About $\frac{1}{3}$ of the fragments cloned from the PCR product have single base differences. Explain this discrepancy.

Answer: The insert Katrina sequenced was obtained from genomic DNA, while the inserts Marina sequenced were obtained from PCR. *Taq* DNA polymerase introduces errors during PCR, so that individual double-stranded molecules that are amplified during PCR may have small amounts of sequence variation. If PCR products are sequenced directly, a population of molecules is sequenced and the amount of variation is small enough that it may not be noticed—at a particular position in the sequence, only a very small number of molecules have an error. However, when PCR products are cloned, each independently isolated plasmid has an insert derived from a different double-stranded DNA PCR product, so errors will be apparent.

17

APPLICATIONS OF RECOMBINANT DNA TECHNOLOGY

REVIEW OF KEY TERMS, SYMBOLS, AND CONCEPTS

In your own words, write a brief, precise definition of each term in the groups below. Check your definitions using the text. Then develop a concept map using the terms in each list.

1	2	3
recombinant DNA technology	DNA polymorphism, marker	DNA marker
random mutagenesis	gene, allele, locus	SNP, STR, VNTR
site-directed mutagenesis	genetic map	PCR
knockout mice	SNP, STR, VNTR	PCR–RFLP
PCR, RT-PCR	cSNP, ncSNP	ASO hybridization
transcriptional regulation	RFLP	reverse-ASO hybridization
alternative mRNA splicing	PCR, multiplex PCR	genetic testing
regulatory gene	PCR–RFLP analysis method	DNA molecular testing
northern blotting	ASO hybridization analysis	prenatal diagnosis
yeast two-hybrid system	reverse ASO hybridization	newborn screening
interaction trap assay	high stringency	carrier detection
yeast *GAL, GAL1, GAL4* genes	DNA microarray	DNA typing
glucose repression	DNA chip, GeneChip® array	DNA fingerprinting
Drosophila sexual behavior	oligonucleotide array	DNA profiling
fruitless (*fru*) gene	probe array	paternity
UAS$_G$ sequence	target nucleic acid	inclusion result
DNA binding domain (BD)	Cy3, Cy5	exclusion result
activation domain (AD)	microsatellite, SSR	forensics
reporter gene, bait	minisatellite	
fusion plasmid, fusion protein	monolocus, single-locus probe	
PEX genes, peroxisome	multilocus probe	

4	5
positional cloning	somatic cell therapy
RFLP	germ-line cell therapy
genetic linkage	genetically modified organism
in situ hybridization	biotechnology
chromosome walking	transgene
chromosome jumping	transgenic cell
zoo blot, Noah's ark blot	transfection
CpG island	transformation
candidate gene	electroporation
Southern blotting	gene gun
northern blotting	*Agrobacterium tumefaciens*
	crown gall disease
	monocot, dicot
	Ti plasmid

4	5
	T-DNA
	transforming DNA
	vir region
	antisense RNA
	Roundup™
	EPSPS
	edible vaccine
	pharming

THINKING ANALYTICALLY

Much of the excitement in modern genetics comes as genetic principles and recombinant DNA technology are combined and applied to address fundamental biological questions as well as practical human problems. The material in this chapter introduces many examples—methods used to investigate biological processes, applications of technology for improving human health and agriculture, determination of paternity, and applications in forensics. As you read through the examples, look for how our understanding of genetic principles and recombinant DNA technology allows for inventive, pragmatic application. Analyze the example by asking

- What questions are being asked?
- Why are they important? How do they relate to a fundamental biological question or practical problem?
- What methods are being used to address the questions?
- What genetic principles or recombinant DNA applications underline a manipulation or investigation? (What did we need to know before being able to do this?)
- Are multiple investigatory approaches being combined?
- How is the method effective in answering the question?

After you have read through the examples, go back and reflect more generally on the types of questions and hypotheses that can be investigated and whether there are limitations to the method or application.

Here is an example, using the yeast two-hybrid system.

- *What are the questions being asked?*
    › Do two proteins physically (directly) interact in the cell?
    › How can we identify which proteins interact with a known protein?
- *Why are they important?*
    › It has been difficult to show that two proteins physically interact within cells.
    › This system allows assessment of interactions in cells, not just in the test tube.
    › We can better understand how a protein functions in cells.
    › We can identify which other proteins, and infer which processes, a protein is involved with. This can be done in an unbiased manner and with no previous knowledge of the other proteins(s) or processes.
    › We can gain insights into disease processes.

- *What methods are being used to address the questions?*
  - › Goal: Develop an in vivo system to monitor whether two proteins interact.
  - › Construct a fusion protein where the cDNA for a known protein of interest is fused to a cDNA for the binding domain of Ga14. Use this as a bait.
  - › Construct a library where cDNAs for genes expressed in a certain tissue are fused to a cDNA for the activation domain of Ga14.
  - › Construct a reporter gene (*lacZ*) with an upstream UAS_G.
  - › Transform yeast with the bait, clones from the library, and the reporter gene construct.
  - › Screen for interactions between proteins of the cDNA library and the bait. If two proteins interact, they will place the binding domain and activation domain of Ga14 in proximity, and the *lacZ* reporter gene will be expressed.
- *What genetic principles or recombinant DNA applications underlie a manipulation or investigation?*
  - › A detailed understanding of *GAL* gene transcriptional regulation
  - › A detailed understanding of molecular cloning and library construction
- *Are multiple investigatory approaches being combined?*
  Yes. This system combines investigations that address protein-protein interactions and transcriptional regulation with experimental strategies used by yeast geneticists.
- *How is the method effective in answering the question?*
  - › Normal interactions during *GAL4* gene expression are being modified to allow these interactions to serve as indicators of interactions between two unrelated proteins. Our detailed understanding of *GAL4* transcriptional regulation is being exploited to provide a more general tool to address protein-protein interaction.
- *What type of questions and hypotheses does this system allow us to investigate? What limitations might this method have?*
  - › We can identify new genes whose products interact with a known protein causing disease, providing new targets for therapeutic intervention.
  - › We can test if two proteins interact, expanding our understanding of a protein's function.
  - › We can test if a mutation inhibits a particular type of protein-protein interaction, thereby causing a disease.
  - › What is the sensitivity of the method? Will the method ever give spurious results?

QUESTIONS FOR PRACTICE

Multiple Choice Questions

1. Positional cloning
   a. requires knowledge of the gene product before the gene can be cloned
   b. generates a transgenic organism that expresses a gene only in certain tissues
   c. isolates a disease gene based on its approximate chromosomal location
   d. is when a cDNA has been cloned into a specific orientation in an expression vector

2. Chromosome walking is
   a. used to obtain a set of overlapping clones from a genomic library
   b. used to obtain a set of overlapping clones from a cDNA library
   c. used to jump between chromosomal locations without cloning intervening DNA
   d. impossible in eukaryotes because of the amount of interspersed repetitive DNA

3. During the positional cloning of a disease gene, four candidate genes are identified. What would provide the most convincing evidence that one of them is responsible for the disease?
    a. A zoo blot shows that one gene is also found in other organisms.
    b. Polymorphisms are present in one of the genes in affected individuals.
    c. One of them is expressed in the tissue affected by the disease.
    d. Mutational changes are present in one of the genes in affected individuals.

4. What is the difference between an STR and a VNTR?
    a. An STR is the same as a microsatellite, while a VNTR is the same as a minisatellite
    b. Both are tandem repeats: STRs are 2 to 6 bp long; VNTRs are 7 to 10s of base pairs long.
    c. Both are simple tandem repeated STRs are 7 to 10s of bp long; VNTRs are 2 to 6 bp long.
    d. STRs are simple tandem repeated sequences found only in one location, while VNTRs are always dispersed in different regions of the genome.

5. What is the difference between a monolocus probe and a multilocus probe?
    a. A monolocus probe detects alleles at one gene, while a multilocus probe detects alleles at different genes.
    b. A monolocus probe detects one of two alleles at one gene, while a multilocus probe detects both alleles at one gene.
    c. A monolocus probe detects STR or VNTR sequences at one locus (site) in the genome, while a multilocus probe detects STR or VNTR sequences at a number of loci (sites) in the genome.
    d. A monolocus probe detects STR sequences, while a multilocus probe detects VNTR sequences.

6. An ASO hybridization is a method
    a. to assess genotypes: allele-specific oligonucleotide hybridization. PCR products differing by one base are distinguished by hybridization with labeled oligonucleotides.
    b. used to screen jumping libraries: "all similar objects" hybridization. A probe is used to identify clones containing similar sequences.
    c. to distinguish between two similar transgenic animals: almost the same organisms. Blood taken from each is typed for a set of DNA markers, and the results are compared.
    d. for testing for the presence of GMOs in foods: agriculturally sensitive oligonucleotide hybridization. Since transgenic plants are modified using expression vectors, an oligonucleotide probe can be used to test if foods contain it and hence contain GM plant material.

7. Which one is *not* an example of a product produced by a biotechnology company?
    a. human growth hormone
    b. DNase
    c. bacteria that can accelerate the degradation of oil pollutants
    d. aspirin

8. How would you identify proteins that interact with PEX1, a protein important in peroxisomal biogenesis?
    a. Use cDNA clones for the *PEX1* gene to screen a cDNA expression library.
    b. Fuse a cDNA clone for the *PEX1* gene to the gene containing the GAL4 DNA binding domain, and use this as bait in a yeast two-hybrid screen.
    c. Fuse a cDNA clone for the *PEX1* gene to a reporter gene such as *lacZ,* and use it as bait in a yeast two-hybrid screen.
    d. Use a cDNA clone for the *PEX1* gene to screen a genomic library.

9. Suppose DNA typing is used in a paternity case. How do exclusion results differ from inclusion results?
    a. Exclusion results are easier to prove; one just needs to show that the male in question has no alleles in common with the baby.

   b. Inclusion results require positive identity to be established and usually require testing for alleles at many genes.

   c. Inclusion results require calculation of the relative odds that an allele came from the accused or from another person and require knowing the frequencies of VNTR and STR alleles in many ethnic groups.

   d. all of the above

10. A researcher wants to evaluate whether single nucleotide changes to a promoter placed upstream from a reporter gene will increase the level of reporter gene expression. What method should she use?

   a. chromosome walking

   b. site-specific mutagenesis

   c. PCR-RFLP

   d. ASO hybridization

11. In a hybridization used to detect SNPs using a DNA microarray, what will be the probe and what will be the target?

   a. The probe consists of fluorescently labeled genomic DNA, the target is a set of oligonucleotides on the DNA microarray.

   b. The target consists of fluorescently labeled genomic DNA, the probe is a set of oligonucleotides on the DNA microarray.

   c. The probe consists of Cy3-labeled cDNA and Cy5-labeled cDNA, the target is a set of oligonucleotides on the DNA microarray.

   d. The target consists of Cy3-labeled cDNA and Cy5-labeled cDNA, the probe is a set of oligonucleotides on the DNA microarray.

12. In a hybridization used to evaluate differences in gene expression using a DNA microarray, what will be the probe and what will be the target?

   a. The probe is fluorescently labeled genomic DNA, the target is a set of oligonucleotides on the DNA microarray.

   b. The target is fluorescently labeled genomic DNA, the probe is a set of oligonucleotides on the DNA microarray.

   c. The probe consists of Cy3-labeled cDNA and Cy5-labeled cDNA, the target is a set of oligonucleotides on the DNA microarray.

   d. The target consists of Cy3-labeled cDNA and Cy5-labeled cDNA, the probe is a set of oligonucleotides on the DNA microarray.

13. What are the differences between the methods used to transform dicots and monocots?

   a. Dicots can be transformed using electroporation while monocots can only be transformed using T-DNA based vectors.

   b. Monocots and dicots can be transformed using T-DNA based vectors, but monocots can also be transformed using electroporation and the gene gun.

   c. Only dicots and not monocots can be transformed using T-DNA based vectors. Monocots must be transformed using electroporation or the gene gun.

   d. No methods exist to transform dicots. Monocots can only be transformed using T-DNA based vectors.

Answers: 1c, 2a, 3d, 4b, 5c, 6a, 7d, 8b, 9d, 10b, 11b, 12d, 13c

Thought Questions

1. When cells are subjected to a heat shock (a transient high-temperature pulse), they become stressed and stop transcribing most genes. They start transcribing others that help them deal

with the stress of the heat shock. How would you assess if transcription of a gene you have cloned is repressed or activated by heat shock?

2. After developing a GMO, some companies secure patents on it. Suppose a company sells seed for a GMO to a farmer under the condition that the farmer may plant the seed but may not retain seed stock from any crop that is harvested. The seed is expensive, and the farmer buys seed from the company only once. How would the company check if the farmer has used seed harvested from the first crop?

3. Suppose the farmer described in Question 2 is quite poor and also no longer retains the seed stock he used prior to planting seed for the GMO. What ethical issues arise?

4. While many mutations causing diseases result in proteins that are nonfunctional or that have altered function (such as in sickle-cell anemia), some are caused by mutant alleles that fail to make a protein product. Therapeutic approaches will differ in this situation. Therefore, when a disease gene is cloned, it is important to know if the disease phenotype is caused by an abnormal protein or by the absence of a protein. Suppose you have cloned a gene for a disease and notice that a restriction site for *ECORI* has been altered in a transcribed region of the gene. How would you experimentally address this issue if you are given a tissue biopsy from a patient with the disease?

5. In what different ways are DNA markers used?

6. Should DNA typing be generally accepted to prove parenthood or guilt? Should it be accepted as providing evidence that excludes an accused individual? Why or why not?

7. Some indigenous peoples have taken positions against the use of GMOs based on a cultural viewpoint that they alter the spiritual integrity of living organisms. Some Maori (an indigenous people in New Zealand), for example, have asked that transgenic sheep that produce milk with pharmaceutically important compounds not be raised on their land and that genetically modified trees harvested for lumber not be grown on their land. How do you respond to these types of concerns, and should these concerns be raised more generally?

8. Even though CF is the most common lethal genetic disease in the United States today, it is not tested for routinely in newborns. Why might this be the case?

9. What is the difference between a genetic test and a diagnostic test if each is also a DNA molecular test? How do the ethical and legal issues surrounding these two tests differ?

10. Somatic gene therapy holds great promise for treating genetic disease. What is the underlying basis for somatic gene therapy? What ethical and legal issues are associated with its implementation?

11. How can site-specific mutagenesis and knockout mice be used to develop experimental models for human disease? Why are these models important?

Thought Questions for Media

After reviewing the media on the *iGenetics* CD-ROM, try to answer these questions.

1. What is meant by a "bait gene"? How does the "bait" allow you to "fish" for interacting proteins?

2. When a yeast two-hybrid screen is done, what is the genotype of the transformed yeast strain? Why are both uracil *and* tryptophan left out of the medium on which the transformed yeast are grown?

3. Suppose a cDNA for the Y gene is fused to the Gal4 AD. This cDNA is expressed in a yeast cell that also expresses a cDNA for the X gene that is fused to the Gal4 BD, and which has a UAS$_G$ promoter upstream of the *lacZ* gene. A blue colony is seen when the

yeast is plated on a medium with X-gal. What control could you do to ensure that the Y-Gal4 AD fusion protein activates *lacZ* expression specifically through its interaction with the X-Gal4 BD fusion protein, and not through some other mechanism?

4. How is a yeast two-hybrid screen done?

5. In the animation on plant genetic engineering, what features in the T plasmid define the DNA sequences that will be transferred to the plant nucleus? Why aren't the *vir* genes transferred?

6. What different types of *vir* genes are there, and how does each function in transferring T DNA into the plant genome? What is the purpose of placing an antibiotic resistance gene in the *vir* region of T vectors?

7. What is the basis for Roundup™ resistance in crop plants?

8. What different types of DNA sequences need to be engineered into T DNA to confer Roundup resistance? How should these be arranged?

9. Normally, *Agrobacterium tumefaciens* infections are harmful to plants. Why are infections by strains carrying T-vector plasmids not harmful?

10. Outline the steps you would take to obtain a stock of seeds for Roundup-resistant plants from a plant whose leaf has been infected with an appropriate recombinant *A. tumefaciens* strain.

11. In the animation on DNA molecular testing, what properties of SNPs distinguish them from other types of DNA polymorphisms?

12. In the example shown in the animation, an SNP led to a single amino acid change that led to disease. Are all, or even most, SNPs likely to be associated with a phenotype?

13. Are all SNPs also RFLPs?

14. How can SNPs be detected if they affect a restriction site? If they do not affect a restriction site?

1. For one primer pair, how many different PCR products can be obtained in one individual? How many different alleles can be detected with one primer pair?

2. For five primer pairs, how many different PCR products can be obtained in one individual? How many different alleles can be detected with five primer pairs?

3. Are alleles detected as STRs dominant or recessive or neither? Why?

4. For one primer pair, how many different patterns of PCR products can be obtained in two individuals? For five primer pairs, how many different patterns of PCR products can be obtained in two individuals?

5. Why is it important in forensics to perform DNA typing on loci that are highly polymorphic?

6. How is the profile probability used to determine the likelihood that two matching samples come from the same individual?

7. What is the advantage of using fluorescent primers and multiplex PCR? If you want to test multiple loci using PCR, what information would you need to know about each pair of primers before labeling them and using them in a multiplex PCR?

SOLUTIONS TO TEXT PROBLEMS

17.1   What modifications are made to the polymerase chain reaction (PCR) to use this method for site-specific mutagenesis?

Answer: Site-specific mutagenesis uses primers that have been modified to incorporate a mutation at a particular site. As shown in text Figure 17.1, p. 457, two separate PCR amplifications are performed to amplify two overlapping DNA segments that together span one region. The products of each PCR reaction overlap by the length of a primer in the center of the region. In the area where the two DNA segments overlap, the PCR primers do not exactly match the template sequence, but incorporate a mutation at one site. The resulting two PCR products are mixed, denatured, and allowed to reanneal. In some cases, the overlapping ends of the different PCR products will anneal. DNA polymerase can then extend the 3' ends to generate a double-stranded DNA molecule that spans the entire region. The entire region, now containing an alteration in DNA sequence at one central site, can then be amplified by using PCR and primers that flank the region. The product with the altered site can then be cloned and used to replace the wild-type sequence in a cell.

17.2   Metalloproteases are enzymes that require a metal ion as a cofactor when they cleave peptide bonds. Members of one family of metalloproteases share the following consensus amino acid sequence in their catalytic site: His-Glu-X-Gly-His-Asp-X-Gly-X-X-His-Asp (X is any amino acid). Structural models of the catalytic site developed from X-ray crystallographic data suggest that the second amino acid, glutamate, is essential for proteolytic activity. Outline the experimental steps you would take to test this hypothesis. Assume you possess a cDNA encoding a metalloprotease having the consensus sequence, and can measure metalloprotease activity in a biochemical assay.

Answer: First use site-specific mutagenesis to introduce a single-base change into the cDNA to alter the primary structure of the metalloprotease gene product. For example, change the codon for the second amino acid in the consensus sequence, glutamate (GAA or GAG, see Figure 14.7, p. 362 in the text), to the codon for alanine (GCA or GCG). While glutamate is an acidic amino acid, alanine is a neutral, nonpolar amino acid (see Figure 14.2, p. 357 in the text). If glutamate is essential for proteolytic activity, the altered protein should not have proteolytic activity. After completing the site-specific mutagenesis (described in Figure 17.1 of the text, p. 457), replace a segment of the wild-type cDNA sequence with the altered sequence. Then express the protein product of the wild-type cDNA (as a positive control) and that of the mutant cDNA and purify the proteins that are produced. Measure the metalloprotease activity of each protein in the biochemical assay to determine whether the second amino acid must be glutamate for the protein to have proteolytic activity.

17.3   Positional cloning is used to identify the gene for an autosomal dominant human disease. Sequence analysis of a mutant allele reveals it to be a missense mutation. Two alternate hypotheses are proposed for how the mutant allele could cause disease. In one hypothesis, the missense mutation alters a critical amino acid in the protein so that the protein is no longer able to function: Heterozygotes with just one copy of the normal allele develop the disease because they have half of the normal dose of this protein's function. In the second hypothesis, the missense mutation alters the protein so that it interferes with a normal process: Heterozygotes develop the disease because the mutant allele actively disrupts a required function. How could you gather evidence to support one of these alternate hypotheses using knockout mice?

Answer: A knockout mouse is a mouse in which the gene has been made nonfunctional, for example, by deleting part of it. Under the first hypothesis, heterozygotes develop the disease because they have only half the normal dose of the gene's function. If this hypothesis is correct, then heterozygous mice having a knockout allele and a normal allele will also have only half of the normal dose of the gene's function, and so these mice should develop symptoms of the disease. If they do not develop symptoms of the disease, this hypothesis would not be supported. The second hypothesis, that the missense mutation alters the protein to interfere with a normal process, should then be investigated further.

17.4   What different types of DNA polymorphisms exist and what different methods can be used to detect them?

Answer: There are three major classes of DNA polymorphisms: single-nucleotide polymorphisms (SNPs), short tandem repeats (STRs), and a variable number of tandem repeats (VNTRs).

SNPs are single base changes. If they alter a restriction site, they can be detected as restriction fragment length polymorphisms (RFLPs). RFLPs can be assessed in a Southern blot made using genomic DNA and using a probe that spans the SNP (see text Figure 17.2, p. 459) or using the PCR-RFLP method (see text Figure 17.3, p. 459). Chromosomal aberrations can also result in an RFLP (for example, a deletion that removes the site, a chromosomal rearrangement the repositions the site). Not all SNPs will affect a restriction site. Independent of whether they affect a restriction site, SNPs can be detected using allele-specific oligonucleotide (ASO) hybridization analysis (see text Figure 17.4, p. 460) or using DNA microarrays (see text Figure 17.5, p. 460).

VNTRs and STRs are tandemly repeated sequences. STRs have 2–6-bp DNA sequences tandemly repeated a few times to about 100 times. Because the overall length of an STR is relatively short, PCR is the preferred method for analyzing STR polymorphic loci. This is illustrated in text Figure 17.6, p. 462. VNTRs are similar to STRs, but have repeats from 7 to a few tens of base pairs long. The length of VNTRs precludes the use of PCR as a convenient way to analyze them. More often, they are detected by restriction enzyme digestion (using enzymes that flank the repeat), Southern blotting, and probing with a labeled DNA fragment containing several copies of the repeating sequence. There are two types of VNTR loci, unique loci and multicopy loci. Probes that detect only one VNTR locus are monolocus probes, while probes that detect VNTRs at multiple sites in the genome are multilocus probes.

17.5   Do all DNA polymorphisms lead to an alteration in phenotype? Explain why or why not.

Answer: Not all DNA polymorphisms lead to an alteration in phenotype. Some are silent. For example, most SNPs are noncoding SNPs (ncSNPs occur in regions of the genome that do not code for gene products). Many of these are silent, although those that occur in gene regulatory regions can affect gene function. Of the less frequent coding SNPs (cSNPs), about half are silent and do not cause missense mutations. STRs and VNTRs that occur in noncoding regions may also be silent.

17.6   Abbreviations used in genomics typically facilitate the quick and easy representation of longer tongue-twisting terms. Explore the nuances associated with some abbreviations by stating whether an RFLP, VNTR, or STR could be identified as an SNP? Explain your answers.

Answer: A SNP is a single nucleotide polymorphism. Since a single base change can alter the site recognized by a restriction endonuclease, a SNP can also be a RFLP, or restriction fragment length polymorphism. Since simple tandem repeats (STRs) and variable number of tandem repeats (VNTRs) are based on tandemly repeated sequences (2 to 6 base-pair repeats for STRs, 7 to tens of base pairs for VNTRs), they will not usually be SNPs.

17.7   The frequency of individuals in a population with two different alleles at a DNA marker is called the marker's heterozygosity. Why would an STR DNA marker with nine known alleles and a heterozygosity of 0.79 be more useful for mapping and DNA fingerprinting studies than a nearby STR having three alleles and a heterozygosity of 0.20?

Answer: If an individual is homozygous for an allele at an STR, all of their gametes have the same STR allele. The STR cannot be used as a marker to distinguish the recombinant- and parental-type gametes of the individual, and so will not be useful for mapping studies. In a population, individuals will be more often heterozygous for STRs with more alleles and higher levels of heterozygosity. The recombinant- and parental-type offspring may be distinguished in individuals heterozygous for an STR, making crosses informative for mapping studies.

   If an STR has few alleles and a low heterozygosity, many individuals in a population will share the same STR genotypes. Therefore, there will be many individuals in the population who, by chance alone, will share the same genotype as a test subject and the STR will not be very useful for DNA fingerprinting studies.

17.8   DNA was prepared from small samples of white blood cells from a large number of people. These DNAs were individually digested with *Eco*RI, subjected to electrophoresis and Southern blotting, and the blot was probed with a radioactively labeled cloned human sequence. Ten different patterns were seen among all of the samples. The following figure shows the results seen in 10 individuals, each of whom is representative of a different pattern.

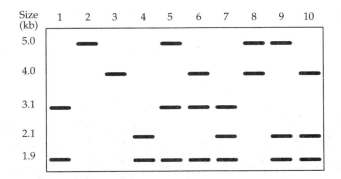

a. Explain the hybridization patterns seen in the 10 representative individuals in terms of variation in *Eco*RI sites.

b. If the individuals whose DNA samples are in lanes 1 and 6 on the blot were to produce offspring together, what bands would you expect to see in DNA samples from these offspring?

Answer:

a. The probe hybridizes to the same genomic region in each of the 10 individuals. Different patterns of hybridizing fragments are seen because of polymorphism of the *Eco*RI sites in the region. If a site is present in one individual but absent in another, different patterns of hybridizing fragments are seen. This provides evidence of restriction fragment length polymorphism. To distinguish between sites that are invariant and those that are polymorphic, analyze the pattern of bands that appear. Notice that the sizes of the hybridizing bands in individual 1 add up to 5 kb. This is also the size of the band in individual 2 and the largest hybridizing band. This suggests that there is a polymorphic site within a 5-kb region. This is indicated in the diagram below, where the asterisk over site *b* depicts a polymorphic *Eco*RI site:

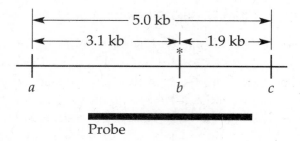

Notice also that the size of the band in individual 3 equals the sum of the sizes of the bands in individual 4. Thus, there is an additional polymorphic site in this 5-kb region. Since the 1.9-kb band is retained in individual 4, the additional site must lie within the 3.1-kb fragment. This site, denoted x, is incorporated into the diagram below. Notice that, because the 1.0-kb fragment flanked by sites a and x is not seen on the Southern blot, the probe does not extend into this region.

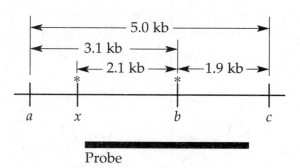

Depending on whether x and/or b are present, one will see either 5-kb, 3.1-kb and 1.9-kb, 2.1-kb and 1.9-kb, or 4-kb bands. In addition, if an individual has chromosomes with different polymorphisms, one can see combinations of these bands. Thus, individual 5 has one chromosome that lacks sites x and b and one chromosome that has site b. The chromosomes in each individual can be tabulated as follows:

Individual	Sites on Each Homologue	Homozygote or Heterozygote?
1	*a, b, c*	homozygote
2	*a, c*	homozygote
3	*x, c*	homozygote
4	*x, b, c*	homozygote
5	*a, c/a, b, c*	heterozygote
6	*x, c/a, b, c*	heterozygote
7	*a, b, c/x, b, c*	heterozygote
8	*a, c/x, c*	heterozygote
9	*a, c/x, b, c*	heterozygote
10	*x, c/x, b, c*	heterozygote

b. Since individual 1 is homozygous, chromosomes with sites at a, b, and c will be present in all of the offspring, giving bands at 3.1 and 1.9 kb. Individual 6 will contribute chromosomes of two kinds, one with sites at x and c and one with sites at a, b, and c. Thus, if this analysis is performed on their offspring, two equally frequent patterns will be observed: a pattern of bands at 3.1 and 1.9 kb and a pattern of bands at 4, 3.1, and 1.9 kb. This is just like the patterns seen in the parents.

17.9   The maps of the sites for restriction enzyme R in the wild type and the mutated cystic fibrosis genes are shown schematically in the following figure:

Samples of DNA obtained from a fetus (F) and her parents (M and P) were analyzed by gel electrophoresis followed by the Southern blot technique and hybridization with the radioactively labeled probe designated "CF probe" in the previous figure. The autoradiographic results are shown in the following figure:

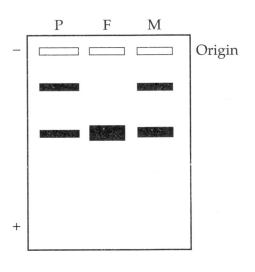

Given that cystic fibrosis is a recessive mutation, will the fetus be affected? Explain.

Answer: Chromosomes bearing CF mutations have a shorter restriction fragment than do chromosomes bearing wild-type alleles. Both parent lanes (M and P) have two bands, indicating that each parent has a normal and a mutant chromosome. The parents are therefore heterozygous for the CF trait. The fetus lane (F) shows only one (lower-molecular-weight) band. The size of the band indicates that the fetus has only mutant chromosomes. The intensity of the band is about twice that of the same-sized band in the parent lanes. This is because the diploid genome of the fetus has two copies of the fragment, while the diploid genome of each parent only has one. Since the fetus is homozygous for the CF trait, it will have CF. (Do not confuse the black bands with the open boxes, which represent the origin, the location where the DNA was loaded before electrophoresis.)

17.10   The enzyme *Tsp*45I recognizes the 5′-bp site 5′-G-T-(either C or G)-A-C-3′. This site appears in exon 4 of the human gene for α-synuclein, where, in a rare form of Parkinson disease, it is altered by a single G-to-A mutation. (Note: Not all forms of Parkinson disease are caused by genetic mutations.)

a. Suppose you have primers that can be used in PCR to amplify a 200-bp segment of exon 4 containing the *Tsp*45I site, and that the *Tsp*45I site is 80 bp from the right primer. Describe the steps you would take to determine if a parkinsonian patient has this α-synuclein mutation.

b. What different results would you see in homozygotes for the normal allele, homozygotes for the mutant allele, and in heterozygotes?

c. How would you determine, in heterozygotes, if the mutant allele is transcribed in a particular tissue?

Answer:

a. Use the PCR-RFLP method: Isolate genomic DNA from the parkinsonian individual, and use PCR to amplify the 200-bp segment of exon 4; purify the PCR product, digest it with *Tsp*45I, and resolve the digestion products by size using gel electrophoresis. The normal allele will contain the *Tsp*45I site, and so produce 120- and 80-bp fragments. The mutant allele will not contain the *Tsp*45I site, and so produce only a 200-bp fragment.

b. Homozygotes for the normal allele will have 120- and 80-bp fragments; homozygotes for the mutant allele will have a 200-bp fragment; heterozygotes will have 200-, 120-, and 80-bp fragments.

c. Use RT-PCR to amplify a DNA copy of the mRNA, and digest the RT-PCR product with *Tsp*45I. First isolate RNA from the tissue. Then make a single-stranded cDNA copy using reverse transcriptase and an oligo(dT) primer. Then amplify exon 4 of the cDNA using PCR, digest the product with *Tsp*45I, and separate the digestion products by size using gel electrophoresis. If a 200-bp fragment is identified in a heterozygote, then the mutant allele is transcribed. If only 120- and 80-bp fragments are identified, then the mutant allele is not transcribed. Note that to accurately assess expression of either allele, it is essential that the RT-PCR reaction is performed on a purified RNA template without contaminating genomic DNA.

17.11   Pathologists categorize different types of leukemia, a cancer that affects cells of the blood, using a set of laboratory tests that assess the different types and numbers of cells present in blood. Patients classified into one category using this method had very different responses to the same therapy: Some showed dramatic improvement while others showed no change or worsened. This finding raised the hypothesis that two (or more) different types of leukemia were present in this set of patients, but that these types were indistinguishable using existing laboratory tests. How would you test this hypothesis using DNA microarrays and mRNA isolated from blood cells of these leukemia patients?

Answer: Use microarray analysis to determine if patients who respond to therapy have a different pattern of gene expression in their blood cells than do patients who fail to respond to therapy. Prepare cDNA from the mRNA isolated from the blood cells of individual leukemia patients, label the cDNAs with fluorescent dyes, and use them in a microarray analysis. For example, label cDNA from a patient who responds to the therapy with Cy3 and label cDNA from a patient who fails to respond to the therapy with Cy5. Mix the labeled cDNAs together and allow them to hybridize to a probe array containing oligonucleotides from many different genes as shown in text Figure 17.5, p. 461. Then identify the set of genes whose pattern of expression differs in the two patients. Repeat the experiment using different pairs of patients to identify the set of genes that shows consistently greater (or lesser) expression in patients who respond to therapy. The hypothesis that two (or more) different types of leukemia are present in this patient population would be supported if there are consistent differences in the gene expression patterns between patients who respond to therapy and patients who fail to respond to therapy. The pattern of gene expression could be further evaluated as a clinical marker.

17.12   Some features that we commonly associate with racial identity, such as skin pigmentation, hair shape, and facial morphology, have a complex genetic basis. However, it turns out that these features are not representative of the genetic differences between racial groups—individuals assigned to different racial categories share many more DNA polymorphisms that not—supporting the contention that race is a social and not a biological construct. How could you use DNA chips to quantify the percentage of SNPs that are shared between individuals assigned to different racial groups?

Answer: Compare the SNPs present in individuals assigned to different racial groups using DNA chips (probe arrays) containing thousands of SNPs that are representative of SNPs throughout the entire genome. That is, use a probe array that has, for each of several thousand representative SNPs, a set of oligonucleotides that match the common allele and all possible variant alleles. Each hybridization will assess the SNP alleles present in one individual, performed by labeling fragments of genomic DNA from one individual as target DNA and hybridizing the labeled DNA to a SNP probe array. An individual's SNP alleles (and whether they are homozygous or heterozygous) will be determined by observing the pattern of fluorescence on the probe array and comparing this to the locations of the oligonucleotides for each SNP allele. The percentage of SNPs that are shared between individuals assigned to different racial groups can be determined by comparing the results of hybridizations with target DNA from individuals from different racial groups.

17.13   What is DNA fingerprinting and what different types of DNA markers are used in DNA fingerprinting? How could this method be used to establish parentage? How is it used in forensic science laboratories?

Answer: DNA fingerprinting is the characterization of an individual in terms of the set of DNA markers the person has. For DNA fingerprinting, DNA markers are usually chosen so that they are highly polymorphic in a population, and they may include RFLPs, simple tandem repeat polymorphisms of 2 to 6 bp in length (STRs, microsatellites), and variable numbers of tandem repeat polymorphisms of 7 to 10s of bp in length (VNTRs, minisatellites). VNTRs can be derived either from just one locus or more than one locus (monolocus or multilocus probes). Since all individuals except for identical (monozygotic) twins have different genomes, individuals of a species differ in terms of their DNA fingerprints.

Each parent donates one allele at a DNA marker to their offspring. Therefore, the set of DNA markers a child has come from his/her parents. If two parents and a child are tested, all of the DNA markers present in the child must also be present in the set of markers in the two parents. However, the parents can have additional markers, since they donate only one of their two alleles at each marker. If a child has no alleles in common with a suspected parent, the suspected individual can be excluded from consideration as the child's parent. If alleles are shared, the likelihood that the set of alleles came from the suspected parent must be calculated and compared with the likelihood that the set of alleles came from another person. If the alleles present in the child and one known parent are known, then one can infer at least some of the alleles that were contributed by the other parent. Usually, multiple DNA markers are used to minimize possible inaccuracy and provide for a better statistical evaluation of the results, showing that the results obtained are not based on sampling variation. The calculation of the relative odds that a set of alleles derives from a suspected parent or from another person depends on knowing the frequencies of the marker alleles in the child's ethnic population. These types of data should be used together with other non-DNA-based evidence to support a claim of parentage.

In forensic science laboratories, DNA fingerprinting is used to match or exclude the DNA fingerprint of a suspected individual with the DNA fingerprint provided by physical evidence (blood, skin, hair, semen, saliva, etc.) gathered at the scene of a crime. As in the analysis for parentage, DNA fingerprinting is very useful for excluding suspects who do not share alleles with that found in the physical evidence. Inclusion results, results where an individual is positively identified as being responsible for a crime, are more difficult to obtain. As for parentage, a calculation of the relative odds that a set of alleles derive from a suspect or from another person must be made, and this requires good population statistics. In addition, there must be evidence that there were no errors in collecting or processing samples. Thus, DNA evidence is often more useful for excluding suspects than for proving their guilt, and when used to support guilt, must usually be supported by additional, non-DNA-based evidence.

17.14   One application of DNA fingerprinting technology has been to identify stolen children and return them to their parents. Bobby Larson was taken from a supermarket parking lot in New Jersey in 1978, when he was 4 years old. In 1990, a 16-year-old boy called Ronald Scott was found in California, living with a couple named Susan and James Scott, who claimed to be his parents. Authorities suspected that Susan and James might be the kidnappers and that Ronald Scott might be Bobby Larson. DNA samples were obtained from Mr. and Mrs. Larson and from Ronald, Susan, and James Scott. Then DNA fingerprinting was done, using a polymorphic probe for a particular VNTR family, with the results shown in the following figure. From the information in the figure, what can you say about the parentage of Ronald Scott? Explain.

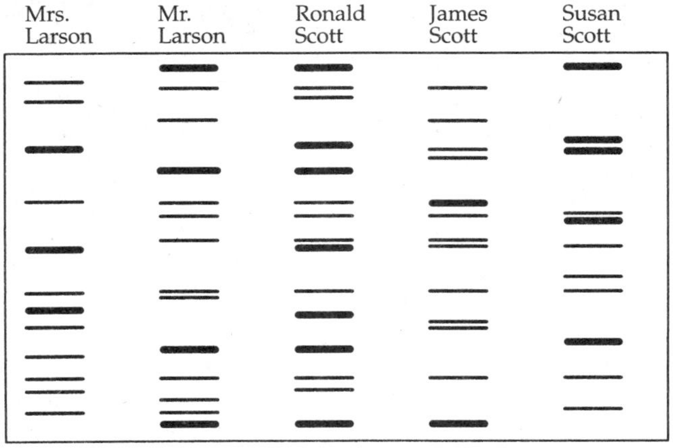

Answer: James and Susan Scott are not the parents of "Ronald Scott." There are several bands in the fingerprint of the boy that are not present in either James or Susan and thus could not have been inherited from either of them (e.g., bands a and b in the figure below). In contrast, whenever the boy's DNA exhibits a band that is missing from one member of the Larson couple, the other member of the Larson couple has that band (e.g., bands c and d). Thus, there is no band in the boy's DNA that he could not have inherited from one or the other of the Larsons. These data thus support an argument that the boy is, in fact, Bobby Larson. These data should be used together with other non-DNA-based evidence to support the claim that the boy is Bobby Larson.

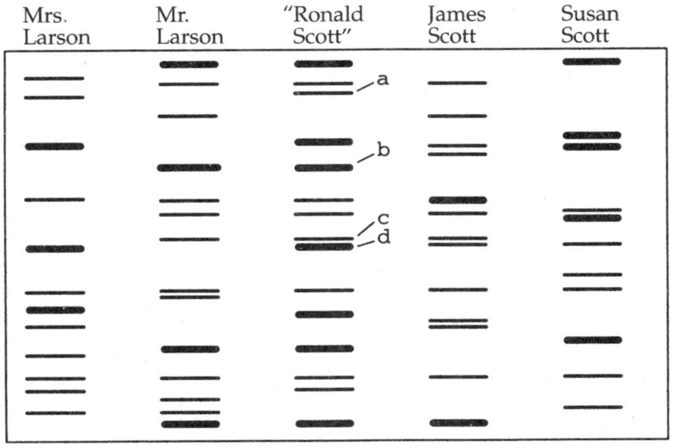

17.15   As described in the text and demonstrated in Question 17.14, VNTRs can robustly distinguish between different individuals. Five well-chosen single-locus VNTR probes used together can almost uniquely identify one individual, as, statistically, they are able to discriminate 1 in 10^9 individuals. However, the use of VNTR markers has largely been supplanted by the use of STR markers. For example, the FBI uses a set of 13 STR markers in forensic analyses. Different fluorescently labeled primers and reaction conditions have been developed so that this marker set can be multiplexed—all of the markers can be amplified in one PCR reaction. The marker set used by the FBI, the number of alleles at each marker, and the probability of obtaining a random match of a marker in Caucasians is listed in the following table:

STR Marker	Number os Allelles	Probability of a Random Match (Based on an analyses of Caucasians)
CSF1PO	11	0.112
FGA	19	0.036
TH01	7	0.081
TPOX	7	0.195
VWA	10	0.062
D3S1358	10	0.075
D5S818	10	0.158
D7S820	11	0.065
D8S1179	10	0.067
D13S317	8	0.085
D16S539	8	0.089
D18S51	15	0.028
D21S11	20	0.039

a. Consider the types of DNA samples that the FBI analyzes and the requirements concerning DNA samples in the methods used to analyze STR and VNTR markers. Why is the use of STR markers preferable to the use of VNTR markers?

b. Why is it advantageous to be able to multiplex the PCR reactions used in forensic STR analyses?

c. Suppose the first four STR markers listed in the table are used to characterize an individual's genotype, and the genotype is an exact match with results obtained from a hair sample found at a crime scene. What is the probability that the individual has been misidentified, that is, what is the chance of a random match when just these four markers are used? About how often do you expect an individual to be misidentified if only these four markers are used?

d. Answer the questions posed in (c) if all 13 STR markers are used.

Answer:

a. The PCR method requires very small (nanograms) amounts of template DNA, and if the primers are designed to amplify only small regions, the DNA can be used even if it is partially degraded. In contrast, VNTR methods require larger amounts (micrograms) of intact DNA, as restriction digests are used to produce relatively large (kb-size fragments) that are then detected by Southern blotting. Some of the DNA samples used in forensic analysis are found in crime scenes and may be stored for years, so that they may often be degraded and not be present in large amounts. STR methods can still be used on such samples, while VNTR methods cannot.

b. Multiplexing PCR reactions ensures that (1) the different STR results obtained in the reaction are all derived from a single DNA sample (laboratory labeling and pipetting errors are minimized), and (2) limited amounts of DNA samples are used efficiently.

c. P(random match) = $(0.112 \times 0.036 \times 0.081 \times 0.195) = 6.4 \times 10^{-5}$. About 1 person in 15,702 would be misidentified by chance alone using just these four markers.

d. P(random match) = $(0.112 \times 0.036 \times 0.081 \times 0.195 \times 0.062 \times 0.075 \times 0.158 \times 0.065 \times 0.067 \times 0.085 \times 0.089 \times 0.028 \times 0.039) = 1.7 \times 10^{-15}$. About 1 person in 5.94×10^{14} would be misidentified by chance alone using all 13 markers.

17.16 About midnight on Saturday, the strangled body of a regular patron of the Seedy Lounge is found in an alleyway near the bar. The police interview the workers and patrons remaining in the bar. A few of the patrons indicate that several individuals, including the bartender, owed money to the deceased. The police notice that the bartender and patrons A, C, D, F, K, L, O, and R all have recent cuts and scratches on their faces and the back of their necks, but are told that these happened during mud-wrestling matches earlier in the evening. DNA samples are obtained from the bartender and the bar's patrons, from the deceased and from scrapings of her fingernails. STR analyses are performed on the DNA samples using three of the markers described in Question 17.15: THO1, D18S51, and D21S11. The sizes of the PCR products obtained in each DNA sample for each marker are shown in the following table:

DNA Sample	STR		
	THO1	**D21S11**	**D18S51**
Victim	162, 170	221, 239	292, 304
Victim's fingernail scraping	162, 170, 174	221, 225, 233, 239	280, 292, 300, 304
Patron A	159, 174	221, 225	292, 316
Patron B	162	221, 235	296, 304
Patron C	162, 174	225, 233	280, 300
Patron D	170, 174	229, 231	300, 304
Patron E	170, 174	225, 233	288, 292
Patron F	162, 166	229, 243	284, 288
Patron G	174	225, 235	292, 308
Patron H	159, 174	221, 233	296
Patron I	159, 174	233, 235	300, 308
Patron J	170, 174	225	284, 296
Patron K	170, 174	231, 235	288, 292
Patron L	159	237, 239	276, 304
Patron M	159, 170	221, 229	304, 308
Patron N	166, 174	229, 239	292, 304
Patron O	170	221, 225	288, 308
Patron P	162, 170	221	296, 300
Patron Q	159, 174	235, 239	284, 304
Patron R	170, 174	225, 233	288, 292
Bartender	170, 174	221, 231	300, 308

a. How many different alleles are present at each marker in these samples and how does this compare to the total number of alleles that exist? How do you explain the appearance of only one marker allele in some individuals? How do you explain the appearance of three and four marker alleles in the DNA sample obtained from the victim's fingernails?

b. Who should the police investigate further if they consider the results obtained using only the D21S11 marker? Explain your reasoning.

c. Who should the police investigate further if they consider the results obtained using all three STR markers? Explain your reasoning.

Answer:

a. This relatively small sample possesses many of the known alleles at these three loci: It has 5 of the 7 known THO1 alleles, 9 of the 20 known D21S11 alleles, and 10 of the 15 known D18S51 alleles. Individuals that have only one marker allele are homozygous for that allele. For example, Patron B is homozygous for the 162-bp allele of THO1. Since one individual can have at most two alleles at one locus, the presence of three and four marker alleles in the DNA sample obtained from the victim's fingernails indicates the presence of two different DNA samples—that of the victim and a person she had scratched.

b. The victim has the 221 and 239-bp alleles at the D21S11 locus. The DNA sample from her fingernail scraping has these alleles and the 225 and 233 alleles. Assuming that she scratched her assailant and not her mud-wrestling opponent, the police should investigate further individuals with the 225 and 233-bp alleles at D21S11, patrons C, E, and R.

c. The victim has the 162 and 170-bp alleles at the THO1 locus. The DNA sample from her fingernail scraping has these alleles and the 174 allele. This means that the assailant has the 174 allele. The assailant could be homozygous for the 174 allele or also have the 162 or the 170 allele. Patrons C, E, and R all have the 174 allele and either the 162 or 170 allele. Thus, the information from the THO1 locus does not allow us to distinguish between patrons C, E, and R.

The victim has the 292 and 304-bp alleles at the D18S51 locus. The DNA sample from her fingernail scraping has these alleles and the 280 and 300 alleles. Only patron C has the 280 and 300 alleles. Therefore, the police should investigate patron C further.

17.17   For rare genetic disorders that have only one mutant allele, genetic tests can be tailored to specifically detect the mutant and normal alleles. However, for more prevalent genetic disorders, such as anemia caused by mutations in α- and β-globin, Duchene muscular dystrophy caused by mutations in the dystrophin gene, and cystic fibrosis caused by mutations in CFTR, there are many different alleles at one gene that can lead to different disease phenotypes. These diseases present a challenge to genetic testing, as for these diseases, a genetic test that identifies only a single type of DNA change is inadequate. How can this challenge be overcome?

Answer: One approach to evaluating the presence of a defined set of alleles is to use reverse ASO hybridization and multiplex PCR. In this approach, several different regions of the gene are amplified in one test tube using PCR and radioactive dNTPs. The radioactively labeled PCR product is used as a probe for hybridization with many different ASOs (allele-specific oligonucleotides) bound to a membrane filter. Each ASO bound to the filter contains a different sequence variant (wild-type or mutant). On the autoradiograms, the dot to which the PCR product binds indicates the allele that the individual has. Inspection of the signal on the autoradiogram allows us to infer whether the individual has any of the mutant alleles used in the test.

If the region being evaluated is very large, an alternate approach is to design a probe array that has oligonucleotides that match the wild-type sequences and known sequence variants (polymorphisms and mutations) throughout the gene that is being evaluated, fluorescently label target DNA from an individual, and hybridize the DNA to the probe array. Since the oligonucleotide locations of the probes for different segments and mutant variants of the gene are known on the probe array, the observed pattern of fluorescence indicates which DNA sequences are present in the individual. This information can be used to determine whether the individual has a mutation in the gene, and if so, where it is and whether it is the same as a known mutation.

17.18   A research team interested in social behavior has been studying different populations of laboratory rats. By using a selective breeding strategy, they have developed two populations of rats that differ markedly in their behavior: One population is abnormally calm and placid, while the second population is hyperactive, nervous, and easily startled. Biochemical analyses of brains from each population reveal different levels of a catacholamine neurotransmitter, a molecule used by neurons to communicate with each other. Relative to normal rats, the hyperactive population has increased levels while the calm population has decreased levels. Based on these results, the researchers have hypothesized that the behavioral and biochemical differences in the two populations are caused by variations in a gene that encodes an enzyme used in the synthesis of the catacholamine. Suppose you have a set of SNPs that are distributed throughout the rat genomic region containing this gene, including its promoter, coding region, enhancers, and silencers. How could you use these SNPs to test this hypothesis?

Answer: If the hypothesis proposed by these researchers is correct, the selective breeding strategy has resulted in two rat populations that differ in DNA sequence at a gene for the synthesis of a catacholamine. Consider as simplest case that there is a one SNP—it may be considered either a mutation or a polymorphism—at this gene that is responsible for the behavioral and biochemical differences. Since SNPs are common, occurring at a frequency of about 1 per 350 bp, the SNP that causes the phenotype is tightly linked to other nearby SNPs. Since crossing-over will not separate the tightly linked SNPs very often, the selective breeding strategy will have selected for the causal SNP as well as the SNPs that are tightly linked to it, including some of the SNPs that we can assay. Under the researchers' hypothesis then, some of the SNPs that we can assay should be more frequent in one of the two rat populations. Identifying such SNPs will localize a region of the gene for further study to identify the causal DNA sequence change.

Therefore, we should detect which SNPs are present in the genomic DNA (isolated from a drop of blood or a tail snipping) of representative individuals of the two rat populations using PCR-RFLP (see text Figure 17.3, p. 459) or the ASO hybridization method (see text Figures 17.4, p. 460, and 17.9, p. 465) and then determine whether any of the SNPs are present at a different frequency in the two rat populations. If we find a region of the gene where SNPs consistently differ between the two rat populations, our data will support the researchers' hypothesis.

The hypothesis might also be evaluated by sequencing the gene in representative individuals from each rat population. However, it would be cumbersome to sequence the potentially large region containing the gene and its promoter, coding region, enhancers, and silencers from multiple individuals. Sequencing would almost certainly reveal differences, as SNPs are frequent, but we would be unable to evaluate their significance without data from many individuals in each population. The method presented here is much more efficient, as it allows us to screen the entire region in multiple individuals drawn from each population.

17.19 When geneticists sought to identify the gene responsible for cystic fibrosis, they did not know the identity of its protein product. Even though they knew nothing about the nature of the defective protein, they did know that something about salt transport was disrupted. The geneticists used a powerful combination of classic and molecular approaches and in 1989 located the cystic fibrosis gene. What steps did they take to identify the CFTR gene?

Answer: Cloning of the gene responsible for cystic fibrosis involved sequential homing in on the gene, using both genetic and molecular data.

1. *Linkage analysis.* First, RFLP markers linked to the CF gene were identified. As the patterns of inheritance of CF were tracked in families, DNA from family members was analyzed using Southern blot analysis to identify RFLP markers that showed genetic linkage to the CF locus.

2. *Chromosomal localization.* The RFLP marker was used as a probe in experiments using in situ hybridization to identify the chromosome on which the CF gene was located — chromosome 7. Additional nearby RFLPs were then identified. These were used in genetic linkage analysis to identify flanking markers that defined which region of chromosome 7 harbored the CF gene.

3. *Cloning genomic DNA.* A combination of chromosomal walking and jumping was used to identify a large number of overlapping genomic DNA clones spanning about 500 kb in the CF region.

4. *Identifying CF candidate gene.* To identify the CF gene among these DNA sequences, subcloned segments of the region were used as probes on zoo blots to detect if any were conserved among other organisms. Five clones with conserved gene sequences were identified. Linkage analysis excluded two from being responsible for CF. A third was excluded because it was a pseudogene. A fourth was excluded because it was not expressed in tissues affected in CF patients. The fifth subclone contained a CpG island. CpG islands are often found near promoters of expressed genes. This clone was used to synthesize a probe that was used to screen a cDNA library made from mRNA expressed in cultured sweat gland tissue cells. These cells were chosen since they might express genes involved in salt transport. The probe identified a single 6.5-kb band on a northern blot and allowed the identification of a candidate cDNA clone. The cDNA clone was used to analyze the genomic clones in more detail.

5. *Confirming that candidate gene is responsible for CF.* The DNA sequences of the candidate CF gene were compared in normal and affected individuals. After a mutational change (a 3-bp deletion) was found in the gene from one CF patient, DNA changes were identified in the gene in additional CF patients.

6. *Characterizing the protein produced by the CF gene.* Computerized analyses of the predicted CF protein provided a basis to propose a structure for the CF protein. In turn, this was used to develop testable hypotheses about its function.

17.20   Chromosome walking and chromosome jumping can both be used to find a gene between flanking markers. What is the difference between chromosome walking and chromosome jumping? Given that these can proceed in either of two opposite directions, what experimental method could you use to verify that a chromosomal walk or jump is going toward the gene you are seeking?

Answer: Several different methods can be helpful to orient a chromosomal walk that has proceeded through several steps, or that includes one or more jumps. After restriction mapping is used to establish how the clones of the walk overlap, DNA markers such as RFLPs that have been positioned relative to each other using genetic mapping studies may be used to identify the orientation of a walk. If a walk has extended far enough, and includes two RFLP markers that have been previously oriented relative to the gene you are seeking, the orientation of the walk can be determined by using Southern blotting to identify which clone contains each marker. An alternative method is to use in situ hybridization to chromosomes. Probes can be made from unique sequences at the ends of a partially completed walk, and hybridized to chromosomes.

The orientation of the walk relative to the centromere and telomere of the chromosome can then be determined.

17.21 A positional cloning approach has been used to identify a chromosomal region for a disease gene, and the region is found to contain several different candidate genes. What is meant by a candidate gene? What types of criteria can be used to prioritize the evaluation of different candidate genes?

Answer: A candidate gene for a disease is a gene that if mutated could conceivably cause the disease, but has not yet been proved to cause the disease. There can be several candidate genes for a disease. Based on the example of the CF candidate gene, potential criteria for prioritizing the evaluation of different candidate genes include their chromosomal location and their expression in a tissue affected by the disease. Other criteria could include the gene product's biochemical and molecular characteristics (for example, in CF, the protein is involved in chloride transport), an altered pattern of expression in tissue from diseased individuals, and the identification of mutations or polymorphisms that are found in diseased but not normal individuals.

17.22 Sexual behavior in *Drosophila* (fruit fly) is under the control of several regulatory genes. Recently the *fru* gene has been molecularly cloned. Explain why male mutants homozygous for *fru* are considered sterile even though they have normal sexual organs and produce normal sperm.

Answer: Drosophila males and females display a stereotyped set of mating behaviors. Male *Drosophila* that are mutant for the *fru* gene are unable to perform male courtship behaviors, and so are unsuccessful at courting and mating with females. Even though they are morphologically normal, they are behaviorally abnormal. They can't persuade a female to allow copulation and so are functionally sterile.

17.23 The *fru* gene has been cloned, and both genomic and cDNA clones for *fru* are available. One means to more fully understand its function in male sexual behavior is to identify genes for proteins that interact with the *fru* gene's protein product. Describe the steps you would take to accomplish this goal.

Answer: Use an interaction trap assay (the yeast two-hybrid system). Fuse the *fru* coding region (obtained from a cDNA) to the sequence of the Gal4p BD, and cotransform this plasmid into yeast with a plasmid library containing the GAL4p AD sequence, which is fused to protein sequences encoded by different cDNAs from the *Drosophila* brain. Purify colonies that express the reporter gene (see text Figure 17.15, p. 474). In these colonies, the transcription of the reporter gene was activated when the AD and BD domains were brought together by the interactions of the *fru* protein with an unknown protein encoded by one of the brain cDNAs. Isolate and characterize the brain cDNA found in these yeast colonies.

17.24 Genetic variability is important for maintaining the ability of a species to adapt to different environments. Therefore, it is important to understand how much genetic variation there is in an endangered species, as this type of information can be used to design better strategies to help the species from becoming extinct. Listed below are four strategies that have been proposed for detecting a SNP in a known DNA sequence in several hundred individuals from an endangered species. Critically evaluate them, and explain why each is, or is not, a good strategy for this purpose.

a. Sacrifice each of the animals or plants in the name of science. Isolate their genomic DNA, prepare libraries from each, and screen for clones containing the sequence. Sequence each clone individually. Then compare the sequences of the different clones.

b. Isolate a few cells (e.g., by using a cheek scraping or leaf sampling) from each of the individuals. Prepare DNA from the samples and use the ASO hybridization method.

c. Isolate a few cells from each of the individuals. Prepare DNA from each of the samples and then use the yeast two-hybrid system.

d. Search the literature to find a restriction enzyme that cleaves the sequence containing the SNP and that cleaves the site when only one SNP allele is present. Use the restriction enzyme to measure the site as an RFLP marker. After isolating a few cells from the individuals, prepare DNA from the cells, digest it with the restriction enzyme, separate it by size using electrophoresis, make a Southern blot, and perform an RFLP analysis.

Answer:

a. This strategy is problematic for two reasons. First, this strategy is likely to push the species further toward extinction, as it eliminates several hundred individuals. Second, it is labor intensive and costly.

b. This is the best strategy: It is essentially noninvasive and does not harm the organisms, and it is efficient and cost effective.

c. This strategy would not accomplish the experimental objective. The yeast two-hybrid system is used to identify proteins that interact with a cloned protein, which is not the objective here.

d. This strategy may not work for several reasons. First, the SNP may not be able to be assessed using the RFLP method, as not all SNPs alter restriction sites. Second, Southern blot analysis usually requires DNA from more than just a few cells. If a restriction enzyme were identified that cleaved the sequence containing the SNP, PCR-RFLP would be a more reasonable choice. See text Figure 17.3, p. 459.

17.25   A scientist is interested in understanding the physiological basis of alcoholism. She hypothesizes that the levels of the enzyme alcohol dehydrogenase, which is involved in the degradation of ethanol, are increased in individuals who routinely consume alcohol. She develops a rat model system to test this hypothesis. What steps should she take to determine if the transcription of the gene for alcohol dehydrogenase is increased in the livers of rats who are chronically fed alcohol compared to a control, abstinent population?

Answer: Isolate RNA from the livers of the alcohol-fed and control rats. Measure the levels of mRNA for alcohol dehydrogenase by either (1) separating the RNA by size using gel electrophoresis, preparing a northern blot, and hybridizing it with a probe made from a cDNA for the alcohol dehydrogenase gene; or (2) using RT-PCR or quantitative RT-PCR.

17.26   In 1990, the first human gene therapy experiment on a patient with adenosine deaminase deficiency was done. Patients who are homozygous for a defective gene for this enzyme have defective immune systems and risk death from diseases as simple as a common cold. Which cells were involved, and how were they engineered?

Answer: In this gene therapy experiment, T cells were isolated from the patient and a viral vector was used to introduce the normal adenosine deaminase gene. The cells were then reintroduced into the patient.

17.27 What methods are used to introduce genes into plant cells, and how are these methods different from those used to introduce genes into animal cells?

Answer: DNA can be introduced into animal cells using transfection methods, using virus-related vectors, and using electroporation. DNA can also be injected into embryos. A variety of vectors with eukaryotic origins of replication, selectable markers, and regulatory elements can be used to direct foreign gene expression in animal cells.

In contrast, the transformation of dicotyledonous plant cells is mediated by the Ti plasmid of *Agrobacterium tumefaciens*. The Ti plasmid has a 30-kb T-DNA region flanked by two repeated 25-bp sequences that are involved in T-DNA excision. When *A. tumefaciens* interacts with the host plant cell, it excises the T-DNA and transfers it to the nucleus of the plant cell. There, the T-DNA integrates into the nuclear genome. To transform plant cells, genes are placed in transformation vectors derived from Ti plasmids and between the 25-bp border sequences of the T-DNA, transformed into *A. tumefaciens*, and the plant cells are infected with *A. tumefaciens*.

To transform monocotyledonous plants, electroporation and gene gun methods are used to introduce DNA into plant cells. Transformed cells are selected (using co-transformed markers) and used to regenerate whole plants.

17.28 The ability to place cloned genes into plants raises the possibility of engineering new, better strains of crops such as wheat, maize, and squash. It is possible to identify useful genes, isolate them by cloning, and insert them directly into a plant host. Usually these genes bring out desired traits that allow the crops in question to flourish. Why, then, is there such concern by consumers about this process? Do you feel that the concern is justified? Defend your answer.

Answer: There is clearly a huge potential for success in crop modification. In addition to the examples presented in the text, visit the Web sites listed below, two of many sites that describe the potential value of a modified, vitamin-A-enhanced rice (golden rice) or mustard (golden mustard). Such products could, in principle, help millions of people suffering from vitamin A deficiency, which is associated with blindness and increased disease susceptibility.

> *http://www.disasterrelief.org/Disasters/000808gmofoods/*
> *http://www.monsanto.com/monsanto/layout/media/00/12–07–00c.asp*

A quick Internet search using "golden rice" or "GMO" identifies mostly organizations opposing the use of GMOs. There are multiple reasons consumers have concerns about GMOs. First, some individuals find that this type of genetic alteration is in conflict with their culturally held views of the spiritual integrity of living organisms (this issue is raised on Thought Question 7). Second, some individuals feel that the dangers associated with genetically modified organisms are unknown or not well enough understood. For example, a gene placed in a particular crop species might be transferred laterally, perhaps via some infectious agent, to another species, where it is undesirable; the gene may confer properties that are deleterious to the environment, which only appear years after introduction and which would not be apparent in controlled testing environments; the altered strain may deleteriously alter existing ecological balances; the altered gene may negatively impact human health in unknown ways; there may be significant concerns about food safety. Third, there are concerns about the maintenance of genetic diversity in crops. What will be the effect on seed stocks if only a few highly desirable strains of some crops are grown? Will this result in increased susceptibility to potential pathogens in the future? Fourth, there are economic concerns: How ethically sound are the business practices of companies involved in these efforts? Should science be financed by corporations? Are these developments in the best interests of underdeveloped nations? Finally, it is a new and largely unknown area. There is much that is unknown, and regulatory agencies are themselves entering a new territory and are being offered conflicting "expert" opinions.

18
GENOMICS

CHAPTER OUTLINE

STRUCTURAL GENOMICS
> Sequencing Genomes
> Whole-Genome Shotgun Sequencing of Genomes
> Selected Examples of Genomes Sequenced
> Bacterial Genomes
> Eukaryotic Genomes
> Insights from Genome Analysis: Genome Sizes and Gene Densities

FUNCTIONAL GENOMICS
> Identifying Genes in DNA Sequences
> Sequence Similarity Searches to Assign Gene Function
> Assigning Gene Function Experimentally
> Describing Patterns of Gene Expression

COMPARATIVE GENOMICS

ETHICS AND THE HUMAN GENOME PROJECT

THINKING ANALYTICALLY

Analysis of genomes is complex. Not only are there multiple approaches to mapping and sequencing a genome, but also a huge amount of information has been gathered through sequence analysis, and an entirely new area—bioinformatics—has emerged to analyze and apply this information. As in previous chapters, organizing information into a contextual framework will help you understand the logic behind the different approaches and retain information obtained from sequence analyses. It will help you to organize the information into separate units.

After reading through the chapter, sketch out answers to questions addressing its main topics, such as, How is a genome mapped? What different types of markers are used? Why is mapping helpful to sequencing a genome? What is a direct shotgun approach? In what different ways are sequences analyzed? What similarities and differences have been seen in the genomes of organisms sequenced to date?

Once you can answer questions like these that relate to structural genomics, identify how sequence information is used in answering questions related to functional and comparative

REVIEW OF KEY TERMS, SYMBOLS, AND CONCEPTS

In your own words, write a brief, precise definition of each term in the groups below. Check your definitions using the text. Then develop a concept map using the terms in each list.

1	2
structural genomics	functional genomics
functional genomics	bioinformatics
comparative genomics	annotation
Human Genome Project (HGP)	sequence similarity search
ELSI program	homology, homologous gene
mapping approach	BLAST
whole-genome shotgun (WGS) approach	ORF
genetic map	*FUN* gene
sequence-tagged site (STS)	orphan family
physical map	single orphan
restriction map	gene knockout
clone contig map, contig	loss-of-function, null allele
expressed sequence tag (EST)	transcriptome
BAC-, YAC-contig	proteome
shotgun approach	probe array
computer algorithm	gene expression profile
sequencing, assembly, finishing	transcriptional fingerprint
dideoxy sequencing	pharmacogenomics
annotation	proteomics
Arabidopsis 2010 project	protein array, microarray, chip
fold-sequence coverage	capture array
draft sequence	large-scale protein array
data mining	NCBI, TIGR, HUGO
NCBI, TIGR, NHGRI, Celera	
gene density	
gene-rich region	
gene desert	
Bacteria, Archaea, Eukarya	

genomics. What are the aims of functional genomics? What methods are used in functional genomics? What new technologies have been developed that allow simultaneous analysis of many genes at once? What are the practical applications of these technologies? What are the aims of comparative genomics? What human problems can it help solve? What biological questions can it address? What tools are used in comparative genomics?

Finally, step back and consider that modern genetics increasingly interfaces with many different aspects of human health, agriculture, and human society. What ethical, legal, and social issues arise with the advent of genomics and its related technologies? Ask yourself if all of these are unquestionably for the good of society or whether some need to be carefully monitored and/or regulated. Step back and think outside the box wherein scientists only pursue information and research for its own sake, justified by the potential for positive benefit to society. What dangers are associated with genomics? How, as a society, can we ensure that the "good" of genomics is utilized while minimizing its dangerous aspects?

QUESTIONS FOR PRACTICE

Multiple Choice Questions

1. Which of the following is *not* a goal of the Human Genome Project?
   a. Identify all the genes in human DNA.
   b. Store DNA sequence and gene-related information in public databases.
   c. Transfer technologies related to genomics to the private sector.
   d. Identify the gene(s) associated with each human disease.

2. Genetic crosses are used for all of the following *except*
   a. to establish the locations of genetic markers on chromosomes
   b. determine the genetic distance between markers
   c. determine the number of base pairs between markers
   d. determine the recombination frequency between markers

3. Which of the following is *not* a DNA marker?
   a. RFLP
   b. STS
   c. ECE
   d. EST

4. What is the best way to generate a BAC clone-contig map?
   a. Use fluorescent in situ hybridization (FISH) to determine the relative chromosomal location of the inserts in each of a set of BAC clones.
   b. Identify which of a set of STSs are present in a set of BAC clones using PCR-derived probes, compare the maps to identify overlapping regions, and then align the clones.
   c. Sequence each of a set of BAC clones using a random shotgun approach, then align the BACs based on the overlapping sequences.
   d. Determine what genes are present in each BAC clone by identifying which of a set of ESTs are present in each BAC clone using PCR-derived probes. Then infer the overlap between different BACs clones from a previously completed genetic map of the genes identified by the ESTs.

5. Which of the following most precisely describes a gene that is a single orphan?
   a. It lacks even a single ORF.
   b. It has an ORF with an unknown function similar to just one gene in a different species.
   c. It has an ORF with an unknown function and is found only in one species.
   d. It has an ORF with an unknown function similar to multiple genes found in other species.

6. DNA microarrays are
   a. aligned sets of printed DNA sequences, ordered according to their map positions
   b. ordered grids of DNA molecules of known sequence fixed at known positions on a solid substrate
   c. aligned sets of printed STRs (microsatellites) in a database
   d. ordered grids of homologous genes from different organisms

7. Which of the following laboratory investigations is most likely to benefit from using a large-scale protein array?
   a. identifying the protein targets of a drug
   b. comparing liver proteins in normal rats to those in rats fed large doses of ethanol
   c. diagnosing the presence of an infectious agent
   d. determining the protein expression profile of yeast cells at different stages of sporulation

8. A genome of a new bacterial species has just been sequenced. You have been assigned the task of identifying all of the organism's genes and assigning each a function. What should you do first?
   a. Perform a sequence similarity search.
   b. Use a probe array to identify the gene expression profile in a bacterial culture.
   c. Use a PCR-based deletion strategy to create a collection of mutant strains that have 500 base-pair segments deleted at 1-kilobase intervals throughout the genome.
   d. Search for ORFs using a computerized algorithm.

9. An EST is an STS marker. Even though it is a unique DNA sequence and tags a specific site, it is not always an exact copy of genomic DNA. Why?
   a. An EST contains vector sequences.
   b. An EST is derived from cDNA, and so introns present in genomic DNA have been removed.
   c. An EST is a fusion of 500 bp from the two ends of a 2-kb plasmid insert used for sequencing genomic DNA.
   d. An EST is a clone of genomic DNA that has been modified to delete repetitive DNA sequences nearby unique sequences.

10. Which of the following can be done using computerized algorithms for prokaryotic, but not eukaryotic, genomes?
    a. Identify the locations of all protein-coding regions.
    b. Identify the locations of known repeated sequences.
    c. Identify the locations of known transposable elements.
    d. Identify the locations of all rRNA genes.

11. Which of the following is *not* true about archaeon genomes?
    a. Some have large main circular chromosomes as well as multiple circular extrachromosomal elements.
    b. The genes involved in energy production, cell division, and metabolism are similar to their counterparts in bacteria.
    c. The genes involved in DNA replication, transcription, and translation are similar to their counterparts in eukaryotes.
    d. They have a higher GC content compared with either bacteria or eukaryotes.

12. Why could Celera Genomics obtain only the sequence of about 97 percent of the genomes of *Drosophila* and humans?
    a. In both organisms, simple telomeric sequences are very difficult to sequence, and these make up about 3 percent of the genome.
    b. In both organisms, long ribosomal DNA repeats make up about 3 percent of the genome. Since these are identical, it is impossible to sequence each individual repeat.
    c. In both organisms, approximately 3 percent of the sequences were in error.
    d. In both organisms, heterochromatic regions mostly near the centromeres are unclonable, making sequences in these regions impossible to obtain.

13. Arguments can be made for using simpler organisms such as *Drosophila* to study complex diseases such as cancer. Which one reason below is the most compelling?

a. *Drosophila* flies do not get cancer, and it would be useful to understand why.

b. It has been impossible to find genes associated with cancer in humans.

c. *Drosophila* flies have homologs for about 61 percent (177 of 289) of human disease genes, making them good models to study the functions of genes involved in human disease.

d. *Drosophila* flies are usually found near human populations and may cause cancer.

14. What is BLAST?

a. the Basic Local Alignment Search Tool, used for homology searches when a DNA or amino acid sequence is compared to other sequences in a database

b. the Big Linear Accelerator Sequencing Tool, a device funded by the Department of Energy that is able to read 10,000 bp of DNA from one sequencing reaction

c. the Best Legal Analysis of Sequence Technology, a document from the American Bar Association that legally analyzes situations faced by biotechnology companies that introduce genetically modified organisms into the environment

d. the Beer Legal Age Screening Tool, a DNA chip used to detect the length of simple telomeric repeats (Since these shorten with each cell division in somatic cells, this can be used to quickly determine whether an individual is over the legal drinking age.)

15. What is the ELSI program?

a. the world's largest bioethics program

b. a program funding studies of the ethical, legal, and social issues related to the availability of genetic information

c. a program that is likely to impact legislation concerning the privacy of genetic information, the safe and effective introduction of genetic information in the clinical setting, the fair use of genetic information, and the education of professionals and the public

d. all of the above

Answers: 1d, 2c, 3c, 4b, 5c, 6b, 7a, 8d, 9b, 10a, 11d, 12d, 13c, 14a, 15d

Thought Questions

1. What are the arguments for and against the ability of states or the federal government to collect genetic data on their populace?

2. Huntington disease is a neurodegenerative disorder inherited as an autosomal dominant disease and is caused by a single gene mutation. It affects about 30,000 Americans. A child of an affected parent has a 50 percent chance of inheriting the disease allele. While there is intensive research into treatments for Huntington disease, there is at present no cure. Carriers of the disease allele will eventually get the disease. Why might someone who has a first-degree relative with the disease want to be tested for the disease allele? Why might the person not want to be tested? Should there be testing for a genetic disease for which there is no cure?

3. Unlike Huntington disease, many different genes appear to be associated with other neurodegenerative diseases, such as Alzheimer disease. Having mutations at one of these genes increases the risk of getting the disease, but does not predict with certainty that an individual will get the disease. Other differences from Huntington disease are that there are disease forms that are apparently not caused by gene mutations, and prevention and treatment options have been proposed. Alzheimer disease affects over 4 million Americans. How do these differences lead you to modify your answers to the questions posed in Thought Question 2? Under what circumstances, if any, would genetic testing for genes related to Alzheimer disease be warranted?

4. How are the challenges faced by proteomics different from the challenges faced by genomics?

5. What is pharmacogenomics? How can it, when combined with the use of DNA chips, be useful to develop drugs that are tailored to the needs of individuals?

6. DNA microarrays can be used to distinguish between two similar cancers affecting the same tissue. How is this done, and why might this be therapeutically helpful?

7. What impact does the completion of a high-resolution map and detailed sequence of the human genome have on positional cloning of disease genes?

8. What is the difference between hypothesis-driven science and descriptive science? In what types of circumstances is descriptive science necessary? How does descriptive science enhance hypothesis-driven science? Are there any circumstances where descriptive science is unwarranted?

9. What is meant when two genes are referred to as homologs? For a gene encoding a protein product, is it better to search for homologs based on DNA sequence or amino acid sequence? Why?

10. Explore the NCBI website for BLAST at *http://www.ncbi.nlm.nih.gov/BLAST/*. What different forms of BLAST have been developed, and what are they used for? Why is it important statistically to evaluate matches between an obtained sequence and sequences in a database as similar sequences are identified?

11. After the human genome sequence was completed, searches were performed to identify possible genes. About 35,000 were found. What different criteria might be used to quickly provide evidence that a DNA sequence is part of a functional gene? (Hint: think about the value of homology searches.)

12. Within different ethnic groups, some alleles are more common than others. Consequently, for diseases influenced by genetic factors, different ethnic groups vary in their frequency of disease alleles. What challenges does this present for genetic testing and for preventative medicine based on the identification of genetically at-risk populations?

Thought Questions for Media

After reviewing the media on the *iGenetics* CD-ROM, try to answer these questions.

1. After shearing the DNA to fragments that are approximately 2 kb in size, why are the DNA fragments treated with SI nuclease?

2. After the 2-kb DNA fragments are cloned into plasmid vectors, they are transformed into bacteria, the plasmids isolated, and their inserts sequenced. Since only about 500 bp of insert DNA can be obtained from each end of an insert, how is the remaining 1 kb of DNA obtained?

3. How are overlapping clones identified to construct a clone contig?

4. Why might there be gaps in an assembled sequence?

5. What are the differences and similarities between a Southern blot and a colony blot?

6. The animation illustrated how microarrays can be designed to identify mutations in a gene as well as to analyze differences in patterns of gene expression.
   a. What differences are there in the design of these two types of microarrays?
   b. What differences are there in labeling the target DNA samples for these two types of analyses?

7. Based on the animation illustrating how gene expression is analyzed using DNA microarrays, discuss how characterization of the transcriptome allows us to gain a more complete understanding of normal cellular events.

8. How are cDNAs labeled in a microarray experiment?

9. Suppose that in a microarray experiment, 97 percent of the spots are yellow. Does this indicate that the experiment is unlikely to yield valuable information? Explain why or why not.

1. Why is it important to expose the microarray to two differently labeled probes simultaneously, instead of using two microarrays with a single probe on each?

2. What are the sources of the tissues for the mRNA samples used to develop the probes for the microarrays? How important is it for the tumor tissue sample to be composed only of tumor cells and not to be a mixture of tumor and normal cells?

3. Is the selection of drugs for Maeve's therapy based on a genotype? Tumor cells usually harbor multiple somatic mutations and often also contain chromosomal rearrangements. If the drug selection is based on a genotype, is it based on her genotype or the genotype of the tumor? If not, what is it based on?

4. Specific genes are placed on microarrays. What information was needed to design the microarray used in this investigation?

5. How could you use microarrays to identify additional genes that might be helpful in identifying more effective therapies for breast cancer?

SOLUTIONS TO TEXT PROBLEMS

18.1   What are STS markers, and in what different ways are they used in structural genomics?

Answer: An STS marker is a sequence tagged site, a sequence of DNA that is unique in the genome and able to tag the genomic site for that sequence. Multiple types of STS markers exist, including polymorphic STRs, nonpolymorphic STRs, ESTs (expressed sequence tags), and random nongene sequences. STS markers are used in making different types of high-density genetic maps of the genome, including YAC and BAC clone contigs and radiation hybrid maps. For example, a YAC contig map can be assembled by first developing STS maps of individual YACs and then comparing these maps to find the clones that share some STSs, indicating overlap of the DNA inserted in each YAC. (See text Figure 18.1, p. 492.)

18.2   Discuss the relationship between STSs and ESTs, addressing whether all STSs are capable of being ESTs, and whether all ESTs are capable of being STSs.

Answer: An STS, or sequence-tagged site, is a unique, known DNA sequence. Some STS markers are derived from genes that are transcribed, and these will also be ESTs (expressed sequence tags). However, other STS markers may be from nongene regions (for example, they may be an STR in a nongene region), and so these will not be ESTs. Providing that an EST is derived from a cDNA for a unique gene, it will also be an STS.

18.3   The average size of fragments, in base pairs, observed after genomic DNA from eight different species was individually cleaved with each of six different restriction enzymes, is shown in Table 18.A on the next page:

a. Under the assumption that each genome has equal amounts of A, T, G, and C, and that on average these bases are evenly distributed, what average fragment size is expected following digestion with each enzyme?

b. How might you explain each of the following?

i.   There is a large variation in the average fragment sizes when different genomes are cut with the same enzyme.

Table 18.A

Species	Enzyme and Recognition Sequence					
	*Apa*I GGGCCC	*Hind*III AAGCTT	*Sac*I GAGCTC	*Ssp*I AATATT	*Srf*I GCCCGGGC	*Not*I GCGGCCGC
Escherichia coli	68,000	8,000	31,000	2,000	120,000	200,000
Mycobacterium tuberculosis	2,000	18,000	4,000	32,000	10,000	4,000
Saccharomyces cerevisiae	15,000	3,000	8,000	1,000	570,000	290,000
Arabidopsis thaliana	52,000	2,000	5,000	1,000	no sites	610,000
Caenorhabditis elegans	38,000	3,000	5,000	800	1,110,000	260,000
Drosophila melanogaster	13,000	3,000	6,000	900	170,000	83,000
Mus musculus	5,000	3,000	3,000	3,000	120,000	120,000
Homo sapiens	5,000	4,000	5,000	1,000	120,000	260,000

ii. There is a large variation in the average fragment sizes when the same genome is cut with different enzymes that recognize sites having the same length (e.g., *Apa*I, *Hin*dIII, *Sac*I, and *Ssp*I).

iii. Both *Srf*I and *Not*I, which each recognize an 8-bp site, cut the *Mycobacterium* genome more frequently than *Ssp*I and *Hin*dIII, which each recognize a 6-bp site.

c. Based on these data, which enzymes would be good choices for constructing a restriction map of a chromosome (or a large segment of a chromosome) in each organism? Explain your choices.

Answer:

a. In a random sequence that is 25 percent each A, G, C, and T, the chance of finding a 6-bp site is $(\frac{1}{4})^6 = 1/4{,}096$, and the chance of finding an 8-bp site is $(\frac{1}{4})^8 = 1/65{,}536$. In such a sequence, *Apa*I, *Hin*dIII, *Sac*I, and *Ssp*I should produce fragments that average 4,096 bp in size, and *Srf*I and *Not*I should produce fragments that average 65,536 bp in size.

b. i. The large variation in average fragment sizes when one restriction enzyme is used to cleave different genomes could reflect (1) the nonrandom arrangements of base pairs in the different genomes (e.g., there is variation in the frequencies of certain sequences that are part of the restriction site in the different genomes), and/or (2) the different base compositions of the genomes (e.g., genomes that are rich in A-T base pairs should have fewer sites for enzymes recognizing sites containing only G-C base pairs).

ii. The large variation in fragment sizes when the same genome is cut with different enzymes that recognize sites having the same length could reflect (1) the nonrandom arrangement of base pairs *in that genome,* and/or (2) the base composition *of that genome.*

iii. If the sequence of *Mycobacterium tuberculosis* was random and contained 25 percent each of A, G, T, and C, enzymes recognizing a 6-bp site should produce fragments that are about 16-fold smaller than enzymes recognizing an 8-bp site. That this is not the case here suggests that at least one of these assumptions is incorrect. Two possibilities are that the genome of *Mycobacterium tuberculosis* is very rich in G-C base pairs and poor in A-T base pairs, and/or that there is a nonrandom arrangement of base pairs so that 5'-AA-3', 5'-TT-3', 5'-AT-3', and/or 5'-TA-3' sequences are rare. (The data given for *Sac*I suggest that the sites 5'-AG-3' and 5'-CT-3', which are part of the *Hin*dIII site, are not rare.)

c. Choose an enzyme that cuts infrequently, so that it becomes tractable to construct a map from the fragments that are produced, but not so infrequently that there is little information gained from the map. The answer will depend in part on the size of the fragment being mapped and the objectives of constructing the map. For example, if a 4-Mb region of the *Caenorhabditis elegans* genome is being mapped, cleavage with *Hin*dIII would produce about 1,333 fragments while cleavage with *Srf*I would produce about four fragments. It will be difficult to construct a map from 1,333 fragments, while the map constructed from just four fragments might not be very helpful. Cleavage with *Not*I would produce about 15 fragments, a number more tractable for constructing a map and that might provide the desired organizational information. Reasonable choices to consider are: for *E. coli, Apa*I, *Srf*I, or *Not*I; for *M. tuberculosis, Hin*dIII or *Ssp*I; for *A. thaliana* and *C. elegans, Not*I; for *S. cerevisiae, D. melanogaster, M. musculus,* and *H. sapiens, Srf*I or *Not*I.

18.4   STS mapping has been useful to generate clone contig maps. Perform the following exercise to consider the logistics of locating STSs using PCR.

    a.   A plasmid library contains 500-bp inserts generated from randomly sheared mouse DNA. How would you identify clones harboring STRs with the dinucleotide repeat $(AT)_N$ or the trinucleotide repeat $(CAG)_N$?

    b.   How would you use these STRs as STSs in a mapping experiment?

    c.   Nusbaum and colleagues generated a YAC-based physical map of the mouse genome by localizing 8,203 STSs onto 960 YAC clones. If each of the STSs were assayed in each of the 960 YAC clones, how many different PCRs would need to be analyzed?

    d.   Although PCRs can be performed robotically, each reaction consumes time and material resources, and with each there is a certain chance of a false positive result or other error. It is therefore advantageous to reduce the number of PCR reactions by pooling individual YACs together. First, yeast colonies containing the YACs are grown individually in the wells of ten 96-well plates. Suppose the plates are numbered I, II, . . . X and the wells of each plate are arrayed in an 8-row × 12-column grid. The rows are coded A–H and the columns coded 1–12. This allows the position of a single YAC to be specified uniquely by a code (e.g., II-C6 specifies the YAC from plate II, row C, column 6). In one pooling scheme, all YACs from each row of a plate are pooled into one well (e.g., those on plate II, A1–A12 are pooled together into a well designated II-A) and all YACs from each column of a plate are pooled into one well (those on plate II, A7–H7 are pooled together into a well designated II-7).

      i.   How many different pools would now have to be screened with each STS?

      ii. How many PCRs would have to be performed?

      iii. If the II6 and IIF pools had a positive result with STS #6239, what is the code of the YAC containing this STS?

      iv. How would you interpret a result where only the IV3 pool was positive for a particular STS?

    e.   Construct a YAC contig based on the results in the following table.

    f.   Devise a method to combine the YACs pooled in (c) to further reduce the number of PCRs. In your method, how many pools are there and how many PCRs must be performed?

STSMarker	Positive YAC Pools
63	II-6, II-A
210	II-6, II-A, IV-C, IV-3
522	VII-E, VII-12, X-G, I-C, I-8
713	I-C, I-8
714	VII-E, VII-12
719	X-H, X-9, IV-C, IV-3
991	X-H, X-9, VII-E, VII-12
1071	II-6, II-A, IV-C, IV-3, X-H, X-9
2631	II-6, II-A
3097	VII-E, VII-12, I-C, I-8
4630	VII-E, VII-12, I-C, I-8
5192	X-H, X-9, IV-C, IV-3
6193	X-H, X-9, VII-E, VII-12
6892	II-6, II-A, IV-C, IV-3

Answer:

a. Synthesize an oligonucleotide of, say, 30 bases, that is either $(AT)_N$ or $(CAG)_N$. Label one end of the oligonucleotide with ^{32}P, using polynucleotide kinase. Then use this as a probe in a colony hybridization to screen the colonies. Sequence the inserts of the positive clones using universal primer sites near the multiple cloning site of the plasmids.

b. After sequencing the positive clones, design primers (based on the unique sequences flanking the repetitive DNA sequences of each insert) for PCR. Have these synthesized and use them in PCR amplifications of genomic DNA. The sizes of the PCR products will indicate the length of the amplified segment. If an STS is polymorphic, multiple alleles corresponding to different lengths of repeats will be found in different individuals. Polymorphic STSs can be used as markers in mapping experiments. The genotypes of individuals in a cross would be determined using PCR, and linkage between two different STSs or between an STS and a phenotype (e.g., disease) can be assessed.

c. $960 \times 8,203 = 7,874,880$ PCR reactions.

d. i. Under this scheme, the 96 wells on each plate are pooled into 12 (rows) + 8 (columns), or 20 samples. For 10 plates, there would be 200 samples.

ii. For 8,203 STSs, there would be $200 \times 8,203 = 1,640,600$ PCRs.

iii. Plate II, row F, column 6.

iv. It is probably a false-positive result, an artifact.

e. STS markers that are shared by a set of YACs define the regions of the YACs that overlap. Therefore, the process of assembling the contig involves identifying which YACs share the same markers and then drawing a possible contig.

To analyze the data in this example, first infer [see (d-iii)] which YACs are positive for specific STS markers:

STSMarker	Positive YACs
63	II-6A
210	II-6A, IV-C3
522	VII-E12, I-C8 (X-G is a false positive)
713	I-C8
714	VII-E12
719	X-H9, IV-C3
991	X-H9, VII-E12
1071	II-6A, IV-C3, X-H9
2631	II-6A
3097	VII-E12, I-C8
4630	VII-E12, I-C8
5192	X-H9, IV-C3
6193	X-H9, VII-E12
6892	II-6A, IV-C3

STS markers that are shared by a set of YACs define the regions of the YACs that overlap. Rearrange the data to indicate which YACs share the same markers:

						MARKER							
YAC	63	210	522	713	714	719	991	1071	2631	3097	4630	5192	6193	6892
II-6A	X	X					X	X						X
IV-C3		X				X	X					X		X
VII-E12			X		X		X			X	X		X	
I-C8		X	X							X	X			
X-H9						X	X	X				X	X	

Rearrange the columns of this table so that markers in a single YAC are in columns positioned close together. Now the overlap between YACs becomes apparent.

							MARKER							
YAC	2631	63	210	6892	1071	5192	719	6193	991	714	3087	4630	522	713
II-6A	X	X	X	X	X									
IV-C3			X	X	X	X	X							
VII-E12								X	X	X	X	X	X	
I-C8											X	X	X	X
X-H9						X	X	X	X	X				

Now use these data to draw the contig. Note that the exact order of some sets of markers cannot be determined from these data. Some sets of markers may be in different orders than those shown below: 2631 and 63; 210 and 6892; 5192 and 719; 6193 and 991; 3097, 4630 and 522. Parentheses are shown around markers with ambiguous locations.

			MARKER				
YAC	(2631 63)	(210 6892) 1071	(5192 719)	(6193 991)	714	(3087 4630 522)	713
II-6A	———————————————						
IV-C3		———————————————					
X-H9			———————————————				
VII-E12					———————————————		
I-C8						———————————————	

f. Multiple pooling schemes are possible, but not all ensure that positive YACs can be identified unambiguously. Try combining similarly positioned wells in each of the 10 different microtiter plates: Combine IA1, IIA1, . . . XA1 into pool A1, combine IB1, IIB1, . . . XB1 into pool B1, etc., to give 96 pools. Also use the plate-row pools described in the question. This would require $96 + 80 = 176$ pools ($176 \times 8{,}203 = 1{,}443{,}728$ PCRs), almost the same as the 200 pools previously described.

18.5 BACs I-VI have been aligned in the contig shown in the following figure after screening a BAC library with a series of numbered STS and EST markers.

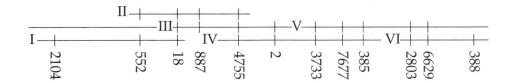

a. During the construction of BAC libraries, genomic DNA segments from different chromosomal regions sometimes are cloned together into a single BAC clone. This generates a chimeric BAC. How would you verify that none of the BACs in this contig was chimeric?

b. How would you identify the chromosomal region from which the contig derived?

c. How would you identify the orientation of the contig on a chromosome (that is, which end of the contig is closer to the telomere or centromere)?

d. How would you identify the distance (in bp) of the markers in each BAC from each other and from the end of the insert?

Answer:

a, b.   Chimeric BACs will hybridize to more than one site. Use FISH to verify that a fluorescently labeled BAC clone hybridizes to a single chromosomal site and identify the chromosomal location of this site.

c.   Label fragments at each end of the BAC-clone contig with a different fluorochrome, and perform FISH on the same chromosome preparation. The orientation of the different signals relative to the centromere and telomere will reveal the orientation of the BAC-clone contig.

d.   A laborious but precise approach is to sequence each BAC and use computerized algorithms to search for the sequences of the STSs relative to the entire BAC sequence. A faster but approximate approach is to generate a restriction map of each BAC using rare-cutting restriction enzymes and identify which restriction fragments contain an STS. This can be done by testing each restriction digest fragment for the presence of an STS using PCR or using a labeled STS as a probe in a Southern blot hybridization.

18.6   Genomes can be sequenced using a whole-genome shotgun approach or a mapping approach.

a. What is the difference between these approaches?

b. Estimate the number of sequencing reactions needed to obtain 97 percent of the sequence of the human genome using the approach taken by the Celera group.

Answer:

a. In a direct shotgun approach, the genome is broken into partially overlapping fragments, each fragment is cloned and sequenced, and the genome sequence is assembled using computerized algorithms. The Celera group used a modification of this approach to sequence the human genome. They robotically sequenced 300 to 500 bp at the ends of many clones, keeping track of which sequences were from the ends of the same clone. In total, they obtained about five times the amount of sequence of the entire genome. This ensured that they would have sufficient overlap among sequences so that they could assemble a nearly complete (97 percent) sequence.

In a mapping approach, high-resolution genetic and physical maps of the genome are obtained and integrated. This allows the position of DNA markers to be mapped onto clone contigs and gives an ordered assembly, without any gaps, of partially over-lapping clones that make up the DNA of each chromosome. For the human genome, this was done using BAC clones. The sequence of each BAC clone is obtained using a shotgun approach: The insert of the clone is cut out, sheared mechanically into a partially overlapping set of fragments appropriate for sequencing, cloned into a plasmid, the plasmid inserts sequenced, and the BAC clone sequence assembled using computerized algorithms. The contiguous sequence of each chromosome is obtained by assembling the sequences of the BAC contig.

b. Estimate the number of required sequencing reactions using estimates of the size of a haploid human genome (3×10^9 bp), the average number of bases that can be read in a sequencing reaction (500) and an aim of about 5-fold sequence coverage. In this case, obtaining 97 percent of the sequence would require $5 \times 0.97 \times (3 \times 10^9)/500 = 29.1 \times 10^6$ sequencing reactions. In fact, the assembled sequence of the Celera group consisted of 3.12×10^9 bp assembled from 26.4×10^6 sequence reads of 550 bp each, a total of 14.5×10^9 bp or 4.6-fold sequence coverage.

18.7 When Celera Genomics sequenced the human genome, they obtained 13,543,099 reads of plasmids having an average insert size of 1,951 bp, and 10,894,467 reads of plasmids having an average insert size of 10,800 bp.

a. Dideoxy-chain termination sequencing provides only about 500–550 nucleotides of sequence. About how many nucleotides of sequence did they obtain from sequencing these two plasmid libraries?

b. Why did they sequence plasmids from two libraries with different-sized inserts?

c. They only sequenced the ends of each insert. How did they determine the sequence lying between the sequenced ends?

Answer:

a. Approximately, $500 \times (13,543,099 + 10,894,467) = 1.22 \times 10^{10}$ nucleotides.

b. If a plasmid with a 2-kb insert has a unique sequence at one end but repetitive sequence at the other end, it will not be possible to continue to assemble the sequence past this plasmid, since many clones in the library have the same repetitive sequence and they come from all over the genome. Since many repetitive sequences are about 5 kb in length, sequencing plasmids with 10-kb inserts circumvents this problem. Some of the 10-kb inserts will have a sequence at one end that overlaps with the unique sequence in the plasmid with the 2-kb insert, as well as the unique sequence at their other end that lies past the repetitive element, and that can be assembled with unique sequence from other plasmids.

c. The sequence of the central region is obtained from the sequence of overlapping clones during sequence assembly.

18.8 Eukaryotic genomes differ in their repetitive DNA content. For example, consider the typical euchromatic 50-kb segment of human DNA that contains the human β T-cell receptor. About 40 percent of it is composed of various genome-wide repeats, about 10 percent encodes three genes (with introns), and about 8 percent is taken up by a pseudo-gene. Compare this to the typical 50-kb segment of yeast DNA containing the *HIS4* gene. There, only about 12 percent is composed of a genome-wide repeat, and about 70 percent encodes genes (without introns). The remaining sequences in each case are untranscribed and either contain regulatory signals or have no discernible information. Whereas some repetitive sequences can be interspersed throughout gene-containing euchromatic regions, others are abundant near centromeres. What problems do these repetitive sequences pose for sequencing eukaryotic genomes? When can these problems be overcome, and how?

Answer: Repetitive sequences pose at least two problems for sequencing eukaryotic genomes. Highly repetitive sequences associated with centromeric heterochromatin consist of short, simple repeated sequences. These are unclonable, making it impossible to obtain the complete genome sequence of organisms with them. More complex repetitive sequences such as those found within euchromatic regions can be cloned and sequenced. However, since they can originate from different genomic locations and since sequencing reads are relatively short (500 bp), the assembly of overlapping sequences can be ambiguous. In the Celera approach where each end of a clone is matched with the other, repetitive sequences are especially problematic for assembling sequences from clones with small inserts (~2 kb). Here, one end of a clone can be unique sequence while the other may be repetitive sequence. It is therefore not possible to identify an overlapping clone from the same region of the genome, since many clones from different genomic regions have the same repetitive sequence. Some of the ambiguities can be resolved by comparing these sequences to overlapping sequences generated from sequencing clones with larger inserts, say, of about 10 kb. Since many repetitive sequences are only about 5 kb in length, a larger insert is likely to contain the repetitive sequence flanked by unique sequence. Another approach is to use sequence information from the clones of an assembled clone contig (e.g., a BAC clone contig).

18.9   How has genomic analysis provided evidence that Archaea is a branch of life distinct from Bacteria and Eukarya?

Answer: Sequencing of archaeon genomes has shown that their genes are not uniformly similar to those of Bacteria or Eukarya. While most of the archaeon genes involved in energy production, cell division, and metabolism are similar to their counterparts in Bacteria, the genes involved in DNA replication, transcription, and translation are similar to their counterparts in Eukarya.

18.10   What is bioinformatics, and what is its role in structural, functional, and comparative genomics?

Answer: Bioinformatics is the use of computers and information science to study genetic information and biological structures. It results from the fusion of biology, mathematics, computer science, and information science. It has become an important field with the widespread use of databases holding sequence, genetic, functional, structural (e.g., protein structure), and bibliographic information. In structural genomics, it is used to assemble genome (and other) sequences, find genes within databases containing sequence and genetic information, align sequences to determine their degree of matching, predict gene structure, and identify the locations of putative genes. In functional genomics, it is used to predict the function of genes based on homology to DNA or amino acid sequences and to analyze information on gene function (e.g., databases containing information on how the function of a gene can alter the transcriptome or proteome). In comparative genomics, it is used to compare homologous gene sequences and compare genomes in terms of similarities and differences between gene and nongene sequences.

18.11   What is the difference between a gene and an ORF? How might you identify the functions of ORFs whose functions are not yet known?

Answer: Physically, a gene is a sequence of DNA that includes a transcribed DNA sequence and the regulatory sequences that direct its transcription (e.g., the promoter). Genes produce RNA (mRNA, rRNA, snRNA, and tRNA) and protein products. Functionally, genes can be identified by the phenotypes of mutations that alter or eliminate the functions of their products. In contrast, an ORF is a potential open reading frame, the segment of a mRNA (mature mRNA in eukaryotes) that directs the synthesis of a polypeptide by the ribosome. Therefore, genes have features that ORFs do not including transcribed and untranscribed sequences and introns. We have experimental evidence for some ORFs (cDNA sequence, detected protein products) but the existence of other ORFs is predicted only from genomic sequence information.

Two general approaches can be used to determine the function of ORFs having unknown functions. First, computerized homology searches can be performed where comparisons are made between the ORF sequence and all sequences (DNA or protein) in a database. The extent of sequence similarity is used to infer whether the ORF encodes a protein with the same or similar function to that of a gene in a database. Since proteins can have multiple functional domains (e.g., a catalytic domain and a DNA-binding domain), sometimes this approach gives only partial insight into the function of the ORF's protein product. Second, an experimental approach can be used to investigate the function of the gene identified by the ORF. In some organisms such as yeast or mice, this may involve analyzing knockout mutations and analyzing the resulting mutant phenotype. In organisms such as humans where this experimental approach would be unethical, a gene in a model organism that encodes its homolog can be investigated instead. This approach may include demonstrating that the ORF encodes a protein, characterizing its protein product, and the proteins it interacts with (e.g., using the yeast two-hybrid system described in Chapter 17).

18.12   Once a genomic region is sequenced, computerized algorithms can be used to scan the sequence to identify potential ORFs.

a.   Devise a strategy to identify potential prokaryotic ORFs by listing features assessable by an algorithm checking for ORFs.

b.   Why does the presence of introns within transcribed eukaryotic sequences preclude direct application of this strategy to eukaryotic sequences?

c.   The average length of exons in humans is about 100–200 base pairs while the length of introns can range from about 100 to many thousands of base pairs. What challenges do these findings pose for identifying exons in uncharacterized regions of the human genome?

d.   How might you modify your strategy to overcome some of the problems posed by the presence of introns in transcribed eukaryotic sequences?

Answer:

a.   Since prokaryotic ORFs should reside in transcribed regions, they should follow a bacterial promoter containing consensus sequences recognized by a sigma factor. For example, promoters recognized by σ^{70} would contain -35 (TTGACA) and -10 (TATAAT) consensus sequences. Within the transcribed region, but before the ORF, there should be a Shine-Dalgarno sequence (UAAGGAGG) used for ribosome binding.

Nearby should be an `AUG` (or `GUG`, in some systems) initiation codon. This should be followed by a set of in-frame sense codons. The ORF should terminate with a stop (`UAG`, `UAA`, `UGA`) codon.

b. Eukaryotic introns are transcribed but not translated sequences in the RNA-coding region of a gene. They will be spliced out of the primary mRNA transcript before it is translated. If not accounted for, they could introduce additional amino acids, frameshifts, and chain-termination signals.

c. The small average size of exons relative to the range of sizes for introns makes it challenging to predict whether a region with only a short set of in-frame codons is used as an exon. Such regions could have arisen by chance or be the remnants of exons that are no longer used due to mutation in splice site signals.

d. Eukaryotic introns typically contain a `GU` at their 5′ ends, an `AG` at their 3′ ends, and a `YNCURAY` branch-point sequence 18 to 38 nucleotides upstream of their 3′ ends. To identify eukaryotic ORFs in DNA sequences, scan sequences following a eukaryotic promoter for the presence of possible introns by searching for sets of these three consensus sequences. Then try to translate sequences obtained if potential introns are removed, testing whether a long ORF with good codon usage can be generated. Since alternative mRNA splicing exists at many genes, more than one possible ORF may be found in a given DNA sequence.

18.13   Annotation of genomic sequences makes them much more useful to researchers. What features should be included in an annotation, and in what different ways can they be depicted? For some examples of current annotations in databases, see the following Web sites:

http://www.yeastgenome.org/	(*S. cerevisiae*)
http://www.flybase.org and http://flybase.net/annot/	(*Drosophila*)
http://www.tigr.org/tdb/e2k1/ath1/	(*Arabidopsis*)
http://www.ncbi.nlm.nih.gov/genome/guide/human/	(*humans*)
http://genome.ucsc.edu/cgi-bin/hgGateway	(*humans*)
http://www.h-invitational.jp/	

Answer: An annotation can vary in its level of completeness, depending on what can be inferred based on homology and what experimental data are available. An annotation could include:

1. The location of a transcribed region depicted graphically with links embedded within the symbols used to depict transcripts
   a. Physical map coordinates
      i. Links to clones, ESTs, STSs, and other DNA markers in the region
   b. Genetic map coordinates
      i. Links to genes, mapping data in the region
   c. Location of introns, exons, alternative splice sites, alternative promoters, poly(A) addition sites, etc.
      i. Physical map coordinates
      ii. Graphical depictions of transcript structures

2. The types of evidence for a transcribed region
   a. Prediction based on computer algorithms that assess possible sequences of a promoter, conceptual ORF(s), appropriate splicing sites, homology with other genes, etc.

      b. Analysis of ESTs

      c. Documentation of splice sites, alternative transcript forms by comparison of cDNA and genomic sequences, etc.

  3. The inferred function of the gene product

      a. Links to reports, database entries, publications

      b. Evidence supporting the inferred function

          i. Inferences based on homology

         ii. Experimental evidence

      c. Information on pathways and processes involving the gene

  4. Information on gene expression levels

      a. Information on where the gene is expressed

      b. Information on levels of expression in different tissues

  5. Information about gene regulation

      a. Genes regulated in a similar manner

      b. Unique features of the gene and its regulation

18.14 One powerful approach to annotate genes is to compare the structures of cDNA copies of mRNAs to the genomic sequences that encode them. However, during the synthesis of cDNA (see Figure 16.8, p. 430), reverse transcriptase may not always copy the entire length of the mRNA and so a cDNA that is not full-length can be generated. This approach to gene annotation often uses ESTs that are not full-length cDNA copies of mRNA. Recently, a large collaboration involving 68 research teams analyzed 41,118 full-length cDNAs to annotate the structure of 21,037 human genes (see *http://www.h-invitational.jp/*).

  a. What types of information can be obtained by comparing the structures of cDNAs with genomic DNA?

  b. Why is it desirable, when possible, to use full-length cDNAs in these analyses?

  c. The research teams characterized the number of loci per Mb of DNA for each chromosome. Among the autosomes, chromosome 19 had the highest ratio of 19 loci per Mb while chromosome 13 had the lowest ratio of 3.5 loci per Mb. Among the sex chromosomes, the X had 4.2 loci per Mb while the Y had only 0.6 loci per Mb. What does this tell you about the distribution of genes within the human genome? How can these data be reconciled with the idea that chromosomes have gene-rich regions as well as gene deserts?

  d. The research teams were able to map 40,140 cDNAs to the current human genome sequence. Of the 978 cDNAs that could not be mapped, 907 could be roughly mapped to the mouse genome. Why might some (human) cDNAs be unable to be mapped to the current human genome sequence while they could be mapped to the mouse genome sequence? (Hint: consider where errors and limited information might exist.)

Answer:

  a. Comparison of cDNA and genomic DNA sequences can define the structure of transcription units by elucidating the location of the intron-exon boundaries, poly(A) sites, and the approximate locations of promoter regions. Comparison of different full-length cDNAs representing the same gene can identify the use of alternative splice sites, alternative poly(A) sites, and alternative promoters.

  b. The analysis of full-length cDNAs provides information about an entire open reading frame, information about the site at which transcription starts and where the promoter lies, and the location of the poly(A) site. Partial-length cDNAs might provide some but not all of this information. While multiple ESTs could be compared and assembled to obtain more information, assembling ESTs is challenging as alternative splice sites, alternative promoters and/or alternative poly(A) sites can be used.

c. Genes are not uniformly distributed among different chromosomes, and some chromosomes have more genes than others. While consistent with the finding that chromosomes have gene-rich regions and gene deserts, more data is needed to infer the relationship between the density of genes on a chromosome and how gene-rich it is. For example, a chromosome with many small genes could still have regions classified as gene deserts.

d. Two possible explanations are that (1) some regions of the genome sequence are incorrectly assembled (e.g., due to the large numbers of repetitive sequences they contain), so that the cDNAs are unable to be mapped to just one region, and (2) some of the genes are in regions that have not yet been assembled (e.g., because they are difficult to clone or sequence using current technologies). As the genome sequence is revised, these issues should be resolved.

18.15 A central theme in genetics is that an organism's phenotype results from an interaction between its genotype and the environment. Because some diseases have strong environmental components, researchers have begun to assess how disease phenotypes arise from the interactions of genes with their environments, including the genetic background in which the genes are expressed. (See *http://pga.tigr.org/desc.shtml* for additional discussion.) How might DNA microarrays be useful in a functional genomic approach to understanding human diseases that have environmental components, such as some cancers?

Answer: One approach is to use model organisms (e.g., transgenic mice) that have been developed as models to study a specific human disease. Expose them and a control population to specific environmental conditions, and then simultaneously assess disease progression and alterations in patterns of gene expression using microarrays. This would provide a means to establish a link between environmental factors and patterns of gene expression that are associated with disease onset or progression.

18.16 How does a cell's transcriptome compare with its proteome?

a. For a specific eukaryotic cell, can you predict which has more total members? Can you predict which has more unique members?

b. Suppose you are interested in characterizing changes in the pattern of gene expression in the mouse nervous system during development. Describe how you would efficiently assess changes in the transcriptome from the time the nervous system forms during embryogenesis to its maturation in the adult.

c. How would your analyses differ if you were studying the proteome?

Answer: The transcriptome is the set of RNAs present in a cell at a particular stage and time, while the proteome is the set of proteins present in the cell at that stage and time.

a. It is likely that the proteome has both more total as well as unique members. It is likely that it has more total members since multiple copies of a protein can be translated from a single mRNA transcript. It is also likely that it has more unique members since a given transcript can in principle give rise to different protein isoforms. Once translated, proteins can be posttranslationally or cotranslationally modified in different ways: phosphorylation, glycosylation, methylation, proteolytic processing, etc. If proteins translated from a single transcript are modified in different ways, multiple protein isoforms are produced.

b. This analysis could be performed using DNA chips. RNA could be isolated from the nervous system of different developmental stages, and the transcriptional profile in each stage could be assessed using DNA chips. To do this type of analysis, RNA from one developmental stage would be labeled with one fluorochrome—say, a fluorescent green dye. RNA from a second developmental stage would be labeled with another fluorochrome—say, a fluorescent red dye. The labeled probes would be hybridized to the DNA chip, and the relative red:green fluorescence bound to a single site on the chip would be used to infer the relative amounts of gene expression of the gene located at that site on the chip. This could be done for many different developmental stages.

c. For the proteome, one would need to assess changes in the proteins produced over time. Thus, one would isolate proteins from the nervous system of different developmental stages and assess the relative abundance of individual proteins using an analytical method such as quantitative two-dimensional gel electrophoresis.

18.17   When cells are exposed to short periods of heat (heat shock), they alter the set of genes they transcribe as part of a protective response.

a. What steps would you take to characterize alterations to the yeast transcriptome following a heat shock?

b. Suppose the transcriptome analyses identify a set of genes whose transcript levels increase following heat shock. How might you experimentally determine which of these genes are required for a protective response following heat shock?

Answer:

a. Use an approach similar to that taken by Pat Brown and Ira Hershowitz to follow changes in gene expression during yeast sporulation—see text discussion on p. 505 and Figure 18.11, p. 506. Innoculate two yeast cultures—one will be a reference sample and grown at the normal temperature while the other will be an experimental sample and subjected to a heat shock. Remove aliquots of the yeast cells from each culture at time points just before and after exposure of the experimental culture to a short heat shock. Isolate mRNA from each aliquot of yeast cells, label the mRNA with Cy3 (reference samples) and Cy5 (experimental samples), and allow these targets to hybridize to a probe array (a DNA chip) onto which has been spotted an entire set of yeast ORF probes generated by PCR. Scan the chip and extract the data to determine which genes show altered patterns of expression following the heat shock.

b. This question raises an important issue. Just because a gene's transcription is altered in response to heat shock does not mean that the gene's function (or absence of function) is necessary to survive the heat shock. For example, a gene could be down regulated in response to heat shock as part of a more general cellular response to limit the synthesis of new proteins that are not required to survive the heat shock. You can evaluate whether a particular gene is required to survive a heat shock by determining if a mutant lacking the gene's function can survive a heat shock.

You are fortunate to be working with yeast, as knockout mutations exist for each of the approximately 6,200 genes. See Figure 18.10, p. 504 for a description of how these mutations were induced. About 4,200 of these are viable and can be used in this approach. For each gene you identify in (a), obtain a knockout mutant strain, subject it to heat shock, and determine whether the mutant remains viable. If the mutant cannot survive a heat shock, the normal gene is required for a protective response following heat shock. If the mutant can survive a heat shock, the normal gene may not be needed for the protective response or may play only a modest role in the protective response. For genes where knockout mutations are inviable (because they eliminate essential gene functions), this approach will not work. An alternate approach is to isolate a conditional or a partial loss-of-function mutation that can be cultured and then assess whether heat shock alters the viability of these cells.

18.18   Mutations in the dystrophin gene can lead to Duchenne muscular dystrophy. The dystrophin gene is among the largest known: It has a primary transcript that spans 2.5 Mb, and produces a mature mRNA that is about 14 kb. Many different mutations in the dystrophin gene have been identified. What steps would you take if you wanted to use a DNA microarray to identify the specific dystrophin gene mutation present in a patient with Duchenne muscular dystrophy?

Answer: Isolate DNA from the blood of the patient and label it with a red fluorescent dye and label DNA from a normal individual with a green fluorescent dye. Prepare a DNA chip consisting of oligonucleotides that collectively represent the entirety of the normal dystrophin gene, and hybridize the chip with the labeled DNAs. The normal, green-labeled DNA will hybridize on all sites. The normal sequences in the patient's red-labeled DNA will hybridize to the oligonucleotides on the chip but mutated sequences will not. Consequently, normal hybridization will be seen as a yellow (red + green) spot, while a mutation will be recognized as a green spot. Identify the region of the gene corresponding to the oligonucleotide spots that have green hybridization signals to determine the site(s) of the mutation.

18.19   Distinguish between structural, functional, and comparative genomics by completing the following exercise. The following list describes specific activities and goals associated with genome analysis. Indicate the area associated with the activity or goal by placing a letter (S, structural; F, functional; C, comparative) next to each item. Some items will have more than one letter associated with them.

_____   Aligning DNA sequences within databases to determine the degree of matching

_____   Annotation of sequences within a sequenced genome

_____   Characterizing the transcriptome and proteome present in a cell at a specific developmental stage or in a particular disease state

_____   Comparing the overall arrangements of genes and nongene sequences in different organisms to understand how genomes evolve

_____   Describing the function of all genes in a genome

_____   Determining the functions of human genes by studying their homologs in nonhuman organisms

_____   Developing a comprehensive two-dimensional polyacrylamide gel electrophoresis map of all proteins in a cell

_____   Developing a physical map of a genome

_____   Developing DNA microarrays (DNA chips)

_____ Identifying homologs to human disease genes in organisms suitable for experimentation

_____ Identifying a large collection of simple tandem repeat or microsatellite sequences to use as DNA markers within one organism

_____ Identifying expressed sequence tags

_____ Making gene knockouts and observing the phenotypic changes associated with them

_____ Mapping a gene in one organism using the lod score method

_____ Sequencing individual BAC clones aligned in a contig using a shotgun approach

_____ Using oligonucleotide hybridization analysis to type an SNP

Answer:

S, F, C	Aligning DNA sequences within databases to determine the degree of matching
F, C	Annotation of sequences within a sequenced genome
F	Characterizing the transcriptome and proteome present in a cell at a specific developmental stage or in a particular disease state
C	Comparing the overall arrangements of genes and nongene sequences in different organisms to understand how genomes evolve
F	Describing the function of all genes in a genome
F, C	Determining the functions of human genes by studying their homologs in non-human organisms
F	Developing a comprehensive two-dimensional polyacrylamide gel electrophoresis map of all proteins in a cell
S	Developing a physical map of a genome
S, F, C	Developing DNA microarrays (DNA chips)
F, C	Identifying homologs to human disease genes in organisms suitable for experimentation
S	Identifying a large collection of simple tandem repeat, or microsatellite sequences to use as DNA markers within one organism
S, F, C	Identifying expressed sequence tags (ESTs)
F, C	Making gene knockouts and observing the phenotypic changes associated with them
S	Mapping a gene in one organism using the lod score method
S	Sequencing individual BAC or PAC clones aligned in a contig using a shotgun approach
S	Using oligonucleotide hybridization analysis to type an SNP

18.20 Genomic analyses of *Mycoplasma genitalium* have not only identified its 523 genes but also allowed identification of the genes that the organism needs for life. The analyses are presented in detail at the Web site of the Institute for Genomic Research (*http://www.tigr.org/minimal/*). Explore this Web site and then address the following questions.

a. How many protein-coding genes have unknown functions?

b. How do we know which genes are required for life?

c. Are any of the protein-coding genes with unknown functions required for life?

d. What functional genomics approaches might be helpful to discern their function?

e. What practical value might there be identifying the minimal gene set needed for life?

f. Is it possible to synthesize DNA sequences containing all the genes identified as essential for life and assemble these on one chromosome? Would this be enough to have a template for a new life form? If not, what additional sequence information would be needed to generate a new life form?

g. What is the difference between the life form generated in (f) and *Mycoplasma genitalium*?

h. What ethical issues arise from these analyses? Who should decide how to address them?

Answer:

a. There are about 180 genes with no known homologs outside of the mycoplasmas, which makes it difficult to assign their function.

b. Both theoretical and experimental paradigms can be used to provide evidence for which genes are required for life.

   A theoretical approach is to estimate how many genes are conserved across large phylogenetic distances, making the assumption that highly conserved genes provide essential functions. Comparison of the *Haemophilus influenzae* and *M. genitalium* genomes suggest that about 265 genes are needed. Comparison of two *Mycoplasma* species (*M. pneumoniae* and *M. genitalium*) shows that *M. pneumoniae* has all 480 of the *M. genitalium* protein-coding genes, plus an additional 197 genes.

   To test which of these 480 are essential for life an experimental approach can be used. One paradigm is to ask whether a mutation that "knocks out" a gene causes a visible mutation, affects the organism's ability to grow, or affects its viability (ability to live). Hutchison et al. used transposons (DNA segments capable of moving and inserting themselves in other sites) to estimate the size of the minimal mycoplasma genome. Their work suggested that between 265 and 300 of the 480 protein-coding genes were essential under laboratory conditions for growth. It is unclear whether any of the genes not essential under laboratory conditions might still be essential for growth in a human host. Thus, the approach begs the question—what "quality" of life?—of "life" under what conditions?

c. Yes. There are 180 genes without a functional assignment. While 69 of these could be disrupted by a transposon insertion without affecting the viability of the organism, 111 could not. This suggests that at least some of them are not dispensable, since if they were dispensable, transposon disruptions in them would have been recovered. The inference from this experiment is that at least some of them must encode essential cellular functions.

d. Multiple functional genomic approaches could be helpful to infer their function. Note that while the knock-out approach using transposon disruption suggested that these genes have an essential function, the results do *not* provide clues as to what they are. To understand the functions of individual genes, it might be helpful to understand when they are transcribed and whether they are coordinately transcribed with other genes during the life cycle of the organism. This could provide some information about the cellular pathways utilizing these genes and so indirectly allow inferences as to their function. For this purpose, gene chips could be used to assess their expression profiles during growth or under different experimental conditions. For example, one might glean some information about the function of a gene by learning that its expression was up- or down-regulated during a particular growth phase, or in a mutant strain where a known function was disrupted. An alternate approach would be to generate mutations in a single gene (whose function is unknown) that did not disrupt its function entirely (e.g., by using a chemical mutagen) and then assess which genes had altered expression patterns as a result.

e. In principle, this would allow the design of cells that could be used as bioreactors to synthesize specific complex biological products that could be useful in improving health, crops, or some other aspect of human existence. It might also provide a basis for designing cells that could be used in implementing some biological or chemical process (e.g., as agents to mediate gene therapy, as agents that could be designed with features to use in toxic waste cleanup, as agents for germ warfare).

f. It is not likely that the assembly of the sequences, by itself, would be sufficient for generating a template for a new life form. Encoded in even a simple genome would have to be instructions for regulating the expression of its genes. Even in the smallest bacteriophage, there are enough instructions encoded to allow for the ordered synthesis of a set of genes and the propagation of a life cycle. In a "cell" there must be at least these types of instructions, which would occur minimally at the level of transcriptional control.

g. *M. genitalium* is able to respond to its environment and grow as an autonomous cell. It has a set of genes that are "fitted" into regulatory circuits that interact. The "life form" generated in (f) may make a set of gene products, but this may not "make" a cell. The cell is more than just the sum of the proteins it possesses.

h. There are multiple ethical issues. One can list some of them as a series of unanswered questions.

i. Is there significant potential for the generation of a biohazard that could adversely affect existing life?

ii. How does the possibility of generating a "living" cell affect our religious and moral beliefs about the origin and sanctity of life?

iii. What are the benefits of pursuing this line of inquiry, relative to our goals for improving human health or the environment?

iv. Do we know enough about the safety, benefits, and risks to venture into such an undertaking?

v. Since it is difficult to understand what the risks of this venture are, how does one evaluate how to proceed?

It would be important that scientists, ethicists, the public, and representative groups who might be affected by this work be part of a team that discussed how to proceed and how to address ethical and other (legal, economic, biohazard) issues that might arise from these investigations. Even with this varied level of input, it might not be possible to anticipate all of the consequences of these types of investigations. There are both pragmatic as well as more philosophical concerns.

18.21 Comparative genomics offers insights into the relationship between homologous genes and the organization of genomes. When the genome of *C. elegans* was sequenced, it was striking that some types of sequences were distributed nonrandomly. Consider the data obtained for chromosome V and the X chromosome shown here. The following figure shows the distribution of genes, the distribution of inverted and tandem repeat sequences, and the location of ESTs in *C. elegans* that are highly similar to yeast genes.

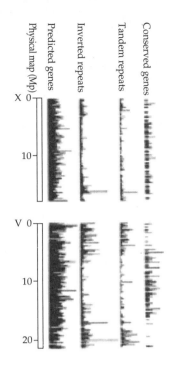

a. How do the distributions of genes, inverted and tandem repeat sequences, and conserved genes compare?

b. Based on your analysis in (a), what might you hypothesize about the different rates of DNA evolution (change) on the arms and central regions of autosomes in *C. elegans*?

c. Curiously, meiotic recombination (crossing-over, discussed in Chapter 6) is higher on the arms of autosomes, with demarcations between regions of high and low crossing-over at the boundaries between conserved and nonconserved genes seen in the physical map. Does this information support your hypothesis in (b)?

Answer:

a. On chromosome V and the X chromosome, genes are distributed uniformly. However, especially on chromosome V, conserved genes are found more frequently in the central regions. In contrast, inverted and tandem-repeat sequences are found more frequently on the arms. It appears that at least on chromosome V, there is an inverse relationship between the frequency of inverted and tandem-repeats and the frequency of conserved genes.

b. Since there are fewer conserved genes on the arms, there appears to be a greater rate of change on chromosome arms than in the central regions.

c. Yes, since increased meiotic recombination provides for greater rates of exchange of genetic material on chromosomal arms.

18.22 What is the difference between a DNA chip and a protein chip? What different types of protein arrays are there, and how are they used to analyze the proteome?

Answer: A DNA chip has DNA probes (oligonucleotides, PCR-amplified cDNA products) bound to a solid substrate (a glass slide, membrane, microtiter well, or silicon chip), while a protein chip has proteins immobilized on solid substrates. Protein arrays are probed by labeling target proteins with fluorescent dyes, incubating the labeled target with the probe array, and measuring the bound fluorescence using automated laser detection.

One type of protein chip is a capture array, where a set of antibodies is bound to a solid substrate and used to evaluate the level and presence of target molecules in cell or tissue extracts. A capture array can be used in disease diagnosis (to evaluate whether a specific protein associated with a disease state is present) and in protein expression profiling (evaluation of the proteome qualitatively and quantitatively). Another type of protein chip is the large-scale protein array, where a large number of purified proteins are spotted onto an array substrate and used to assay one of a wide range of biochemical functions, including protein-protein interactions and drug-target interactions. In contrast to the antibodies that are spotted onto capture arrays, large-scale protein arrays are spotted with proteins produced and purified from an expression library.

19

REGULATION OF GENE EXPRESSION IN BACTERIA AND BACTERIOPHAGES

CHAPTER OUTLINE

THINKING ANALYTICALLY

This chapter presents how the expression of bacterial and bacteriophage genes is regulated. While the regulation of a particular set of genes can be complex, it is invariably based on a general paradigm: The structural organization of a set of genes is related to both their induction by effector molecules and subsequent expression (as diagrammed in Figure 16.1). As you study the regulation of each set of genes, keep this general principle in mind. Reconsider it with each of the operons; analyze how it is used and elaborated upon.

Once you have a solid understanding of the general principle underlying prokaryotic gene expression, the challenge is to understand and retain the complex details of the regulatory circuitry. It will help to first study the text figures and then diagram (eventually, from memory) the structure of each operon and the role of regulatory factors within it. This is especially valuable when considering the regulation of genes in phage λ. This will take time, as it requires repeated practice and comparative review.

REVIEW OF KEY TERMS, SYMBOLS, AND CONCEPTS

In your own words, write a brief, precise definition of each term in the groups below. Check your definitions using the text. Then develop a concept map using the terms in each list.

1	2	3
regulated genes	*lac, trp* operons	lambda (λ) phage
constitutive genes	*lacA, -Z, -Y, -I* genes	lytic pathway
inducible, repressible genes, operons	negative, positive control	lysogenic pathway
inducers and effectors	allolactose	genetic switch
controlling site	partial diploids	lambda repressor
housekeeping genes	cis-dominance	*cro* gene
induction	trans-dominance	*recA* gene
coordinate induction	$O^c, I^-, I^s, I^{-d}, I^Q, I^{SQ}$ mutations	antiterminator
operon	superrepressor	$P_R, P_{RE}, P_{RM}, P_L, P_I$
polycistronic mRNA	allosteric shift	O_R, O_L
polygenic mRNA	catabolite repression	antitermination
effector molecule	glucose effect	readthrough transcription
operator	CAP-cAMP complex	integrase
repressor molecule	CAP site	*cI, cII, cIII* genes
promoter	adenylate cyclase	Q, Cro, N, O, P proteins
polarity	III^{Glc}	induction by UV light
polar effects	effector	
polar mutations	aporepressor	
missense mutation	attenuation	
nonsense mutation	attenuator site (*att*)	
	pause signal	
	antitermination	
	feedback inhibition	
	allosteric shift	
	inducible, repressible operon	

You can strengthen your understanding of the principles underlying prokaryotic gene expression and your detailed knowledge of the regulatory circuitry of operons by examining the consequences of mutations in different operon elements. First, address why mutations in regulatory elements such as the promoter or operator will be cis-dominant. Second, address why some mutations in the genes for diffusible regulatory factors *can* be trans-dominant. Third, consider how the properties of different mutations in a specific operon can provide you with a basis to develop a model of its structure. Finally, relate the regulation of the operon to how it functions within a biochemical or growth pathway.

QUESTIONS FOR PRACTICE

Multiple Choice Questions

1. Genes that respond to the needs of a cell or organism in a controlled manner are known as
   a. inducer genes
   b. effector genes
   c. regulated genes
   d. constitutive genes

2. A mutation that causes a gene always to be expressed, irrespective of what the environmental conditions may be, is known as a(n)
   a. inducer mutation
   b. effector mutation
   c. regulator mutation
   d. constitutive mutation

3. A gene that is stimulated to undergo transcription in response to a particular molecular event that occurs at a controlling site near that gene is said to be
   a. inducible
   b. constitutive
   c. a promoter
   d. an inducer

4. An operon that is inducible may be under
   a. positive control only
   b. negative control only
   c. constitutive control only
   d. both positive and negative control

5. The effector molecule that induces the *lac* protein-coding genes is
   a. lactose
   b. allolactose
   c. glucose
   d. β-galactosidase

6. Before transcription of the *lac* operon can occur, RNA polymerase must bind strongly to the promoter. This happens when
   a. CAP binds to the CAP site in the promoter
   b. a CAP-cAMP complex binds to the CAP site in the promoter
   c. catabolite repression occurs
   d. $lacI^+$ is mutated to $lacI^-$

7. Which of the following is an example of an effector molecule acting via positive control?
   a. lactose inducing the *lac* operon
   b. glucose causing catabolite repression
   c. tryptophan attenuating the *trp* operon
   d. the λ cI gene product, at a certain cellular level, favoring the lysogenic pathway

8. Bacterial operons containing protein-coding genes for the synthesis of amino acids, such as the tryptophan operon, are customarily classified as
   a. negatively controlled
   b. positively controlled
   c. repressible operons
   d. inducible operons

9. Ultraviolet light induces integrated phage λ to enter the lytic pathway by
   a. cleaving repressor monomers
   b. initiating transcription of the cro gene
   c. converting the RecA protein to a protease
   d. all of the above, directly or indirectly

10. Which of the following is *not* critical to the genetic switch controlling the choice between the lysogenic and lytic pathways in phage λ?
   a. the competition between the products of the *cI* gene and the *cro* gene
   b. transcription of the genes *J, K, L,* and *M*
   c. the concentration of λ repressor protein dimers
   d. whether RNA polymerase binds to P_{RM} or P_R

Answers: 1c, 2d, 3a, 4d, 5b, 6b, 7b, 8c, 9d, 10b

Thought Questions

1. What three proteins are synthesized when lactose is the sole carbon source in *E. coli*? What does each do?

2. What are the roles of *lacA, lacI, lacO, lacY,* and *lacZ* in *E. coli* carbohydrate metabolism? In what order are they arranged in the DNA molecule?

3. Describe the sequence of events that occurs in the *lac* operon when *E. coli* is grown in the presence of both glucose and lactose.

4. Distinguish between cis-dominant mutations and trans-dominant mutations. Why is *lacO*^c cis-dominant but *lacI*^s trans-dominant? Would a mutation that led to constitutive expression of the *cI* gene be cis- or trans-dominant?

5. How is attenuation related to the coupling of transcription and translation in prokaryotes?

6. What are the fundamental differences between the *lac* and *trp* operons in *E. coli*?

7. How do structural changes in DNA or RNA conformation play a role in:
   a. catabolite repression by glucose and CAP-cAMP?
   b. antitermination and termination in attenuation of the *trp* operon?

8. What is catabolite repression?

9. In what different or similar ways are allosteric shifts used in the regulation of inducible and repressible operons?

10. Speculate as to what phenotype each of the following λ mutations would have: (a) a mutation in the *cI* gene that resulted in repressor molecules that bound O_L and O_R less well; (b) a mutation in the *cro* gene that resulted in no Cro protein being produced; (c) a mutation in O_R that resulted in decreased affinity for repressor molecules.

Thought Questions for Media

After reviewing the media on the *iGenetics* CD-ROM, try to answer these questions.

1. In the *lac* operon is the operator wholly within, wholly outside of, or partially within the promoter?

2. How could you distinguish between promoter and operator mutations?

3. An operon is defined as a cluster of genes along with an adjacent promoter and operator that controls the transcription of those genes.
   a. Must all operons be polycistronic?
   b. Must all of the genes in one operon be transcribed together?

4. How does lactose mediate de-repression of the *lac* operon? Does it act directly?

5. Which type of mutation in the *lac* operon causes the production of β-galactosidase whether or not lactose is present?

6. Why does it make sense for *E. coli* to transcribe the *lac* operon at a very slow rate in the presence of glucose? (Consider the products of β-galactosidase enzymatic activity.)

7. What phenotype would you expect of a mutation in the CAP site in the *lac* promoter?

1. In the iActivity, the strain containing the O^c mutation ($I^+ O^c Z^+$) showed constitutive expression both by itself and as a merozygote of the form $I^+ O^c Z^+/F' I^+ O^+ Z^+$. Would the phenotype be different if the merozygote had a genotype that was either $I^+ O^c Z^+/F' I^+ O^+ Z^-$ or $I^+ O^c Z^+/F' I^- O^+ Z^+$? For each case, why or why not?

2. In the iActivity, the strain containing the I^s mutation ($I^s O^+ Z^+$) showed repressed expression both by itself and as a merozygote of the form $I^s O^+ Z^+/F' I^+ O^+ Z^+$. Would the phenotype be different if the merozygote had a genotype that was either $I^s O^+ Z^+/F' I^+ O^+ Z^-$ or $I^s O^+ Z^+/F' I^- O^+ Z^+$? For each case, why or why not?

3. In the iActivity, you analyzed data showing that the *lac* operon was inducible and under negative control.
   a. In theory, if you can experimentally demonstrate that an operon is inducible, must it also be under negative control?
   b. Can an inducible operon be under both negative and positive control?
   c. Can an inducible operon be under only positive control?

SOLUTIONS TO TEXT PROBLEMS

19.1   What is meant by constitutive gene expression? How is constitutive gene expression unlike regulated gene expression?

Answer: Genes that are always active in growing cells are constitutive, or housekeeping genes. Thus, constitutive gene expression refers to gene expression that is constantly present, while regulated gene expression refers to gene expression that is controlled in response to the needs of a cell or organism. We tend to think of genes that show constitutive gene expression as also having a fixed, unchanging level of expression, although this may not always be true. In fact, all genes are regulated at some level.

19.2   Give two examples of effector molecules, and discuss how effector molecules function to regulate gene expression.

Answer: Allolactose and tryptophan are effector molecules that regulate the *lac* and *trp* operons, respectively. Effectors cause allosteric shifts in repressor proteins to alter their affinity for operator sites in DNA. When allolactose binds to the *lac* repressor, it loses its affinity for the *lac* operator, inducing transcription at the *lac* operon. When tryptophan interacts with the *trp* aporepressor, it is converted to an active repressor that can bind the *trp* operator, repressing transcription at the *trp* operon.

19.3   Operons produce polygenic mRNA when they are active. What is a polygenic mRNA? What advantages, if any, does it confer in terms of the cell's function?

Answer: Polygenic mRNAs contain coding information for more than one protein. These mRNAs are transcribed from operons that contain several genes encoding related functions, such as catalyzing steps of a biosynthetic pathway. One advantage conferred by utilizing such mRNAs is that cells can regulate all of the steps of a pathway coordinately. By using a polygenic mRNA, the synthesis of each of a set of enzymes acting in one pathway can be produced by a single regulatory signal.

19.4   How does lactose trigger the coordinate induction of the synthesis of β-galactosidase, permease, and transacetylase? Why does the synthesis of these enzymes not occur when glucose is also in the medium?

Answer: The addition of lactose to *E. coli* cells brings about a rapid synthesis of these three enzymes by the induction of a single promoter that lies upstream from the genes for these three enzymes. The three genes are part of a *lac* operon that is transcribed as a single unit. When lactose is added, it is metabolized (isomerized) to allolactose, which binds to a repressor protein. Without bound allolactose, the repressor protein blocks transcription from the *lac* promoter. Hence, the *lac* operon is normally under a negative control mechanism. When allolactose is bound, the repressor protein is inactivated and is unable to bind to the operator site to block transcription. As a result, RNA polymerase binds to the promoter and initiates transcription of a single mRNA that encodes all three proteins.

One of the enzymes that is synthesized, β-galactosidase, cleaves lactose to produce glucose and galactose (which is converted to glucose in a subsequent enzymatic step). Consequently, if glucose is present in the medium, it is redundant to induce the *lac* operon. Glucose blocks induction of the *lac* operon by utilizing a positive control mechanism. In this catabolite repression, glucose causes a great reduction in the amount of cAMP in the cell. For normal induction of the *lac* operon, cAMP must complex with a CAP (catabolite gene activator protein) that in turn binds to a CAP site upstream of the *lac* promoter and activates transcription. In the absence of cAMP, the cAMP-CAP complex is absent, and so transcription cannot be activated.

19.5   An *E. coli* mutant strain synthesizes β-galactosidase whether or not the inducer is present. What genetic defect(s) might be responsible for this phenotype?

Answer: There are two possibilities. First, the repressor protein bound by the inducer could be unable to bind to the operator (i.e., the repressor is nonfunctional and the strain is I^-), so that its presence or absence makes no difference. Second, there could be base-pair alterations in the operator region that make it unrecognizable by the repressor protein (i.e., the operon is constitutively expressed and the strain is O^c).

19.6   How did the discovery of polarity and polar mutations contribute to the formulation of the operon hypothesis? How do polar mutations prevent the initiation of translation at downstream genes in a polycistronic mRNA?

Answer: In genes that are not organized into operons, nonsense mutations affect only the function of the product of the gene with the mutation. Polar mutations are nonsense mutations that affect the function of genes in the cluster that lie downstream from the mutation. For example, nonsense mutations in *lacZ* knock out the function not only of β-galactosidase, but also that of permease and transacetylase. The discovery of polarity indicated that all three genes—β-galactosidase, permease, and transacetylase—are clustered together and transcribed into a single RNA molecule—a polycistronic mRNA—rather than three separate mRNAs.

When a ribosome encounters a nonsense mutation in an upstream gene from a polycistronic mRNA, it stops translation and releases the partially completed, nonfunctional protein. Although it will continue to slide along the polycistronic mRNA for a short while, it will typically dissociate from the mRNA before the next start codon. Thus, the downstream gene(s) will not be translated and a polar effect will be observed.

19.7   Distinguish the effects you would expect from (a) a missense mutation and (b) a nonsense mutation in the *lacZ* (β-galactosidase) gene of the *lac* operon.

Answer:

a. A missense mutation results in partial or complete loss of β-galactosidase activity, but no loss of permease and transacetylase activities.

b. Unless the nonsense mutation is very close to the normal chain-terminating codon for β-galactosidase, it is likely to have polar effects. Therefore, permease and transacetylase activities would be lost in addition to the loss of β-galactosidase activity.

19.8   Elucidation of the regulatory mechanisms associated with the enzymes of lactose utilization in *E. coli* was a landmark in our understanding of regulatory processes in microorganisms. In formulating the operon hypothesis as applied to the lactose system, Jacob and Monod found that results from particular partial diploid strains were invaluable. In terms of their operon hypothesis, what specific information did analyses of partial diploids provide that analyses of haploids could not?

Answer: The use of partial diploids allowed observation of the consequences of placing sequences in *trans* and in *cis*. Partial diploids were used to show that some regulatory sequences must lie in cis to *lacZ* (the *lacO* region upstream of the *lacZ* gene) to exert a regulatory effect. For example, O^c mutations are cis-acting: They cause constitutive activation of the *lac* promoter that they lie upstream from, but do not activate any other promoter. Partial diploids also were used to show that the *lacI* gene encoded a trans-acting factor (a diffusible product that could bind to the *lacO* region) and that promoter function did not require a diffusible substance. For example, a *lacI*$^+$ gene on a plasmid could function in trans to regulate a *lac* promoter in cis to a *lacI*$^-$ gene.

19.9   For the *E. coli lac* operon, write the partial diploid genotype for a strain that will produce β-galactosidase constitutively and permease by induction.

Answer: To observe such a phenotype, there are three requirements: (1) A functional *lacZ* and a nonfunctional *lacY* gene must both lie downstream from a functional operator and promoter; (2) a nonfunctional *lacZ* and a functional *lacY* gene must lie downstream from an inducible (i.e., O^+) promoter; and (3) a functional repressor gene must be present in the cell (I^+). A genotype that satisfies these requirements is the partial diploid *lacI*$^+$ *lacO*^{c} *lacP*$^+$ *lacZ*$^+$ *lacY*$^-$/*lacI*$^+$ *lacO*$^+$ *lacP*$^+$ *lacZ*$^-$ *lacY*$^+$. Only one *lacI*$^+$ gene is required, so one may be *lacI*$^-$.

19.10   Mutants were instrumental in elaborating the model for regulation of the *lac* operon.
a. Discuss why $P_{lac}-$ and *lacO*^{c} mutants are cis-dominant but not trans-dominant.
b. Explain why *lacI*^{s} and *lac*$^{-d}$ mutants are trans-dominant to the wild-type *lacI*$^+$ allele but *lacI*$^-$ mutants are recessive.
c. Discuss the consequences of mutations in the repressor gene promoter as compared with mutations in the structural gene promoter.

Answer:
a. Mutants that are *lacO*^{c} are mutants in the operator region that is normally bound by the repressor protein. The *lacO*^{c} mutants result from a DNA alteration that precludes

the repressor protein from binding the operator. Since the repressor normally blocks transcription initiation from a downstream promoter, *lacO*^{c} mutants result in constitutive transcription at that promoter. Since the operator acts only on an adjacent, and not any other, promoter, mutations in the operator are only cis-dominant. The *lacO*^{c} has no effect on other lactose operons in the same cell because *lacO*^{c} does not code for a product that could diffuse through the cell and affect other DNA sequences.

b. In wild-type strains, *lacI* encodes a repressor that can block transcription at the *lac* operon by binding to an operator region. By binding the operator, the repressor blocks RNA polymerase from binding to the promoter. This activity of the repressor can be altered if allolactose is present, which binds to the repressor and inhibits it from binding the operator. The superrepressor (*lacI*^{s}) mutation results in a repressor protein that can bind the operator, but cannot bind allolactose. Once it binds to the operator, it cannot leave the operator, and transcription is always blocked. This results in a dominant mutation, since even if normal repressor molecules (made by *lacI*$^+$) are present, the superrepressor molecules do not vacate the operator region, and the operon cannot be induced. On the other hand, *lacI*$^-$ mutants either do not make repressor protein or make repressor that is unable to bind to the operator. In a partial diploid that has both *lacI*$^-$ and *lacI*$^+$ genes, repressor proteins are made (by *lacI*$^+$) and are capable of diffusing to any *lac* operator region to regulate a *lac* operon. Hence, *lacI*$^-$ is recessive to *lacI*$^+$ because the defect caused by the absence of the repressor proteins in *lacI*$^-$ mutants can be overcome by the synthesis of diffusible repressor protein from the *lacI*$^+$ gene.

c. The *lacI* gene is constitutively expressed at a low level. Mutations in the repressor gene promoter result in either an increase or a decrease in the level of expression of the repressor gene. They will not affect the structure of the repressor molecule, but will affect its cellular concentration. If the mutation causes a complete loss of expression of the repressor protein, it would have the same phenotype as an *I*$^-$ mutant, since no functional repressor would be produced. These mutants would result in constitutive expression of *lacZ*, *lacY*, and *lacA* and be recessive to *I*$^+$ in partial diploids. If the promoter mutation only partially decreases transcription below normal levels, there will be increased expression of *lacZ*, *lacY*, and *lacA* in the absence of inducer. If the promoter mutation increases transcription above normal levels (the *lacI*^{Q} and *lacI*^{SQ} mutations presented in the text), a large number of repressor molecules will be produced. Such mutants will reduce the efficiency of induction of the *lac* operon by allolactose, and will be trans-dominant.

19.11   This question involves the *lac* operon of *E. coli*, where *I* = *lacI* (the repressor gene), *P* = *P*$_{lac}$ (the promoter), *O* = *lacO* (the operator), *Z* = *lacZ* (the β-galactosidase gene), and *Y* = *lacY* (the permease gene). Complete Table 19.A, using + to indicate that the enzyme in question will be synthesized and − to indicate that the enzyme will not be synthesized.

Answer: The answer is given in the table on page 463.

Table 19.A

		Inducer Absent		Inducer Present	
	Genotype	β-Galactosidase	Permease	β-Galactosidase	Permease
a.	$I^+ P^+ O^+ Z^+ Y^+$				
b.	$I^+ P^+ O^+ Z^- Y^+$				
c.	$I^+ P^+ O^+ Z^+ Y^-$				
d.	$I^- P^+ O^+ Z^+ Y^+$				
e.	$I^S P^+ O^+ Z^+ Y^+$				
f.	$I^+ P^+ O^c Z^+ Y^+$				
g.	$I^S P^+ O^c Z^+ Y^+$				
h.	$I^+ P^+ O^c Z^+ Y^-$				
i.	$I^{-d} P^+ O^+ Z^+ Y^+$				
j.	$\dfrac{I^- P^+ O^+ Z^+ Y^+}{I^+ P^+ O^+ Z^- Y^-}$				
k.	$\dfrac{I^- P^+ O^+ Z^+ Y^-}{I^+ P^+ O^+ Z^- Y^+}$				
l.	$\dfrac{I^S P^+ O^+ Z^+ Y^-}{I^+ P^+ O^+ Z^- Y^+}$				
m.	$\dfrac{I^+ P^+ O^c Z^- Y^+}{I^+ P^+ O^+ Z^+ Y^-}$				
n.	$\dfrac{I^- P^+ O^c Z^+ Y^-}{I^+ P^+ O^+ Z^- Y^+}$				
o.	$\dfrac{I^S P^+ O^+ Z^+ Y^+}{I^+ P^+ O^c Z^+ Y^+}$				
p.	$\dfrac{I^{-d} P^+ O^+ Z^+ Y^-}{I^+ P^+ O^+ Z^- Y^+}$				
q.	$\dfrac{I^+ P^- O^c Z^+ Y^-}{I^+ P^+ O^+ Z^- Y^+}$				
r.	$\dfrac{I^+ P^- O^+ Z^+ Y^-}{I^+ P^+ O^c Z^- Y^+}$				
s.	$\dfrac{I^- P^- O^+ Z^+ Y^+}{I^+ P^+ O^+ Z^- Y^-}$				
t.	$\dfrac{I^- P^+ O^+ Z^+ Y^-}{I^+ P^- O^+ Z^- Y^+}$				

	Genotype	Inducer Absent		Inducer Present	
		β-Galactosidase	Permease	β-Galactosidase	Permease
a.	$I^+ P^+ O^+ Z^+ Y^+$	−	−	+	+
b.	$I^+ P^+ O^+ Z^- Y^+$	−	−	−	+
c.	$I^+ P^+ O^+ Z^+ Y^-$	−	−	+	−
d.	$I^- P^+ O^+ Z^+ Y^+$	+	+	+	+
e.	$I^S P^+ O^+ Z^+ Y^+$	−	−	−	−
f.	$I^+ P^+ O^c Z^+ Y^+$	+	+	+	+
g.	$I^S P^+ O^c Z^+ Y^+$	+	+	+	+
h.	$I^+ P^+ O^c Z^+ Y^-$	+	−	+	−
i.	$I^{-d} P^+ O^+ Z^+ Y^+$	+	+	+	+
j.	$\dfrac{I^- P^+ O^+ Z^+ Y^+}{I^+ P^+ O^+ Z^- Y^-}$	−	−	+	+
k.	$\dfrac{I^- P^+ O^+ Z^+ Y^-}{I^+ P^+ O^+ Z^- Y^+}$	−	−	+	+
l.	$\dfrac{I^S P^+ O^+ Z^+ Y^-}{I^+ P^+ O^+ Z^- Y^+}$	−	−	−	−
m.	$\dfrac{I^+ P^+ O^c Z^- Y^+}{I^+ P^+ O^+ Z^+ Y^-}$	−	+	+	+
n.	$\dfrac{I^- P^+ O^c Z^+ Y^-}{I^+ P^+ O^+ Z^- Y^+}$	+	−	+	+
o.	$\dfrac{I^S P^+ O^+ Z^+ Y^+}{I^+ P^+ O^c Z^+ Y^+}$	+	+	+	+
p.	$\dfrac{I^{-d} P^+ O^+ Z^+ Y^-}{I^+ P^+ O^+ Z^- Y^+}$	+	+	+	+
q.	$\dfrac{I^+ P^- O^c Z^+ Y^-}{I^+ P^+ O^+ Z^- Y^+}$	−	−	−	+
r.	$\dfrac{I^+ P^- O^+ Z^+ Y^-}{I^+ P^+ O^c Z^- Y^+}$	−	+	−	+
s.	$\dfrac{I^- P^- O^+ Z^+ Y^+}{I^+ P^+ O^+ Z^- Y^-}$				
t.	$\dfrac{I^- P^+ O^+ Z^+ Y^-}{I^+ P^- O^+ Z^- Y^+}$	−	−	+	−

19.12   A new sugar, sugarose, induces synthesis of two enzymes from the *sug* operon of *E. coli*. Some properties of deletion mutations affecting the appearance of these enzymes are as follows (here, "+" = enzyme induced normally, i.e., synthesized only in the presence of the inducer; C = enzyme synthesized constitutively; 0 = enzyme cannot be detected):

Mutation of	Enzyme 1	Enzyme 2
Gene *A*	+	0
Gene *B*	0	+
Gene *C*	0	0
Gene *D*	C	C

a. The genes are adjacent, in the order *A B C D*. Which gene is most likely to be the structural gene for enzyme 1?

b. Complementation studies using partial diploid (*F'*) strains were made. The extrachromosomal element (*F'*) and chromosome each carried one set of *sug* genes. The results were as follows:

Genotype of F'	Chromosome	Enzyme 1	Enzyme 2
$A^+ B^- C^+ D^+$	$A^- B^+ C^+ D^+$	+	+
$A^+ B^- C^- D^+$	$A^- B^+ C^+ D^+$	+	0
$A^- B^+ C^- D^+$	$A^+ B^- C^+ D^+$	0	+
$A^- B^+ C^+ D^+$	$A^+ B^- C^+ D^-$	+	+

From all the evidence given, determine whether the following statements are true or false:
i.   It is possible that gene *D* is a structural gene for one of the two enzymes.
ii. It is possible that gene *D* produces a repressor.
iii. It is possible that gene *D* produces a cytoplasmic product required to induce genes *A* and *B*.
iv. It is possible that gene *D* is an operator locus for the *sug* operon.
v.   The evidence is also consistent with the possibility that gene *C* could be a gene that produces a cytoplasmic product required to induce genes *A* and *B*.
vi. The evidence is also consistent with the possibility that gene *C* could be the controlling end of the *sug* operon (the end from which mRNA synthesis presumably commences).

Answer: First consider the data in general terms. Mutations in genes *A* or *B* result in a loss of one, but not both, enzyme activities. These are likely to be structural genes for the enzymes (*B* = enzyme 1, *A* = enzyme 2). The mutations in genes *C* and *D* result in loss or constitutive expression of both enzyme activities, suggesting that these genes or regions regulate or affect both genes.

a. Gene *B* is likely to be the gene for enzyme 1, since only a mutation in gene *B* produced a loss of enzyme 1 activity with no effect on enzyme 2 activity. (By similar logic, gene *A* codes for enzyme 2.)

b. i. False. A mutation in D leads to the constitutive synthesis of both enzymes 1 and 2, so D cannot be a structural gene for either enzyme.

ii. True. D could encode a repressor. Suppose the repressor acted as the *lac* repressor does in the *lac* operon. If mutations in D inactivated the repressor, then an absence of repressor would lead to constitutive activation of the operon. In this model, D^- mutants would be recessive to D^+ mutants, which is seen in the analysis of partial diploids ($A^- B^+ C^+ D^+/A^+ B^- C^+ D^-$ shows inducible activity of both A^+ and B^+).

iii. False. If D was needed to induce the *sug* operon, D^- mutants should produce no enzymes. This is not observed.

iv. False. One does see that D^- mutants are constitutive, as would be expected if a D^- operator region could not be bound by a repressor to repress transcription. However, not all of the results support this view. If D was an operator, consider what phenotype would be expected in the partial diploid $A^- B^+ C^+ D^+/A^+ B^- C^+ D^-$. If D^- was a defective operator, this partial diploid would express the A^+ gene (enzyme 2) constitutively, in a cis-dominant manner, and not in an inducible manner. This is not seen. What is seen is a trans-dominant effect where the A^+ gene is inducible. Thus, D is not an operator.

v. False. Since the products of genes A and B are not inducible in C^- mutants, one might speculate that gene C produced a cytoplasmic product that was required to induce genes A and B. However, consider the partial diploid data. In two of the partial diploids ($A^+ B^- C^- D^+/A^- B^+ C^+ D^+$ and $A^- B^+ C^- D^+/A^+ B^- C^+ D^+$), the wild-type genes on the C^- chromosome are not expressed, even though a C^+ gene is on another chromosome. This indicates that C shows cis-dominance and not trans-dominance. If C^+ encoded a cytoplasmic factor, it would diffuse and be capable of acting in a trans-dominant fashion. Since it does not, C does not encode a cytoplasmic, trans-acting factor.

vi. True. The cis-dominant effects that C^- mutants show in partial diploids could be explained if the mutations were in the controlling end of the *sug* operon, in a region such as the promoter.

19.13 Four different polar mutations, *1*, *2*, *3*, and *4*, in the *lacZ* gene of the *lac* operon were isolated after mutagenesis of *E. coli*. Each caused total loss of β-galactosidase activity. Two revertant mutants, due to suppressor mutations in genes unlinked to the *lac* operon, were isolated from each of the four strains: Suppressor mutations of polar mutation *1* are *1A* and *1B*, those of polar mutation *2* are *2A* and *2B*, and so on. Each of the eight suppressor mutations was then tested, by appropriate crosses, for its ability to suppress each of the four polar mutations; the test involved examining the ability of a strain carrying the polar mutation and the suppressor mutation to grow with lactose as the sole carbon source. The results follow (+ = growth on lactose and − = no growth):

Polar Mutation	Suppressor Mutation							
	1A	1B	2A	2B	3A	3B	4A	4B
1	+	+	+	+	+	+	+	+
2	+	−	+	+	+	+	−	−
3	+	−	+	−	+	+	−	−
4	+	+	+	+	+	+	+	+

A mutation to a UAG codon is called an amber nonsense mutation, and a mutation to a UAA codon is called an ochre nonsense mutation. Suppressor mutations allowing reading of UAG and UAA are called amber and ochre suppressors, respectively.

a. Which of the polar mutations are probably amber? Which are probably ochre?

b. Which of the suppressor mutations are probably amber suppressors? Which are probably ochre suppressors?

c. How would you explain the anomalous failure of suppressor *2B* to permit growth with polar mutation *3*? How could you test your explanation most easily?

d. Explain precisely why ochre suppressors suppress amber mutants but amber suppressors do not suppress ochre mutants.

Answer:

a. An ochre suppressor mutation will allow reading of the UAA codon, and have a 3'-AUU-5' anticodon. Because the 5'-U in the anticodon can wobble base-pair with either a 3'-A or 3'-G in a codon, the ochre suppressor will also allow reading of the UAG codon, and be an amber mutant suppressor. An amber suppressor mutation will allow reading of a UAG codon, and will have a 3'-AUC-5' anticodon. It will be able to suppress only amber mutants. Thus, amber mutants will be those suppressible by all given suppressors, while ochre mutants will be those suppressible by only a subset of suppressors. Mutants *1* and *4* appear to be amber mutants, while mutants *2* and *3* appear to be ochre mutants.

b. An ochre suppressor will allow growth of all four polar mutations, while an amber suppressor will allow growth of only *1* and *4*. Therefore, *1B*, *4A*, and *4B* are amber suppressors, while *1A*, *2A*, *3A*, and *3B* are ochre suppressors. *2B* is unlike either of these two groups, and will be discussed in part (c).

c. Since *2B* suppresses *1*, *2*, and *4*, it is probably an ochre suppressor. *3* might not be suppressed because of the mechanism of suppression. Any of a number of tRNA molecules could have been mutated to have an anticodon that will base-pair with an ochre triplet (UAA). It is likely that this tRNA will bring an amino acid to the ochre triplet that is different from that coded by the wild-type message. In some cases, this could result in a nonfunctional β-galactosidase enzyme. In turn, this would lead to the cell being unable to grow on lactose as a sole carbon source, so that suppression would not be seen. To test the hypothesis that a nonfunctional β-galactosidase protein is made, generate antibodies to the wild-type β-galactosidase protein, and use them to immunoprecipitate β-galactosidase protein from mutant cells. If protein can be precipitated even though no enzyme activity can be detected, the hypothesis is supported.

d. The fact that ochre suppressors suppress amber mutants but amber suppressors do not suppress ochre mutants can be explained by the wobble hypothesis (see text p. 363 and Figure 14.8, p. 362). A 5'-U in the tRNA anticodon can pair with either a 3'-A or 3'-G in the codon. Thus, a 5'-UAG-3' amber mutant codon and a 5'-UAA-3' ochre mutant codon can both be read by a 3'-AUU-5' ochre suppressor anticodon. This allows ochre suppressors to suppress both ochre and amber nonsense mutants. On the other hand, 5'-C in the tRNA anticodon can pair only with a 3'-G in the codon. Thus, a 3'-AUC-5' amber suppressor anticodon can pair only with a 5'-UAG-3' amber mutant codon.

19.14 What consequences would a mutation in the catabolite activator protein (*CAP*) gene of *E. coli* have for the expression of a wild-type *lac* operon?

Answer: The *CAP*, in a complex with cAMP, is required to facilitate RNA polymerase binding to the *lac* promoter. The RNA polymerase binding occurs only in the absence of glucose, and only if the operator is not occupied by repressor (i.e., lactose is also absent). A mutation in the *CAP* gene, then, would render the *lac* operon incapable of expression because RNA polymerase would be unable to recognize the promoter.

19.15 The presence of glucose in the medium along with lactose leads to catabolite repression. Explain why catabolite repression is considered to be a form of positive control, while repression by the *lac* repressor is considered to be a form of negative control.

Answer: Whether the form of control is positive or negative reflects whether transcription is stimulated or repressed. Catabolite repression is considered a form of positive control because in the absence of glucose, CAP-cAMP is produced and binds to the *lac* promoter to stimulate transcription of the *lac* operon. In constrast, the *lac* repressor is considered a form of negative control because when lactose is absent, the repressor binds the *lac* operator and blocks transcription.

19.16 DNase protection experiments were helpful to elucidate the functions of different DNA sequences in the *lac* promoter.
   a. What is a DNase protection experiment and how does it provide this information?
   b. How are the binding sites for the *lac* repressor, RNA polymerase, and CAP-cAMP arranged at the 5'end of the *lac* operon?
   c. What effects would you expect each of the following mutations to have on the coordinate induction of the *lac* operon by lactose (in the absence of glucose)? Explain your reasoning. (The base-pair coordinates used here are those specified in Figure 19.14.)
      i.   a deletion of base pairs from +3 to +18
      ii. a T-A to G-C transversion at −12
      iii. a T-A to G-C transversion at −69
      iv. a G-C to A-T transition at +28
      v.   a G-C to A-T transition at +9
   d. Would any of the mutations listed in (c) affect catabolite repression of the *lac* operon?

Answer:
   a. A DNase protection experiment is an in vitro method to identify DNA sites that are bound by a protein. After a purified protein is allowed to bind a DNA segment, the complex is treated with DNase and sequences unprotected by protein binding are digested. Then, the sequence of the protected region is determined. DNase protection experiments defined the location of the operator, promoter, and CAP-cAMP-binding site.
   b. See Figure 19.14, p. 528.

c. i. This deletion disrupts the operator, so the operon would be expressed constitutively.

ii. A −12 transversion alters the −10 promoter consensus sequence, possibly decreasing the efficiency of transcription initiation. The operon may still be coordinately induced, but there will be diminished levels of β-galactosidase, permease, and transacetylase activity.

iii. A −69 transversion alters the consensus sequence for the CAP-binding site. If CAP-cAMP is unable to bind the CAP site, RNA polymerase will not be recruited to the promoter and the operon will not be coordinately induced.

iv. A +28 transition alters the Shine-Dalgarno sequence and, by affecting translation initiation, could result in diminished or absent β-galactosidase and, due to polar effects, diminished or absent expression of permease and transacetylase.

v. A +9 transition alters the operator. It could either have no effect, cause the repressor to have more affinity for the operator (preventing coordinate induction in a cis-dominant manner), or cause the repressor to have less affinity for the operator (leading to constitutive expression).

d. None of the mutants will prevent catabolite repression.

19.17   The *lac* operon is an inducible operon, whereas the *trp* operon is a repressible operon. Discuss the differences between these two types of operons.

Answer: Both inducible and repressible operons allow sensitive control of transcription. They differ from each other in the details of how such control is achieved. In the *lac* operon, a repressor protein bound to an operator blocks transcription unless lactose is present. Thus, if lactose is absent, the system is OFF. When lactose is added, it is converted to allolactose and acts as an effector molecule to release the repressor from the operator, so that RNA polymerase can transcribe the operon. In the *trp* operon, the control strategy is the opposite. When tryptophan is abundant in the medium, the operon is turned off, as it is unnecessary to synthesize the enzymes needed to build tryptophan. Tryptophan also acts as an effector molecule. It binds to an aporepressor protein and converts it into an active repressor. This repressor is capable of binding the *trp* operator to reduce transcription of the *trp* operon protein-coding genes by RNA polymerase. Transcription of the *trp* operon is reduced by about 70 percent in the presence of tryptophan, while the aporepressor has no affinity for the operator in the absence of tryptophan.

The *trp* operon also can be regulated by attenuation, a mechanism that controls the ratio of the transcripts that include the five structural genes to those that are terminated before the structural genes. Under conditions where some tryptophan is present in the medium, short 140-bp transcripts are produced, and transcription is attenuated. Under conditions of tryptophan starvation or limitation, full-length transcripts are produced. A model for how attenuation occurs is described in text Figures 19.16, p. 530, and 19.17, p. 531.

19.18   Transcription of the *trp* operon can be reduced through a combination of repression using an aporepressor and attenuation.

a. How much of a reduction in transcription can be achieved using the aporepressor, and how much of a reduction in transcription can be achieved using attenuation? Speculate why the aporepressor might be unable to completely silence expression of the *trp* operon, and why this might be advantageous to *E. coli*.

b. Explain how the mechanism of attenuation is dependent on translation of transcripts at the *trp* operon.

Answer:

a. When the cell has an abundance of tryptophan, the aporepressor interacts with tryptophan and becomes an active repressor able to repress transcription of the *trp* operon by about 70-fold. Attenuation can further reduce transcription of the *trp* operon by a factor of eight to ten.

There are at least three possible reasons that the active repressor could be unable to completely repress transcription at the *trp* operon: (1) It may not be very abundant in the cell; (2) it may not have a very high affinity for the operator and so may not be bound to it constantly (transcription initiation could occur when it is not bound); and/or (3) it may have a relatively low affinity for tryptophan (which it must bind to become and remain an active repressor).

Tryptophan is an amino acid present in many proteins, and is very likely to be present in the enzymes used for its own biosynthesis. If transcription at the *trp* operon were completely silenced when tryptophan is abundant, a cell growing in the presence of tryptophan would cease to make even low levels of the biosynthetic enzymes needed to produce tryptophan. Over time, the biosynthetic enzymes would be degraded (assuming they have a set half-life). Should tryptophan levels become depleted, such a cell would be unable to synthesize new enzymes for tryptophan biosynthesis (since the tryptophan needed for incorporation into the polypeptide chain of the enzyme would be unavailable) and also be unable to synthesize any other protein containing tryptophan. This would lead to the cell's death. By allowing some transcription of the *trp* operon, the cell ensures that it can survive should tryptophan ever become unavailable, as it retains an ability to synthesize tryptophan.

b. In the *trp* operon, the mRNA transcript of the leader region includes a sequence that can be translated into a short polypeptide. Just prior to the stop codon in this transcript are two adjacent codons for tryptophan. Transcription is coupled to translation during the synthesis of the *trp* operon, so that RNA polymerase is synthesizing the *trp* mRNA just ahead of a ribosome that is translating it. If enough tryptophan is present in the cell for the ribosome to translate the two Trp codons, the ribosome continues to the stop codon for the leader peptide. In doing so, it covers part of the newly synthesized mRNA and allows an attenuator structure to form (the 3:4 structure shown in text Figure 19.17b, p. 531). This structure terminates transcription. If very few tryptophan molecules are present, the amount of Trp-tRNA molecules is very low and the ribosome translating the leader transcript stalls at the tandem Trp codons. This allows an antitermination structure to form (the 2:3 pairing shown in text Figure 19.17a, p. 531) so that RNA polymerase can continue and transcribe the structural genes. In this way, transcription of the structural genes of the *trp* operon is dependent on translation.

19.19 In the presence of high intracellular concentrations of tryptophan, only short transcripts of the *trp* operon are synthesized because of attenuation of transcription 5′ to the structural genes. This is mediated by the recognition of two Trp codons in the leader sequence. What effect would mutating these two codons to UAG stop codons have on the regulation of the operon in the presence or absence of tryptophan? Explain.

Answer: For a wild-type *trp* operon, the absence of tryptophan results in antitermination; that is, the structural genes are transcribed and the tryptophan biosynthetic enzymes are made. This occurs because a lack of tryptophan results in the absence of, or at least a very low level of, Trp-tRNA.Trp. In turn, this causes the ribosome translating the leader sequence to stall at the Trp codons (see Figure 19.17a, p. 531). When the ribosome is stalled at the Trp codons, the RNA being synthesized just ahead of the ribosome by RNA polymerase assumes a particular secondary structure. This favors continued transcription of the structural genes by the polymerase. If the two Trp codons were mutated to stop codons, then the mutant operon would function constitutively in the same way as the wild-type operon in the absence of tryptophan. The ribosome would stall in the same place, and antitermination would result in transcription of the structural genes.

For a wild-type *trp* operon, the presence of tryptophan turns off transcription of the structural genes. This occurs because the presence of tryptophan leads to the accumulation of Trp-tRNA.Trp, which allows the ribosome to read the two Trp codons and stall at the normal stop codon for the leader sequence. When stalled in that position, the antitermination signal cannot form in the RNA being synthesized; instead, a termination signal is formed, resulting in the termination of transcription. In a mutant *trp* operon with two stop codons instead of the Trp codons, the stop codons cause the ribosome to stall, even though tryptophan and Trp-tRNA.Trp are present. This results in an antitermination signal and transcription of the structural genes.

In sum, in both the presence and the absence of tryptophan, the mutant *trp* operon will not show attenuation. The structural genes will be transcribed in both cases and the tryptophan biosynthetic enzymes will be synthesized.

19.20   The mutant *E. coli* strains described in the following table are individually inoculated into two different media, one with supplemental tryptophan and one without:

Mutant	Phenotype
1	Aporepressor is unable to bind to tryptophan.
2	A point mutation in the *trp* operator prevents binding by an active Trp repressor.
3	The *trpE* gene has a nonsense mutation.
4	The levels of Trp-tRNA.Trp are decreased due to a mutation in a gene for Trp-aminoacyl synthetase.
5	Three adjacent G–C base pairs in region 4 (see Figures 19.16 and 19.17) are mutated to A–T base pairs.

For each mutant and medium, state whether the level of tryptophan synthetase will be increased or decreased relative to the level found in a wild-type strain and, where possible, by how much, and why.

Answer:

1: If the aporepressor cannot bind to tryptophan, it will not be converted to an active repressor when tryptophan is present. This will lead to constitutive expression of tryptophan synthetase: In medium without tryptophan, expression will be the same as in the wild type; in medium with tryptophan, expression will be reduced only through attenuation, and so it will be about 70-fold more than in the wild type.

2: The *trp* operon will exhibit constitutive expression, so mutant *2* will show the same expression patterns as mutant *1*.

3: The *trpE* gene is the first gene transcribed in the *trp* operon (see Figure 19.15, p. 529). A nonsense mutation could have a polar effect, leading to diminished or absent translation of *trpB* and *trpA*, which encode tryptophan synthetase. Therefore, in a medium without tryptophan where the operon is not repressed, mutant *3* would produce diminished levels of tryptophan synthetase compared to wild-type cells. In a medium with tryptophan, the levels will be the same as in wild-type cells (very low).

4: Trp-tRNA.Trp molecules are needed to attenuate transcription at the *trp* operon. Therefore, if the levels of Trp-tRNA.Trp are always low, transcription of the *trp* operon will not be attenuated even when tryptophan levels are high. Therefore, in medium with tryptophan, mutant *4* will have about 8- to 10-fold higher levels of tryptophan synthetase than will wild-type cells. In medium without tryptophan, attenuation does not occur, so mutant *4* will have levels of tryptophan synthetase that are similar to the wild type.

5: In mutant *5*, the 3:4 attenuator structure shown in Figure 19.17, p. 531, will not form, so attenuation will not occur when tryptophan levels are high. Tryptophan synthetase levels will be the same as in mutant *4*.

19.21 In the bacterium *Salmonella typhimurium*, seven of the genes coding for histidine biosynthetic enzymes are located adjacent to one another in the chromosome. If excess histidine is present in the medium, the synthesis in all seven enzymes is coordinately repressed, whereas in the absence of histidine all seven genes are coordinately expressed. Most mutations in this region of the chromosome result in the loss of activity of only one of the enzymes. However, mutations mapping to one end of the gene cluster result in the loss of all seven enzymes, even though none of the structural genes have been lost. What is the counterpart of these mutations in the *lac* operon system?

Answer: Mutations that cause the loss of activity in just one enzyme are mutations in structural genes, similar to *lacZ* and *lacY* mutations. Mutations that cause loss of all seven enzymes could be mutations in the promoter, similar to $P_{lac}-$ mutations.

19.22 On infecting an *E. coli* cell, bacteriophage λ has a choice between the lytic and lysogenic pathways. Discuss the molecular events that determine which pathway is taken.

Answer: When λ infects a cell and its genome circularizes, phage growth begins when RNA polymerase binds the λ early operon promoters P_L and P_R and transcribes mRNA for the *N* and *cro* genes, respectively. N acts as a transcription antiterminator and extends RNA synthesis to other genes that include the gene *cII*. The cII protein turns on *cI*, the gene for the λ repressor, as well as genes *O* and *P*, whose products are needed for DNA replication, and gene Q, whose product is needed for transcription of the late genes used in lysis and to produce phage particle proteins. Once the λ repressor and Cro are produced, there is competition between them to set a genetic switch that determines whether a lysogenic or a lytic pathway will be taken. A lysogenic pathway is taken if the λ repressor dominates, while a lytic pathway is taken if Cro dominates. The events that occur under Cro or repressor domination are diagrammed in text Figure 19.21, p. 535.

In a cell that will undergo lysogeny, the cII protein (stabilized by the cIII protein) will stimulate transcription of mRNA from P_{RE} and from P_I. This results in synthesis of repressor and of integrase. The λ repressor binds to two operators: O_R and O_L. This binding blocks transcription from P_R and P_L and so blocks production of Cro and N. It also stimulates transcription from P_{RM} and the production of additional λ repressor molecules. The stable repression of transcription from P_R together with the integrase-catalyzed integration of the λ chromosome into the *E. coli* host chromosome leads to the establishment of lysogeny. If the λ repressor is not present in high enough concentration or is cleaved and inactivated, the absence of repressor at O_R leads to the binding of RNA polymerase at P_R and transcription of the *cro* gene. When the Cro protein is produced in increasing amounts, it acts to decrease transcription from P_R and P_L, which reduces synthesis of the cII protein and block synthesis of repressor from P_{RM}. By this means, an increase in Cro protein blocks production of the λ repressor. It allows enough transcription from P_R for Q proteins to accumulate and stimulate the late gene transcription that is needed for starting the lytic pathway. In this way, the competition between the Cro and λ repressor proteins for the operator sites determines how the genetic switch is set.

19.23   How do the λ repressor protein and the Cro protein regulate their own synthesis?

Answer: The λ repressor binds to the operator O_R. When the λ repressor is bound, it stimulates synthesis of more repressor mRNA from the promoter P_{RM}. Eventually, however, at very high concentrations of λ repressor, O_R will be bound by the λ repressor in a way that blocks further transcription from the promoter P_{RM}. This maintains repressor concentrations in the cell. In the absence of repressor binding to O_R, RNA polymerase transcribes *cro* mRNA from the promoter P_R. As the concentration of Cro protein produced from this mRNA increases, Cro binds to the operator O_R and decreases synthesis from P_R and P_L. Not only does this serve to block synthesis of λ repressor mRNA from P_{RM}, but it will eventually block synthesis of *cro* mRNA as well. Thus, just as the λ repressor regulates the ability of RNA polymerase to transcribe its own *cI* gene from P_{RM}, Cro regulates the ability of RNA polymerase to transcribe *cro* from P_R. See Figure 19.21.

19.24   A mutation in the phage λ *cI* gene results in a nonfunctional *cI* gene product. What phenotype would you expect the mutant phage to exhibit?

Answer: The *cI* gene product is a repressor protein that acts to keep the lytic functions of the phage repressed when λ is in the lysogenic state. A *cI* mutant strain would lack the repressor and be unable to repress lysis, so that the phage would always follow a lytic pathway.

19.25   Bacteriophage λ can form a stable association with the bacterial chromosome because the virus manufactures a repressor. This repressor prevents the virus from replicating its DNA and making lysozyme and all the other tools used to destroy the bacterium. When you induce the virus with UV light, you destroy the repressor, and the virus goes through its normal lytic cycle. The repressor is the product of a gene called the *cI* gene and is a part of the wild-type viral genome. A bacterium that is lysogenic for λ^+ is full of repressor protein, which confers immunity against any λ virus added to these bacteria. Added viruses can inject their DNA, but the repressor from the resident virus prevents replication, presumably by binding to an operator on the incoming virus. Thus, this system has many elements analogous to the *lac* operon. We could diagram a virus as shown in the following figure. Several mutations of the *cI* gene are known. The c_i mutation results in an inactive repressor.

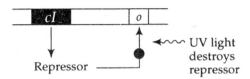

a. If you infect *E. coli* with λ containing a c_i mutation, can it lysogenize (form a stable association with the bacterial chromosome)? Why or why not?

b. If you infect a bacterium simultaneously with a wild-type c^+ and a c_i mutant of λ, can you obtain stable lysogeny? Why or why not?

c. Another class of mutants called c^{IN} makes a repressor that is insensitive to UV destruction. Will you be able to induce a bacterium lysogenic for c^{IN} with UV light? Why or why not?

Answer:

a. No. The repressor is necessary to keep the lytic function of the phage repressed and allow the phage to enter lysogeny. In the presence of a c_i mutation, only the lytic pathway can be taken.

b. Yes. A normal repressor will be made from the wild-type c^+ gene. This repressor is diffusible and will act in a trans-dominant manner to repress lytic growth.

c. No. UV irradiation of lysogenic bacteria destroys repressor function, which in turn leads to induction, including excision of λ and lytic growth. In a c^{IN} mutant, the repressor would not be destroyed following UV irradiation, so that the prophage would not be excised and lysogeny would be retained.

19.26   Five λ mutants have the molecular phenotypes shown in the left column of the following table:

Mutant	Molecular Phenotype	Lytic Growth	Lysogenic Growth	Inducible by UV Light
1	The Cro protein is unable to bind DNA.			
2	The N protein does not function.			
3	The cII protein does not function.			
4	The Q protein does not function.			
5	P_{RM} is unable to bind RNA polymerase.			

Fill in the table to indicate whether each mutant will be able to undergo lytic or lysogenic growth. For mutants able to follow a lysogenic pathway, state whether they can be induced by UV-light.

Answer:

Mutant	Molecular Phenotype	Lytic Growth	Lysogenic Growth	Inducible by UV Light
1	The Cro protein is unable to bind DNA.	no	yes	no
2	The N protein does not function.	no	no	no
3	The cII protein does not function.	yes	no	yes
4	The Q protein does not function.	no	yes	no
5	P_{RM} is unable to bind RNA polymerase.	yes	no	yes

20

REGULATION OF GENE EXPRESSION IN EUKARYOTES

CHAPTER OUTLINE

OPERONS IN EUKARYOTES

LEVELS OF CONTROL OF GENE EXPRESSION IN EUKARYOTES

CONTROL OF TRANSCRIPTION INITIATION

Chromatin Remodeling

Activation of Transcription by Activators and Coactivators

Blocking Transcription with Repressors

Combinatorial Gene Regulation

Case Study: Regulation of Galactose Utilization in Yeast

Regulation of Gene Expression by Steroid Hormones

GENE SILENCING AND GENOMIC IMPRINTING

POSTTRANSCRIPTIONAL CONTROL

RNA Processing Control

mRNA Transport Control

mRNA Translational Control

mRNA Degradation Control

Protein Degradation Control

RNA INTERFERENCE: A MECHANISM FOR SILENCING GENE EXPRESSION

THINKING ANALYTICALLY

Eukaryotic gene regulation is fundamentally different from prokaryotic gene regulation, as the packaging of eukaryotic DNA into chromatin establishes a new basis for the regulation of gene expression: Unless the transcription machinery can access DNA, genes are not transcribed. Thus, the remodeling of chromatin structure is the first factor to be considered when thinking about how eukaryotes regulate gene expression. Additional factors contribute to the complexity of eukaryotic gene regulation as well: Eukaryotic genes are organized differently than their prokaryotic counterparts; the transfer of information from DNA to protein is more complicated in eukaryotes and provides more opportunity for cells to regulate gene expression; and eukaryotic organisms are capable of substantially more cellular interactions, including those that lead to development and differentiation. As you approach this material, keep these factors in mind.

REVIEW OF KEY TERMS, SYMBOLS, AND CONCEPTS

In your own words, write a brief, precise definition of each term in the groups below. Check your definitions using the text. Then develop a concept map using the terms in each list.

1	2	3
prokaryotic operon	transcriptional control	RNA processing control
C. elegans operon	gene silencing	mRNA transport control
polygenic mRNA	genomic imprinting	alternative polyadenylation
monogenic mRNA	telomere position effect	alternative splicing
intergenic region	Sir complex	differential splicing
trans-splicing	*SIR2, SIR3, SIR4* genes	protein isoform
spliced leader (SL)-RNA	histone methylation	*CALC* gene
	histone methyl transferase	CGRP, calcitonin
	*Hpa*II, *Msp*I	spliceosome retention
	CpG island	snRNP
	insulator	nuclear pore
	Prader-Willi, Angelman syndrome	hnRNA

4	5	6
mRNA transcriptional control	mRNA transcription control	mRNA translation control
chromatin remodeling	*GAL1, GAL3, GAL4, GAL7, GAL10, GAL80* genes	mRNA degradation control
DNase I hypersensitive site	UAS$_G$	protein degradation control
histone acetyl transferase	divergent transcription	stored mRNA
histone deacetylase	steroid hormone	polyadenylation
nucleosome remodeling complex	polypeptide hormone	deadenylation
activator	second messenger	cytoplasmic polyadenylation
helix-turn-helix, zinc finger, leucine zipper DNA motif	signal transduction	adenylate/uridylate (AU)-rich element (ARE)
coactivator	steroid hormone receptor, response element	deadenylation-dependent, -independent decay pathway
mediator complex	chaperone	PAB-dependent poly(A) nuclease
enhancer	effector molecule	yeast *PAN1, DCP1* genes
silencer	plant hormone	decapping
repressor	gibberellins, auxins, cytokinins, ethylene, abscisic acid	5'→3' exonuclease
combinatorial gene regulation		endonuclease
quenching		proteolysis
		ubiquitin
		N-end rule

4	5	6
		RNA interference (RNAi) short interfering RNA (siRNA) antisense RNA Dicer, RISC knock-down vs. knock-out

While gene regulation in eukaryotes can occur at many more levels than in prokaryotes, one of the fundamental principles of gene regulation that we discovered from studying prokaryotes is retained—gene expression can be controlled through positive and negative regulation. Therefore, as you examine the strategies used by eukaryotes to regulate their gene expression, consider whether a parallel situation exists in prokaryotes and, if it does, how the mechanism used by eukaryotes differs.

The multiple levels used by eukaryotes to regulate gene expression require you to pay very close attention to detail. Start by reviewing the material in Chapters 10, 12, and 13, so that you have a solid grasp of how eukaryotes package their DNA and how they produce functional gene products. Then consider each level of regulation separately. Refer frequently to the text figures, redrawing them to solidify your understanding. Finally, think about how different levels of gene regulation can be integrated. For example, a gene may first be regulated at the level of transcriptional activation and then be posttranscriptionally or posttranslationally regulated. At the level of transcriptional activation, it will be important to consider how the gene's organization and regulatory factors, and alterations in chromatin structure, lead to the activation or quenching of transcription. Then, it will be important to understand how at subsequent points of regulation, different aspects of gene expression are targeted by distinct sets of regulatory proteins.

QUESTIONS FOR PRACTICE

Multiple Choice Questions

1. In eukaryotes, cell- and tissue-specific gene expression is achieved via
   a. the use of operons
   b. a cell- and tissue-specific set of activators and repressors
   c. DNase I sensitivity
   d. selective deletion of genes not active in differentiated cells
2. Transcription in eukaryotes is activated if activators are bound to
   a. enhancer elements
   b. promoter elements
   c. the operator
   d. both a and b
3. Histones act in eukaryotic gene regulation primarily as
   a. enhancers of gene expression
   b. repressors of gene expression
   c. promoters
   d. proteins preventing DNase I digestion

4. A gene lying in chromatin that is actively transcribed by RNA polymerase II is likely to
   a. have DNase I-hypersensitive sites upstream from the protein coding region
   b. have DNase I-hypersensitive sites within the protein coding region
   c. be relatively insensitive to DNase I
   d. be heavily methylated

5. The receptors for steroid hormones
   a. lie on the cell surface and act via second messengers
   b. lie on the cell surface and when bound by hormone are transported into the nucleus and bind response elements to activate transcription
   c. lie in the cytoplasm and act via second messengers
   d. lie in the cytoplasm and when bound by hormone are transported into the nucleus and bind response elements to activate transcription

6. Receptors bound by polypeptide hormones can alter gene expression by
   a. acting through a second messenger such as cAMP
   b. directly binding DNA sequences in the 5' end of genes
   c. directly binding DNA sequences in the 3' end of genes
   d. affecting signal transduction
   e. both a and d
   f. both b and c

7. In *C. elegans*, polygenic mRNAs are *not*
   a. transcribed under the control of a single promoter
   b. processed by trans-splicing
   c. sequentially translated
   d. capped and polyadenylated

8. How is ATP hydrolysis used in chromatin remodeling?
   a. to catalyze histone acetylation by HATS so that chromatin forms a looser structure
   b. to catalyze histone deacetylation by HDACs so that chromatin forms a tighter structure
   c. to methylate DNA
   d. to alter the position of nucleosomes by nucleosome remodeling complexes

9. What is the relationship between trans-activators and coactivators?
   a. Trans-activators bind DNA with the assistance of coactivators.
   b. Trans-activators activate transcription by interacting with coactivators, which in turn interact with transcription factors.
   c. Coactivators prevent repressors from binding DNA so that trans-activators can bind DNA.
   d. Coactivators bind DNA using helix-turn-helix, zinc-finger, and leucine-zipper domains. Trans-activators bind to the coactivators and interact with transcription factors to activate transcription.

10. The *GAL1*, *GAL7*, and *GAL10* genes are located near each other and are coordinately expressed. How is their expression prevented by the Gal80p protein?
    a. In the absence of an inducer, Gal80p binds to the Gal4p activation domain and prevents it from activating transcription.
    b. Gal80p blocks transcription by binding to an operator in the GAL operon.
    c. In the absence of an inducer, Gal80p binds a silencer near these genes.
    d. In the presence of galactose, an inducer is formed that binds to Gal80p. The inducer-bound Gal80p binds to Gal4p and prevents it from binding to the UAS_G operator. This allows RNA polymerase to bind the promoter.

11. Which type of histone modification is able to prevent the spreading of heterochromatin?
    a. deacetylation
    b. acetylation
    c. demethylation
    d. methylation

12. Which of the following could be most readily explained by genomic imprinting?
    a. the transcriptional silencing of only the paternal copy of a gene
    b. the liver-specific transcriptional activation of both alleles of a gene
    c. the silencing of a gene containing an expanded CpG island near its promoter
    d. the activation of a gene containing a HRE in the presence of a steroid hormone

13. When a cDNA library is screened using a probe from a single-copy gene, two different cDNAs are isolated. When their sequences are analyzed, two different, unrelated ORFs are identified. Based on this observation, which of the following types of gene regulation might occur at this gene?
    a. alternative polyadenylation
    b. alternative splicing
    c. mRNA degradation control
    d. alternative protein phosphorylation
    e. both a and b
    f. all of the above

14. What role does RISC have in RNAi?
    a. RISC cleaves double-stranded RNA into 21–23-bp fragments having 3′ overhangs.
    b. RISC binds short double-stranded RNA, unwinds it, pairs one strand with a complementary mRNA, and then cleaves that mRNA.
    c. RISC binds short double-stranded RNA, unwinds it, pairs one strand with a complementary mRNA, and, in doing so, blocks transcription.
    d. RISC amplifies the interference signal.
    e. all of the above except a

Answers: 1b, 2d, 3b, 4a, 5d, 6e, 7c, 8d, 9b, 10a, 11d, 12a, 13e, 14e

Thought Questions

1. Suppose you have cloned gene X. (a) Design an experiment to determine if gene X *is available for transcription* in liver and in brain tissue. (b) Now design an experiment to determine if gene X *is transcribed* in liver and/or in brain tissue. How do these experiments differ?

2. Present evidence that argues for or against the following statement: "Histone and nonhistone proteins function as repressors of gene expression in eukaryotic cells."

3. What is the evidence that methylation plays a role in transcriptional control in some eukaryotes? What evidence is there that methylation does not entirely determine the transcriptional activity of some genes?

4. Consider how hormones act. (a) Since hormones are diffusible, why are certain cells, but not others, targets of a particular hormonal signal? (b) How is the response of a cell to a peptide hormone fundamentally different from the response of a cell to a steroid hormone? (c) Propose a hypothesis to explain why it is that steroid hormone receptors are located inside the target cells, whereas the receptors for peptide hormones are located on the surfaces of target cells.

5. The primary transcripts of many genes undergo tissue-specific alternative mRNA splicing. Indeed, about 15 percent of human disease mutations are associated with mRNA splicing alterations. (a) For genes that are alternatively spliced, do you expect the proteins that are produced to have identical functions? Why or why not? (b) How might spliceosomes process primary mRNAs differently in different tissues to achieve alternative mRNA processing?

6. What is the mechanism that underlies the ability of a short double-stranded RNA molecule to block gene expression? What different roles does an activated RISC complex have in this process?

7. In what different ways is RNAi a valuable experimental tool? For example, how might RNAi be helpful to identify the function of orphan genes, or to elucidate the functions of a gene that produces different protein isoforms?

8. Why is it important to regulate the degradation of proteins? In what different ways can protein degradation be regulated?

9. In Chapter 14, we saw that the 5′ cap was important for translation initiation. How is it also important in mRNA stability and degradation?

10. Mammalian cells have between 10,000 and 100,000 SHR molecules. Why might so many SHR molecules be necessary?

Thought Questions for Media

After reviewing the media on the *iGenetics* CD-ROM, try to answer these questions.

1. In neuronal cells, the calcitonin/CGRP transcript is cleaved and polyadenylation occurs at the site called pA_2, producing a pre-mRNA with exons 1, 2, 3, 4, and 5. In the thyroid gland, a transcript with exons 1, 2, 3, and 4 is produced and calcitonin, a functional peptide hormone encoded by exon 4, is produced. Why is calcitonin not produced in neurons?

2. What role do chaperones play in hormone receptor activation?

3. Where is the GRE found, and what is its role in triggering a specific transcriptional response to the presence of glucocorticoids?

4. Why is the transcriptional response to glucocorticoid exposure cell-type specific?

1. Explain, in general terms, how the regulatory networks affected by a specific gene can be identified by comparing the gene expression profiles in normal cells and in cells where that gene's function has been knocked out.

2. In these types of experiments, can we infer whether the gene acts directly to activate or repress the expression of genes in a regulatory network?

SOLUTIONS TO TEXT PROBLEMS

20.1   How common is it for eukaryotic organisms to have genes organized into operons? How are these operons different from those found in prokaryotic organisms? In what ways are they similar?

Answer: About 15 percent of the genes in *C. elegans* are organized into operons, and operons have also been found in other nematodes. With the rare exceptions of the operons of nematodes, genes in eukaryotes are not clustered into operons. Like the prokaryotic operons, nematode operons are transcribed from a single promoter. However, unlike the prokaryotic operons, the mRNAs are not translated sequentially. Rather, the pre-mRNA is processed to generate monogenic mRNAs using trans-splicing and polyadenylation. In trans-splicing, an approximately 100-nucleotide capped SL-(spliced leader) RNA in an snRNP recognizes each 3′ splice site, cleaves the pre-mRNA near the 5′ end of each gene, and splices the first 22 nucleotides of the SL-RNA to its 5′ end. Simultaneous cleavage at the polyadenylation site and addition of the poly(A) tail generates the 3′ end of each gene. See text Figure 20.1, p. 545.

20.2   A nonsense mutation occurs in the first transcribed gene in a eukaryotic operon. If the mutation is close to the 5′ end of the gene's open reading frame, do you expect it to show polarity? Why or why not?

Answer: In prokaryotes, a nonsense mutation in an upstream gene of a polygenic transcript can prevent reinitiation of translation at downstream genes, leading to polarity. The polygenic transcripts of eukaryotic operons are processed by 5′ trans-splicing, which adds a capped, spliced leader RNA, and 3′ cleavage and polyadenylation, so that each gene is separately translated. Consequently, nonsense mutations in an upstream gene will not affect translation of downstream genes and they will not show polarity.

20.3   Three genes *a, b,* and *c* in the nematode *C. elegans* are very tightly linked in the order *a-b-c* and have unrelated functions. An analysis of cDNA and genomic DNA sequences for each gene reveals that their mRNAs share a common 22-base-pair 5′ region, but that these 22 base pairs are not present in the genomic sequences of these genes. Furthermore, the *b* and *c* genes lack promoters. Explain the mechanism that results in each mRNA gaining the same 22-bp 5′ sequence, and how *b* and *c* can be transcribed without promoters.

Answer: These three genes are organized into an operon whose expression is regulated by one promoter and that is transcribed into a single pre-mRNA. The pre-mRNA is then processed by trans-splicing and polyadenylation to produce three monogenic mRNAs, one for each gene. The 22 base 5′ end of each cDNA is not present in the genomic sequences of these genes because it results from trans-splicing. During trans-splicing, the same capped SL-RNA is spliced onto the 5′ end of each mRNA.

20.4   Critically evaluate the following contention: Prokaryotes and eukaryotes use fundamentally different mechanisms to control gene expression.

Answer: Prokaryotic promoters are transcribed unless they are silenced by the binding of repressors that block transcription initiation. For example, RNA polymerase can initiate transcription of the *lac* operon whenever a repressor protein is not bound to the *lac* operator. Prokaryotic genes are able to be transcribed without altering how they are packaged—they are accessible for transcription. In eukaryotes, the nucleosome organization of chromosomes has a generally repressive effect on gene expression—it impedes the transcription machinery from accessing genes. For a eukaryotic gene to be activated, the chromatin structure must be remodeled in the vicinity of the core promoter. This aspect of eukaryotic gene regulation is fundamentally different from the regulation of gene expression in prokaryotes.

20.5   In *Drosophila*, pulses of the steroid-hormone ecdysone trigger molting between the larval stages and then, at the end of the larval stages, trigger the formation of a pupa, where the larva will metamorphose into an adult fly. Immediately after the ecdysone pulse at the end of the larval stages, transcription of several genes, including *Eip93F*, is dramatically increased. To investigate how ecdysone regulates *Eip93F*, chromatin is isolated from staged wild-type animals just before and just after the ecdysone pulse at the end of the larval stages. The chromatin is distributed to separate test tubes where it is digested for two minutes with different concentrations of DNase I. DNA is then purified from each sample and digested with *Eco*RI. The resulting DNA fragments are then resolved by size using gel electrophoresis and a Southern blot is made. The Southern blot is probed with two *Eco*RI fragments from the *Eip93F* gene: a 4.0-kb fragment from its promoter and a 3.0-kb fragment from its protein-coding region. The following figure shows the results, where the thickness of the band corresponds to the intensity of hybridization signal.

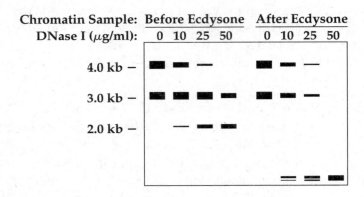

a. Explain why the 4-kb band, but not the 3-kb band, diminishes in intensity when chromatin isolated before the pulse of ecdysone is treated with increasing concentrations of with DNase I. How do you explain the increasing amounts of the 2-kb band in these samples?

b. Explain why both the 4-kb and 3-kb bands diminish in intensity when chromatin isolated after the pulse of ecdysone is treated with increasing amounts of DNase I. How do you interpret the increasing amounts of low-molecular-weight digestion products in these samples?

Answer:

a. The disappearance of the 4-kb band indicates that the promoter region has a DNase I hypersensitive site—a less highly coiled site where DNA is more accessible to

DNase I for digestion. Since DNase I digestion produces a 2-kb band, the site lies near the middle of the 4-kb *Eco*RI fragment. The 3-kb band does not diminish in intensity except at the highest DNase I concentration, so the region of the gene containing the 3-kb *Eco*RI fragment is more highly coiled by nucleosomes.

b. Following the ecdysone pulse, increased concentrations of DNase I lead to the disappearance of both the 4- and 3-kb bands, indicating that DNase I has increased access and the gene is less tightly coiled during transcription. The appearance of low-molecular-weight digestion products indicates that DNase I cannot access some regions of the 4- and 3-kb *Eco*RI fragments. These regions may be bound by proteins such as general transcription factors.

20.6 Chromatin remodeling is essential for gene activation and can be achieved using different mechanisms.

a. What different types of enzymes are used to modify histones, and how do these enzymatic modifications lead to chromatin remodeling?

b. In what other ways can chromatin be remodeled?

c. What phenotype(s) would you expect to see in a mutant where a protein involved in chromatin remodeling failed to function?

Answer:

a. Histone acetyl transferases (HATs) and histone deacetylases (HDACs) modify nucleosomes by acetylating or deacetylating core histones. As positively charged acetyl groups are added to the negatively charged histones by HATs, the histones slowly lose their affinity for negatively charged DNA, and the 30-nm chromatin fiber loses histone H1 and changes conformation to a 10-nm chromatin fiber. This makes promoters more accessible for the activation of transcription. Conversely, as HDACs remove acetyl groups from histones, the histones will slowly gain more affinity for DNA. The 30-nm chromatin fiber will reform and make promoters less accessible for the activation of transcription.

b. Chromatin can also be remodeled using ATP-dependent nucleosome remodeling complexes. They use the energy of ATP hydrolysis to alter nucleosome position or structure and facilitate the binding of the transcription machinery to the core promoter. See text Figure 20.3, p. 548.

c. Mutants in which a protein involved in chromatin remodeling fails to function will exhibit marked decreases in the expression of many genes in different, unrelated pathways. They are likely to grow very slowly and may be unable to complete some cellular processes altogether. See the text discussion of the discovery of the SWI/SNF genes that affect mating-type switching in yeast, p. 548.

20.7 DNA, histones, promoter-binding proteins, and enhancer-binding proteins are mixed together in the following orders:

a. first histones and DNA, then promoter-binding proteins

b. first histones and promoter-binding proteins, then DNA

c. first DNA and promoter-binding proteins, then histones

d. first histones, promoter-binding proteins and enhancer-binding proteins, then DNA

For each case, state whether transcription can occur. Explain your answers.

Answer:

a. Histones repress gene expression, so if they are present on DNA, promoter-binding proteins cannot bind promoters and transcription cannot occur.

b. Histones will compete more strongly for promoters than will promoter-binding proteins, so transcription will not occur.

c. If promoter-binding proteins are already assembled on promoters, nucleosomes will be unable to assemble on these sites, so transcription will occur.

d. Enhancer-binding proteins will help promoter-binding proteins to bind promoters even in the presence of histones, so transcription will occur.

20.8   Promoters, enhancers, general transcription factors, activators, coactivators, and repressors that regulate the expression of one gene often have structural features that are similar to those regulating the expression of other genes. Nonetheless, the transcriptional control of a gene can be exquisitely specific: It will be specifically transcribed in some tissues at very defined times. Explore how this specificity arises by addressing the following questions:

a. Distinguish between the functions of promoters and enhancers in transcriptional regulation.

b. Distinguish between the functions of general transcription factors, activators, coactivators, and repressors in transcriptional regulation.

c. What structural features are found in activators, and what role do these play in transcriptional activation?

d. How is the mechanism by which eukaryotic repressors function different from that by which prokaryotic repressors function?

e. How can the same enhancer stimulate as well as quench transcription?

f. Given that several different genes may contain the same types of promoter and enhancer elements, and a number of the proteins that bind these elements contain the same structural features, how is transcriptional specificity generated?

Answer:

a. Promoters and enhancers are DNA elements that bind specific regulatory proteins. Promoter elements are located just upstream from the site at which transcription begins, while enhancers are usually some distance away, either upstream or downstream. Some promoter elements—those in the core promoter—are required for transcription to begin, while others—those in the promoter proximal region—are specialized for the gene they control and determine whether transcription can occur. In contrast to promoter elements that are crucial for determining whether transcription can occur, enhancer elements ensure maximal transcription of the gene. Specific regulatory proteins bound to enhancer elements can activate transcription through their interaction with multiprotein complexes interacting with proteins bound to the promoter elements.

b. General transcription factors (GTFs) are proteins that are required for basal levels of transcription but are unable, by themselves alone, to influence the rate of transcription. TBP and TFIID, which bind to core promoter elements, are examples of GTFs. Activators (also known as trans-activators) influence transcription at a distance.

They bind to enhancers and simulate transcription through interactions with coactivators. Coactivators are multiprotein complexes that interact both with activators and with transcription factors. The coactivators do not bind DNA directly, but rather mediate interactions between activators and transcription factors, the proteins that do bind to DNA. Interactions between activators and a coactivator complex, between a coactivator complex and RNA polymerase II, and between RNA polymerase II and the GTFs cause the DNA to loop back upon itself and lead to the activation of transcription. Repressors are similar to activators in that they bind to both DNA elements and coactivators. However, they quench transcription rather than stimulate it. For this reason, DNA sites bound by repressors are referred to as silencers.

c. Activators have two functional domains. One domain is a DNA-binding domain that binds to an enhancer. This domain has a structural motif important for DNA-binding such as a helix-turn-helix, zinc finger, or leucine zipper motif. The second domain is a transcription activation domain that stimulates transcription by recruiting a coactivator protein complex.

d. Repressors in eukaryotes are unlike those in prokaryotes in that they do not directly bind to DNA to prevent RNA polymerase from binding a promoter. Unlike the situation in prokaryotes, in eukaryotes, the packaging of DNA by nucleosomes establishes a repressed state that must be overcome by activators. Eukaryotic repressors counteract the action of activators to block transcription. Like activators, they have two domains, in this case a DNA-binding domain and a repressor domain. Repressors can work in several different ways: (1) They can bind to DNA sites near an activator's binding site and block the action of the activator's activation domain through their repressor domain; (2) they can bind to a DNA site that overlaps the site bound by the activator and prevent the activator from binding; and (3) they can recruit a histone deacetylase complex to bring about chromatin compaction.

e. Whether a particular DNA sequence serves to stimulate or quench transcription depends first on the type of regulatory protein bound to it. If it is bound by an activator that stimulates transcription, the sequence serves as an enhancer, while if it is bound by a repressor that quenches transcription, it acts as a silencer. The magnitude of its effect on transcription also depends on the strength of that protein's effect compared with that of regulatory proteins bound to other DNA elements in the gene. Therefore, the function of a DNA element able to bind a regulatory protein depends on the type of regulatory proteins present in a cell and the types of other DNA elements present in the gene.

f. Transcriptional specificity is generated through the combinatorial use of a set of GTFs, activators, and repressors. Some of these are common to many cells, while others have a more restricted spatial and/or temporal distribution. The transcription of defined subsets of genes in different cells at different times is controlled by the constellation of regulatory proteins present in a eukaryotic cell at a particular time.

20.9   Eukaryotic organisms have a large number of copies (usually more than a hundred) of the genes that code for ribosomal RNA, yet they have only two copies (one on each of two homologs) of genes that code for a ribosomal protein. Explain how eukaryotes can produce the same number of ribosomal RNAs and ribosomal proteins, given this disparity in gene copy number.

Answer: The final product of an rRNA gene is an rRNA molecule. Therefore, a large number of genes are required to produce the large number of rRNA molecules required for ribosome biosynthesis. In contrast, ribosomal proteins are the end products of the translation of mRNAs. The mRNAs can be translated over and over to produce the large number of ribosomal protein molecules required for ribosomal biosynthesis.

20.10   Both peptide and steroid hormones can affect gene regulation of a targeted population of cells.

    a. What is a hormone?

    b. Distinguish between the mechanisms by which a peptide and a steroid hormone affect gene expression.

    c. What role does each of the following have in a physiological response to a peptide or a steroid hormone?

       i.   steroid hormone receptor (SHR)

       ii.   chaperone

       iii. steroid hormone response element (HRE)

       iv.   second messenger

       v.   cAMP and adenylate cyclase

    d. How can the same steroid hormone simultaneously activate distinct patterns of gene expression in two different cell types and have no affect on a third cell type?

Answer:

    a. A hormone is typically a low-molecular-weight chemical messenger synthesized in low concentrations in one tissue or cell and transmitted in body fluids to another part of the organism. Hormones act as effector molecules to produce specific effects after binding to a receptor in target cells that may be remote from the hormone's point of origin. Hormones function to regulate gene activity, physiology, growth, differentiation, or behavior.

  b, c. Steroid hormones and peptide hormones employ two fundamentally different mechanisms to regulate gene expression. Steroid hormones traverse the cell membrane and bind to a cytoplasmic steroid hormone receptor (SHR). SHRs have three domains, a DNA-binding domain, an activation or repression domain, and a specific, steroid hormone binding domain. When the steroid hormone is absent the SHR is inactive, and the SHR complexes with chaperones to maintain its funtionality. After the steroid hormone enters the cell, it complexes with the SHR and displaces Hsp90, a chaperone. The complex of the hormone and SHR then enters the nucleus and acts directly to regulate gene expression. The DNA-binding domain of the SHR binds to specific steroid hormone response elements (HREs) near genes to regulate their transcription.

      Polypeptide hormones, such as insulin and certain growth factors, bind to receptors that reside on cell surfaces. The activated, bound receptor then transduces a signal via a second-messenger system inside the cell. Some activated receptors increase the activity of the membrane-bound enzyme adenylate cyclase, which results in an increase in cAMP levels. Increased intracellular levels of cAMP send an intracellular signal (a second messenger) to activate other cellular processes that ultimately lead to an alteration in patterns of gene expression.

d. Each hormone acts on specific target cells that have receptors capable of recognizing and binding that particular hormone. Polypeptide hormones bind to receptors on the cell surface, while steroid hormones bind to receptors inside the cell. Two distinct cell types that possess receptors for a hormone can be activated by the same hormone signal while a third cell type that lacks receptors for the hormone will fail to respond. However, even the two cell types that possess hormone receptors may respond in different ways. Though the cells have the same SHR, they may differ in the types of other regulatory proteins they have. Since the pattern of genes that is activated by the hormone is dependent on the array of other regulatory proteins that are present, two receptor-bearing cell types may respond differently to the same hormone signal.

20.11 The following figure shows the effect of the hormone estrogen on ovalbumin synthesis in the oviduct of 4-day-old chicks. Chicks were given daily injections of estrogen ("Primary stimulation") and then after 10 days the injections were stopped. Two weeks after withdrawal (25 days), the injections were resumed ("Secondary stimulation").

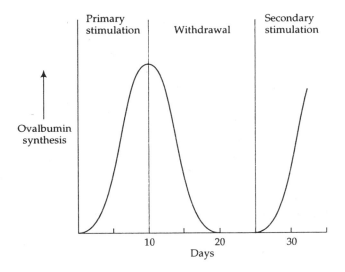

Provide possible explanations of these data.

Answer: The data indicate that the synthesis of ovalbumin is dependent upon the presence of the hormone estrogen. These data do not address the mechanism by which estrogen achieves its effects. Theoretically, it could act (1) to increase transcription of the ovalbumin gene by binding to an intracellular receptor that, as an activated complex, stimulates transcription at the ovalbumin gene; (2) to stabilize the ovalbumin precursor mRNA; (3) to increase the processing of the precursor ovalbumin mRNA; (4) to increase the transport of the processed ovalbumin mRNA out of the nucleus; (5) to stabilize the mature ovalbumin mRNA once it has been transported into the cytoplasm; (6) to stimulate translation of the ovalbumin mRNA in the cytoplasm; or (7) to stabilize (or process) the newly synthesized ovalbumin protein. Experiments in which the levels of ovalbumin mRNA were measured have shown that the production of ovalbumin mRNA is primarily regulated at the level of transcription.

20.12   In what different ways are the tails of histones chemically modified? What are the biological consequences of each type of modification?

Answer: Histone tails can be acetylated by histone acetyl transferases (HATs) and methylated by histone methyl transferases (HMTs). Acetylation, as described in the solution to Question 20.6, results in chromatin remodeling—loosening of chromatin structure—that allows for transcriptional activation, while deacetylation by histone deacetylases (HDACs) can result chromatin remodeling—tightening of chromatin structure—that results in the silencing of transcription. Methylation of histone tails is important to delimit regions of transcriptional silencing. For example, the Sir silencing complex—a protein complex that includes the HDAC Sir2p—is recruited to telomeric regions by Rap1p, a telomere-repeat-binding protein. As Sir2p catalyzes the local removal of acetyl groups from histone tails, the deacetylated histones can be recognized directly by the Sir silencing complex, so that waves of binding and deacetylation occur during the spreading of heterochromatic structure along the chromosome. Methylation of histone H3 tails blocks this process so that heterochromatin forms over a limited distance.

20.13   A cloned DNA sequence was used to probe a Southern blot. There were two DNA samples on the blot, one from white blood cells and the other from a liver biopsy of the same individual. Both samples had been digested with *Hpa*II. The probe bound to a single 2.2-kb band in the white blood cell DNA but bound to two bands (1.5 and 0.7 kb) in the liver DNA.

a. Is this difference likely to result from a somatic mutation in a *Hpa*II site? Explain.

b. How would it affect your answer if you knew that white blood cell and liver DNA from this individual both showed the two-band pattern when digested with *Msp*I?

Answer:

a. A somatic mutation is rather unlikely. Humans are diploid, so finding a single 2.2-kb band in the white blood cell DNA indicates that both homologous chromosomes have the same restriction sites. The liver cells have two bands whose sizes add up to the size of the band in white blood cells. Thus, it would appear that the liver has an additional site on *both* chromosomes, not just on one chromosome (heterozygous cells would have three bands). For this to occur by a mutational mechanism, two different somatic mutations would have had to occur in the same site in the progenitor cell that gave rise to all of the cells being examined (possibly to all of the cells of the tissue), an unlikely event.

b. *Msp*I recognizes the same CCGG sequence as *Hpa*II, but unlike *Hpa*II, it cleaves the methylated sequence C^mCGG as well as the unmethylated CCGG. That both liver cells and white blood cells show the same 2-band pattern when digested with *Msp*I, but not *Hpa*II, indicates that there is a central methylated site in the white blood cells, but not in the liver cells. Increased DNA methylation has, in some instances, been correlated with a decrease in transcriptional activity. However, not all methylated DNA is transcriptionally silent.

20.14   In what different ways can DNA methylation affect gene expression?

Answer: In some organisms, DNA methylation results in the transcriptional silencing of specific DNA sequences. One example illustrating how DNA methylation can affect gene expression is the abnormal methylation associated with an expanded triplet repeat in the *FMR-1* gene. This results in transcriptional silencing at *FMR-1* and the development of fragile-X syndrome.

A second example is DNA imprinting. In DNA imprinting, specific DNA sequences are methylated in just one parent prior to their inheritance. The methylation imprint is maintained following DNA replication in zygotic cells so that the paternally and maternally contributed chromosomes of the zygote retain their different methylation patterns. Therefore, imprinting results in different patterns of gene silencing on maternally and paternally contributed chromosomes. An illustration of this for the linked *Igf2* and *H19* genes is shown in text Figure 20.16, p. 560. A single enhancer downstream from both genes controls their expression. However, an activator bound to this enhancer activates the upstream *Igf2* gene only if its action is not blocked by the binding of the CTCF protein to an insulator element lying in between the genes. If the CTCF protein binds the insulator element, only the downstream *H19* gene is transcribed. Methylation of the paternally contributed chromosome—imprinting—alters this situation. Methylated of the insulator element and the *H19* promoter in the paternally contributed chromosome prevents binding of the insulator by the CTCF protein and transcription of the *H19* gene. Consequently, the paternal copy of the *Igf2* gene, but not the *H19* gene, is expressed. In contrast, the maternally contributed chromosome is not methylated in this region. Thus, the insulator element functions, allowing expression of the maternal *Igf2* gene and preventing expression of the maternal *H19* gene.

20.15   When male mice heterozygous for a small deletion on chromosome 2 are mated to normal females, deletion-bearing offspring have thin bodies and are slow moving, while non-deletion-bearing offspring are normal. However, when females heterozygous for the same deletion are mated to normal males, all offspring are normal.

a. How can these findings be explained in terms of imprinting?

b. When, and in what cell types, does imprinting occur?

c. *Neuronatin* is a gene that lies within the deleted region. A DNA polymorphism exists in the 3'-UTR of the *Neuronatin* gene that can be distinguished using PCR-RFLP (see Chapter 17 for a discussion of PCR-RFLP). How would you determine if the *Neuronatin* gene is expressed in a manner consistent with its being imprinted in embryos produced by the cross? Explain what results you would expect if it is imprinted, and what results you would expect if it is not imprinted.

Answer:

a. When an abnormal phenotype is associated with imprinting, it is seen when a nonfunctional imprinted allele is inherited and a functional normal allele is not inherited. Here, offspring inheriting a deletion from the male but not the female parent exhibit an abnormal phenotype. Therefore, in a normal individual the paternally contributed allele of a gene within the deletion must function while the maternally contributed allele must be imprinted.

b. Imprinting is associated with the methylation of DNA sequences that results in gene silencing. It occurs during the formation of gametes—during spermatogenesis and oogenesis.

c. The DNA polymorphism lies in an expressed region of the gene, so the polymorphism can be evaluated in cDNAs made from *Neuronatin* mRNA by using RT-PCR-RFLP—isolate mRNA from the tissue where the *Neuronatin* gene is normally expressed, prepare cDNA by reverse transcription, and use PCR-RFLP to characterize the polymorphism. RT-PCR-RFLP allows us to determine if one or both alleles of the *Neuronatin* gene are expressed. Contrast this with PCR-RFLP using a genomic DNA template, which would tell us about the genotype of an individual—whether they are homozygous or heterozygous. Based on the reasoning in (a), we can hypothesize that the maternally contributed copy of the *Neuronatin* gene is imprinted and the paternally contributed copy of the gene is expressed. Suppose that a normal male homozygous for one polymorphism is crossed to a normal female homozygous for another polymorphism. Under the imprinting hypothesis, RT-PCR-RFLP using mRNA prepared from the offspring should identify the polymorphism present in the father and not the mother. If the hypothesis is incorrect and both the maternal and paternal alleles are expressed—the *Neuronatin* gene is not imprinted—then RT-PCR-RFLP should identify the polymorphisms present in each parent. Whether or not the gene is imprinted, PCR-RFLP used with a genomic DNA template should reveal that the individual is heterozygous for the polymorphism.

If we use RT-PCR-RFLP to examine the hemizygous individuals described in the problem and the *Neuronatin* gene is imprinted, we should find that the affected, deletion-bearing offspring produced from a cross of a deletion-bearing male with a normal female do not express the *Neuronatin* gene. The only copy of the gene they received—the one from their mother—is imprinted. In contrast, the normal offspring produced from a cross of a deletion-bearing female with a normal male should express only the paternal allele.

20.16 Both fragile X syndrome and Huntington disease are caused by trinucleotide repeat expansion. Individuals with fragile X syndrome have at least 200 CGG repeats at the 5' end of the *FMR-1* gene. In contrast, individuals with Huntington disease have 36 or more in-frame CAG repeats within the protein-coding region of the *huntingtin* gene.

a. Do you expect gene expression at the two genes to be affected in the same way by these repeat expansions? Explain your answer.

b. Based on your answer to (a), why might fragile X syndrome be recessive, whereas Huntington disease is dominant?

c. Generate a hypothesis to explain why is the number of trinucleotide repeats needed to cause a disease phenotype is different at each gene.

Answer:

a. In fragile X syndrome, the expanded CGG repeat results in hypermethylation and transcriptional silencing. A CAG codon specifies glutamine, so in Huntington disease, the expanded CAG repeat results in the inclusion of a polyglutamine stretch within the huntingtin protein. This causes it to have a novel, abnormal function.

b. A heterozygote with a CGG repeat expansion near one copy of the *FMR-1* gene will still have one normal copy of the *FMR-1* gene. The normal gene can produce a normal product, even if the other is silenced. (The actual situation is made somewhat more complex by the process of X inactivation in females, but in general one would expect that a mutation that caused transcriptional silencing of one allele would not affect a normal allele on a homolog.) In contrast, a novel, abnormal protein is produced by the CAG expansion in the disease allele in Huntington disease. Since the disease phenotype is due to the presence of the abnormal protein, the disease trait is dominant.

c. Transcriptional silencing may require significant amounts of hypermethylation, and so require more CGG repeats for an effect to be seen. In contrast, protein function may be altered by a stretch of more than 36 glutamines.

20.17   Although the primary transcript of a gene may be identical in two different cell types, the translated mRNAs can be quite different. Consequently, in different tissues, distinct protein products can be produced from the same gene. Discuss two different mechanisms by which the production of mature mRNAs can be regulated to this end; give a specific example for each mechanism.

Answer: Two mechanisms are those controlling the choice of poly(A) site and choice of splice site. An example of regulation by choice between alternative poly(A) sites is the production of IgM in B lymphocytes. In two different developmental stages of B lymphocytes, two different poly(A) sites are used, leading to IgM molecules that are either secreted or membrane bound. This results from the developmental repression of one component (CstF-64) of the three-polypeptide CstF (cleavage stimulation factor) complex that functions in poly(A) addition. Its repression causes the formation of a CstF complex with a lower affinity for one poly(A) site, and the use of an alternative site.

An example of gene regulation by differential mRNA splicing is the regulation cascade that governs sex determination in *Drosophila*. Here, an X:A ratio of 1.0 leads to female development as a result of a cascade of alternative female-specific pre-mRNA splicing at a set of regulatory genes that includes *Sxl, tra,* and *dsx*. An X:A ratio of 0.5 leads to male development when a male-specific pattern of splicing occurs at these genes.

20.18   How are pre-mRNAs prevented from being transported into the cytoplasm until after their introns have been removed?

Answer: For mRNAs to be transported out of the nucleus, they must be capped and bound by proteins that interact with other proteins at the nuclear pore complexes that direct them out of the nucleus. In the spliceosome retention model, the binding of unspliced mRNAs to spliceosomes prevents this interaction. After mRNAs are spliced and no longer interact with spliceosomes, they are able to interact with these proteins and can be transported out of the nucleus. In contrast, intronic sequences that remain associated with spliceosomes cannot be transported.

20.19   Although many mRNAs are present in the cytoplasm of unfertilized vertebrate and invertebrate embryos, the rate of protein synthesis is very low. After fertilization, the rate of protein synthesis increases dramatically without new mRNA transcription.

a. What differences are seen in the length of poly(A) tails between inactive, stored mRNAs and actively translated mRNAs?

b. What role does cytoplasmic polyadenylation have in this process?

c. What signals are present in mRNAs that control polyadenylation and deadenylation?

d. In what way is deadenylation also important for controlling mRNA degradation?

Answer:

a. Inactive stored mRNAs have shorter poly(A) tails (15–90 As) while actively translated mRNAs have longer poly(A) tails (100–300 As).

b. At least some of the mRNAs that will be stored prior to translation are initially processed with a longer poly(A) tail of 300–400 As. Their poly(A) tail is rapidly deadenylated to a length of 40–60 As in the mature stored message, and later, when activated for translation, increased in length by about 150 A nucleotides by a cytoplasmic polyadenylation enzyme.

c. A sequence in the 3′ UTR, the adenylate/uridylate (AU)-rich element (ARE, UUUUUAU), that lies upstream from the polyadenylation sequence (AAUAAA) provides a signal for deadenylation and polyadenylation.

d. In addition to the rapid deadenylation pathway discussed above, mRNAs can undergo a default, slow decrease in poly(A) length. If the poly(A) tails are deadenylated and become too short (5–15 As) to bind PAB (poly(A) binding protein), the 5′ cap is removed enzymatically (decapping) and the mRNA is degraded from the 5′ end by a 5→3′ exonuclease. This deadenylation-dependent pathway is one of two major mRNA decay pathways. In the other major pathway, decapping and 5′→3′ degradation or internal endonuclease cleavage occurs without deadenylation.

20.20   Although most eukaryotes lack operons such as those found in prokaryotes, the exceptional conserved organization of the *ChAT/VAChT* locus in *Drosophila* is reminiscent of a prokaryotic operon. *ChAT* is the gene for the enzyme choline acetyltransferase, which synthesizes acetylcholine, a neurotransmitter released by one neuron to signal another neuron. *VAChT* is the gene for the vesicular acetylcholine transporter protein, which packages acetylcholine into vesicles before its release from a neuron. Both *ChAT* and *VAChT* are expressed in the same neuron.

Part of the *VAChT* gene is nested within the first intron of the *ChAT* gene, and the two genes share a common regulatory region and a first exon. The structure of a primary mRNA and two processed mRNA transcripts produced by this locus are diagrammed in the following figure. The common regulatory region important for transcription of the locus in neurons is shown in the DNA, black rectangles in RNA represent exons, lines connecting the exons represent spliced intronic regions, and AUG indicates the translation start codons within the *ChAT* and *VAChT* mRNAs. Polyadenylation sites are not shown.

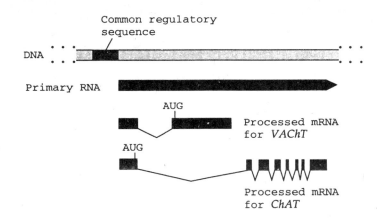

a. In what ways is the organization of the *VAChT/ChAT* locus reminiscent of a bacterial operon?

b. Why is the organization of this locus not structurally equivalent to a bacterial operon?

c. Based on the transcript structures shown, what modes of regulation might be used to obtain two different protein products from the single primary mRNA?

Answer:

a. In bacterial operons, a common regulatory region controls the production of single mRNA from which multiple protein products are translated. These products function in a related biochemical pathway. Here, two proteins that are involved in the synthesis and packaging of acetylcholine are both produced from a common primary mRNA transcript.

b. Unlike the proteins translated from an mRNA synthesized from a bacterial operon, the protein products produced at the *VAChT/ChAT* locus are not translated sequentially from the same mRNA. Here, the primary mRNA appears to be alternatively processed to produce two distinct mature mRNAs. These mRNAs are translated starting at different points, producing different proteins.

c. At least two mechanisms are involved in the production of the different ChAT and VAChT proteins: alternative mRNA processing and alternative translation initiation. After the first exon, an alternative 3′ splice site is used in the two different mRNAs. In addition, different AUG start codons are used.

20.21   Four different cDNAs were identified when a cDNA library was screened with a probe from one gene. The locations of introns and exons in the gene were determined by comparing the cDNA and genomic DNA sequences. The results are summarized in the following figure: Exons are represented by filled rectangles with protein-coding regions shaded black and 5′- and 3′-UTR regions shaded grey; introns are represented by thin lines.

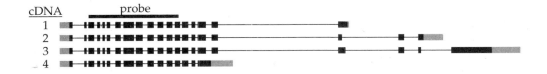

a. How many different protein isoforms are encoded by this gene?

b. Carefully inspect these data and generate a specific hypothesis about the type(s) of posttranscriptional control that could generate these different protein isoforms.

Answer:

a. Four different protein isoforms differing in their C terminus are produced.

b. The cDNA structures indicate that alternative mRNA splicing or alternative poly(A) site choice is used to generate the different protein isoforms. This leads to different 5′ splice sites being used: The last exon of cDNA 4 contains a 5′ splice site that is used by cDNAs 1, 2, and 3; the last exon of cDNA 1 contains a 5′ splice site that is used by cDNAs 2 and 3; and the last exon of cDNA 2 contains a 5′ splice site that is used by cDNA 3. Each cDNA also uses a different poly(A) site. From these data, we can't tell whether alternative splicing leads to the use of different poly(A) sites, or alternative poly(A) site choice leads to the use of different 5′ splice sites.

21

GENETIC ANALYSIS OF DEVELOPMENT

CHAPTER OUTLINE

BASIC EVENTS OF DEVELOPMENT

MODEL ORGANISMS FOR THE GENETIC ANALYSIS OF DEVELOPMENT

DEVELOPMENT RESULTS FROM DIFFERENTIAL GENE EXPRESSION
Constancy of DNA in the Genome During Development
Examples of Differential Gene Activity During Development
Exception to the Constancy of Genomic DNA During Development: DNA Loss
    in Antibody-Producing Cells

CASE STUDY: SEX DETERMINATION AND DOSAGE COMPENSATION IN MAMMALS AND *DROSOPHILA*
Sex Determination in Mammals
Dosage Compensation Mechanism for X-linked Genes in Mammals
Sex Determination in *Drosophila*
Dosage Compensation in *Drosophila*

CASE STUDY: GENETIC REGULATION OF THE DEVELOPMENT OF THE *DROSOPHILA* BODY PLAN
Drosophila Developmental Stages
Embryonic Development
Microarray Analysis of *Drosophila* Development

THINKING ANALYTICALLY

This chapter considers how gene expression can be controlled in a temporal and spatial context during development. It is important that you have a solid grasp on the mechanisms that eukaryotes use to regulate gene expression––it may be helpful to review the material in Chapter 20. Thus far, we have considered gene expression in defined situations and in fairly short timelines. In eukaryotic development, timelines can be much longer, there can be multiple modes of regulation acting simultaneously, and more interactions between genes, and the spatial dimension is added. Therefore, the fundamental question being asked in developmental genetics, "How do different cell types and tissues differentiate from a common precursor cell?" has a complex and detailed answer.

REVIEW OF KEY TERMS, SYMBOLS, AND CONCEPTS

In your own words, write a brief, precise definition of each term in the groups below. Check your definitions using the text. Then develop a concept map using the terms in each list.

1

development
determination
differentiation
totipotent
morphogenesis
fate map
model organism
animal cloning
regeneration
nuclear transplantation
G_0 phase
cell lineage
Saccharomyces cerevisiae
Drosophila melanogaster
Caenorhabditis elegans
Arabidopsis thaliana
Danio rerio
Mus musculus

2

differential gene activity
α-, β-, δ-, γ-, ε-, ζ-globin genes
embryonic, fetal, adult hemoglobin
polytene chromosome
puff
ecdysone
early, late genes
lymphocyte, T, B cell
antigen
antibody
humoral immune response
clonal selection
immunoglobulin
light (L), heavy (H) chain
C, V, J, D segments
somatic recombination
antibody diversity

3

sex determination
dosage compensation
mammal
Drosophila
testis-determining factor, TDF
sex reversal individual
SRY, Sry
X inactivation center (XIC)
Xce, Xist genes
histone methylation
regulation cascade model
Sxl, tra, dsx genes
alternative vs. default splicing
sis-a, sis-b, dpn genes
P_E, P_L promoters at *Sxl*
DSX-F, DSX-M
mle, msl-1, msl-2, msl-3, mof genes
MSL complex
chromatin remodeling

4

Drosophila development
polar cytoplasm
syncitium
cellular blastoderm
parasegments
segments
imaginal discs
maternal effect gene
morphogen
bicoid, caudal, nanos, hunchback genes
segmentation gene
gap, pair rule, segment polarity genes
homeotic gene
selector gene
homeotic mutation
biothorax complex
Antennapedia complex
homeobox
homeodomain
Hox gene

It is useful to maintain a biological perspective as you consider how differential gene expression is regulated during development. For example, it will help to have a thorough understanding of the biology of the system that is developing. Focus on gaining an understanding of how the organism (e.g., *Drosophila, C. elegans*) or population of cells (e.g., the immune system) develops. Then, consider how genes are regulated during its development, and how regulatory events at one stage of development impact on gene expression in subsequent stages. It will be helpful to construct flowcharts that illustrate how genes are regulated along a timeline or in a developmental context: Show how genes are turned on and off, what their targets are, and whether they affect the expression of other genes positively or negatively.

QUESTIONS FOR PRACTICE

Multiple Choice Questions

1. Sex-type in *Drosophila* is controlled by
   a. the ratio of X chromosomes to autosomes
   b. a cascade of regulated, alternative RNA splicing
   c. sex-specific selection of polyadenylation sites
   d. sex-specific transport control
   e. a and b
   f. all of the above

2. A wide diversity of immunoglobulin molecules is obtained via
   a. new gene synthesis
   b. combinatorial gene regulation
   c. alternative RNA splicing
   d. somatic recombination of DNA

3. Evidence for the totipotency of nuclei in eukaryotic cells is provided by
   a. the existence of homeotic mutations
   b. nuclear transplantation experiments
   c. the transcription of fetal globin genes during fetal development but not during adult life
   d. finding that nuclei in different tissues have the same amount of DNA

4. In *Drosophila*, maternal effect genes are genes that
   a. are expressed in the mother during oogenesis and whose products will specify spatial organization in the developing embryo
   b. cause females to lay eggs
   c. affect the number or polarity of body segments
   d. affect the identify of a segment
   e. b, c, and d

5. The homeotic mutations found in *Drosophila*
   a. cause transformation of one segment to another
   b. alter the function of proteins that are regulators of transcription
   c. alter highly conserved functions found in nearly all organisms
   d. do all of the above

6. How does the *dsx* gene in *Drosophila* function in sex determination?
   a. It is active only in XY animals where it activates male differentiation. XX animals lack its function so that a default pathway of female differentiation ensues.
   b. It is active only in XX animals where it activates female differentiation. XY animals lack its function so that a default pathway of male differentiation ensues.

    c. It functions to control transcription at key X-linked genes. In XX animals, these genes are transcribed twice as much as in XY animals. The doubled dose of key X-linked gene products results in female differentiation.

    d. Its pre-mRNA is alternatively spliced in XX animals to make a female-specific protein that represses male differentiation. Its pre-mRNA undergoes default splicing in XY animals to make a male-specific protein that represses female differentiation.

7. Why might an individual have female secondary sexual characteristics but an XY karyotype?

    a. One of their X chromosomes has been inactivated—they are really XXY.

    b. They have a point mutation in the *SRY* gene.

    c. They have a point mutation in the *SXL* gene.

    d. They have a mutation in the *XIST* gene.

8. In *Drosophila*, which of the following types of chromatin modifications is associated with dosage compensation?

    a. Histone H3 methylation

    b. Histone H4 acetylation

    c. Histone H1 demethylation

    d. Histone H2A deacetylation

9. Your neighbor adores his calico cat and, being totally unaware of the availability of stray cats at the humane society, hires a firm to clone her. He receives a report with DNA analyses showing that the clone and his cat are genetically identical and a cat carrier containing the cloned cat. When he releases the clone from the carrier, he is outraged. Not only does the cat look nothing like his beloved original, it hisses at him incessantly. What is the most likely explanation?

    a. X inactivation contributes to the distribution of pigment in calico cats. Since X inactivation is a random process in different cells, the two cats are unlikely to be alike.

    b. Environmental factors in the research laboratory have influenced the cloned cat's personality.

    c. The two cats do not have identical patterns of gene expression because of the incomplete reprogramming of the donor nucleus.

    d. Your neighbor is uninformed about the capabilities of cloning technology.

    e. all of the above

10. In which tissue is fetal hemoglobin containing two α and two γ polypeptides synthesized?

    a. embryonic yolk sac

    b. fetal liver

    c. fetal spleen

    d. fetal bone marrow

    e. b and c

Answers: 1e, 2d, 3b, 4a, 5d, 6d, 7b, 8b, 9e, 10e

Thought Questions

1. Consider the cascade of alternative splicing events that are used to regulate sex type in *Drosophila*. From an evolutionary perspective, what advantages might there be to having a cascade of events regulate such an important process? (Hint: Consider the nature of the initial signal for sex type, the X:A ratio—a ratio of either 1:2 or 2:2—and how a splicing cascade amplifies this signal.)

2. Suppose Gurdon's experiments on totipotency were repeated using differentiated B lymphocytes as the source of nuclei for transplantation. To what extent would you expect these nuclei

to be totipotent? If mature organisms developed from cells with transplanted B cell nuclei, what characteristics would their immune system have?

3. In *Drosophila*, the polar cytoplasm contains factors determining that the nuclei migrating into this region will become germ-line cells. (a) Design an experiment to gather evidence supporting this conclusion. (b) What phenotype would be associated with mutants lacking these factors? (c) Would the mutants from (b) identify maternal effect genes?

4. What are homeoboxes, and what is their role in the regulation of development in eukaryotes?

5. In vertebrates, including mammals such as humans, homeotic gene complexes have been identified using heterologous probes. As discussed in the text, the gene complexes show similar structural organization, even though vertebrates are not segmented in the same way as invertebrates. What significance might this finding have? How might analyzing mutations in these genes in vertebrates help demonstrate the function of these genes and test this significance?

6. Multiple model organisms have been studied intensively with the aim of understanding the genetic control of differential gene expression during development. Describe the features of *Drosophila*, *Caenorhabditis*, and *Arabidopsis* that make them useful for such analysis. Why is it important to study more than one model organism?

7. The primary transcripts of many genes undergo tissue-specific alternative mRNA splicing. Indeed, about 15 percent of human disease mutations are associated with mRNA splicing alterations. (a) For genes that are alternatively spliced, do you expect the proteins that are produced to have identical functions? Why or why not? (b) How might spliceosomes process pre-mRNAs differently in different tissues to achieve alternative mRNA processing? (Hint: Consider the regulation cascade that controls sex determination in *Drosophila*. Are there sex-specific splicing factors?)

8. Dolly is the lamb that was generated by fusing adult mammary epithelial cells with enucleated oocytes. (a) How does this experiment demonstrate totipotency? (b) How could you use molecular methods to experimentally demonstrate that Dolly had the same genotype as the adult mammary epithelial cells? (c) Suppose you sampled Dolly's DNA when she was 1 year old. Do you expect Dolly's simple telomeric repeats to be the same length as other sheep that were 1 year old? (d) What impact might imprinting and incomplete genome reprogramming have on our ability to successfully clone eukaryotic organisms?

9. Exposure of *Drosophila* salivary glands to the steroid hormone ecdysone induces a set of early puffs that are followed by a set of late puffs. What mechanism underlies the differential gene expression associated with this pattern of puffs? How would you experimentally gather evidence to support this mechanism?

10. Compare the processes that lead to chromatin remodeling during dosage compensation in mammals and in *Drosophila*.

Thought Questions for Media

After reviewing the media on the *iGenetics* CD-ROM, try to answer these questions.

1. How is the X:A ratio sensed in *Drosophila*? What molecules constitute the numerator and denominator in this ratio?

2. How does an X:A ratio of 1:1 lead to sex-specific transcriptional activation of *Sxl*?

3. Later in development, *Sxl* is transcribed in both sex-types from a constitutive promoter. Why then is functional SXL protein made only in females?

4. Use diagrams to explain what is meant by a *cascade of alternative splicing events.*

5. In what sense does male development in *Drosophila* represent a default state?

6. What leads to the production of different DSX proteins in males and females?

7. Loss of function mutants at *dsx* result in intersexual development—animals in which male and female development proceeds simultaneously. Explain the origin of this phenotype based on your understanding of the molecular function of the DSX proteins.

8. What is meant by a syncytium?

9. At what point during *Drosophila* development is the germ line established? How is the germ line formed separately from somatic cells?

10. At what point during development do you expect each of the following sets of genes to be transcribed? To be translated?
    a. maternal effect genes
    b. segmentation genes
    c. homeotic genes

11. How is cell fate gradually restricted by the expression of the genes in Thought Question 10?

12. Why might some homeotic mutations be lethal, while others are not?

1. In the iActivity, microarrays were used to identify genes that were differentially expressed between differentiating tissues (mesoderm and endoderm) and stem cells. Suppose two genes, *A* and *B*, are upregulated in differentiating mesoderm relative to stem cells. Gene A is expressed 13-fold more and Gene B is expressed 2.5-fold more. Neither gene is upregulated in differentiating endoderm.
    a. From this data alone, can you infer anything about the *relative* importance of these genes' expression for mesodermal differentiation?
    b. Suppose mutants were available that caused the rate of each gene's transcription to be halved. Can you predict which mutation might have a more severe effect on mesodermal differentiation?
    c. Based on the results of the microarray analyses, can you infer which gene produces more mRNA copies (the total number of transcripts per gene) in differentiating mesodermal cells?

2. What is meant by a gene that can *serve as a marker* for endoderm differentiation? What features must such a gene have?

3. A gene that encodes a transcription factor is expressed at high levels in differentiating endoderm and differentiating mesoderm, but not in stem cells. What types of genes might be targets of this transcription factor? Would it invariably act through positive regulation of these targets?

4. A gene that encodes a transcription factor is expressed at stem cells, but not in differentiating cells. What role might this gene play in keeping stem cells from differentiating? Would it invariably act through negative regulation of target genes?

SOLUTIONS TO TEXT PROBLEMS

21.1   Distinguish between the terms *development*, *determination*, and *differentiation*.

   Answer: Development is a process of regulated growth and cellular change. It results from the interaction of the genome with the cytoplasm and external environment. It involves a programmed sequence of phenotypic changes that are typically irreversible. Initially, cells of the zygote are totipotent—they have the potential to develop into any cell type of the complete organism. As the cells of the zygote divide and interact, they may lose their totipotency and become *determined*—they follow a genetic program that sets their fate. *Differentiation* refers to the process of cellular change in development that leads to the formation of distinct types of cells, tissues, and organs through the regulation of gene expression. Determination and differentiation are thus a part of development and lead to cells that have characteristic structural and functional properties.

21.2   What is totipotency? Give an example of the evidence for the existence of this phenomenon. What two mechanisms are used to restrict a cell's totipotency during development?

   Answer: Totipotency refers to the capacity of a nucleus to direct a cell through all of the stages of development. A nucleus taken from the cell of a differentiated tissue is said to be totipotent if, when it is injected into an enucleated egg, it can direct the development of the organism to the adult stage. If the egg is able to develop into an adult, the differentiated nucleus must have retained all of the genetic information needed to direct development again from the start. In 1975, Gurdon demonstrated that the nucleus of a skin cell of an adult frog was at least partly totipotent. When injected into an enucleated egg, such a nucleus could direct development to the tadpole stage. Then, in 1997, Wilmut and his colleagues demonstrated that a mammalian nucleus was totipotent. They showed that the nucleus of a mammary epithelium-derived cell could direct the development of a sheep to the adult stage. Since then, other mammals, including mice, rats, goats, cattle, horses, cats, and monkeys have been successfully cloned.

   A cell's totipotency is restricted as it becomes determined to adopt different cell fates. This occurs either by induction where an inductive signal produced by one cell or group of cells affects the development of another cell or group of cells, or by asymmetric cell division which leads to a new distribution of cell-determining molecules in daughter cells.

21.3   It is possible to excise small pieces of early embryos of the frog, transplant them to older embryos, and follow the course of development of the transplanted material as the older embryo develops. A piece of tissue is excised from a region of the late blastula or early gastrula that would later develop into an eye and is transplanted to three different regions of an older embryo host (see part a of Figure 21.A). If the tissue is transplanted to the head region of the host, it will form eye, brain, and other material characteristic of the head region. If the tissue is transplanted to other regions of the host, it will form organs and tissues characteristic of those regions in normal development (e.g., ear, kidney). In contrast, if tissue destined to be an eye is excised from a neurula and transplanted into an older embryo host to exactly the same places as used for the blastula or gastrula transplants, in every case the transplanted tissue differentiates into an eye (see part b of Figure 21.A). Explain these results.

Figure 17.A   (a) **Tissue from late blastula**   (b) **Tissue from neurula**
                        or early gastrula

Area that will later
become neural tissue

Excise tissue
that would later
become an eye

Excise tissue
that would later
become an eye

Transplant into older
host embryo

Transplant into older
host embryo

Develops
into a kidney

Develops
into an eye

Develops into   Develops
head material   into an ear
(eye, brain)

Develops   Develops
into an eye   into an eye

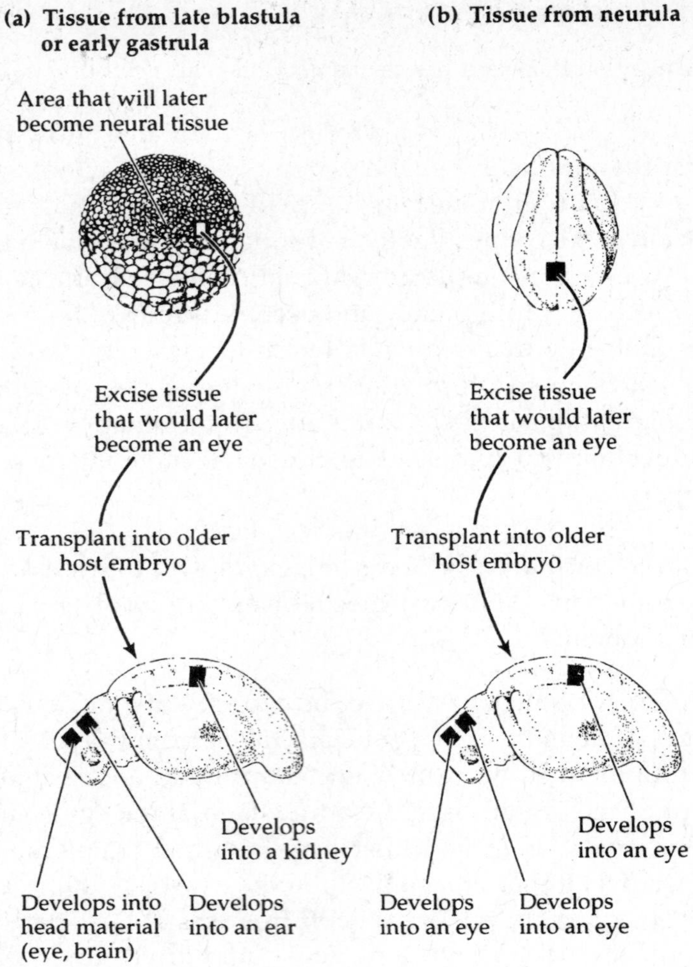

Answer: This experiment demonstrates the phenomenon of *determination* and when it occurs during development. The tissue taken from the blastula or gastrula has not yet been committed to its final differentiated state in terms of its genetic programming; that is, it has not yet been *determined*. Thus, when the tissue is transplanted into the host, it adopts the fate of nearby tissues and differentiates in the same way as they do. Presumably, cues from the tissue surrounding the transplant determine its fate. In contrast, tissues in the neurula stage are stably determined. By the time the neurula developmental stage has been reached, a developmental program has been set. In other words, the fate of neurula tissue transplants is *determined*. Upon transplantation, they will differentiate according to their own set genetic program. Tissue transplanted from a neurula to an older embryo cannot be influenced by the surrounding tissues. It will develop into the tissue type for which it has been determined, in this case, an eye.

21.4   With respect to the genetic analysis of development, what is meant by a *model* organism? Describe the features that model organisms possess to make them attractive for the genetic analysis of development, using specific examples.

Answer: A model organism for the genetic analysis of development must have two fundamental attributes: It must develop, and mutants affecting developmental processes must be available for use in developmental genetic analyses. Model organisms that are attractive for the genetic analysis of development share many properties: They are often easy to culture, develop over relatively short time periods, have a rich array of mutants that affect developmental processes, have well-characterized genetics so that crosses can be performed and analyzed, and have characterized and sequenced genomes that facilitate molecular analyses. Organisms such as *Saccharomyces cerevisiae*, *Drosophila melanogaster*, *Caenorhabditis elegans*, *Arabidopsis thaliana*, *Danio rerio*, and *Mus musculus* all have these properties to different degrees. The choice among organisms depends on the developmental process that is being studied. For example, the yeast *S. cerevisiae*, even with its limited developmental repertoire, is ideal for investigating fundamental questions such as the role of cellular signaling in differentiation. The fly *D. melanogaster* has a rich array of mutants, making it well suited for detailed analyses of developmental mechanisms such as those underlying sex-determination and pattern formation. The nematode *C. elegans* has a completely defined fate map, so it is well suited to investigations addressing how cell fate is determined. Finally, *A. thaliana*, *D. rerio*, and *M. musculus* serve as models for studying plant, vertebrate embryogenesis, and mammalian development, respectively.

21.5 In the set of experiments used to clone Dolly, six additional live lambs were obtained. Why is the production of Dolly more significant than the production of the other lambs? What is the evidence that Dolly resulted from the fusion of a nucleus from one cell with the cytoplasm of another?

Answer: The production of Dolly was significant because it demonstrated the apparent complete totipotency of a nucleus from a mature mammary epithelium cell. The other six lambs developed from the transplanted nuclei of embryonic or fetal cells, which one might expect to be less determined and have a higher degree of totipotency due to their younger developmental age.

Two types of evidence indicate that Dolly resulted from the fusion of a nucleus of a mature mammary epithelium cell with the cytoplasm of a donor egg. First, Dolly had the same phenotype as the ewe who donated the mammary epithelium cell—Finn Dorset (whiteface)—and not that of the surrogate mother—Scottish blackface. Second, analysis of four microsatellite markers demonstrated that Dolly's DNA matched the DNA of the donor mammary epithelial cells but not the DNA of the recipient ewe.

21.6 In Woody Allen's 1973 film *Sleeper*, the aging leader of a futuristic totalitarian society has been dismembered in a bomb attack. The government wants to clone the leader from his only remaining intact body part, a nose. The characters Miles and Luna thwart the cloning by abducting the nose and flattening it under a steam roller.

a. In light of the 1996 cloning of the sheep Dolly, how should the cloning have proceeded if Miles and Luna had not intervened?

b. If methods like those used for Dolly had been successful, in what genetic ways would the cloned leader be unlike the original?

c. Suppose that instead of a nose, only mature B cells (B lymphocytes of the immune system) were available. What genetic deficits would you expect in the "new leader"?

d. Based on what has been discovered about cloned cats and mice, in what nongenetic ways might the cloned leader differ from the original? What (constructive) advice would you give the totalitarian government, based on these findings?

e. If the cloning of the leader had succeeded, can you make any prediction about whether the "cloned leader" would be interested in perpetuating the totalitarian state?

Answer:

a. Based on the work of Wilmut and his colleagues, the nose cells would first be dissociated and grown in tissue culture. The cells would be induced into a quiescent state (the G_0 phase of the cell cycle) by reducing the concentration of growth serum in the medium. Then they would be fused with enucleated oocytes from a donor female and allowed to grow and divide by mitosis to produce embryos. The embryos would be implanted into a surrogate female. After the establishment of pregnancy, its progression would need to be maintained.

b. While the nuclear genome would generally be identical to that in the original nose cell, cytoplasmic organelles presumably would derive from those in the enucleated oocyte. Therefore, the mitochondrial DNA would not derive from the original leader. In addition, because telomeres in an older individual are shorter, one might expect the telomeres in the cloned leader to be those of an older individual.

c. In mature B cells, DNA rearrangements at the heavy- and light-chain immunoglobulin genes have occurred. One would expect the cloned leader to be immunocompromised, as he would be unable to make the wide spectrum of antibodies present in a normal individual.

d. It is likely that the cloning process will be very inefficient, with most clones dying before or soon after birth. The surviving clones are also likely to differ in body shape and personality, and are unlikely to be normal since a nucleus donated from the differentiated nose cell is unlikely to be completely reprogrammed. One suggestion is for the government to hire a good plastic surgeon to alter the appearance of a good actor able to assume the role of the totalitarian leader.

e. There is no way to predict the psychological profile of the cloned leader based on his genetic identity. Even identical twins, who are genetically more identical than such a clone, do not always share behavioral traits.

21.7   Discuss some of the evidence for differential gene activity during development. How have microarray analyses enhanced our understanding of this process?

Answer: The evidence for differential gene activity during development is vast. Classic lines of evidence stem from (1) studies on the differential expression of the α, β, γ, δ, ε, and ζ classes of globin genes during development, (2) differential puffing patterns in the polytene chromosomes in Dipteran insects, and (3) the studies on genes that are expressed in a temporal and spatially specific manner during the development of *Drosophila*. See the text for additional discussion.

Microarray analysis was used to study changes in *Drosophila* gene expression patterns brought about by the hormone ecdysone during the metamorphosis of the larva into a pupa. About 40 percent of the genes in *Drosophila* were analyzed (N = 6,240). Of these, about 8 percent (N = 534) exhibited differential gene expression. This study provided researchers with a catalog of genes that are induced or repressed by ecdysone during this developmental window. For example, it showed that four hours prior to the formation of the pupa, a set of genes involved in the differentiation of the nervous system is induced, while another set of genes required for muscle formation is repressed in anticipation of the future breakdown of larval muscle tissues.

21.8   Discuss the expression of human hemoglobin genes during development.

Answer: Separate loci code for α-like and β-like globin polypeptides which form distinct types of hemoglobin at different times during human development. In the embryo, hemoglobin is initially made in the yolk sac and consists of two ζ polypeptides and two ε polypeptides. ζ polypeptides are α-like, while ε polypeptides are β-like. At about three months of gestation, hemoglobin synthesis switches to the fetal liver and spleen. There, hemoglobin is made that consists of two α polypeptides and two β-like polypeptides, either two γA polypeptides or two γG polypeptides. Just before birth, hemoglobin synthesis switches to the bone marrow, where predominantly α polypeptides and β polypeptides are made along with some β-like δ polypeptides.

21.9   How are the hemoglobin genes organized in the human genome, and how is this organization related to their temporal expression during development?

Answer: All of the α-like genes (the ζ and α genes and three pseudogenes) are located in a gene cluster on chromosome 16, while all of the β-like genes (ε, γG, γA, δ, and β genes and one pseudogene) are located in a gene cluster on chromosome 11. At a very general level, the organization of the two gene clusters is similar. Both sets of genes are arranged on the chromosome in an order that exactly parallels the time when the genes are transcribed during human development. In both gene clusters, the genes transcribed in the embryo are at the left end of the cluster; the genes transcribed in the fetus are to the right of these; and the genes transcribed in the adult are farthest to the right. This is intriguing since the genes are also transcribed in different tissues.

21.10   In humans, β-thalassemia is a disease caused by failure to produce sufficient β-globin chains. In many cases, the mutation causing the disease is a deletion of all or part of the β-globin structural gene. Individuals homozygous for certain of the β-thalassemia mutations are able to survive because their bone marrow cells produce γ-globin chains. The γ-globin chains combine with α-globin chains to produce fetal hemoglobin. In these people, fetal hemoglobin is produced by the bone marrow cells throughout life, whereas normally it is produced in the fetal liver. Use your knowledge about gene regulation during development to suggest a mechanism by which this expression of γ-globin might occur in β-thalassemia.

Answer: There are a number of possibilities. One is that the γ-globin genes in bone marrow are under negative regulation by β-globin (or some metabolite of it). When β-globin is not formed, the γ-globin gene is derepressed.

21.11   What are polytene chromosomes? Discuss the molecular nature of the puffs that occur in polytene chromosomes during development.

Answer: Polytene chromosomes occur in Dipteran insects such as *Drosophila*. Polytene chromosomes are formed by endoreduplication, in which repeated cycles of chromosome duplication occur without nuclear division or chromosome segregation. Since they can be 1,000 times as thick as the corresponding chromosomes in meiosis or in the nuclei of normal cells, they can be stained and viewed at the light microscope level. Distinct bands, or chromomeres, are visible, and genes are located both in bands and in interband regions. A puff results when a gene in a band or interband region is expressed at very high levels in a particular developmental stage. Puffs are accompanied by a loosening of the chromatin structure that allows for efficient transcription of a particular DNA region. When increased transcriptional activity at the gene ceases at a later developmental stage, the puff disappears, and the chromosome resumes its compact configuration. In this way, the appearance and disappearance of puffs provides a visual representation of differential gene activity.

21.12   Puffs of regions of the polytene chromosomes in salivary glands of *Drosophila* are surrounded by RNA molecules. How would you show that this RNA is single stranded and not double stranded?

Answer: That RNA molecules are present in puffs can be demonstrated by feeding or injecting *Drosophila* larvae with radioactive uridine, an RNA precursor. After this treatment, salivary glands are dissected from larvae and a spread of the polytene chromosomes is subjected to autoradiography to detect the location of the incorporated uridine. Silver grains are evident over puffed regions, indicating that they contain RNA. Evidence that the RNA is single stranded might be obtained by treating such spreads with an RNase that was capable of digesting only single-stranded RNA. The radiolabel should be recovered in the solution, and not remain bound to the salivary gland chromosome. A lack of signal on puffs following autoradiographic detection (compared to controls) would provide evidence that the RNA in puffs is single stranded.

21.13   In experiment A, ^{3}H-thymidine (a radioactive precursor of DNA) is injected into larvae of *Chironomus*, and the polytene chromosomes of the salivary glands are later examined by autoradiography. The radioactivity is seen to be distributed evenly throughout the polytene chromosomes. In experiment B, ^{3}H-uridine (a radioactive precursor of RNA) is injected into the larvae, and the polytene chromosomes are examined. The radioactivity is first found only around puffs; later, radioactivity is also found in the cytoplasm. In experiment C, actinomycin D (an inhibitor of transcription) is injected into larvae and then ^{3}H-uridine is injected. No radioactivity is found associated with the polytene chromosomes, and few puffs are seen. The puffs that are present are much smaller than the puffs found in experiments A and B. Interpret these results.

Answer: Experiment A results in all of the DNA becoming radioactively labeled. The distribution of radioactive label throughout the polytene chromosomes indicates that DNA is a fundamental and major component of these chromosomes. The even distribution of label suggests that each region of the chromosome has been replicated to the same extent. This provides support for the contention that band and interband regions are the result of different types of packaging, not different amounts of DNA replication. Experiment B results in the radioactive labeling of RNA molecules. The finding that label is found first in puffs indicates that these are sites of transcriptional activity that arise from molecules that are in the process of being synthesized. The later appearance of label in the cytoplasm reflects the completed RNA molecules that have been processed and transported into the cytoplasm, where they will be translated. Experiment C provides additional support for the hypothesis that transcriptional activity is associated with puffs. The inhibition of RNA transcription by actinomycin D blocks the appearance of signal over puffs, indicating that it blocks the incorporation of ^{3}H-uridine into RNA in puffed regions. The fact that the puffs are much smaller indicates that the puffing process itself is associated with the onset of transcriptional activity for the genes in a specific region of the chromosome.

21.14 Explain how it is possible for both of the following statements to be true: The mammalian genome contains about 10^5 genes. Mammals can produce about 10^6 to 10^8 different antibodies.

Answer: The ability to make 10^6 to 10^8 different antibodies arises from the combinatorial way in which antibody genes are generated in different antibody-producing cells during their development, and not the existence of this many separate antibody genes in each and every mammalian cell. A template that exists in germ-line cells is processed differently during the development of different antibody-producing cells to generate antibody diversity.

 Antibody molecules consist of two light (L) chains and two heavy (H) chains. The amino acid sequence of one domain of each type of chain is variable, and generates antibody diversity. In the germ-line DNA of mammals, coding regions for these immunoglobulin chains exist in tandem arrays of gene segments. For light chains, there are many variable (V) region-gene segments, a few joining (J) segments, and one constant (C) gene segment. Somatic recombination during development results in the production of a recombinant V-J-C DNA molecule that, when transcribed, produces a unique functional L chain. From a particular gene in one cell, only one L chain is produced. A large number of L chains are obtained by recombining the gene segments in many different ways. Diversity in these L chains results from variability in the sequences of the multiple V segments, variability in the sequences of the four J segments, and variability in the number of nucleotide pairs deleted at the V-J joints. H chains are similar, except that several D (diversity) segments can be used between the V and J segments, increasing the possible diversity of recombinant H chain genes. The type of C gene segment chosen for the constant domains of the H chain determines whether the antibody is IgM, IgD, IgG, IgE, or IgA.

21.15 Antibody molecules (Ig) are composed of four polypeptide chains (two of one light-chain type and two of one heavy-chain type) held together by disulfide bonds.
   a. If for the light chain there were 300 different V_κ segments and four J_κ segments, how many different light chain combinations would be possible?
   b. If for the heavy chain there were 200 V_H segments, 12 D segments, and 4 J_H segments, how many heavy chain combinations would be possible?
   c. Given the information in (a) and (b), what would be the number of possible types of IgG molecules (L + H chain combinations)?

Answer:
   a. Not considering the variability in the number of nucleotide pairs deleted at the V_κ-J_κ joint, there are 300 (V_κ segments) $\times$ 4 (J_κ segments) = 1,200 different light chain combinations. This number is a lower bound for an estimate, since significantly more variability can be obtained by using imprecise V-J joining.

b. Not considering the imprecise joining of gene segments that comprise the chain's variable region, there are 200 (V_H segments) $\times$ 12 (D segments) $\times$ 4 (J_H segments) = 9,600 different heavy chain combinations for each of the eight types of Ig (IgM, IgD, IgG_3, IgG_1, IgG_{2b}, IgG_{2a}, IgE, or IgA; see p. 583 and Fig. 21.13, p. 584). If all possible types of Ig are considered, there are 9,600 $\times$ 8 (C_H segments) = 76,800 heavy chain combinations. As in part (a), this is a lower bound for an estimate of heavy chain variability.

c. IgG has four C segments (γ_3, γ_1, γ_{2b}, γ_{2a}; see p. 585 and Fig. 21.13, p. 584). An estimate of the minimal number of combinations for IgG molecules would be 1,200 $\times$ 9,600 $\times$ 4 = 4.608 $\times$ 10^7.

21.16 How was the testis-determining factor gene (*TDF*) identified, and what evidence is there to support the contention that the *SRY* gene is the TDF gene?

Answer: The gene for TDF was identified by analyzing rare sex-reversal individuals—males who are XX and females who are XY. Cytogenetic analysis of these individuals revealed that XX males had attached to one of their X chromosomes a small fragment from near the tip of the small arm of the Y chromosome, and that XY females had deletions of the same region. Molecular analyses identified a male-specific gene—the *SRY* gene—lying within the translocated region and identified an equivalent gene, *Sry*, in mice. Three lines of evidence support the view that the *SRY* gene is the gene for TDF: (1) The mouse *Sry* gene is expressed in the undifferentiated genital ridges of the embryo just before the formation of the testes, that is, where it would be expected to be expressed if it was the gene for TDF; (2) transgenic XX mice carrying a 14-kb DNA fragment with the *Sry* gene develop testes and have male secondary sexual development; and (3) some rare XY human females have simple mutations in the *SRY* gene, suggesting loss-of-function mutations at *SRY* lead to sex reversal.

21.17 A male mouse cell line has been generated in which the *gfp* (*green fluorescent protein*) gene has been inserted onto the X chromosome under the control of a constitutively expressed promoter. Cells expressing the *gfp* gene exhibit bright-green fluorescence under UV light.

a. What pattern of green fluorescence do you expect to see in this cell line?

b. The cell line is modified by introducing a segment of DNA containing *Xic* into an autosome. How do you expect the pattern of green fluorescence to change? Why?

c. A cell of the modified cell line described in (b) exhibits green fluorescence. Which copy of *Xist*—the one on the X chromosome or the one on the autosome—is being expressed? On which chromosome does its expression lead to chromatin remodeling, and how?

Answer:

a. Each cell should exhibit green fluorescence, since the gene is constitutively expressed.

b. About half the cells will exhibit green fluorescence. If more than one *Xic* is present, X inactivation will occur on one *Xic*-containing chromosome. Either the X chromosome or the *Xic*-bearing autosome will be inactivated, at random. If the X chromosome is inactivated, the *gfp* gene will not be expressed.

c. The *Xist* gene on the autosome is being expressed. Since the cell exhibits green fluorescence, the X chromosome with the *gfp* gene is not inactivated and the *Xic*-bearing autosome is inactivated. The *Xist* gene on the autosome is transcribed and its RNA coats the autosome to trigger the methylation of histone H3. This initiates chromatin remodeling to silence genes on the *Xic*-bearing autosome.

21.18   In *Drosophila,* sex type is determined by the X:A ratio.

   a. How is this ratio detected early in development, and how does it lead to sex-specific transcription at *Sxl?*

   b. A mutation in *Sxl* affects P_E so that early transcription of *Sxl* does not occur. The upstream P_L promoter is unaffected, so constitutive transcription from this late promoter occurs in all cells regardless of their X:A ratio. What phenotype do you expect this mutation to have in animals with an X:A ratio of 1:2? In animals with an X:A ratio of 2:2?

   c. A *tra* mutant has a nonsense mutation into exon 2. What phenotype do you expect this mutant to have in animals with an X:A ratio of 1:2? In animals with an X:A ratio of 2:2?

   d. The TRA protein targets the *dsx* pre-mRNA for alternatively splicing. However, if TRA is not present, male differentiation ensues. Why then do animals with knockout mutations in *dsx* have both male and female characteristics?

Answer:

   a. The X:A ratio is detected by interactions of the protein products of three X-linked numerator genes (*sis-a, sis-b, sis-c*) and one autosomal denominator gene (*dpn*). The numerator gene products can either form homodimers or heterodimers with the denominator gene product. When the X:A ratio is 2:2, an excess of numerator gene products leads to the formation of many homodimers. These serve as transcription factors to activate *Sxl* transcription from P_E. When the X:A ratio is 1:2, most numerator subunits are found in heterodimers, so *Sxl* transcription is not activated. Therefore, activation of *Sxl* at P_E by the homodimers serves to detect the X:A ratio and leads to the early sex-specific synthesis of SXL protein.

   b. Transcription at P_E is essential to generate a functional SXL protein in animals with an X:A ratio of 2:2. It is not used in individuals with an X:A ratio of 1:2, so these animals will be unaffected, and differentiate as males. In individuals with an X:A ratio of 2:2, SXL initiates a cascade of alternative mRNA splicing at *Sxl, tra,* and *dsx* that leads to the implementation of female differentiation. If there is no transcription from P_E, no functional SXL protein will be produced in these animals, and a default set of splice choices at *Sxl, tra,* and *dsx* will be used. In principle, this would lead to male differentiation in individuals with an X:A ratio of 2:2. However, SXL also prevents the translation of *msl-2* transcripts so that dosage compensation does not normally occur in individuals with an X:A ratio of 2:2. Without SXL, *msl-2* transcripts will be translated so dosage compensation will occur, leading to four doses of X-linked gene products. The imbalance in X and autosomal gene product dosage is likely to be lethal to these animals.

   c. This *tra* mutation will eliminate functional TRA protein. Since functional TRA is not normally present in animals with an X:A ratio of 1:2, this mutation will have no effect on these animals—they will differentiate normally into males. TRA is normally present in individuals with an X:A ratio of 2:2, where it functions to regulate alternative splicing at *dsx* and produce DSX-F, which implements female differentiation by repressing male-specific gene expression. Without TRA, default splicing will occur at *dsx* and produce DSX-M, which implements male differentiation by repressing female-specific gene expression. Thus, animals with an X:A ratio of 2:2 will be males.

d. Animals with knockout mutations at *dsx* will produce neither DSX-M, which represses female-specific gene expression, nor DSX-F, which represses male-specific gene expression. Therefore, neither male- nor female-specific gene expression will be repressed, and both male and female differentiation pathways will proceed.

21.19 The SXL protein binds to its own mRNA as well as the mRNAs of *tra* and *msl-2*. How does it regulate its own expression through mRNA binding? Is this mechanism the same as, or different from, the mechanism by which it regulates the expression of *tra* and *msl-2*?

Answer: The early SXL protein binds to the *Sxl* pre-mRNA to cause alternative splicing: Exons E1 and 3 are skipped and exons L1, 2, 4, 5, 6, 7, and 8 are included. The resulting mRNA produces a functional late SXL protein. In the absence of the early SXL protein, default splicing occurs to produce a transcript that includes exon 3. This exon has a stop codon in frame with the start codon at the beginning of exon 2, so no functional SXL protein is produced. The SXL protein regulates *tra* in a similar manner: SXL binds to the *tra* pre-mRNA to produce an mRNA that encodes an active TRA protein. In the absence of SXL, a default stop codon containing exon is included and an active TRA protein is not produced.

In contrast to its role in alternative mRNA splicing at *Sxl* and *tra*, the late SXL protein serves to block translation of *msl-2* transcripts. In XX animals (females), the SXL late protein binds to the transcript of *msl-2*. This blocks its translation so that no MSL2 protein is produced. As a result, dosage compensation does not occur. In XY animals (males) where SXL protein is not produced, the *msl-2* transcript is translated and dosage compensation occurs.

21.20 In *Drosophila*, mutations at five genes (*mle, msl-1, msl-2, msl-3,* and *mof*) lead to male-specific lethality during the larval stages due to defective dosage compensation.

a. What common process does each of these genes function in, and how does it lead to dosage compensation in *Drosophila* males?

b. Explain why females with these mutations develop normally.

c. How is the mechanism by which dosage compensation occurs in *Drosophila* related to the molecular steps regulating sex determination?

d. Why do *Sxl* gain-of-function mutations cause male lethality, while *Sxl* loss-of-function mutations have no effect on male development?

Answer:

a. Each of these genes functions in dosage compensation, the twofold increase in the transcriptional activity of the male's single X chromosome that occurs to match the transcriptional activity of the female's two X chromosomes. Initially, the protein products of these five genes form a complex that binds to about 35 chromatin entry sites (CES) on the male's X chromosome. Then, the MSL complexes spread from those sites in both directions into the flanking chromatin. The MOF protein of the MSL complex is a histone acetyltransferase, and during the spreading of the complexes along the X chromosome, it remodels chromatin to allow for the higher level of transcription of the X chromosome genes in males.

b. In females, the MLE, MSL-1, MSL-3, and MOF proteins are produced. The MSL-2 protein is not produced (see the solution to Question 21.19). Without the MSL-2 protein, the MSL complex is unable to bind to the X chromosome. Since none of the proteins function in females, mutations in their genes do not affect females.

c. The mechanisms for sex determination and dosage compensation are interrelated through the function of the *Sxl* gene. In XX animals (females), SXL protein is produced. SXL initiates a cascade of regulated splicing at the *tra* and *dsx* genes that culminates in female sexual differentiation. SXL also blocks the translation of *msl-2* transcripts so that dosage compensation does not occur in XX animals. In XY animals (males), active SXL protein is not produced. This results in default splicing at the *tra* and *dsx* genes and male sexual differentiation, and allows the translation of *msl-2* transcripts so that dosage compensation can occur.

d. Gain-of-function mutations at *Sxl* will result in XY animals having an active SXL protein. The SXL protein binds to *msl-2* transcripts to prevent their translation, so dosage compensation will not occur. Therefore, these males will have their single X chromosome transcribed only half as much as normal males or normal females and have an imbalance in gene function that is lethal. Loss-of-function mutations at *Sxl* will not affect males because no SXL protein is made in normal males.

21.21   The following figure shows the percentage of ribosomes found in polysomes in unfertilized sea urchin oocytes (0 hours) and at various times after fertilization:

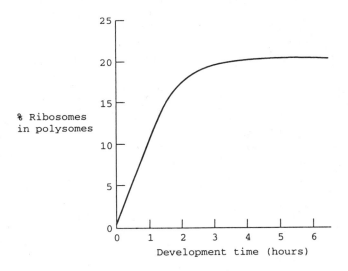

In the unfertilized egg, less than 1 percent of ribosomes are present in polysomes, and at 2 hours postfertilization about 20 percent of ribosomes are present in polysomes. It is known that no new mRNA is made during the time period shown. How can these data be interpreted?

Answer: Preexisting, maternally packaged mRNAs that have been stored in the oocyte are recruited into polysomes as development begins following fertilization.

21.22   Define *imaginal disc* and *homeotic mutant*.

Answer: An *imaginal disc* is a larval structure in *Drosophila* and other insects that undergo a metamorphic transformation from a larva to an adult. During metamorphosis in the pupal period, it will differentiate into an adult structure. It is essentially a sac of cells, a disc, that is set aside early in embryonic development. The disc increases in cell number by mitosis, but remains undifferentiated during larval growth. Although the disc remains undifferentiated up until the time hormonal signals activate the differentiation of adult structures during the pupal period, its fate is *determined* very early in development. Thus, separate imaginal discs are destined to differentiate into different adult structures, such as legs, eyes, antennae, and wings.

Homeotic mutants alter the identity of particular segments, transforming them into copies of other segments. Homeotic mutations thus affect the determination, or fate, of a disc. Mutations in the *Antennapedia* gene complex can, for example, transform the fate of cells of the antenna disc so that a leg is formed where an antenna should be. The general inference from such observations is that homeotic genes normally play a pivotal role in establishing the segmental identity of an undifferentiated cell.

21.23   Both *bicoid* and *nanos* are maternal effect genes whose mRNAs are transcribed in the mother and then localized in the cytoplasm of developing embryos.

    a. How does maternal deposition of these mRNAs lead to gradients of morphogens along the anterior-posterior axis of the developing embryo?

    b. What would be the effect of loss-of-function mutations in *bicoid* on the distribution of CAUDAL protein, and how does this contribute to the phenotype of these *bicoid* mutations?

    c. What would be the effect of loss-of-function mutations in *nanos* on the distribution of HUNCHBACK protein, and how does this contribute to the phenotype of these *nanos* mutations?

    d. What is the function of the morphogen gradients established by these two genes?

Answer:

    a. The *bicoid* mRNA is localized to the anterior pole of the egg cytoplasm by other anterior maternal effect genes. When translated, its protein product diffuses to form an anterior-to-posterior gradient—the highest concentration of BICOID is at the anterior end of the egg and little to no BICOID protein is in the posterior third of the egg. In contrast, the *nanos* mRNA is localized to the posterior pole of the egg cytoplasm by a group of posterior maternal effect genes and, when translated, the NANOS protein forms a high-posterior-to-low-anterior gradient.

        The BICOID protein functions both as a transcription factor, activating and repressing genes along the anterior-posterior axis of the embryo, and as a translational repressor, blocking translation of the evenly distributed *caudal* mRNA. Thus, production of the BICOID protein leads to a high-posterior-to-low-anterior gradient of CAUDAL protein. The NANOS protein also functions to repress translation, specifically translation of *hunchback* mRNAs. Since NANOS is distributed in a high-posterior-to-low-anterior gradient, HUNCHBACK protein will be distributed in a high-anterior-to-low-posterior gradient.

    b. One of the functions of the BICOID protein is to block the translation of the *caudal* mRNA. In loss-of-function *bicoid* mutations, *caudal* mRNA would be translated uniformly throughout the embryo. Since the CAUDAL protein functions later in the segmentation phase of development to activate genes needed for the formation of posterior structures, one might expect that this would lead to the formation of posterior structures at both ends of the embryo, not just the posterior end. Indeed, this is the phenotype seen in *bicoid* mutants.

c. In loss-of-function *nanos* mutations, the evenly distributed *hunchback* mRNAs will be translated so that HUNCHBACK protein is uniformly distributed in the egg. HUNCH-BACK is important for defining segments within the developing embryo. The uniform distribution of HUNCHBACK would fail to define segments properly. This is one aspect of loss-of-function *nanos* mutations, which have a no-abdomen phenotype.

d. The morphogenetic gradients established by *bicoid* and *nanos* direct the formation of gradients of CAUDAL and HUNCHBACK proteins, and, more generally, set the stage for the formation of anterior and posterior structures through the direct or indirect activation or repression of segmentation genes—genes that determine segments and have specific roles in specifying regions of the embryo.

21.24   Imagine that you observed the following mutants (*a–e*) in *Drosophila*.

a. Mutant *a:* In homozygotes, phenotype is normal, except wings are oriented backward.

b. Mutant *b:* Homozygous females are normal but produce larvae that have a head at each end and no distal ends. Homozygous males produce normal offspring (assuming the mate is not a homozygous female).

c. Mutant *c:* Homozygotes have very short abdomens, which are missing segments A2 through A4.

d. Mutant *d:* Affected flies have wings growing out of their heads in place of eyes.

e. Mutant *e:* Homozygotes have shortened thoracic regions and lack the second and third pair of legs.

Based on the characteristics given, assign each of the mutants to one of the following categories: maternal effect gene, segmentation gene, or homeotic gene.

Answer:

a. As the *polarity* of an individual, differentiated structure is altered, *a* is likely to be a segmentation gene. In particular, it is likely to be a member of the subclass of segmentation genes known as segment polarity genes.

b. The abnormal appearance of progeny of homozygous mothers, but not homozygous fathers, suggests that homozygous mothers produce abnormal oocytes. This view is bolstered by the phenotype of the abnormal progeny. Progeny with defect(s) at one end could arise if a maternally produced gradient in the oocyte was abnormal. Thus, *b* is likely to be a maternal effect gene.

c. The absence of a set of segments is characteristic of the phenotypes of gap gene mutants, a subclass of segmentation gene mutants.

d. It would appear that the fate of the eye disc has been changed to that of a wing disc. The segmental *identity* of cells is controlled by homeotic genes, and so *d* is likely to be a homeotic mutation.

e. In mutant *e*, a set of segments is missing (T2 and T3). Thus, like *c*, *e* is likely to be a mutant in a gap gene, a subclass of segmentation gene mutants.

21.25   What is the evidence that homeotic genes specify not only the invertebrate body plan, but also the vertebrate body plan?

Answer: Homeotic gene complexes are found in all major animal phyla except for sponges and coelenterates. They share a similar, clustered gene organization and have highly conserved homeobox sequences. The most compelling evidence that the homeotic genes specify the vertebrate body plan is provided by data on their patterns of gene expression during vertebrate development, an analysis of the effects of mutations, and embryological analyses. See also text Figure 21.29, p. 600.

21.26   If actinomycin D, an antibiotic that inhibits RNA synthesis, is added to newly fertilized frog eggs, there is no significant effect on protein synthesis in the eggs. Similar experiments have shown that actinomycin D has little effect on protein synthesis in embryos up until the gastrula stage. After the gastrula stage, however, protein synthesis is significantly inhibited by actinomycin D, and the embryo does not develop further. Interpret these results.

Answer: Preexisting mRNA that was made by the mother and packaged into the oocyte prior to fertilization is translated up to the gastrula stage. After gastrulation, new mRNA synthesis is necessary for the production of proteins needed for subsequent embryonic development.

22

GENETICS OF CANCER

CHAPTER OUTLINE

THINKING ANALYTICALLY

Although cancer genetics is complex, focusing on two key issues should help put the topics of this chapter in perspective. Since cancer results from unconstrained cellular growth, the first key issue is to understand how each of the processes discussed in this chapter affects cellular growth. It will be helpful to review how cells proceed through a cell cycle. As you read the text, pay particular attention to how progression through the cell cycle is controlled. Notice how mutations in oncogenes and tumor suppressor genes, both of which impact the cell cycle, can result in cancer through different mechanisms.

A second key issue is that cancer typically occurs in somatic cells, and so it is important to consider the effect of accumulating mutations in somatic cells and not only germline cells. To understand the relationship between hereditary and sporadic forms of cancer, it will help to review the relationship between somatic and germ-line cells. To understand how mutations accumulate in cancerous cells, it is necessary to consider the role of DNA repair within the cell cycle, the consequences of errors in DNA repair, and the effects of carcinogens and ionizing radiation.

REVIEW OF KEY TERMS, SYMBOLS, AND CONCEPTS

In your own words, write a brief, precise definition of each term in the groups below. Check your definitions using the text. Then develop a concept map using the terms in each list.

1	2	3
tumor	tumor virus	multistep nature of cancer
neoplasm	oncogene	cell cycle
transformation	DNA vs. RNA tumor virus	pRB, E2F
monolayer	sarcoma	G_0 phase
contact inhibition	retrovirus	cyclin Cdk complex
benign	RSV, HIV-1	phosphorylation
malignant	*gag, pol, env, src* genes	p53, *TP53*, Mdm2, *ARF*
cancer	nononcogenic retrovirus	p21, *WAF1*
metastasis	insertional mutagenesis	G_1-to-S, G_2-to-M, M
oncogenesis	AIDS	checkpoints
terminal differentiation	cytolytic virus	*START*
stem cell	transducing retrovirus	BRCA1, BRCA2
self-renewal	nontransducing retrovirus	apoptosis
germ-line cell	replication-competent	cell cycle arrest
somatic cell	transformation-competent	mutator gene
growth factor	viral oncogene, v-*onc*	tumor suppressor
growth-inhibiting factor	proto-oncogene	oncogene
signal transduction	cellular oncogene, c-*onc*	mismatch repair
cell cycle	growth factor	HNPCC, hereditary FAP
dedifferentiation	protein kinase	*hMSH2, hMLH1, hPMS1,*
two-hit model	proviral DNA	*hPMS2, mutS, mutL* genes
familial cancer	tumor suppressor gene	telomerase
sporadic cancer	signaling cascade	replicative senescence
tumor suppressor gene	gene amplification	carcinogen
oncogene, proto-oncogene		direct-acting carcinogen
mutator gene		ultimate carcinogen
multistep nature of cancer		procarcinogen
sporadic vs. hereditary		UVA, UVB
retinoblastoma		ionizing, non-ionizing
loss of heterozygosity		radiation
unilateral, bilateral tumor		

QUESTIONS FOR PRACTICE

Multiple Choice Questions

1. About what percentage of breast cancer is hereditary?
   a. 1%
   b. 5%

   c. 10%

   d. 50%

2. Familial retinoblastoma appears to be inherited as a dominant trait. What factor(s) accounts for this?

   a. A dominant mutation is transmitted through the germ line.

   b. Somatic mutation occurs after a recessive mutation is transmitted through the germ line.

   c. Heterozygous individuals are infected with a retrovirus.

   d. About half of the homozygous, normal individuals in an affected family are infected with a retrovirus at birth.

3. RNA tumor viruses cause cancer

   a. by the same mechanism as DNA tumor viruses

   b. only if their proviral DNA integrates near a proto-oncogene

   c. if they carry an oncogene

   d. if they infect germ-line cells

4. Which of the following is *not* a feature of a viral oncogene?

   a. It lacks introns.

   b. Its expression is regulated identically to the corresponding cellular proto-oncogene.

   c. Its structure and location in the viral genome varies.

   d. It is not conserved between viral strains.

5. Which of the following is untrue about the product of the *TP53* gene?

   a. It is a transcription factor.

   b. It is a tumor suppressor.

   c. It can induce cell death.

   d. When present, it activates the cell cycle from G_1 to S.

6. What is the cellular function of the unphosphorylated product of the *RB* gene, pRB?

   a. pRB binds to transcription factors E2F to maintain cells in G_1 or G_0.

   b. pRB binds to p53 to stimulate apoptosis.

   c. pRB is a growth factor, stimulating cellular proliferation.

   d. pRB is involved in DNA mismatch repair.

7. Which type of protein phosphorylates the product of the *RB* gene, pRB?

   a. a receptor kinase

   b. a phosphorylase

   c. a cyclin/cyclin-dependent kinase

   d. p53

8. What is the relationship between a proto-oncogene and its corresponding viral oncogene?

   a. Only the viral oncogene contains introns.

   b. If the viral oncogene encodes a transcription factor, the corresponding proto-oncogene will encode a growth factor.

   c. The proto-oncogene encodes a product essential to the normal development and function of the organism. The viral oncogene encodes an altered product that has aberrant function.

   d. Usually, the proto-oncogene product is produced in greater amounts. The viral oncogene, being under the control of the retroviral promoter, enhancer, and poly(A) signals, produces smaller amounts of a product.

9. What is apoptosis?

   a. the process by which cancerous cells extend cellular processes in an amoeba-like manner

   b. the process of uncontrollable cellular growth caused by overexpression of an oncogene

   c. cell death that results from the multiple mutations that accumulate in a cancerous cell

   d. the process of programmed cell death

10. A genetic predisposition to hereditary nonpolyposis colon cancer (HNPCC) can result from mutations in any of the four genes *hMSH2, hMLH1, hPMS1,* and *hPMS2.* All of these genes are homologous to *E. coli* and yeast genes known to be involved in DNA repair. From this information, one might speculate that HNPCC is caused by a mutation
    a. in a mutator gene
    b. in an oncogene
    c. in a proto-oncogene
    d. in a tumor suppressor gene

Answers: 1b, 2b, 3c, 4b, 5d, 6a, 7c, 8c, 9d, 10a

Thought Questions

1. What is the relationship of the cell cycle to cancer?

2. What is the two-hit mutation model for cancer? How is it related to the multistep nature of cancer?

3. In all retroviral RNA genomes, specific sequences are found at the left (R and U_5) and right (R and U_3) ends. In proviral DNA, long terminal repeats (LTRs) are found at each end. How are the LTRs produced?

4. What is meant by a nononcogenic retrovirus? Can infection by a nononcogenic retrovirus lead to increased susceptibility to cancer? Give an example.

5. Distinguish between a viral oncogene, a cellular oncogene, and a proto-oncogene. Are each of these always associated with a transducing retrovirus?

6. There are at least six functional classes of oncogene, products. What are these functional classes? How can alteration in such diverse functions lead to a similar oncogenic phenotype?

7. Does the HIV virus itself have viral oncogenes? Why are individuals infected with HIV more susceptible to cancer and infection?

8. What information would you want to know about the UV lights used in tanning salons? Do you think they can be safe?

9. In what ways is programmed cell death a normal process?

10. What is the most lethal common carcinogen you know of?

11. Cancers are staged according to how far they have progressed toward matastasis. This is illustrated in text Figure 22.13 for the development of hereditary adenomatous polyposis, a colorectal cancer. What genetic changes are associated with the staged progression of tumors? Is there a specific order of changes that *always* occurs? Once a tumor has progressed to a certain stage, can it revert to an earlier stage?

Thought Questions for Media

After reviewing the media on the *iGenetics* CD-ROM, try to answer these questions.

1. How do growth-stimulating factors and growth-inhibiting factors differ, if at all, in terms of being able to
    a. turn on pathways?
    b. bind to specific cell membrane receptors?
    c. transduce signals to the nucleus?
    d. trigger the expression of one or more genes for one or more proteins that stimulate cell division?

2. Many growth-stimulating and growth-inhibiting factors have been identified. Is it the presence or absence of specific factors, the shear number of stimulating or inhibiting factors, or the balance between factors that determines whether a normal, healthy cell in the G_1 phase of the cell cycle is triggered to proceed into the S, G, and M phases?

3. After DNA damage occurs, how are the levels of p53 protein altered?

4. What is the immediate consequence of altering the level of p53 protein?

5. Why do normal cells undergo apoptosis after extensive DNA damage? Why don't cells with mutations in each of their *TP53* genes undergo apoptosis?

6. How is a cyclin-Cdk complex involved in controlling the ability of a cell to proceed from G_1 into S phase? What is the role of the p21 protein in regulating this process?

7. Why are cells that proceed into S phase with unrepaired DNA damage more likely to become cancerous?

8. The animation illustrated how the p53 protein, following DNA damage, triggers a cellular pathway with inhibitory and stimulatory events.
   a. Which cellular events are inhibited by p53?
   b. Which cellular events are stimulated by p53?
   c. How are these inhibitory and stimulatory events connected?

SOLUTIONS TO TEXT PROBLEMS

22.1   Progression of cells through the cell cycle is tightly regulated, and cancerous cells fail to respond to signals that normally regulate cellular proliferation.

    a. Which stages of the cell cycle are subject to regulation?

    b. What types of proteins regulate progression through the cell cycle, and what is the role of protein phosphorylation in this process?

Answer:

    a. The cell cycle is regulated at cell cycle checkpoints that occur between G_1 and S (the START checkpoint in yeast), between G_2 and M, and during M just before the separation of chromatids.

    b. The cyclins and cyclin-dependent kinases (Cdks) regulate progression through the cell cycle. The levels of the cyclins oscillate in a regular pattern through the cell cycle. At each checkpoint, a cyclin binds to a Cdk to form a complex. When the cell is ready to pass the checkpoint, the complex is activated. For example, at the G_2-M checkpoint, the cyclin-Cdk complex is activated by dephosphorylation of the Cdk. The activated Cdk then phosphorylates other cell cycle control proteins to affect their function, thereby leading to the cell's transition into the next phase of the cell cycle.

22.2   Explain why HIV-1, the causative agent of AIDS, is considered a nononcogenic retrovirus, even though numerous types of cancers are frequently seen in patients with AIDS.

Answer: HIV-1 is a retrovirus that infects and then kills cells of the immune system. This leads to a disabling of the immune response and an increase in susceptibility to certain types of cancer. In contrast to the mechanism by which cancer occurs following infection with HIV-1, an oncogenic retrovirus transforms the growth properties of infected cells to neoplastic growth.

22.3   Distinguish between a transducing retrovirus and a nontransducing retrovirus.

Answer: Retroviruses are viruses that have two copies of an RNA genome within a protein core surrounded by a membranous envelope. Transducing retroviruses carry an oncogene from a host cell, while nontransducing retroviruses do not. Consequently, transducing retroviruses may be capable of transforming cells to a cancerous state.

22.4   Cellular proto-oncogenes and viral oncogenes are related in sequence, but they are not identical. What is the fundamental difference between the two?

Answer: Cellular proto-oncogenes have important roles in regulating normal cellular processes such as cell division and differentiation. Viral oncogenes are mutated, abnormally expressed forms of proto-oncogenes that cause neoplastic growth. Viral oncogenes lack introns and, as shown in text Figure 22.7, p. 615, are often fused to viral genes.

22.5   An autopsy of a cat that died from feline sarcoma revealed neoplastic cells in the muscle and bone marrow but not in the brain, liver, or kidney. To gather evidence for the hypothesis that the virus FeSV contributed to the cancer, Southern blot analysis (see Chapter 16, pp. 439–441) was performed on DNA isolated from these tissues and on a cDNA clone of the FeSV viral genome. The DNA was digested with the enzyme *Hind*III, separated by size on an agarose gel, and transferred to a membrane. The resulting Southern blot was hybridized with a ^{32}P-labeled probe made from a 1.0-kb *Hind*III fragment of the feline *fes* proto-oncogene cDNA. The autoradiogram revealed a 3.4-kb band in each lane, with an additional 1.2-kb band in the lanes with muscle and bone marrow DNA. Only a 1.2-kb band was seen in the lane loaded with *Hind*III-cut FeSV cDNA. Explain these results, including the size of the bands seen. Do these results support the hypothesis?

Answer: If FeSV contributed to the feline sarcoma, FeSV should be found in the neoplastic tissues (muscle and bone marrow). The Southern blot provides this evidence: A 1.2-kb DNA fragment hybridizes to the *fes* cDNA probe in the lanes with DNA from muscle and bone marrow, and in the control lane with FeSV cDNA. The size difference between the 1.2-kb hybridizing fragment and the 1.0-kb *fes* proto-oncogene *Hind*III-cut cDNA probe reflects their different origins. The *fes* proto-oncogene is found normally in a cat, while the FeSV *fes* oncogene is found in a retrovirus. The size of the fragment in the retrovirus may reflect a polymorphic *Hind*III site and/or a gene rearrangement. The *fes* proto-oncogene normally functions in the cat, so its DNA should be present in all tissues. The 3.4-kb DNA fragment is found in all of the cat tissues, so is likely to be the genomic sequence. Since the *fes* proto-oncogene cDNA has a 1.0-kb *Hind*III fragment, the mRNA of this gene is very likely spliced to remove a 2.4-kb intron.

22.6   The sequences of proto-oncogenes are highly conserved among a large number of animal species. Based on this fact, what hypothesis can you make about the functions of the proto-oncogenes?

Answer: The high degree of conservation of proto-oncogenes suggests that they function in normal, essential, conserved cellular processes. Given the relationship between oncogenes and proto-oncogenes, it also suggests that cancer occurs when these processes are not correctly regulated.

22.7   Proto-oncogenes produce a diverse set of gene products.
   a. What types of gene products are made by proto-oncogenes? Do these gene products share any features?
   b. Which of the following mutations might result in an oncogene?
      i.   a deletion of the entire coding region of a proto-oncogene
      ii.   a deletion of a silencer that lies 5′ to the coding region
      iii. a deletion of an enhancer that lies 3′ to the coding region
      iv.   a deletion of a 3′ splice site acceptor region
      v.   the introduction of a premature stop codon
      vi.   a point mutation (single base-pair change in the DNA)
      vii. a translocation that places the coding region near a constitutively transcribed gene
      viii. a translocation that places the gene near constitutive heterochromatin

   Answer:
   a. Proto-oncogenes encode a diverse set of gene products that include growth factors, receptor and nonreceptor protein kinases, receptors lacking protein kinase activity,

membrane-associated GTP-binding proteins, cytoplasmic regulators involved in intracellular signaling, and nuclear transcription factors. These gene products all function in intercellular and intracellular pathways that regulate cell division and differentiation.

b. In general, mutations that activate a proto-oncogene convert it into an oncogene. Since (i), (iii), and (viii) cause a decrease in gene expression, they are unlikely to result in an oncogene. Since (ii) and (vii) could activate gene expression, they could result in an oncogene. Mutations (iv), (v), and (vi) cannot be predicted with certainty. The deletion of a 3' splice site acceptor would alter the mature mRNA and possibly the protein produced and may or may not affect the protein's function and regulation. Similarly, it is difficult to predict the effect of a nonspecific point mutation or a premature stop codon. The text presents examples where these types of mutations have caused the activation of a proto-oncogene and resulted in an oncogene.

22.8   Explain the likely mechanism underlying the formation of an oncogene in each of the following mutations:

a. a mutation in the promoter for platelet-derived growth factor that leads to an increase in the efficiency of transcription initiation

b. a mutation affecting a regulatory domain of a nonreceptor tyrosine kinase that causes it always to be active

c. a mutation affecting the structure of a membrane-associated G protein that eliminates its ability to hydrolyze GTP.

Answer:

a. Increased transcription of the mRNA will lead to increased levels of the growth factor, which in turn will stimulate fibroblast growth and division.

b. Constitutive activation of a nonreceptor tyrosine kinase could lead to abberant, unregulated phosphorylation and activation of many different proteins, including growth factor receptors, that are involved in signaling cascades used in regulating cellular growth and differentiation.

c. When the growth factor EGF binds its membrane-bound receptor, it stimulates its autophosphorylation, which allows Grb2 to bind and recruit SOS. SOS displaces GDP from Ras, a membrane-associated G protein, so that it can bind GTP. This allows Ras to recruit and activate Raf-1 to initiate the cytoplasmic MAP kinase signaling cascade. In turn, this activates transcription factors such as Elk-1 to induce transcription of cell cycle–specific target genes. Therefore, if a membrane-associated G protein were unable to hydrolyze GTP, it would be constitutively active and lead to constant expression of genes needed for cell cycle progression.

22.9   Describe a pathway leading to altered nuclear transcription following the binding of a growth factor to a surface-membrane-bound receptor. How can mutations in genes involved in this pathway result in oncogenes?

Answer: One such pathway is illustrated in text Figure 22.8, p. 617. As shown in this figure, the binding of EGF to its surface-membrane-bound receptor stimulates a signaling cascade that leads to transcriptional activation of cell cycle–specific target genes. First, EGF binding stimulates autophosphorylation of the receptor, which allows binding of

GRB2. In turn, GRB2 binding recuits SOS to the plasma membrane, which displaces GDP from Ras. This allows Ras to bind GTP, which recruits and activates Raf-1. The activated Raf-1 stimulates the MAP-kinase cascade—a cascade of cytoplasm-based protein phosphorylations—to produce phosphorylated ERK. Phosphorylated ERK moves from the cytoplasm into the nucleus where it activates, through phosphorylation, several transcription factors, including Elk-1. The activated transcription factors then turn on the transcription of defined sets of cell cycle–specific target genes.

Mutations that lead to the constitutive or aberrant activation of the signaling pathway result in oncogenes. Three specific examples are mutations that activate EGF-receptor autophosphorylation in the absence of EGF binding, mutations that abolish the ability of Ras to hydrolyze GTP to GDP, and mutations that lead to the constitutive activation of Elk-1. All of these mutations are oncogenic, because they lead to the constitutive transcriptional activation of cell cycle–specific target genes.

22.10   What are the three main ways in which a proto-oncogene can be changed into an oncogene?

Answer: A proto-oncogene can be changed into an oncogene if there is an increase in the activity of its gene product or an increase in gene expression that leads to an increased amount of gene product. This can result from point mutation, deletion, or gene amplification.

22.11   List two ways in which cancer can be induced by a retrovirus.

Answer: One mechanism by which a transducing retrovirus causes cancer is for the viral oncogene it carries to be expressed under the control of retroviral promoters. A second mechanism involves insertional mutagenesis that results when the proviral DNA of a retrovirus integrates near a proto-oncogene. In this event, the expression of the proto-oncogene can be controlled by the promoter and enhancer sequences in the retroviral LTR. As these sequences do not respond to the environmental signals that normally regulate proto-oncogene expression, overexpression of the proto-oncogene is induced.

22.12   After a retrovirus that does not carry an oncogene infects a particular cell, northern blots indicate that the amount of mRNAs transcribed from a particular proto-oncogene became elevated approximately 13-fold compared with uninfected control cells. Propose a hypothesis to explain this result.

Answer: One hypothesis is that the proviral DNA has integrated near the proto-oncogene, and the expression of the proto-oncogene has come under the control of promoter and enhancer sequences in the retroviral LTR. This could be assessed by performing a whole-genome Southern blot analysis to determine whether the organization of the genomic DNA sequences near the proto-oncogene has been altered.

22.13   In what ways is the mechanism of cell transformation by transducing retroviruses fundamentally different from transformation by DNA tumor viruses? Even though the mechanisms are different, how are both able to cause neoplastic growth?

Answer: Tumor growth induced by transforming retroviruses results either from the activity of a single viral oncogene or from the activation of a proto-oncogene caused by the nearby integration of the proviral DNA. The oncogene can cause abnormal cellular proliferation via the variety of mechanisms discussed in the text. The expression of a proto-oncogene, normally tightly regulated during cell growth and development, can be altered if it comes under the control of the promoter and enhancer sequences in the retroviral LTR.

DNA tumor viruses do not carry oncogenes. They transform cells through the action of one or more genes within their genomes. For example, in a rare event, the DNA virus can be integrated into the host genome, and the DNA replication of the host cell may be stimulated by a viral protein that activates viral DNA replication. This would cause the cell to move from the G_0 to the S phase of the cell cycle.

For both transducing retroviruses and DNA tumor viruses, abnormally expressed proteins lead to the activation of the cell from G_0 to S and abnormal cell growth.

22.14   You have a culture of normal cells and a culture of cells dividing uncontrollably (isolated from a tumor). Experimentally, how might you determine whether uncontrolled growth was the result of an oncogene or a mutated pair of tumor suppressor alleles?

Answer: Experimentally fuse cells from the two cell lines and then test the resultant hybrids for their ability to form tumors. If the uncontrolled growth of the tumor cell line was caused by a mutated pair of tumor suppressor alleles, the normal alleles present in the normal cell line would "rescue" the tumor cell line defect. The hybrid line would grow normally and be unable to form a tumor. If the uncontrolled growth of the tumor cell line was caused by an oncogene, the oncogene would also be present in the hybrid cell line. The hybrid line would grow uncontrollably and form a tumor.

22.15   What is the difference between a hereditary cancer and a sporadic cancer?

Answer: Hereditary cancer is associated with the inheritance of a germ-line mutation; sporadic cancer is not. Consequently, hereditary cancer runs in families. For some cancers, both hereditary and sporadic forms exist, with the hereditary form being much less frequent. For example, retinoblastoma occurs when both normal alleles of the tumor suppressor gene *RB* are inactivated. In hereditary retinoblastoma, a mutated, inactive allele is transmitted via the germ line. Retinoblastoma occurs in cells of an *RB*/+ heterozygote when an additional somatic mutation occurs. In the sporadic form of the disease, retinoblastoma occurs when both alleles are inactivated somatically.

22.16   Why are mutations in tumor suppressors, such as mutations in *RB* and *TP53*, said to cause recessive disorders when they appear to be dominant in pedigrees?

Answer: Mutations in tumor suppressor genes such as *RB* and *TP53* are recessive because the cancer develops only if both alleles are mutant. The presence of one normal allele in a heterozygote serves to suppress tumor formation, so the normal allele is dominant. The disorder appears dominant in pedigrees because in an individual who has inherited a single mutation—an individual who has inherited a mutant allele from one parent and a normal allele from the other parent—there is a high likelihood that the remaining wild-type allele will be inactivated by mutation. That is, the inheritance of the single gene mutation predisposes a person to the cancer, but does not cause it directly.

22.17   Individuals with hereditary retinoblastoma are heterozygous for a mutation in the *RB* gene; however, their cancerous cells often have two identically mutated *RB* alleles. Describe three different mechanisms by which the normal *RB* allele can be lost. Illustrate your answer with diagrams.

Answer: Two identical mutated alleles can arise by chromosome nondisjunction during mitosis (this is similar in mechanism to nondisjunction during meiosis II, illustrated in Figure 3.20, p. 59); or by mitotic recombination (illustrated in Figures 7.8, p. 175; 7.10, p. 177; and 7.11, p. 178). A third mechanism that would lead to the loss of the normal *RB* allele and result in a loss of heterozygousity would be a second mutational hit at the *RB* locus—a somatic mutation that deleted the *RB* gene. While gene conversion by mismatch repair (illustrated in Figure 11.18, p. 303) generates two identical alleles, it requires pairing of homologous chromosomes during meiosis and so would not be expected to occur in somatic cells such as those of the retina.

22.18   Although there has been a substantial increase in our understanding of the genetic basis for cancer, the vast majority of cases of many types of cancer are not hereditary.
a. How might studying a familial form of a cancer provide insight into a similar, more frequent sporadic form?
b. The incidence of cancer in several members of an extended family might reasonably raise concern as to whether there is a genetic predisposition for cancer in the family. What does the term *genetic predisposition* mean? What might be the basis of a genetic predisposition to a cancer that appears as a dominant trait? What issues must be addressed before concluding that a genetic predisposition for a specific type of cancer exists in a particular family?

Answer:
a. Studies of hereditary forms of cancer have led to insights into the fundamental cellular processes affected by cancer. For example, substantial insights into the important role of DNA repair and the relationship between the control of the cell cycle and DNA repair have come from analyses of the genes responsible for hereditary forms of human colorectal cancer. For breast cancer, studying the normal functions of the *BRCA1* and *BRCA2* genes promises to provide substantial insights into breast and ovarian cancer.
b. *Genetic predisposition* for cancer refers to the presence of an inherited mutation that, with additional somatic mutations during the individual's life span, can lead to cancer. For diseases such as retinoblastoma, a genetic predisposition has been associated with the inheritance of a recessive allele of the *RB* tumor suppressor gene. Retinoblastoma occurs in *RB/rb* individuals when the normal allele is mutated in somatic cells and the pRB protein no longer functions. Because somatic mutation is likely, the disease appears dominant in pedigrees.
   Although there is a substantial understanding of the genetic basis for cancer and the genetic abnormalities present in somatic cancerous cells, there are also substantial

environmental risk factors for specific cancers. Environmental risk factors must be investigated thoroughly when a pedigree is evaluated for a genetic predisposition for cancer.

22.19 Explain how progression through the cell cycle is regulated by the phosphorylation of the retinoblastoma protein pRB. What phenotypes might you expect in cells where

    a. pRB was constitutively phosphorylated?

    b. pRB was never phosphorylated?

    c. a severely truncated pRB protein was produced that could not be phosphorylated?

    d. a normal pRB protein was produced at higher than normal levels?

    e. a normal pRB protein was produced at lower than normal levels?

Answer: When unphosphorylated, pRB inhibits cell cycle progression by binding to a complex of the transcription factors E2F. This ensures that the cell remains in G_1 or in the quiescent G_0 state until it is ready to proceed to S. When progression into S phase is signaled, pRB is phosphorylated by a cyclin/Cdk complex. This prevents it from binding to E2F, which in turn allows E2F to activate transcription at genes needed for cell cycle progression. After S phase, pRB is dephosphorylated.

    a. If pRB were constitutively phosphorylated, it would never bind to E2F, there would be no inhibition of cell cycle progression at G_1/S, and neoplastic growth would occur.

    b. If pRB was never phosphorylated, cells would be stalled in the G_0 phase (and presumably eventually undergo apoptosis).

    c. If a truncated protein were produced, it would be unable to bind to E2F and (regardless of its phosphorylation state) would be unable to inhibit cell cycle progression. Just as in (a), this would result in neoplastic growth.

    d. Higher levels of normal pRB could result in a longer cell cycle. If more pRB were present in the cell, more unphosphorylated pRB would be available to bind to E2F during G_1. In this event, if an unphosphorylated pRB molecule that was bound to a particular E2F molecule were phosphorylated by a cyclin/Cdk complex and the E2F molecule were released, another unphosphorylated pRB molecule present in the cell could bind to the same E2F molecule. This would prevent this particular E2F molecule from participating in activating the gene expression needed for progression into S phase. Only when enough pRB molecules are phosphorylated will a sufficient number of E2F molecules be released to provide for transcriptional activation. Since the number of pRB molecules is increased relative to the number of cyclin/Cdk complexes, it will take longer for the cyclin/Cdk complexes to phosphorylate enough pRB molecules to effect transcriptional activation. If pRB levels are so high that the cyclin/Cdk complex is unable to phosphorylate enough of the pRB molecules present, cells could be stalled in the G_0 phase, just as in (b).

    e. Lower levels of normal pRB might result in a shorter cell cycle.

22.20 Mutations in the *TP53* gene appear to be a major factor in the development of human cancer.

    a. Discuss the normal cellular functions of the *TP53* gene product and how alterations in these functions can lead to cancer.

b. Explain how mutations at both alleles of *TP53* may be involved in 50 percent of all human cancer when familial cancers caused by mutations in *TP53* are rare and associated with a specific type of cancer, Li-Fraumeni syndrome.

c. Suppose cells in a cancerous growth are shown to have a genetic alteration that results in diminished *TP53* gene function. Why can we not immediately conclude that the mutation *caused* the cancer? How would the effect of the mutation be viewed in light of the current, multistep model of cancer?

Answer:

a. The *TP53* gene produces the transcription factor p53. p53 has cellular roles in the repair of DNA damage, in protecting the cell against oncogenes, and in programmed cell death (apoptosis). The p53 protein is regulated by phosphorylation and by interactions with another phosphoprotein, Mdm2. In a normal cell, neither protein is phosphorylated and they are able to bind together. Mdm2 stimulates degradation of p53 so that the amount of p53 is kept low in normal cells.

When DNA damage occurs, p53 initiates a cascade of events leading to cellular arrest in G_1. DNA damage results in the phosphorylation of both p53 and Mdm2 on the domains where they normally interact. The phosphorylation prevents their interaction and p53 degradation, so that p53 accumulates. p53 then activates transcription of DNA repair genes and of *WAF1*, which encodes p21. The p21 protein binds to the G_1-to-S checkpoint Cdk4-cyclin D complexes and inhibits their activity. This results in the failure of pRB phosphorylation, the inhibition of E2F, and cell arrest in G_1.

When viral or cellular oncogenes such as *ras* are expressed, the expression of the *ARF* gene is induced. Its product, p14, binds to Mdm2 in the p53-Mdm2 complex and blocks Mdm2's stimulation of p53 degradation. This leads to the activation of gene transcription by p53 to provide protection against oncogenes.

During apoptosis, p53 does not induce DNA repair genes or *WAF1*, but activates the *BAX* gene. The BAX protein blocks the function of the BCL-2 protein, which is a repressor of the apoptosis pathway. Without the BCL-2 repressor, the apoptotic pathway is activated and the cell commits suicide.

Alterations in *p53* function can lead to cancer in three ways: failure to hold cells in G_0/G_1 arrest, failure to allow DNA repair prior to S (thus allowing mutations to be transmitted to the next generation), or failure to induce apoptosis appropriately—when cells have high levels of DNA damage.

b. Li-Fraumeni syndrome results from the inheritance of one mutant copy of *TP53*. This rare cancer develops when the second copy of *TP53* becomes mutated in somatic cells. In contrast to this rare form of cancer, many cancers accumulate mutations in both alleles of the *TP53* gene during their progression. That is, mutations in *TP53* do not cause them, but *TP53* mutations are among the several genetic changes that are found in them.

c. Cancerous cells typically have a number of genetic abnormalities, or lesions. The *TP53* mutation may not be the initial genetic abnormality that occurred. Rather, it may have occurred subsequent to a mutation that led to the activation of an oncogene, the introduction of a mutator gene, or the inactivation of alleles at a tumor suppressor gene. Hence, the *TP53* mutation may not be the primary genetic lesion. Most cancers develop as a result of the accumulation of mutations in a number of genes, often over extended periods of time (decades). Such mutations accumulate in part due to the lack of successful DNA repair at the G_1-to-S checkpoint in the cell cycle.

22.21   The p53 protein can influence multiple pathways involved in tumor formation.

    a. Explain how the functions of p53 are regulated by phosphorylation.

    b. Through what pathway does the phosphorylation of p53 influence phosphorylation of pRB to control cell cycle progression?

    c. What pathways can be activated by p53 in response to DNA damage? What determines which pathway is activated?

Answer:

    a. In a normal cell, neither p53 nor a protein that interacts with it, Mdm2, is phosphorylated. They bind together and Mdm2 stimulates degradation of p53 so that the amount of p53 is kept low in normal cells. However, when DNA damage occurs, both p53 and Mdm2 are phosphorylated so that their interaction is prevented. This leads to the accumulation of p53.

    b. After the phosphorylation of p53 leads to its accumulation, p53 activates transcription of *WAF1*, which encodes p21. The p21 protein binds to the G_1-to-S checkpoint Cdk4-cyclin D complexes and inhibits their activity. Without the activity of the Cdk4-cyclin D complexes, pRB in the pRB-E2F complex does not become phosphorylated, so that E2F remains inhibited. This leads to cell arrest in G_1.

    c. Under low levels of DNA damage—levels that can be repaired—phosphorylation of p53 leads to the activation of *WAF1* as described in part b. When high levels of DNA damage occur the cell activates a pathway leading to apoptosis. In this pathway, p53 activates the *BAX* gene, whose protein product blocks the function of the BCL-2 protein. The BCL-2 protein normally represses the apoptosis pathway. Without it, the apoptotic pathway is activated.

22.22   What is apoptosis? Why is the cell death associated with apoptosis desirable, and how is it regulated?

Answer: Apoptosis is programmed, or suicidal, cell death. Cells targeted for apoptosis are those that have high levels of DNA damage and so are at a greater risk for neoplastic transformation. (During the development of some tissues in multicellular organisms, cell death via apoptosis is a normal process.) Apoptosis is regulated by p53, among other proteins. In cells with large amounts of DNA damage, p53 accumulates and functions as a transcription factor to activate transcription of DNA repair genes and *WAF1*, whose product, p21, leads cells to arrest in G_1. If very high levels of DNA damage exist, p53 does not induce DNA repair genes and *WAF1*, but activates the *BAX* gene, whose product blocks the BCL-2 protein from repressing the apoptosis pathway. By blocking BCL-2 function, the apoptosis pathway is activated.

22.23   Two alternative hypotheses have been proposed to explain the functions of *BRCA1* and *BRCA2*. One hypothesis proposes that these are tumor suppressor genes while the other proposes that these are mutator genes.

    a. Distinguish between a tumor suppressor gene and a mutator gene.

    b. What cellular roles do *BRCA1* and *BRCA2* have that is consistent with their being tumor suppressor genes? Mutator genes?

Answer:

a. A tumor suppressor gene suppresses uncontrolled cell proliferation that is characteristic of cancer cells, while a mutator gene is any gene that, when mutant, increases the spontaneous mutation frequencies of other genes.

b. Both *BRCA1* and *BRCA2* have been proposed to function in homologous recombination, cellular responses to DNA damage, and mRNA transcription. Their proposed involvement in the repair of DNA damage is consistent with these genes being tumor suppressor genes—*TP53*, a known tumor suppressor gene is also involved in the repair of DNA damage. Indeed, the patterns of inheritance seen for *BRCA1* and *BRCA2* mutations—patterns similar to that of hereditary retinoblastoma—are like those expected of tumor suppressor genes. However, their proposed involvement in the cellular response to DNA damage and in homologous recombination is also consistent with these genes being mutator genes, as the normal (unmutated) forms of mutator genes are often involved in DNA replication and DNA repair.

22.24 Telomerase activity is not normally present in differentiated cells, but is almost always present in cancerous cells. Explain whether telomerase activity alone can lead to cancer. If it cannot, why is it present in most cancerous cells, and what would be the biological consequences of eliminating it from cancerous cells?

Answer: Telomerase activity alone does not lead to cancer—it is normally found in germ-line cells that do not invariably become cancerous. However, the enzyme is reactivated as a secondary event in all major human cancer types. The presence of telomerase enables cancer cells to maintain telomere length. Long telomeres contribute to chromosome stability, which is necessary for cells to proliferate indefinitely. If telomerase were eliminated from cancer cells, the telomeres would shorten and eventually become so short that the complex between telomere sequences and telomere-binding proteins would be disrupted. This would lead to DNA damage that, in normal cells, would trigger apoptosis. However, the apoptotic pathway might not be activated in cancerous cells, especially if they had already lost *TP53* gene function. Nonetheless, cancerous cells without active telomerase are unlikely to be able to divide for many generations because the too-short telomeres create unstable chromosomes.

22.25 Material that has been biopsied from tumors is useful for discerning both the type of tumor and the stage to which a tumor has progressed. It has been known for a long time that biopsied tissues with more differentiated cellular phenotypes are associated with less advanced tumors. Explain this finding in terms of the multistep nature of cancer.

Answer: Tumors result from multiple mutational events that typically involve both the activation of oncogenes and the inactivation of tumor suppressor genes. The analysis of hereditary adenomatous polyposis, an inherited form of colorectal cancer, has shown that the more differentiated cells found in benign, early-stage tumors are associated with fewer mutational events, while the less differentiated cells found in malignant and metastatic tumors are associated with more mutational events. Although the path by which mutations accumulate varies between tumors, additional mutations that activate oncogenes and inactivate tumor suppressor genes generally result in dedifferentiation and increased cellular proliferation.

22.26   Some tumor types have been very frequently associated with specific chromosomal translocations. In some cases, these translocations are found as the only cytogenetic abnormality. In each case examined to date, the chromosome breaks that occur result in a chimeric gene (pieces of two genes fused together) that encodes a fusion protein (a protein consisting of parts of two proteins fused together, corresponding to the coding sequences of the chimeric gene). A partial list of tumor-specific chromosomal translocations in bone and soft tissue tumors is given here.

Tumor Type	Translocation	Characteristic	Genes
Ewing sarcoma	t(11;22) (q24;q12) t(21;22) (q22;q12) t(7;21) (q22;q12)	Malignant	FLI1, EWS, ERD, EWS, ATV1, EWS
Soft tissue clear cell carcinoma	t(12;22) (q13;q12)	Malignant	ATF1, EWS
Myxoid chondro-sarcoma	t(9;22) (q22–31;q12)	Malignant	CHN, EWS
Synocial sarcoma	t(11;22) (q13;q12)	Malignant	SSX1, SSX2, SYT
Lipoma	t(var;12) (var;q13–15)	Benign	HMGI-C
Leiomyoma	t(12;14) (q13–15;q23–24)	Benign	HMGI-C

a. What conclusions might you draw from the fact that in some cases these translocations are found as the only cytogenetic abnormality?

b. How might the formation of a chimeric protein result in tumor formation?

c. Based on the data presented here, can you infer whether the genes near the break points of these translocations are tumor suppressor genes or proto-oncogenes? If so, which?

d. Can you speculate as to how multiple translocations involving the *EWS* gene result in different sarcomas?

e. It is often difficult to diagnose individual sarcoma types based solely on tissue biopsy and clinical symptoms. How might the cloning of the genes involved in translocation break points associated with specific tumors have a practical value in improving tumor diagnosis and management?

Answer:

a. The fact that some translocations are found as the only cytogenetic abnormality in certain cancers probably means that they are a key event in tumor formation. It does not necessarily mean that they are the primary cause of the tumor or the first of many mutational events.

b. A chimeric fusion protein may have different functional properties than either of the two proteins from which it derives. If it results in the activation of a proto-oncogene product into a protein that has oncogenic properties or if it results in the inactivation of a tumor suppressor gene product, it could play a key role in the genetic cascade of events leading to tumor formation.

c. Before drawing conclusions as to whether these chromosomal aberrations inactivate the function of tumor suppressor genes, or activate quiescent proto-oncogenes, it is necessary to have additional molecular information on the effects of the translocation break points on specific transcripts. Finding that the translocation break points result in a lack of gene transcription or in transcripts that encode nonfunctional products would support the hypothesis that the translocation inactivated a tumor suppressor gene. Finding that the translocation break points result in activation of gene transcription or in the production of an active fusion protein would support the hypothesis that the translocation activated a previously quiescent proto-oncogene.

d. One hypothesis is that the various fusion proteins that result from different translocations involving the *EWS* gene somehow result in the transcription activation of different proto-oncogenes, and this leads to the different sarcomas that are seen. (Sarcomas are cancers found in tissues that include muscle, bone, fat, and blood vessels.)

e. If translocation break points are conserved within a tumor type, molecular-based diagnostics can be developed to identify the break points relatively quickly from a tissue biopsy. For example, if the genes at the break points have been cloned, PCR methods can be used to address whether the gene is intact or disrupted, using the DNA from cells of a tumor biopsy. Primers can be designed to amplify different segments of the normal gene. Then PCR reactions containing these primers and either normal control DNA or tumor-cell DNA can be set up to determine if each segment of a candidate gene is intact (a PCR product of the expected size is obtained) or disrupted (no PCR product will be obtained, because the gene has been rearranged).

Such molecular analyses would provide fast, accurate tumor diagnosis. If the different tumor types respond differentially to different regimens of therapeutic intervention, then a more rapid, unequivocal diagnosis of a particular tumor type should allow for the earlier prescription of a more optimized regime of therapeutic intervention. In addition, understanding the nature of the normal gene products of the affected genes may allow for the development of sarcoma-specific therapies.

22.27   What mechanisms ensure that cells with heavily damaged DNA are unable to replicate?

Answer: To proceed from G_1 into S, one or more G_1 cyclins bind to the cyclin-dependent kinase CDC28/cdc2 and activate it. The cyclin-dependent kinase then phosphorylates key proteins that are needed for progression into S. In the presence of heavy DNA damage, p53 is stabilized. The p53 protein acts as a transcription factor to activate *WAF1*, which produces p21. The p21 protein binds to cyclin/Cdk complexes and blocks the kinase activity required to activate the genes needed for the cell to make the transition through the cell cycle checkpoints, for example, from G_1 to S (see the answer to Question 22.21). Thus, stabilization of p53 leads to cell cycle arrest at the G_1-to-S or other cell cycle checkpoints. This arrest allows the cell time to induce the necessary repair pathways and repair the DNA or, if damage is too severe, to undergo apoptosis.

22.28   Distinguish between direct-acting carcinogens, procarcinogens, and ultimate carcinogens in the induction of cancer.

Answer: Direct-acting carcinogens are chemicals that bind to DNA and act as mutagens. Procarcinogens are chemicals that must be converted by normal cellular enzymes to become active carcinogens. These products, most of which also bind DNA and act as mutagens, are referred to as ultimate carcinogens.

22.29   What sources of radiation exist, and how does radiation induce cancer?

Answer: Sources of radiation include the sun, cellular telephones, radioactive radon gas, electric power lines, X-ray machines, and some household appliances. Radiation can be nonionizing, such as the ultraviolet and infrared light from the sun, or ionizing, such as that from X-rays or radon gas. Radiation increases the amount of DNA damage in cells. If this damage is not repaired or is repaired incorrectly, somatic mutations result. Increasing the amount of DNA damage therefore increases the chance of somatic mutation. Mutations in oncogenes or tumor suppressor genes will contribute to the induction of cancer.

23

NON-MENDELIAN INHERITANCE

CHAPTER OUTLINE

THINKING ANALYTICALLY

After developing skills to analyze nuclear inheritance patterns, non-Mendelian patterns of inheritance are quite a twist at first. However, since non-Mendelian inheritance patterns are usually due to the extranuclear inheritance of mitochondrial or chloroplast genomes, the patterns that are seen are well defined. In general, they follow from the cytoplasmic contributions of one (usually female) parent to the offspring.

Consideration of the variation in inheritance patterns and the variation in the structure and function of the mitochondrial and chloroplast genomes provides a valuable perspective on nuclear inheritance patterns and nuclear genome structure and function. After organizing the factual information given in this chapter, take the time to compare and contrast the organization, replication, and expression of mitochondrial and chloroplast genes with that of nuclear

REVIEW OF KEY TERMS, SYMBOLS, AND CONCEPTS

In your own words, write a brief, precise definition of each term in the groups below. Check your definitions using the text. Then develop a concept map using the terms in each list.

1	2	3
extranuclear inheritance, gene	mitochondria	chloroplast
non-Mendelian inheritence, gene	nucleoid region	nucleoid region
uniparental inheritance	H, L strands	SSC, LSC, IRA, IRB regions
maternal inheritance	D loop	23S, 5S, 4.5S, 16S rRNAs
paternal inheritance	continuous replication	cycloheximide
biparental inheritance	ORF, URF	ribulose bisphosphate decarboxylase (rubisco)
cytohet	12S, 16S rRNAs	*rbcL, rbcS*
DNA methylation	mitochondrial genetic code	shoot variegation in four o'clocks
maternal effect	mitochondrial DNA	leukoplast
nuclear genotype	[*poky*] in *Neurospora*	outer, inner membrane
maternal effect gene	*petite, grande* in yeast	stroma
infectious inheritance	*nuclear, netural, suppreseive petites*	thylakoid
endosymbiont theory	protoperithecia	grana
cyanobacterium	controlled cross	
purple nonsulfur photosynthetic bacterium	heteroplasmy	
erythromycin resistance in *Chlamydomanas*	Leber's hereditary optic neuropathy	
heterosis	Kearns-Sayre syndrome	
heterozygote superiority	myoclonic epilepsy and ragged-red fiber disease	
genic, cytoplasmic male sterility		
restorer of fertility (Rf)		
barnase		
killer yeast		
L, M viruses		
shell-coiling direction in *Limnaea peregra*		

genes. The salient features of an important biological process can often be better understood by examining variations in it.

Unlike extranuclear inheritance patterns that are controlled by extranuclear genes, maternal effects and genomic imprinting are controlled by nuclear genes. Keep this distinction clear. In considering maternal effects, it will help to diagram the generations of a cross and illustrate how contributions from the parental generation are important for the development of the progeny. Carefully read the material in the chapter and do the problems to discover how to distinguish a maternal effect from an extranuclear, maternally inherited trait.

QUESTIONS FOR PRACTICE

Multiple Choice Questions

1. DNA is found in
   a. the nucleus of eukaryotes
   b. mitochondria
   c. chloroplasts
   d. all subcellular organelles
   e. all of the above except d

2. In general, extranuclear genes show
   a. maternal inheritance
   b. Mendelian inheritance
   c. maternal effects
   d. dominance

3. Mitochondrial DNA is
   a. typically single stranded and coiled in nucleosomes
   b. uniform in size in all species
   c. uniformly double stranded, supercoiled, and circular
   d. found only in animals and fungi

4. In the D-loop model of replication of mitochondrial DNA, replication
   a. is synchronous with nuclear DNA replication
   b. originates at distinct locations on the H and L strands and is continuous
   c. originates on both strands in the D-loop and is semidiscontinuous
   d. occurs only in the maternal cytoplasm

5. The ribosomal proteins of mitochondrial ribosomes are
   a. coded for in the nucleoid region
   b. coded for in the nuclear genome, and are the same as those used in cytoplasmic ribosomes
   c. coded for in the nuclear genome, and are different from those used in cytoplasmic ribosomes
   d. encoded on genes in between mitochondrial rRNA genes

6. Which of the following is *not* a consequence of the differences in the genetic code used by most mitochondrial and nuclear genes?
   a. Fewer tRNAs are needed in the mitochondria.
   b. Fewer amino acids are used in mitochondrial proteins.
   c. Wobble is used more extensively in translating mitochondrial genes.
   d. Some nonsense codons in mitochondrial genes are sense codons in nuclear genes, and vice versa.

7. Which of the following is true about both mitochondrial and chloroplast genomes?
   a. Both have double-stranded, circular, supercoiled genomes that typically exist in multiple copies per organelle.
   b. Both are small in size, typically only 16–18 kb.
   c. Both have only one copy of rRNA genes.
   d. Both have a similar density in a CsCl density gradient.

8. Which of the following is *not* a typical characteristic of extranuclear traits?
   a. They show uniparental inheritance.
   b. They cannot be mapped relative to nuclear genes.
   c. They show non-Mendelian segregation in crosses.
   d. They are affected by substitution of a nucleus with a different genotype.

9. Two brothers have a hereditary disease associated with a particular lesion in mitochondrial DNA. One brother is more severely affected than the other. Which of the following is *not* a plausible explanation for this observation?
   a. The brothers have different degrees of heteroplasmy.
   b. The brothers have different proportions of two mitochondrial types.
   c. The brothers do not have identical nuclear genomes.
   d. Different mitochondrial genes are affected in the two brothers.

10. Which of the following crosses is the most useful for generating hybrid seed?
   a. female [*CMS*] *rf/rf* × male [*CMS*] *Rf/rf*
   b. male [*CMS*] *rf/rf* × female [*CMS*] *Rf/rf*
   c. female [*CMS*] *Rf/Rf* × male [*CMS*] *rf/rf*
   d. male [*CMS*] *Rf/Rf* × female [*CMS*] *rf/rf*

Answers: 1e, 2a, 3c, 4b, 5c, 6b, 7a, 8d, 9d, 10a

Thought Questions

1. What are the tenets of the endosymbiont hypothesis? What arguments can you marshal to support or detract from this hypothesis?

2. Speculate on the evolutionary origin of plant chloroplasts compared to the presumed origin of mitochondria.

3. Is there any evidence that there are paternal contributions to mtDNA inheritance? What, if any, significance might these contributions have?

4. How might molecular methods be employed to analyze a maternal line of descent? How might this be useful to trace movements of populations over time? Do any significant complications arise because of paternal contributions to mtDNA inheritance?

5. Distinguish between biparental and uniparental inheritance. Which organisms have organelles that show biparental inheritance?

6. Non-Mendelian inheritance is usually associated with genes in mitochondrial or chloroplast DNA. What other examples of non-Mendelian inheritance exist? What is the significance of the examples you cite?

7. In what ways does cpDNA differ from nuclear and mitochondrial DNA?

8. The mtDNAs of most organisms contain similar coding information. Given this, speculate why animal mtDNAs are always relatively small (<20 kb), while mtDNAs of some other organisms can be quite large. (Put another way, what advantages do small extranuclear mtDNA genomes confer?) Why should genes be retained in mtDNA at all, since at least some proteins (e.g., ribosomal proteins) are able to be imported into the mitochondria?

9. Why is it useful to develop male-sterile crops? How does strategy for developing these plants using CMS differ from a strategy that employs nuclear transgenes?

10. How do you account for the maternal effect in shell coiling in the snail *Limnaea peregra*? What features of this phenomenon indicate that it is controlled via a maternal effect and not via extranuclear inheritance?

Thought Questions for Media

After reviewing the media on the *iGenetics* CD-ROM, try to answer these questions.

1. A four o'clock plant shows variegation. That is, part of the plant has only green leaves, part has only white leaves, and part has mixed green and white leaves.

Suppose you fertilized flowers in the green-only region of the plant with pollen from flowers in the white-only part of the plant. Based on the animation, what outcomes would you expect?

2. What is the difference between a leukoplast and a chloroplast?

3. In the material effect animation, dextral (right-handed) coiling is dominant to sinistral (left-handed) coiling, but the coiling pattern is determined by the maternal contribution, not the genotype of the zygote. Suppose you had a population of *Dd* dextral snails.

    a. What crosses would you do to obtain a *Dd* sinistral snail? (Remember that you can tell only the progeny's coiling phenotype and must infer the genotype by testcrossing or selfing!)

    b. What crosses would you do to obtain a *dd* sinistral snail?

    c. Could you ever obtain a *DD* sinistral snail?

4. How does maternal effect differ from maternal inheritance, since both involve cytoplasmic contributions from the mother?

1. The iActivity provided data indicating that individuals with MERFF show one or more symptoms that can include hearing loss, dementia, muscle jumps, seizures, unsteadiness, vision problems, and loss of cognitive skills.

    a. How would you characterize MERFF in terms of penetrance, expressivity, and pleiotrophy?

    b. Based on what you know about the mode of inheritance of MERFF, what hypotheses might you form to explain its symptomatic variability?

    c. How might you use methods similar to those presented in the iActivity to test your hypotheses?

2. The iActivity presented data that allowed you to distinguish between heteroplasmic maternal inheritance and autosomal dominant inheritance with variable expressivity.

    a. What methods were used in the iActivity to distinguish between these modes of inheritance?

    b. Would it ever be possible to distinguish between these modes of inheritance based on pedigree analysis alone?

    c. What challenges do mitochondrial inheritance present for the analyses of pedigrees?

3. Suppose a MERFF-like disorder was caused by a mutation in a nuclear gene whose product was imported into the mitochondrion. What types of inheritance might be associated with this disorder?

SOLUTIONS TO TEXT PROBLEMS

23.1   The endosymbiont theory provides an explanation for the origin of mitochondria and chloroplasts.

a. What are the tenets of this theory?

b. Why are these organelles no longer able to survive as independent organisms?

c. What role did plant photosynthetic activity and changes in the Earth's atmosphere play in the evolution of mitochondria?

d. Which attributes of fundamental cellular processes, such as DNA packaging and translation, provide evidence in support of this theory?

Answer:

a. The endosymbiont theory holds that ancestral eukaryotic cells were anaerobic organisms lacking mitochondria and chloroplasts. They formed symbiotic relationships with a purple nonsulfur photosynthetic bacterium whose oxidative phosphorylation activities benefited them. Then, as atmospheric oxygen increased due to photosynthetic activity, the photosynthetic activity of the intracellular bacterium was lost. Over time, the eukaryotic cell became dependent on the intracellular bacterium for survival and mitochondria were formed. Chloroplasts were formed through a similar symbiotic relationship when eukaryotic cells ingested an oxygen-producing photosynthetic cyanobacterium.

b. During the evolution of mitochondria and chloroplasts, some of their ancestral genes were transferred to the nuclear genome. Mitochondria and chloroplasts can no longer survive as independent organisms without these genes.

c. Photosynthetic activity led to increased atmospheric O_2, so the photosynthetic activity provided by the purple nonsulfur photosynthetic bacterium was no longer essential and was lost.

d. Like many bacterial genomes, many mitochondrial and all chloroplast genomes are circular. Like bacterial genomes, all mitochondrial and chloroplast genomes are supercoiled and packaged without histones in nucleoid bodies. Translation in mitochondria and chloroplasts is not identical to that in prokaryotes, but some features of translation in these organelles are reminiscent of prokaryotic translation. This includes the use of 70S ribosomes with 50S and 30S subunits, the presence of a 16S rRNA in the 30S subunit, the use of fMet-tRNA as the initiator tRNA, and the sensitivity to inhibitors of bacterial but not cytoplasmic ribosome function.

23.2   The nuclear, mitochondrial, and chloroplast genomes are different in size, structure, gene content, and gene organization.

a. Compare the structure of the nuclear genome, the mitochondrial genome, and the chloroplast genome.

b. How does the gene content and organization of the mitochondrial genome compare to that of the chloroplast genome?

c. The mitochondrial genome varies widely in size, ranging from less than 20 kb to more than 2,000 kb. What accounts for this variation?

Answer:

a. The nuclear genome is organized into linear chromosomes that are long double-stranded DNA molecules packaged into chromatin by histone and nonhistone chromosomal proteins. Both mitochondrial and chloroplast genomes are circular, supercoiled, double-stranded DNA molecules and are not complexed with packaging proteins. While there is some variation in the size of chloroplast genomes (80 to 600 kb) and mitochondrial genomes (<20 kb in animals, 80 kb in yeast, 100 to 2,000 kb in plants), these extracellular genomes are not nearly as large as most eukaryotic chromosomes. Unlike nuclear chromosomes that contain diverse genes and noncoding sequences, mitochondria contain similar, unique genes, with larger mitochondrial genomes containing more noncoding sequences. Diploid, mitotically active eukaryotic cells have two copies of each chromosomal homolog that divide in an exquisitely organized mitosis. On the other hand, extranuclear genomes can exist in a variable number of copies per cell and can replicate independently in many phases of the cell cycle. A single yeast mitochondrion, for example, can have 10–30 nucleoid regions, each containing four to five mitochondrial DNA genomes.

b. The mitochondrial genome has fewer genes than does the chloroplast genome. The mitochondrial genome contains genes for tRNAs, rRNAs, and some of the polypeptide components of the proteins cytochrome oxidase, NADH-dehydrogenase, and ATPase, while the chloroplast genome contains genes for rRNA (two copies), tRNAs, and about 100 ORFs of which only about 60 have known functions. These include some of the proteins required for transcription, translation (ribosomal proteins, RNA polymerase subunits, translation factors), and photosynthesis.

   The genes on mitochondrial DNA are found on both strands. In the compact, intronless mammalian mtDNA, genes encoding the rRNAs and the mRNAs are separated by tRNA genes. After each strand is transcribed into a single RNA molecule, the tRNAs are cut out by specific enzymes that simultaneously liberate the mRNAs and rRNAs. In the larger mtDNAs of yeast and plants, tRNA genes do not separate the other genes, introns can be found, and there is separate transcription termination. Like the genes of mtDNA, cpDNA genes are also transcribed on both strands. Unlike mtDNA however, the genome contains two identical inverted repeats that each contain copies of the rRNA as well as other genes. These repeats are arranged in between a short single copy (SSC) and long single copy (LSC) region. See text Figures 23.4, p. 634 and 23.7, p. 638.

c. Large mtDNA genomes contain more DNA that does not code for gene products, either intronic or nontranscribed DNA sequences. Large and small mtDNA genomes do not appear to differ in gene content.

23.3   In what ways is DNA replication in animal mitochondrial genomes different from nuclear DNA replication?

Answer: mtDNA replication differs from nuclear DNA replication in that it uses a mitochondrial set of DNA polymerases, occurs throughout the cell cycle, and is not semi-discontinuous—mtDNA replication on each of the H and L mtDNA strands is initiated at an origin of replication on that strand and proceeds continuously until replication of the strand is completed.

23.4   Imagine that you have discovered a new genus of yeast. In your studies on this organism, you isolate DNA and subject it to CsCl density gradient centrifugation. In addition to a major peak, you observe a minor peak of lighter density. How could you determine whether the minor peak represents organellar (presumably mitochondrial) DNA as opposed to an AT-rich repeated sequence in the nuclear genome?

Answer: Multiple approaches are possible. First, you could purify mitochondria, isolate DNA from them, and use CsCl density gradient centrifugation to determine if this DNA has an identical density to that of the minor peak. Second, you could isolate the minor peak and visualize the DNA under the electron microscope. If the molecules are circular, they are unlikely to be nuclear fragments. Third, you could grow the yeast in the presence of an intercalating agent such as acridine and see whether this treatment causes the minor peak species to disappear. If it does, the minor peak is organellar in origin. Fourth, you could isolate the minor peak, label it (by nick translation, for example), and hybridize this labeled DNA to DNA within suitably prepared yeast cells. Then, using the electron microscope, you could determine whether the label is found over the nucleus or the mitochondria. Fifth, you could isolate the minor peak DNA and study its sequence similarity with mitochondrial DNA from other yeast.

23.5   What genes are present in the human mitochondrial genome, and how was this determined?

Answer: Human mitochondrial DNA contains the genes for the 12S and 16S mitochondrial rRNAs; 22 tRNAs; cytochrome b; and some of the polypeptide subunits of cytochrome c oxidase (COI, COII, COIII), NADH-dehydrogenase (ND1–6), and ATPase (ATPase 6 and 8). See text Figure 23.4, p. 634.

23.6   How is transcription and processing of transcripts in animal mitochondria different from that in eukaryotic nuclei or prokaryotes?

Answer: Mitochondrial transcription in animals is unlike nuclear or prokaryotic transcription in that each strand is transcribed into a single RNA that is then processed to produce the RNAs for individual genes. Most of the genes encoding mRNAs and rRNAs are separated by tRNA genes. The mRNAs and rRNAs are produced as the tRNA genes are recognized and cut out by specific enzymes. The mRNAs also lack 5' caps, introns and chain-terminating codons (mRNAs end in U or UA, polyadenylation completes the missing parts of a UAA stop codon).

23.7   What conclusions can you draw from the fact that most nuclear-encoded mRNAs and all mitochondrial mRNAs have a poly(A) tail at the 3' end?

Answer: That most nuclear and mitochondrial mRNAs are polyadenylated suggests that polyadenylation serves some function that is basic and very likely important, perhaps related to mRNA stability or translation. However, the features unique to polyadenylation of mitochondrial transcripts give little insight into a general function. For example, polyadenylation is necessary in some mitochondrial mRNAs to complete a missing part of a UAA stop codon, but this function is not required for nuclear mRNAs. Since mitochondrial mRNAs do not exit the mitochondrion, the 3' poly(A) tail is presumably not needed for transport between cellular compartments.

Additional insights into the role of polyadenylation in transcript stability and translation have come from experiments discussed in Chapter 20, p. 563–564. Experiments in which *Xenopus* oocytes were injected with globin mRNA with and without a poly(A) tail have suggested that one more general function of the tail is to confer stability on eukaryotic mRNAs. This view has been supported by investigations that have shown that shortening of the poly(A) tail plays a significant part in the control of mRNA degradation. Additional insights into the function of poly(A) tails have come from observations on the control of translation of maternally deposited mRNAs that are stored in oocytes. Here, the length of the poly(A) tail is regulated cytoplasmically, and stored mRNAs have shorter poly(A) tails than actively translated mRNAs. A longer poly(A) tail is able to promote the initiation of translation.

23.8   A substantial body of evidence indicates that defects in mitochondrial energy production may contribute to the neuronal cell death seen in a number of late-onset neurodegenerative diseases, including Alzheimer disease, Parkinson disease, Huntington disease, and amyotrophic lateral sclerosis (ALS, or Lou Gehrig disease). Some of these diseases have been associated with mutations in the nuclear genome. One experimental system that has been developed to evaluate the contributions of the mitochondrial genome to these diseases uses a cytoplasmic hybrid known as a cybid. Cybids are made by repopulating a tissue culture cell line that has been made mitochondria deficient with mitochondria from the cytoplasm of a human platelet cell. The cybids thus have nuclear DNA from the tissue culture cell and mitochondrial DNA from the human platelet cell.

     The mitochondrial protein cytochrome oxidase has subunits encoded by both nuclear and mitochondrial genes. Patients with Alzheimer disease have been reported to have lower levels of cytochrome oxidase than do age-matched controls.

a. What is the experimental evidence that cytochrome oxidase has subunits encoded by both nuclear and mitochondrial genes?

b. Given the means to assay cytochrome oxidase activity, how would you investigate whether the decreased levels of cytochrome oxidase activity in patients with Alzheimer disease could be ascribed to nuclear or mitochondrial genetic defects? What controls would you create?

c. If you can demonstrate that the mitochondrial contribution to cytochrome oxidase is responsible for lowered cytochrome oxidase activity, can you conclude that each mitochondrion of an affected individual has an identical defect?

Answer:

a. Mitochondrial ribosomes are sensitive to most inhibitors of bacterial ribosome function, such as chloramphenicol. However, they are generally insensitive to antibiotics to which cytoplasmic ribosomes are sensitive, such as cycloheximide. Selective use of antibiotics that inhibit translation by either mitochondrial or cytoplasmic ribosomes has shown that components 2, 5, and 6 of cytochrome oxidase are synthesized on mitochondrial ribosomes, while components 1, 3, 4, and 7 are synthesized on cytoplasmic ribosomes. See Figure 23.5, p. 638, and the accompanying text.

b. Compare cytochrome oxidase activity in cybids made with platelets from diseased individuals and in cybids made with platelets from age-matched control individuals. It is important to assess several different enzyme activities associated with mitochondrial proteins to ensure that the deficits in cytochrome oxidase are specific.

c. As discussed in the text (p. 647), the cells of individuals with diseases resulting from mitochondrial DNA defects have a mixture of mutant and normal mitochondria; that is, they show heteroplasmy. Thus, assays in cybids are measurements of the enzyme activity present in a population of mitochondria in a cell. It would be unlikely that each of the mitochondria of an affected individual has an identical defect.

23.9   Discuss the differences between the universal genetic code of the nuclear genes of most eukaryotes and the code found in human mitochondria. Is there any advantage to the mitochondrial code?

Answer: While plant mitochondria use the universal nuclear genetic code, mitochondria from other organisms, including humans, utilize a code that differs in codon designation. In addition, the mitochondrial code typically has more extensive wobble, so that many fewer tRNAs are needed to read all possible sense codons. In humans, only 22 mitochondrial tRNAs are needed, as opposed to the 32 required for nuclear mRNAs. (There is some variability between mitochondrial codes of different organisms, as well as between the nuclear codes of different unicellular organisms. For example, ciliated protozoa do not use the universal nuclear code.) Fewer mitochondrial tRNAs confer the advantage that fewer tRNA genes are needed to read all possible sense codons, allowing for a smaller mitochondrial genome.

23.10   Compare the cytoplasmic, mitochondrial, and chloroplast protein-synthesizing systems.

Answer: Compare these protein-synthesizing systems from three perspectives: the ribosomes used in translation; the mechanisms of translation initiation, elongation, and termination; and the nature of the genetic code.

Mitochondrial and chloroplast ribosomes are different from each other as well as from cytoplasmic ribosomes. Cytoplasmic ribosomes are 80S, with 60S and 40S subunits. Human mitochondrial ribosomes are 60S, with 45S and 35S subunits. Chloroplast ribosomes are 70S, with 50S and 30S subunits, and are similar in size to prokaryotic ribosomes. Mitochondrial ribosomes lack the 5S and 5.8S rRNA components of cytoplasmic ribosomes, and have a 16S rRNA and 12S rRNA in the large and small subunits, respectively. Chloroplast ribosomes have a 23S, 5S, and 4.5S rRNA in the large subunit, and a 16S rRNA in the small subunit. The number of ribosomal proteins found in the chloroplast and mitochondrial ribosomes is not well defined. With a few exceptions (*Neurospora*, yeast), the proteins of mitochondrial ribosomes are distinct from those in cytoplasmic ribosomes. It is known that most mitochondrial ribosomal proteins are encoded by nuclear genes, and chloroplast ribosomal proteins are encoded by both nuclear and chloroplast genes.

Translation initiation in mitochondria (except for plant and yeast mitochondria) is quite different from initiation in the cytoplasm. Animal mitochondrial mRNAs lack a 5′ cap and have virtually no 5′ leader sequence. Therefore, mitochondrial ribosomes bind to mitochondrial mRNAs and orient translation initiation differently than do cytoplasmic ribosomes. Initiation in plant and yeast mitochondria is more like cytoplasmic translation initiation: Although there is no 5′ cap, there is a 5′ leader and translation initiation may be at the first AUG. In chloroplasts, translation initiation is much like in bacteria. A formylmethionyl tRNA is used to initiate all proteins, and the formylation reaction is catalyzed by a transformylase in the chloroplast. Both mitochondria and chloroplasts use their own translation initiation factors, elongation factors, and release factors that are distinct from those used in the cytoplasm. In some ways, translation in both chloroplasts and mitochondria is similar to translation in prokaryotes. Chloroplast and mitochondrial ribosomes are both insensitive to cycloheximide, an inhibitor of cytoplasmic ribosomes, but sensitive to nearly all antibiotics that inhibit prokaryotic translation.

The genetic code used by both chloroplast and cytoplasmic ribosomes is the universal code, while the code used by mitochondria is different. Mitochondria have fewer tRNAs, and there is considerably more wobble.

23.11   Rubisco comprises about half of the protein found in green plant tissue, making it the most common protein on Earth. It consists of eight identical large subunits and eight identical small subunits.

a. What is the function of rubisco?

b. Without resorting to DNA sequencing, how would you experimentally determine that the large subunit gene is in the chloroplast genome and that the small subunit gene is in the nuclear genome?

c. A nonsense mutation occurs near the 5′ end of the protein-coding region in a single copy of the gene for the small subunit. What genetic and phenotypic properties do you expect for this mutation?

d. A nonsense mutation occurs near the 5′ end of the protein-coding region in a single copy of the gene for the large subunit. What genetic and phenotypic properties do you expect for this mutation?

Answer:

a. Rubisco is the first enzyme used in the pathway for fixation of carbon dioxide in photosynthesis.

b. Determine which subunit is made by plant cells grown in tissue culture with an antibiotic that selectively inhibits translation by either mitochondrial (e.g., neomycin) or cytoplasmic (e.g., cyclohexamide) ribosomes.

c. Since the small subunit is encoded by the nuclear genome, a mutation in one of the two gene copies will eliminate about half of the small subunit produced, decreasing functional rubisco levels by about half. These plants might show phenotypic effects (e.g., slower growth), making the mutation dominant. Homozygous mutants lack functional rubisco, so they would be unable to carry out photosynthesis. Such mutants would die soon after seed germination, making the mutation recessive lethal.

d. Each chloroplast contains multiple cpDNA copies with different species having different copy numbers. In species with a high copy number (e.g., *Chlamydomonas*), a mutation in one copy would have a negligible effect on the amount of functional rubisco in the cell, and could have little phenotypic effect. If the cpDNA copy number were small, a phenotypic effect might be seen. Plants with the cpDNA mutation could also have cells with different numbers of mutation-bearing cpDNAs and show variegation.

23.12 What features of non-Mendelian inheritance distinguish it from the inheritance of nuclear genes?

Answer: Traits showing non-Mendelian inheritance exhibit differences in reciprocal crosses that are unrelated to sex, cannot be mapped to nuclear linkage groups, produce progeny ratios atypical of Mendelian segregation, and are indifferent to nuclear substitution.

23.13 Distinguish between maternal effect and non-Mendelian inheritance.

Answer: Maternal effect is the determination of gene-controlled characters by the maternal genotype prior to the fertilization of the egg cell. A maternal effect is seen when nuclear genes of the mother function to specify some characteristic of the zygote. For example, a maternal effect could result from a maternally transcribed gene whose product was localized at one pole of the embryo and was responsible for the polarity of the developing embryo.

In contrast to nuclear genes that can function in the mother, the genes involved in non-Mendelian inheritance are extranuclear and are found in mitochondria and chloroplasts. If the mitochondria are inherited with transmission of the mother's cytoplasm, mitochondrial genes will show maternal inheritance, a type of extranuclear inheritance.

23.14   Reciprocal crosses between two types of the evening primrose, *Oenothera hookeri* and *Oenothera muricata*, produce the following effects on the plastids:

O. hookeri female × *O. muricata* male → Yellow plastids

O. muricata female × *O. hookeri* male → Green plastids

Explain the difference between these results, noting that the chromosome constitution is the same in both types.

Answer: Both offspring have the same hybrid nuclear constitution, but differ in the maternal cytoplasm they inherited. Thus, factors in the cytoplasm are likely to be the cause of the phenotypic difference that is seen. The two possible types of cytoplasmic factors are those that would cause a maternal effect and those that are inherited maternally. If the difference in phenotype is due to a maternal effect, the difference is due to the deposition in the oocyte of gene products from the mother's nuclear genome. For these results to be explained by a maternal effect, one would have to hypothesize (1) that such a maternally contributed gene product affected plastid color and (2) that this product was different in *O. hookeri* and *O. muricata*. If the difference in phenotype is due to maternal inheritance, the plastids contributed by one female parent would have to be different from the plastids contributed by the other. Specifically, yellow plastids would have been contributed by the *O. hookeri* female and green plastids would have been contributed by the *O. muricata* female.

It is noteworthy that in most angiosperms (flowering plants), the plastids are inherited only from the maternal parent. Thus, maternal inheritance of plastids would explain the results presented in this problem. However, in the genus *Oenothera*, biparental inheritance (inheritance from both the male and the female parent) of plastids has been reported (see text, p. 649). If biparental inheritance does not occur all of the time (i.e., maternal inheritance occurs some of the time), many progeny would be just as described in the results presented here. When plastids are contributed biparentally, and the plastids contributed by the male and female parent segregate to different daughter cells during mitosis, a green patch of tissue would be seen in the yellow progeny from the cross *O. hookeri* female × *O. muricata* male, and a yellow patch of tissue would be seen in the green progeny from the cross *O. muricata* female × *O. hookeri* male.

23.15   A series of crosses are performed with a recessive mutation in *Drosophila* called *tudor*. Homozygous *tudor* animals appear normal and can be generated from the cross of two heterozygotes, but a true-breeding *tudor* strain cannot be maintained. When homozygous *tudor* males are crossed to homozygous *tudor* females, both of which appear to be phenotypically normal, a normal-appearing F_1 is produced. However, when F_1 males are crossed to wild-type females, or when F_1 females are crossed to wild-type males, no progeny are produced. The same results are seen in the F_1 progeny of homozygous *tudor* females crossed to wild-type males. The F_1 progeny of homozygous *tudor* males crossed to wild-type females appear normal, and they are capable of issuing progeny when mated either with each other or with wild-type animals.

a. How would you classify the *tudor* mutation? Why?

b. What might cause the *tudor* phenotype?

Answer:

a. The *tudor* mutation is a maternal effect mutation. Homozygous *tudor* mothers give rise to sterile progeny, regardless of their mate.

b. The grandchildless phenotype results from the absence of some maternally packaged component in the egg needed for the development of the F_1's germ line.

23.16 A form of male sterility in corn is maternally inherited. Plants of a male-sterile line crossed with normal pollen give male-sterile plants. Some lines of corn carry a dominant, so-called restorer (*Rf*) gene that restores pollen fertility in male-sterile lines.

a. If a male-sterile plant is crossed with pollen from a plant homozygous for gene *Rf*, what will be the genotype and phenotype of the F_1?

b. If the F_1 plants of (a) are used as females in a testcross with pollen from a normal plant (*rf/rf*), what would be the result? Give genotypes and phenotypes and designate the type of cytoplasm.

Answer:

a. If the normal cytoplasm is [*N*] and the male-sterile cytoplasm is [*Ms*], the cross is [*Ms*] *rf/rf* ♀ × [*N*] *Rf/Rf* ♂, and the F_1 would be [*Ms*] *Rf/rf* and is male-fertile.

b. The cross is [*Ms*] *Rf/rf* ♀ × [*N*] *rf/rf* ♂. Half of the progeny would be [*Ms*] *Rf/rf* and be male fertile, and half would be [*Ms*] *rf/rf* and be male sterile.

23.17 In *Neurospora* a chromosomal gene *F* suppresses the slow-growth characteristic of the [*poky*] phenotype and makes a [*poky*] culture into a *fast*-[*poky*] culture, which still has abnormal cytochromes. Gene *F* in combination with normal cytoplasm has no detectable effect. (Hint: Because both nuclear and extranuclear genes must be considered, it is convenient to use symbols to distinguish the two. Thus cytoplasmic genes are designated in square brackets, e.g., [*N*] for normal cytoplasm, [*poky*] for *poky*.)

a. A cross in which *fast*-[*poky*] is used as the female (protoperithecial) parent and a normal wild-type strain is used as the male parent gives half [*poky*] and half *fast*-[*poky*] progeny ascospores. What is the genetic interpretation of these results?

b. What would be the result of the reciprocal cross of the cross described in (a), that is, normal female × *fast*-[*poky*] male?

Answer:

a. Parents: [*poky*]*F* × [*N*] + ; progeny: [*poky*]*F* and [*poky*] + in equal numbers. Standard [*poky*] ([*poky*] +) is found among the offspring, so gene *F* must not effect a permanent alteration of [*poky*] cytoplasm. The 1:1 ratio of [*poky*] to *fast*-[*poky*] indicates that all progeny have the [*poky*] mitochondrial genotype (by maternal inheritance) and that the *F* gene must be a nuclear gene segregating according to Mendelian principles. Therefore, the [*poky*] progeny are [*poky*] + , and the *fast*-[*poky*] are [*poky*]*F*.

b. Parents: [*N*] + × [*poky*]*F* ; progeny [*N*] + and [*N*]*F* in equal numbers. These two genotypes are phenotypically normal and indistinguishable.

23.18 Distinguish between *nuclear (segregational)*, *neutral*, and *suppressive petite* mutants of yeast.

Answer: Petites are a class of mutations that affect mitochondrial function. *Neutral petites* are able to grow on a medium that will support fermentation, but cannot grow on a medium that supports only aerobic respiration because they lack nearly all of their mitochondrial DNA. *Nuclear petites* result from a nuclear mutation that affects mitochondrial function (e.g., in a nuclear gene that encodes a subunit of a mitochondrial protein). These mutations, if crossed to a wild-type (*grande*) strain, will show a typical 2:2 Mendelian segregation pattern. *Neutral* and *suppressive petites* result from mutations in the mitochondrial genome. They show uniparental inheritance (but not maternal inheritance): When crossed to a normal cell, and (*grande*) diploids go through meiosis, a 0:4 ratio of *petite:grande* is seen. *Suppressive petites*, unlike *neutral petites*, do have an effect on the wild type. A diploid formed from a *suppressive petite* and a normal cell will have respiratory properties intermediate between the *petite* and normal. Mitosis in this diploid results in mostly *petites* (up to 99%) with a respiratory-deficient phenotype, and meiosis results in a 4:0 ratio of *petite:grande*. At a molecular level, the *suppressive petite* mutations result from partial deletions of the mitochondrial DNA.

23.19   In yeast, a haploid *nuclear* (*segregational*) *petite* is crossed with a *neutral petite*. Assuming that both strains have no other abnormal phenotypes, what proportion of the progeny ascospores are expected to be *petite* in phenotype if the diploid zygote undergoes meiosis?

Answer: Nuclear genes show Mendelian segregation patterns, so +/*petite* should show 2:2 segregation and give ½ *petite* and ½ wild-type (*grande*) progeny.

23.20   When grown on a medium containing acriflavin, a yeast culture produces a large number of very small (*tiny*) cells that grow very slowly. How would you determine whether the slow-growth phenotype was the result of a cytoplasmic factor or a nuclear gene?

Answer: This problem is formally very similar to determining the mode of inheritance for yeast *petite* mutants. Acriflavin intercalates between base pairs and might be used to introduce *petite*-like mutations. If the *tiny* phenotypes are due to a nuclear gene, then meiosis in cells formed from a cross of *tiny* × wild type should result in a 2:2 segregation of *tiny*:wild-type ascospores. If, in contrast, an extranuclear gene is involved, one would expect a 0:4 or 4:0 ratio and see only wild-type or *tiny* progeny.

23.21   Mating in *Chlamydomonas* is syngamous with cells of each mating type (mt^+, mt^-) contributing an equal amount of cytoplasm. Erythromycin resistance ([ery^r]) is a chloroplast trait. Streptomycin resistance can be a nuclear (str^r) or cytoplasmic trait ([str^r]).

a. For each of the following crosses, provide the expected zygotic phenotypes (erythromycin resistant or sensitive, streptomycin resistant or sensitive) and their approximate proportions:

  i. mt^+ [ery^r] × mt^- [ery^s]
  ii. mt^- [ery^r] × mt^+ [ery^s]
  iii. mt^+ str^r [str^s] × mt^- str^s [str^s]
  iv. mt^+ str^s [str^r] × mt^- str^s [str^s]
  v. mt^+ str^r [str^r] × mt^- str^s [str^s]
  vi. mt^- str^r [str^s] × mt^+ str^s [str^s]
  vii. mt^- str^s [str^r] × mt^+ str^s [str^s]
  viii. mt^- str^r [str^r] × mt^+ str^s [str^s]

b. Suppose that in cross (ii), about 5 percent of the zygotes are cytohets. What phenotype will these cells have, and what phenotypes do you expect to observe as these cells undergo successive mitotic divisions?

Answer:

a. Keep track of each trait separately, as the traits for nuclear loci (*mt* and *str*) will show 2:2 segregation patterns reflecting Mendelian inheritance while the cytoplasmic traits will reflect the inheritance of cpDNA. In *Chlamydomonas*, both parents contribute equal amounts of cytoplasm to the zygote, but about 95 percent of the progeny show uniparental inheritance and have the cytoplasmic trait of the *mt*$^+$ parent. This occurs because the *mt*$^+$ cpDNA but not the *mt*$^-$ cpDNA becomes highly methylated during zygote maturation and subsequent zygote germination, and because the highly methylated *mt*$^+$ cpDNA replicates much better than the lightly methylated *mt*$^-$ cpDNA (which is also destroyed within hours after mating). Thus, the phenotype of most zygotes reflects the genotype of the cpDNA contributed by the *mt*$^+$ parent. The remaining 5 percent of the zygotes show biparental inheritance of cpDNA and exhibit the extranuclear traits of both parents—they are cytohets. Under the assumptions that the nuclear loci assort independently and that a single *str*^{r} allele, whether nuclear or cytoplasmic, confers resistance to streptomycin, the results of each cross are shown in the following table:

	Cross	Nuclear Trait Segregation	Chloroplast Trait Segregation	Observed Phenotypes
i.	*mt*$^+$ [*ery*^{r}] × *mt*$^-$ [*ery*^{s}]	50% each *mt*$^+$ *mt*$^-$	95% [*ery*^{r}] 5% [*ery*^{s}]	47.5% *mt*$^+$ *ery*^{r} 47.5% *mt*$^-$ *ery*^{r} 2.5% *mt*$^+$ *ery*^{s} 2.5% *mt*$^-$ *ery*^{s}
ii.	*mt*$^-$ [*ery*^{r}] × *mt*$^+$ [*ery*^{s}]	50% each *mt*$^+$ *mt*$^-$	95% [*ery*^{s}] 5% [*ery*^{r}]	47.5% *mt*$^+$ *ery*^{s} 47.5% *mt*$^-$ *ery*^{s} 2.5% *mt*$^+$ *ery*^{r} 2.5% *mt*$^-$ *ery*^{r}
iii.	*mt*$^+$ *str*^{r}[*str*^{s}] × *mt*$^-$ *str*^{s}[*str*^{s}]	25% each *mt*$^+$ *str*^{r} *mt*$^+$ *str*^{s} *mt*$^-$ *str*^{r} *mt*$^-$ *str*^{s}	all [*str*^{s}]	25% *mt*$^+$ *str*^{r} 25% *mt*$^+$ *str*^{s} 25% *mt*$^-$ *str*^{r} 25% *mt*$^-$ *str*^{s}
iv.	*mt*$^+$ *str*^{s}[*str*^{r}] × *mt*$^-$ *str*^{s}[*str*^{s}]	50% each *mt*$^+$ *str*^{s} *mt*$^-$ *str*^{s}	95% [*str*^{r}] 5% [*str*^{s}]	47.5% *mt*$^+$ *str*^{r} 47.5% *mt*$^-$ *str*^{r} 2.5% *mt*$^+$ *str*^{s} 2.5% *mt*$^-$ *str*^{s}
v.	*mt*$^+$ *str*^{r}[*str*^{r}] × *mt*$^-$ *str*^{s}[*str*^{s}]	25% each *mt*$^+$ *str*^{r} *mt*$^+$ *str*^{s} *mt*$^-$ *str*^{r} *mt*$^-$*str*^{s}	95% [*str*^{r}] 5% [*str*^{s}]	47.5% *mt*$^+$ *str*^{r} 47.5% *mt*$^-$ *str*^{r} 2.5% *mt*$^+$ *str*^{s} 2.5% *mt*$^-$*str*^{s}
vi.	*mt*$^-$ *str*^{r}[*str*^{s}] × *mt*$^+$ *str*^{s}[*str*^{s}]	25% each *mt*$^+$ *str*^{r} *mt*$^+$ *str*^{s} *mt*$^-$ *str*^{r} *mt*$^-$ *str*^{s}	all [*str*^{s}]	25% *mt*$^+$ *str*^{r} 25% *mt*$^+$ *str*^{s} 25% *mt*$^-$ *str*^{r} 25% *mt*$^-$ *str*^{s}
vii.	*mt*$^-$ *str*^{s}[*str*^{r}] × *mt*$^+$ *str*^{s}[*str*^{s}]	50% each *mt*$^+$ *str*^{s} *mt*$^-$ *str*^{s}	95% [*str*^{s}] 5% [*str*^{r}]	47.5% *mt*$^+$ *str*^{s}, 47.5% *mt*$^-$ *str*^{s} 2.5% *mt*$^+$ *str*^{r} 2.5% *mt*$^-$ *str*^{r}
viii.	*mt*$^-$ *str*^{r}[*str*^{r}] × *mt*$^+$ *str*^{s}[*str*^{s}]	25% each *mt*$^+$ *str*^{r} *mt*$^+$ *str*^{s} *mt*$^+$ *str*^{r} *mt*$^-$ *str*^{s}	95% [*str*^{s}] 5% [*str*^{r}]	26.25% *mt*$^+$ *str*^{r} 26.25% *mt*$^-$ *str*^{r} 23.75% *mt*$^+$ *str*^{s} 23.75% *mt*$^-$ *str*^{s}

b. Cytohets are the zygotes that show biparental inheritance and have received chloroplasts from each parent. In cross (ii), they are the 5 percent of the progeny that are mt^+ ery^r and mt^- ery^r in phenotype, but note that they have two types of chloroplasts: ery^r and ery^s. Most of the time, the erythromycin-sensitive and erythromycin-resistance traits will segregate in successive mitotic divisions and produce pure $[ery^s]$ and $[ery^r]$ strains. This presumably reflects the segregation of the different chloroplast chromosomes. However, it is also possible that one of the biparental zygotes will not segregate the two traits due to a recombination event between the two types of cpDNA that leads to a cpDNA chromosome carrying both alleles.

23.22 Several investigators have demonstrated that chemical and environmental treatments of plants and animals can lead to abnormalities that persist for several generations before disappearing. For example, Hoffman found that treating the bean *Phaseolus vulgaris* with chloral hydrate led to abnormalities in leaf shape that persisted in the female (but not male) line for almost six generations before disappearing.

   a. In what different ways could you explain the origin of these abnormalities, and their disappearance after several generations?

   b. What broader implications might these findings have?

   Answer:

   a. The persistence of leaf-shape defects in the female but not male line suggests that a maternally contributed component was affected by the chloral hydrate treatment. Such components could include some of the mitochondria and mtDNA genomes, chloroplasts and cpDNA genomes, or other maternal factors such as maternally packaged proteins or mRNAs. Among these possibilities, the data throw suspicion that some, but not all, of the mitochondria or chloroplast genomes were damaged. Not only would this lead to phenotypes that are heritable in the female line, their disappearance after six generations could be explained by the segregation of normal from damaged genomes in successive generations of organellar division and the repopulation of cells with the normal organelles.

   b. If the hypothesis put forth above is correct, environmental agents that alter the genomes in some but not all uniparentally contributed organelles could result in traits that appear to be heritable over multiple generations but that then disappear as organelles with normal genomes repopulate cells. One implication of this is that environmental toxins that affect mtDNA or cpDNA genomes may have effects that last for generations, long after the toxin has been removed from the environment.

23.23 *Drosophila melanogaster* has a sex-linked, recessive mutant gene called *maroon-like (ma-l)*. Homozygous *ma-l* females or hemizygous *ma-l* males have light-colored eyes because of the absence of the active enzyme xanthine dehydrogenase, which is involved in the synthesis of eye pigments. When heterozygous *ma-l$^+$/ma-l* females are crossed with *ma-l* males, all the offspring are phenotypically wild type. However, half the female offspring from this cross, when crossed back to *ma-l* males, give all *ma-l* progeny. The other half of the females, when crossed to *ma-l* males, give all phenotypically wild-type progeny. What is the explanation for these results?

Answer: Since *ma-l* is sex linked, we know it is nuclearly inherited. We can therefore diagram the crosses and their progeny. In the first cross, *ma-l$^+$/ma-l* × *ma-l*/Y gives ¼ *ma-l/ma-l* females, ¼ *ma-l$^+$/ma-l* females, ¼ *ma-l$^+$*/Y males, and ¼ *ma-l*/Y males. To explain why the half of the total F_1 progeny that are *ma-l* hemizygotes or homozygotes do not have light-colored eyes, consider the possibility of a maternal effect. The parental female was *ma-l$^+$/ma-l* and so could have provided normal xanthine dehydrogenase (or stable, normal xanthine dehydrogenase mRNA or the biochemical product of the xanthine dehydrogenase reaction) to the embryo. The results of the crosses indicate that this maternally transmitted product is sufficient for the F$_1$ progeny to have a normal phenotype. When the F$_1$ females are backcrossed to the *ma-l* males, two different crosses are possible. One is *ma-l$^+$/ma-l* × *ma-l*/Y. Just like the initial cross, this will give all wild-type progeny. The other cross is *ma-l/ma-l* × *ma-l*/Y, which will give all *maroon-like* progeny (as the mother can no longer supply the *ma-l$^+$* product to her progeny).

23.24 When females of a particular mutant strain of *Drosophila melanogaster* are crossed to wild-type males, all the- viable progeny flies are females. Hypothetically, this result could be the consequence of either a sex-linked, male-specific lethal mutation or a maternally inherited factor that is lethal to males. What crosses would you perform to distinguish between these alternatives?

Answer: If the male lethality is caused by a sex-linked, male-specific lethal mutation (*l*), the cross can be written as *l/l* ♀ × +/Y ♂, giving *l/+* (♀) and *l/Y* (dead ♂) progeny. A cross of the F$_1$ females to normal males (*l/+* ♀ × +/Y ♂) will give a 2:1 ratio of females to males (¼ *l/+* ♀, ¼ +/+ ♀, ¼ +/Y ♂, and ¼ *l/Y* dead ♂). If the male lethality is caused by a maternally inherited cytoplasmic factor lethal to males, the F$_1$ females will receive this factor in cytoplasm from their mother, so (like their mothers) should have only female offspring when mated to wild-type males.

23.25 Reciprocal crosses between two *Drosophila* species, *D. melanogaster* and *D. simulans*, produce the following results:

> *melanogaster* ♀ × *simulans* ♂ → females only
> *simulans* ♀ × *melanogaster* ♂ → males, with few or no females

Propose a possible explanation for these results.

Answer: When *melanogaster* females are crossed with *simulans* males, *melanogaster* cytoplasm is given to the offspring. Female progeny will have an X chromosome from the *melanogaster* as well as the *simulans* parent, while male progeny will have an X chromosome from the *melanogaster* parent but only a Y chromosome (with few structural genes) from the *simulans* parent. Female progeny survive selectively because of nuclear gene products encoded on the *simulans* X chromosome, which are needed for hybrid survival in *melanogaster* cytoplasm. When a *simulans* female is crossed with a *melanogaster* male, *simulans* cytoplasm is given to the offspring. Female progeny will have a *melanogaster* as well as a *simulans* X chromosome, while male progeny will have a *simulans* X chromosome and a *melanogaster* Y chromosome. Since few or no females are recovered, it appears that the *melanogaster* X chromosome encodes products that (generally) cause lethality in the *simulans* cytoplasm.

23.26 Some *Drosophila* are very sensitive to carbon dioxide; administering it to them anesthetizes them. The sensitive flies have a cytoplasmic particle called sigma that has many properties of a virus. Resistant flies lack sigma. The sensitivity to carbon dioxide shows strictly maternal inheritance. What would be the outcome of the following two crosses: (a) sensitive ♀ × resistant ♂ and (b) sensitive ♂ × resistant ♀?

Answer: In (a), all of the progeny will inherit the *sigma* factor from the (sensitive) female parent. Consequently, all the progeny will be sensitive. In (b), the resistant female parent lacks the *sigma* factor. All of the progeny will also lack the factor and so be resistant.

23.27   A few years ago, Chile allowed its government agents to kidnap, torture, and kill many young adults in opposition to the regime in control. The children of abducted young women were often taken and given to government supporters to raise as their own. Now that the political situation has changed, grandparents of these stolen children are trying to locate and reclaim them as their legitmate grandchildren. Imagine that you are the judge in a trial centering on the custody of a child. Mr. and Mrs. Escobar believe Carlos Mendoza is the son of their abducted, murdered daughter. If this is true, then Mr. and Mrs. Sanchez are the paternal grandparents of the child because their son (also abducted and murdered) was the husband of the Escobars' daughter. Mr. and Mrs. Mendoza claim that Carlos is their natural child. The attorney for the Escobar and Sanchez couples informs you that scientists have discovered a series of RFLPs in human mitochondrial DNA. He tells you that his clients are eager to be tested and asks that you order that Mr. and Mrs. Mendoza and Carlos also be tested.

   a.   Can mitochondrial RFLP data be helpful in this case? In what way?

   b.   Does the mitochondrial DNA of all seven parties need to be tested to resolve the case? If not, whose mitochondrial DNA actually needs to be tested in this case? Explain your choices.

   c.   Assume that the mitochondrial DNA of critical people has been tested and you have received the results. How would the results resolve the question of Carlos's parentage?

   Answer:

   a.   Mitochondrial RFLP data can be helpful to trace the maternal line of descent. Carlos Mendoza will have inherited his mitochondrial DNA from his mother, and she will have inherited it from her mother. If Mrs. Escobar and Mrs. Mendoza have different mitochondrial RFLPs, it can be determined which of them contributed mitochondria to Carlos.

   b.   Only Carlos and individuals who might have maternally contributed his mitochondria (Mrs. Escobar and Mrs. Mendoza) need to be tested. The potential grandfathers need not be tested. Mrs. Sanchez also need not be tested: She may have given mitochondria to Carlos's father, but the father would not have passed them on to Carlos.

   c.   If Mrs. Mendoza and Mrs. Escobar do not differ in mitochondria RFLPs, the data will not be helpful. If the mitochondrial RFLPs do differ, and Carlos matches Mrs. Mendoza, the case should be dismissed. If Carlos matches Mrs. Escobar, then the Escobar and Sanchez couples are indeed the grandparents, and the Mendozas have claimed a stolen child.

23.28   The analysis of mitochondrial DNA has been very useful in assessing the history of specific human populations. For example, a 9-bp deletion in the small intergenic region between the genes for cytochrome oxidase subunit II and tRNA.Lys (see Figure 23.3) has been a very informative marker to trace the origins of Polynesians. The deletion is widely distributed across southeast Asia and the Pacific and is present in 80 to 100 percent of individuals in the different populations within Polynesia. One of the most polymorphic regions of the mitochondrial genome is found in the region between the genes for tRNA.Phe and tRNA.Thr. In Asians with the 9-bp deletion, a specific set of DNA sequence polymorphisms in this region is found. Using the 9-bp deletion and the DNA sequence polymorphisms as markers, comparative analysis of Asian populations has found a genetic trail of mitochondrial DNA variation. The trail begins in Taiwan, winds through the Philippines and Indonesia, proceeds along the coast of New Guinea, and then moves into Polynesia. Based on an estimated rate of mutation in the tRNA.Phe to tRNA.Thr region, this expansion of mitochondrial DNA variants is thought to be about 6,000 years old. This is consistent with linguistic and archaeological evidence that associates Polynesian origins with the spread of the Austronesian language family out of Taiwan between 6,000 and 8,000 years ago.

a. Why are these types of mitochondrial DNA polymorphisms such good markers for tracing human migration patterns?

b. Why is it important to correlate findings from mitochondrial DNA polymorphisms with other (non-DNA) assessment methods?

c. Why might sequences in the tRNA.Phe to tRNA.Thr region be more polymorphic than other sequences in the mitochondrial genome?

d. The 9-bp deletion has also been found in human populations in Africa. What different explanations are possible for this, and how might these explanations be evaluated?

Answer:

a. The analysis of mitochondrial DNA is a powerful tool in analyzing maternal lineages in humans. By following the inheritance of a set of polymorphisms linked to mitochondrial DNA, maternal lineages can be followed.

b. In cases such as this one, it is important to relate these findings to independent assessment methods (language, archaeological evidence). This is because DNA polymorphisms can arise at multiple points in a maternal lineage, and taken alone, are not proof of a historical relationship.

c. This region lacks coding sequences, and is used to regulate mitochondrial DNA replication. Thus, it may have fewer constraints that force conservation of its DNA sequence.

d. There are at least three explanations: a single origin of the deletion in Asia, with migration to Africa; a single origin in Africa, with migration to Asia; or independent, identical mutations in both Africa and Asia. If either of the first two explanations were correct, then DNA in the tRNA.Phe to tRNA.Thr region would be relatively similar in sequence in those African and Asian populations with the deletion. If the third explanation were correct, then DNA in the tRNA.Phe to tRNA.Thr region would be relatively dissimilar in those African and Asian populations with the deletion. (The latter is seen.)

23.29 The pedigree in the following figure shows a family in which a rare inherited disease called Leber hereditary optic atrophy is segregating. This condition causes blindness in adulthood. Studies have recently shown that the mutant gene causing Leber hereditary optic atrophy is located in the mitochondrial genome.

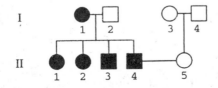

a. What other modes of inheritance (e.g., autosomal dominant, X-linked recessive) are consistent with the inheritance of this rare disease? How could you provide evidence that this disease is not inherited using these modes?

b. Assuming II-5 is normal, what proportion of the offspring of II-4 and II-5 are expected to inherit Leber's hereditary optic atrophy?

c. Assuming that II-2 marries a normal male, what proportion of their sons should be affected? What proportion of their daughters should be affected?

Answer:

a. From an analysis of just this pedigree, we cannot exclude X-linked dominant, autosomal dominant, and autosomal recessive inheritance. These modes of inheritance might be excluded if an expanded kindred—one that included additional branches and generations not shown here—were examined. Since the trait is due to a mutation in mitochondrial DNA, attempts to map the trait to a nuclear linkage group would fail.

b. Since II-4 is a male, he will not contribute any of his mutant mitochondria to his offspring. All of his offspring will have normal mitochondria from their mother and be normal.

c. Since II-2 is a female, all of her progeny will obtain her mutant mitochondria, and hence all of her sons and daughters will be affected.

23.30   The inheritance of shell-coiling direction in the snail *Limnaea peregra* has been studied extensively. A snail produced by a cross between two individuals has a shell with a right-hand twist (dextral coiling). This snail produces only left-hand (sinistral) progeny on selfing. What are the genotypes of the F_1 snail and its parents?

Answer: The F_1 snail gives sinistral offspring when selfed, so it is *dd*. Therefore, both parents had a *d* allele. Since the F_1 has a dextral pattern, its maternal parent was *Dd*. The paternal parent could have been either *Dd* or *dd*.

24

POPULATION GENETICS

CHAPTER OUTLINE

SPECIATION
Barriers to Gene Flow
Genetic Basis for Speciation

REVIEW OF KEY TERMS, SYMBOLS, AND CONCEPTS

In your own words, write a brief, precise definition of each term in the groups below. Check your definitions using the text. Then develop a concept map using the terms in each list.

1	2	3	4
transmission genetics	models of genetic variation	Hardy-Weinberg equilibrium	speciation
molecular genetics	allele frequency cline	random mating	barriers to gene flow
population genetics	classical model	nonrandom mating	postzygotic, prezygotic isolation
quantitative genetics	neutral mutation model	positive assortative mating	hybrid sterility, inviability, breakdown
genetic structure	fitness	negative assortative mating	reinforcement
genotype frequency	polymorphic locus	inbreeding	temporal isolation
allele frequency	proportion of polymorphic loci	mutation	ecological isolation
Mendelian population	heterozygosity (H)	forward, reverse mutation	behavioral incompatibility
gene pool	RFLP	equilibrium frequency	mechanical isolation
gene counting	nucleotide substitution	random genetic drift	gametic isolation
Hardy-Weinberg law	synonymous change	effective population size	Haldane's rule
infinitely large population	nonsynonymous change	sampling error	
migration	coding sequence	founder effect	
natural selection	noncoding sequence	bottleneck effect	
random mating	microsatellite, STR	migration	
mutation		gene flow	
genetic drift		population viability analysis	
		infinite alleles model	
		neutral theory of molecular evolution	
		variance, standard error	
		natural selection	
		Darwinian fitness	
		pleiotropy	
		selection coefficient	
		protected polymorphism	
		heterosis	
		overdominance	
		heterozygote superiority	

THINKING ANALYTICALLY

Population genetics considers patterns of genetic variation found among individuals within groups and how these patterns vary geographically and evolve over time. As you study population genetics, your perspective will shift away from a direct consideration of the molecular and biochemical mechanisms that underlie the inheritance and expression of traits in a single individual or cell to a statistical evaluation of the *effects* of these processes at the level of the group, population, or species.

At the conceptual core of population genetics lies the Hardy-Weinberg law. Initially, focus your efforts on understanding its assumptions and predictions. Practice the examples given in the text and the text problems to become fluent in analyzing whether a population with two alleles at an autosomal locus is in Hardy-Weinberg equilibrium. Then consider extensions of the law to loci with more than two alleles or loci with sex-linked alleles. Finally, investigate the different ways that the assumptions of the Hardy-Weinberg law may be violated and the consequences of each to the genetic structure of a population.

Be aware that the allelic symbolism used in population genetics is often different from that used in Mendelian or molecular genetics. Consider the following example. In earlier chapters, the genotypes for normal and sickle-cell hemoglobin were written either as $\beta^A\beta^A$, $\beta^A\beta^S$, $\beta^S\beta^S$ or Hb^AHb^A, Hb^AHb^S, Hb^SHb^S. Here they are written as *Hb-A/Hb-A*, *Hb-A/Hb-S*, *Hb-S/Hb-S*. Similarly, care should be taken with the letters p and q, which symbolize the frequency within a population of dominant and recessive members of an allelic pair (but do not symbolize the alleles themselves!).

One of the main concerns of population genetics is how to model changes in allele and genotype frequencies under a set of specified conditions. The models are often very elegant and employ equations that identify relationships between a set of variables identified in a population. Memorization of the equations used to analyze changes in populations is by itself insufficient to understand the models. It will be difficult, if not impossible, to solve even moderately challenging word problems if you can only plug numbers into equations. While you will find it helpful to recognize by sight several of the equations, you need to do more. Take the time to understand the model that leads to the equation and how the mathematical and statistical analyses in the model are related to a biological question.

A few hints for solving some of the problems are in order. First, you will often find that the frequency of one class of homozygote (usually the recessive homozygotes, given by q^2) is the only piece of hard data available. As you work back from this information, keep track of the assumptions that you are making. Otherwise, you may enter into the realm of circular reasoning. For example, you can legitimately calculate q as the square root of q^2 and, if only two alleles exist, calculate p by equating it with $1-q$. However, if you then determine that heterozygotes exist at a frequency of $2pq$, you are assuming that random mating is occurring and the conditions of Hardy-Weinberg equilibrium are satisfied. Further work with your calculated values of p^2, $2pq$, and q^2 will continue to reflect that assumption. Often, you will need to look for more information or another approach to prove that the population in question is in Hardy-Weinberg equilibrium. Second, doing the math requires care and, if you are not exceptionally fluent with it, patience. It will sometimes help to factor out common multipliers [e.g., $2pq^2 + q^2 = q^2(2p + 1)$]. At other times, it will be helpful to recognize members of a binomial expansion [e.g., $p^2 + 2pq + q^2 = (p + q)^2$].

After you have mastered the concepts underlying the Hardy-Weinberg law, focus on understanding how genetic variation can be measured. A number of important models have been developed that have been supported by data gathered using a variety of methods, including those of molecular genetics. Models have been proposed to address the substantial amount of genetic variation that exists in a population and the factors that can lead to changes in a population's genetic structure. Some of these employ quantitative analysis and equations. Approach these models just as you did the Hardy-Weinberg law. First become familiar with the conceptual issues of the model, and then relate the variables in an equation to the key factors considered in the model.

QUESTIONS FOR PRACTICE

Multiple Choice Questions

1. In a small population, 30 percent of the individuals have blood type M, 40 percent of the individuals have blood type MN, and 30 percent of the individuals have blood type N. If p equals the frequency of the L^M allele and q equals the frequency of the L^N allele, what are p and q?
   a. $p = 0.30, q = 0.30$
   b. $p = 0.50, q = 0.50$
   c. $p = 0.30, q = 0.70$
   d. $p = 0.50, q = 0.30$

2. Is the population described in Question 1 above in Hardy-Weinberg equilibrium?
   a. Yes; the calculated genotypic frequencies equal the expected genotype frequencies.
   b. Yes; at equilibrium you always have equal numbers of recessive and dominant homozygotes.
   c. No; the frequency of heterozygotes is too large and the frequency of homozygotes is too low.
   d. No; the frequency of heterozygotes is too low and the frequency of homozygotes is too large.

3. Which of the following is *not* an assumption about a population in Hardy-Weinberg equilibrium?
   a. The population is isolated.
   b. Random mating occurs in the population.
   c. The population is free from mutation.
   d. The population is free from natural selection.
   e. The population is free from migration.

4. The frequency of one form of X-linked color blindness varies among human ethnic groups. What can be said about whether each ethnic group is in Hardy-Weinberg equilibrium? (Let q = the frequency of the normal allele and p = the frequency of the color-blind allele.)
   a. None of the ethnic populations can be in equilibrium, since all have different values for p and q.
   b. Only the entire human population is in equilibrium.
   c. Some of the ethnic populations may be in equilibrium, provided that the frequency of the trait in males is p and the frequency in females is p^2.
   d. Some of the ethnic populations may be in equilibrium, provided that the frequency of the trait in both sexes is p^2.
   e. All of the ethnic populations will be in equilibrium, since each satisfies the criteria for a population in Hardy-Weinberg equilibrium.

5. In a large, randomly mating population, 80 percent of the individuals have dark hair and 20 percent are blond. Assuming that hair color is controlled by one pair of alleles, is the allele for dark hair necessarily dominant to the one for blond hair?
   a. Yes, because otherwise the population would not be dominated by dark-haired individuals.

   b. Yes, because more of something (in this case, hair color) is always dominant to less of that thing.

   c. No, because relative frequencies of alleles in a randomly mating population are unrelated to issues of dominance and recessiveness.

   d. No, because although there is a relationship between dominance and allele frequency, that relationship is not seen in this example.

6. The amounts of genetic variation in a population can be explained in a number of ways. In one model, a large amount of genetic variation is explained by recurrent mutation and random changes in allele frequencies. In this model, natural selection selects against some of the variation affecting fitness but does not select for or against much of the genetic variation. This model is termed
   a. the classical model
   b. the neutral mutation model
   c. the random mutation model

7. Which of the following populations are in Hardy-Weinberg equilibrium?

Population	Genotypes		
	AA	*Aa*	*aa*
a	0.72	0.20	0.08
b	0.12	0.80	0.08
c	0.08	0.01	0.91
d	0.25	0.50	0.25

8. How is the rate of forward mutation likely to be related to the rate of reverse mutation?
   a. The rate of forward mutation is generally lower because there is mutational pull back to a specific form.
   b. The rate of forward mutation is generally higher because once an allele has changed, it is nearly as likely that a subsequent change will be to yet another new form.
   c. The rate of forward and reverse mutations are codependent and usually equal.
   d. The relative rates of forward and reverse mutation are so highly variable that one cannot formulate an accurate generalization comparing the two.

9. Which of the following can result in genetic drift?
   a. a sampling error
   b. random factors producing unexpected mortality
   c. the establishment of a population by a small number of breeding individuals
   d. a drastic reduction in the size of a population
   e. a change in environmental conditions that affects selection
   f. all of the above

10. What is the effective population size for a population consisting of 10 breeding males and 2 breeding females?
    a. 12
    b. 6
    c. 7
    d. 2

11. Which of the following is *not* true concerning the effect of migration among populations?

a. Migration tends to increase the effective size of the populations, leading to a reduction in genetic drift.

b. Migration is associated with gene flow and introduces new alleles to the population.

c. If the allele frequencies of migrants and the recipient population differ, migration can lead to the further differentiation of two populations.

d. Migration and genetic drift have opposite effects on size and variability. Migration effectively increases size and variability, whereas drift acts in opposition.

12. What can one generally state about measuring the fitness associated with a specific genotype?

a. It is easy to assess and can be based on the number of offspring of an individual.

b. Since it is based on the reproductive ability of a genotype, it is an absolute term, which requires no assumptions.

c. It will remain the same from generation to generation.

d. It is difficult to measure because of pleiotropic effects.

13. Consider a recessive trait that results in a complete lack of reproductive success of homozygotes. How might such a recessive trait be maintained at a high level in a population?

a. through new mutation

b. through heterosis

c. through overdominance

d. through heterozygote superiority

e. all of the above

14. For a given eukaryotic gene, which rate of DNA change is expected to be the lowest?

a. the relative rate of change in nonfunctional pseudogenes

b. the relative rate of change in introns

c. the relative rate of change for synonymous substitutions in coding sequences

d. the relative rate of change for nonsynonymous substitutions in coding sequences

e. the relative rate of change in leaders and trailers

15. Which of the rates in Question 14 is expected to be the highest?

Answers: 1b, 2d, 3a, 4c, 5c, 6b, 7d, 8b, 9f, 10c, 11c, 12d, 13e, 14d, 15d

Thought Questions

1. Consider a population that is not in Hardy-Weinberg equilibrium for a pair of alleles. Show why equilibrium values for genotype frequencies are reached in one generation after the onset of random mating for autosomal alleles, but more than one generation of random mating is required for sex-linked alleles.

2. Distinguish between heterozygosity and the proportion of polymorphic loci.

3. Mutation pressure is rarely the most important determinant of allelic frequency. What other factors are important?

4. Can a population be in Hardy-Weinberg equilibrium for one, but not another, pair of alleles?

5. What is the most likely explanation if the allelic frequencies agree with those predicted by the Hardy-Weinberg law, but the genotype frequencies do not? (Hint: Consider mating systems and heterosis.)

6. Why do mutation, migration, and drift not necessarily lead to adaptation?

7. What are five different molecular methods that can be used to measure genetic variation? What advantages and disadvantages are there to each?

8. Distinguish between nonrandom mating, positive assortative mating, negative assortative mating, and inbreeding.

9. What are the arguments for and against investing funds and human effort in a population viability analysis for an endangered species?

10. What are the values of preserving and promoting genetic diversity within a species? How does the development of genomics and technologies related to genomics affect our ability to assess and influence genetic diversity, both in our own species and in other species?

Thought Questions for Media

After reviewing the media on the *iGenetics* CD-ROM, try to answer the following questions.

1. What assumptions of the Hardy-Weinberg law were illustrated by the animation?

2. Under these assumptions, what are the predictions of the Hardy-Weinberg law?

3. Which of these assumptions were not valid in the population of Darwin's finches under study by Dr. Grant?

4. What was the purpose of Dr. Grant's study of these finches?

5. The animation indicated that beak shape and size are genetically complex. Some genes will have additive effects, some will be epistatic, and a few will show simple dominance and recessiveness. Why, for the purposes of the study, was it useful to assume the existence of a simple gene that, when homozygous recessive, contributes to a decrease in beak size?

6. Is fitness constant for a given trait such as beak length? If not, according to what conditions does it vary?

7. The animation describes how natural selection for small and large beak size in Darwin's finches is complex. For example, even though birds with larger beaks can crack harder seeds, in some years, there can still be selection for small beak size. Why might this be the case?

SOLUTIONS TO TEXT PROBLEMS

24.1   In the European land snail *Cepaea nemoralis*, multiple alleles at a single locus determine shell color. The allele for brown (C^B) is dominant to the allele for pink (C^P) and to the allele for yellow (C^Y). The dominance hierarchy among these alleles is $C^B > C^P > C^Y$. In one population sample of *Cepaea*, the following color phenotypes were recorded:

Brown	236
Pink	231
Yellow	33
Total	500

Assuming that this population is in Hardy-Weinberg equilibrium (large, randomly mating, and free from evolutionary processes), calculate the frequencies of the C^B, C^Y, and C^P alleles.

Answer: Equate the frequency of each color with the frequency expected in Hardy-Weinberg equilibrium, letting $p = f(C^B)$, $q = f(C^P)$, and $r = f(C^Y)$.

Brown: $f(C^BC^B) + f(C^BC^P) + f(C^BC^Y) = p^2 + 2pq + 2pr = 236/500 = 0.472$
Pink: $f(C^PC^P) + f(C^PC^Y) = q^2 + 2qr = 231/500 = 0.462$
Yellow: $f(C^YC^Y) = r^2 = 33/500 = 0.066$

Now solve for p, q, and r, knowing that $p + q + r = 1$.

$r^2 = 0.066$, so $r = \sqrt{0.066} = 0.26$

There are two approaches to solve for q. First, since $q^2 + 2qr = 0.462$, one can substitute in $r = 0.26$, giving $q^2 + 2q(0.26) = 0.462$. Recognize this as a quadratic equation and set it equal to 0, and solve for q. That is, solve the equation $q^2 + 0.52q - 0.462 = 0$. Solving the quadratic equation for q, one has:

$$q = \frac{-0.52 \pm \sqrt{(0.52)^2 - 4(1)(-0.462)}}{2(1)} = 0.467$$

A second approach to solve for q is to realize that

$q^2 + 2qr = 0.462$ and $r^2 = 0.066$.

Adding left and right sides of the equations together, one has

$q^2 + 2qr + r^2 = 0.066 + 0.462$
$(q + r)^2 = 0.528$
$q + r = 0.726$
$q = 0.726 - r = 0.726 - 0.26 = 0.467$
Since $p + q + r = 1$, $p = 1 - (q + r) = 1 - (0.26 + 0.467) = 0.273$.

24.2 Three alleles are found at a locus coding for malate dehydrogenase (MDH) in the spotted chorus frog. Chorus frogs were collected from a breeding pond, and each frog's genotype at the MDH locus was determined with electrophoresis. The following numbers of genotypes were found:

M^1M^1	8
M^1M^2	35
M^2M^2	20
M^1M^3	53
M^2M^3	76
M^3M^3	62
Total	254

a. Calculate the frequencies of the M^1, M^2, and M^3 alleles in this population.

b. Using a chi-square test, determine whether the MDH genotypes in this population are in Hardy-Weinberg proportions.

Answer:

a. The tally for M^1 alleles is as follows:

Genotype	# Individuals	# M^1 Alleles
M^1M^1	8	16
M^1M^2	35	35
M^1M^3	53	53
Total		104

The total number of individuals is 254; thus the total number of alleles is 254×2, or 508. The frequency of M^1 alleles is $104/508 = 0.205$. The frequency of the other alleles is obtained similarly so that $f(M^1) = 0.20 = p$, $f(M^2) = 0.30 = q$, and $f(M^3) = 0.50 = r$.

b. For three alleles with frequencies p, q, and r, a population in Hardy-Weinberg equilibrium will have $p^2 + 2pq + 2pr + q^2 + 2qr + r^2 = 1$. To test the hypothesis that the population is in Hardy-Weinberg equilibrium, calculate the numbers of individuals expected in each class using this relationship and the values for p, q, and r obtained in (a). Calculate the value of χ^2 as shown in the following table.

Genotype	Observed Value (o)	Expected Frequency	Expected Value (e)	d ($o - e$)	d^2/e
M^1M^1	8	$p^2 = 0.04$	10	−2	0.40
M^1M^2	35	$2pq = 0.12$	30	5	0.83
M^2M^2	20	$q^2 = 0.09$	23	−3	0.39
M^1M^3	53	$2pr = 0.20$	51	2	0.08
M^2M^3	76	$2qr = 0.30$	75	1	0.01
M^3M^3	62	$r^2 = 0.25$	64	−2	0.06
					$\chi^2 = 1.77$

Since the six phenotypic classes are completely specified by three allele frequencies, the number of phenotypes (6) minus the number of alleles (3) determines the degrees of freedom ($6 - 3 = 3$). With 3 degrees of freedom, $0.70 < P < 0.50$. The hypothesis is accepted as possible, and it appears that the population is in Hardy-Weinberg equilibrium.

24.3   In a large interbreeding population, 81 percent of the individuals are homozygous for a recessive character. In the absence of mutation or selection, what percentage of the next generation would be homozygous recessives? Homozygous dominants? Heterozygotes?

Answer: The conditions of this problem meet the requirements for a population in Hardy-Weinberg equilibrium. In such a population, if p equals the frequency of A, and q equals the frequency of a, one expects p^2 (AA), $2pq$ (Aa), and q^2 (aa) genotypes after random mating. Here, $q^2 = 0.81$, so $q = 0.9$. Since $p + q = 1$, $p = 1 - 0.9 = 0.1$. In the next generation, one would expect $p^2 = (0.1)^2 = 0.01$ (or 1%) AA genotypes, $2pq = 2(0.1)(0.9) = 0.18$ (or 18%) Aa genotypes, and $q^2 = (0.9)^2 = 0.81$ (or 81%) aa genotypes.

24.4   Let A and a represent dominant and recessive alleles whose respective frequencies are p and q in a given interbreeding population at equilibrium (with $p + q = 1$).
  a. If 16 percent of the individuals in the population have recessive phenotypes, what percentage of the total number of recessive genes exist in the heterozygous condition?
  b. If 1.0 percent of the individuals were homozygous recessive, what percentage of the recessive genes would occur in heterozygotes?

Answer:
  a. Since $q^2 = 0.16$, $q = \sqrt{0.16} = 0.40$. Since $p + q = 1$, $p = 0.60$. The frequency of heterozygotes is $2pq = 2(0.40)(0.60) = 0.48$. Each of the 48 percent of the heterozygotes has one recessive allele, while each of the 16 percent of the homozygous recessive individuals has two. Thus, the percentage of the total number of recessive alleles in heterozygotes is $(0.48)/[0.48 + 2(0.16)] = 0.48/0.80 = 0.60$, or 60 percent.
  b. If $q^2 = 0.01$, then $q = 0.1$, and $p = 0.9$. $2pq = 2(0.1)(0.9) = 0.18$. The percentage of the total number of recessive alleles in heterozygotes is $(0.18)/[0.18 + 2(0.01)] = 0.18/0.20 = 0.90$, or 90 percent.

24.5   A population has eight times as many heterozygotes as homozygous recessives. What is the frequency of the recessive gene?

Answer: The frequency of heterozygotes in a population in equilibrium is $2pq$, and the frequency of homozygous recessives is q^2. Here, there are eight times as many heterozygotes as homozygous recessives, so $2pq = 8q^2$. Since $p + q = 1$, $p = 1 - q$, and one can substitute $1 - q$ for p. This gives $2(1 - q)q = 8q^2$. Dividing both sides by q and multiplying through, one has $2 - 2q = 8q$. Thus, $2 = 10q$, and $q =$ the frequency of the recessive allele $= 0.20$.

24.6   In a large population of range cattle, the following ratios are observed: 49 percent red (RR), 42 percent roan (Rr), and 9 percent white (rr).

a. What percentage of the gametes that give rise to the next generation of cattle in this population will contain allele *R?*

b. In another cattle population, only 1 percent of the animals are white and 99 percent are either red or roan. What is the percentage of *r* alleles in this case?

Answer:

a. In the red *RR* animals, all of the gametes contain the *R* allele, while in the roan *Rr* animals, half of the gametes contain the *R* allele. Therefore, [49 + (42/2)] = 70 percent of the gametes will contain the *R* allele. Another way to look at this problem is to realize that the frequency of gametes bearing a certain allele is the same as the frequency of the allele in the population. Let *p* equal the frequency of *R* in the population. Since there are 49 percent red animals, $p^2 = 0.49$, so $p = 0.70$ (or 70 percent).

b. If one lets *q* represent the frequency of *r*, since 1 percent of the animals are white, one has $q^2 = 0.01$. Hence, $q = 0.1$.

24.7   In a gene pool the alleles *A* and *a* have initial frequencies of *p* and *q*, respectively. Show that the allelic frequencies and zygotic frequencies do not change from generation to generation as long as there is no selection, mutation, or migration; the population is large; and the individuals mate at random.

Answer: Let the frequency of allele *A* equal *p* and the frequency of allele *a* equal *q*, with $p + q = 1$. Then, in the initial generation, the frequency of *AA* genotypes is p^2, the frequency of *aa* genotypes is q^2. The frequency of the remaining genotypes (i.e., *Aa* heterozygotes) must be $1 - (p^2 + q^2)$. Since $p + q = 1$, $(p + q)^2 = 1^2$, and $p^2 + 2pq + q^2 = 1$. Therefore, $1 - (p^2 + q^2) = 2pq$, and the frequency of *Aa* heterozygotes must be $2pq$.

Assume there is no selection, mutation, or migration and that the individuals mate at random. Since there are three different genotypes in the population, nine crosses are possible. The frequency of each type of cross is determined by the frequency of each parental genotype. The types of crosses, the frequency of each cross, the types of progeny, and the frequency of each progeny class are listed in the table below:

Cross	Cross Frequency	Progeny Ratios			Progeny Frequencies		
		AA	*Aa*	*aa*	*AA*	*Aa*	*aa*
$AA \times AA$	$p^2 \times p^2 = p4$	all			p^4		
$AA \times Aa$	$p^2 \times 2pq = 2p^3q$	½	½		p^3q	p^3q	
$AA \times aa$	$p^2 \times q^2 = p^2q^2$		all			p^2q^2	
$Aa \times AA$	$2pq \times p^2 = 2p^3q$	½	½		p^3q	p^3q	
$Aa \times Aa$	$2pq \times 2pq = 4p^2q^2$	¼	½	½	p^2q^2	$2p^2q^2$	p^2q^2
$Aa \times aa$	$2pq \times q^2 = 2pq^3$		½	½		pq^3	pq^3
$aa \times AA$	$q^2 \times p^2 = p^2q^2$		all			p^2q^2	
$aa \times Aa$	$q^2 \times 2pq = 2pq^3$		½	½		pq^3	pq^3
$aa \times aa$	$q^2 \times q^2 = q^4$			all			q^4

To determine the frequency of a particular zygotic class, add up the frequency of the progeny in that class. Then factor out a common multiplier, and note that $p + q = 1$ and that $(p + q)^2 = p^2 + 2pq + q^2$. One has:

$$\begin{aligned}
\text{frequency } (AA) &= p^4 + 2p^3q + p^2q^2 \\
&= p^2(p^2 + 2pq + q^2) \\
&= p^2(p + q)^2 \\
&= p^2(1)^2 = p^2
\end{aligned}$$

$$\begin{aligned}
\text{frequency } (Aa) &= 2p^3q + 4p^2q^2 + 2pq^3 \\
&= 2pq(p^2 + 2pq + q^2) \\
&= 2pq(p + q)^2 \\
&= 2pq(1)^2 = 2pq
\end{aligned}$$

$$\begin{aligned}
\text{frequency } (aa) &= pq + 2pq^3 + q^4 \\
&= q^2(p^2 + 2pq + q^2) \\
&= q^2(p + q)^2 \\
&= q^2(1)^2 = q^2
\end{aligned}$$

Thus, the zygotic frequencies do not change from one generation to the next.

Since all of the gametes of the AA parents and half of the gametes of the Aa parents will bear the A allele, the frequency of A in the gene pool of the next generation is $p^2 + pq = p(p + q) = p$. Since all of the gametes of the aa parents and half of the gametes of the Aa parents will bear the a allele, the frequency of a in the gene pool of the next generation is $q^2 + pq = q(q + p) = q$. Thus, the allelic frequencies do not change from one generation to the next.

24.8 The S-s antigen system in humans is controlled by two codominant alleles, S and s. In a group of 3,146 individuals the following genotypic frequencies were found: 188 SS, 717 Ss, and 2,241 ss.

a. Calculate the frequency of the S and s alleles.

b. Determine whether the genotypic frequencies are in Hardy-Weinberg equilibrium by using the chi-square test.

Answer:

a. Let p equal the frequency of S and q equal the frequency of s. Since homozygotes have two identical alleles and heterozygotes have one recessive and one dominant allele, one has:

$$p = \frac{2(188)[SS] + 717[Ss]}{2(3,146)} = \frac{1093}{6292} = 0.1737$$

$$q = \frac{717[Ss] + 2(2,241)[ss]}{2(3,146)} = \frac{5199}{6292} = 0.8263$$

b. Remember that in a χ^2 test, one uses the actual numbers of progeny observed and expected, and not the frequencies. With a hypothesis that the population is in Hardy-Weinberg equilibrium, one has:

Class	Observed (*o*)	Expected Frequency	Expected (*e*)	*d* (*o* − *e*)	d^2/e
SS	8	$p^2 = 0.0302$	95	93	91.0
Ss	717	$2pq = 0.287$	903	−186	38.3
ss	2,241	$q^2 = 0.683$	2,148	93	4.0
	3,146	1	3,146	0	133.3

There is only 1 degree of freedom because the three genotypic classes are completely specified by two allele frequencies, namely, *p* and *q*. (d.f. = number of phenotypes − number of alleles = 3 − 2 = 1.) The χ^2 value of 133.3, for 1 degree of freedom, gives $P < 0.0001$. Therefore, the distribution of genotypes differs significantly from that expected if the population were in Hardy-Weinberg equilibrium.

24.9   Refer to Problem 22.8. A third allele is sometimes found at the *S* locus. This allele S^u is recessive to both the *S* and the *s* alleles and can be detected only in the homozygous state. If the frequencies of the alleles *S*, *s*, and S^u are *p*, *q*, and *r*, respectively, what would be the expected frequencies of the phenotypes *S*−, *Ss*, *s*−, and S^uS^u?

Answer: The frequencies are

$$f(S-) = f(SS) + f(SS^u) = p^2 + 2pr$$
$$f(Ss) = 2pq$$
$$f(s-) = f(ss) + f(sS^u) = q^2 + 2qr$$
$$f(S^uS^u) = r^2$$

24.10   In a large interbreeding human population, 60 percent of individuals belong to blood group O (genotype *i/i*). Assuming negligible mutation and no selective advantage of one blood type over another, what percentage of the grandchildren of the present population will be type O?

Answer: The conditions described in the problem indicate that the population is in Hardy-Weinberg equilibrium. Under equilibrium conditions, neither the allele nor the zygotic frequencies change from one generation to the next. Therefore, in two generations there should still be 60 percent type O individuals.

24.11   A selectively neutral, recessive character appears in 40 percent of the males and in 16 percent of the females in a large, randomly interbreeding population. What is the gene's frequency? What proportion of females are heterozygous for it? What proportion of males are heterozygous for it?

Answer: There are several possible explanations for the difference in the frequency of the trait in males and females. Two are sex linkage and autosomal linkage with sex-influenced expression. Sex linkage can be readily examined. If the population is in

Hardy-Weinberg equilibrium (which this one is), the frequency of the recessive allele causing the trait is q, and the gene is X linked, then the frequency in XY males would be q, while the frequency in XX females would be q^2. Since the frequency in males is 0.4 and the frequency in females is $(0.4)^2 = 0.16$, the data fit a model of sex linkage with $q = 0.4$. The frequency of heterozygous XX (female) individuals is $2pq = 2(0.6)(0.4) = 0.48$. Since the trait appears to be sex linked, no heterozygous males exist.

24.12   Suppose you found two distinguishable types of individuals in wild populations of some organism in the following frequencies:

	Type 1	Type 2
Females	99%	1%
Males	90%	10%

The difference is known to be inherited. Are these data compatible with the trait being X linked?

Answer: As in the answer to Problem 24.11, there are several explanations for differences in the frequency of a trait between males and females. Test the possibility of sex linkage first. Suppose the trait is sex linked and recessive and the population is in equilibrium. Let q equal the frequency of the recessive allele, and p equal the frequency of the dominant allele. If the gene is X linked, one would expect XY males to express the recessive trait at a frequency of q, and XX females to express the recessive trait at a frequency of q^2. By inspecting the data given, one can see that the frequency of type 2 individuals in females is 0.01, which is the square of the frequency of type 2 individuals in males: $(0.10)^2 = 0.01$. Thus, this trait appears to be controlled by a sex-linked pair of alleles occurring with allele frequencies of $q = 0.1$ recessive and $p = 0.9$ dominant.

24.13   Red-green color blindness is caused by a sex-linked recessive gene. About 64 women out of 10,000 are color blind. What proportion of men would be expected to show the trait if mating is random?

Answer: Let q equal the frequency of the recessive allele, and p equal the frequency of the dominant allele. One expects homozygotes showing the trait to appear at a frequency of q^2 in a population at equilibrium. If $q^2 = 64/10,000 = 0.0064$, $q = 0.08$. Thus, one would expect 8 percent of XY male individuals to show the trait.

24.14   About 8 percent of the men in a population are red-green color blind (because of a sex-linked recessive gene). Answer the following questions, assuming random mating in the population, with respect to color blindness.
   a. What percentage of women would be expected to be color blind?
   b. What percentage of women would be expected to be heterozygous?
   c. What percentage of men would be expected to have normal vision two generations later?

Answer:

   a. Let q equal the frequency of the recessive allele, and p equal the frequency of the dominant allele. The frequency of males with the trait will be q, and the frequency of females will be q^2. Here, $q = 0.08$, so $q^2 = 0.0064$, or 0.64 percent.

b. Since $q = 0.08$, $p = 0.92$. The frequency of heterozygotes is $2pq = 0.1472$, or 14.72 percent. Only women can be heterozygotes, so the frequency of heterozygous women is 14.72 percent.

c. If the population is in Hardy-Weinberg equilibrium, the frequencies of alleles and the frequency of zygotic phenotypes will not change in two generations. Since $p = 0.92$, 92 percent of the XY males will have normal vision.

24.15 List some of the basic differences in the classical, balance, and neutral mutation models of genetic variation.

Answer: In the classical model, natural populations possess little variation. Within each population, a single allele is strongly favored by natural selection because it "functions the best." This is the wild-type allele, and most individuals are homozygous for this allele. Occasionally, a mutation arises. Usually, this mutation is deleterious, and strong selection occurs against it. In the rare case that a new mutation is advantageous for survival or reproduction, the new allele will increase in frequency and eventually become the new wild type.

In the balance model, there is significant genetic variation within a population. Many alleles exist at each locus, and appear in the population in intermediate frequencies. Members of a population are heterozygous at numerous loci. Natural selection *actively* maintains genetic variation within a population by balancing selection to prevent any single allele from reaching a high frequency.

In a neutral mutation model, there is also significant genetic variation within a population. As in the balance selection model, many alleles exist in intermediate frequencies and members of the population are heterozygous at numerous loci. In the neutral-mutation model, much of this genetic variation is explained by recurrent mutation and random changes in the frequency of alleles that are physiologically equivalent, and not necessarily by natural selection. Hence, in this model, most variation is neutral with regard to selection.

24.16 Two alleles of a locus, A and a, can be interconverted by mutation:

$$A \underset{v}{\overset{u}{\rightleftarrows}} a$$

where u is a mutation rate of 6.0×10^{-7} and v is a mutation rate of 6.0×10^{-8}. What will be the frequencies of A and a at mutational equilibrium, assuming no selective difference, no migration, and no random fluctuation caused by genetic drift?

Answer: Let q equal the frequency of a, and p equal the frequency of A, with $q + p = 1$. As discussed in the text, when the population is at equilibrium, the frequency of p and q is given by

$$q = \frac{u}{u+v} = \frac{6 \times 10^{-7}}{(6 \times 10^{-7}) + (6 \times 10^{-8})} = \frac{6 \times 10^{-7}}{(6 \times 10^{-7}) + (0.6 \times 10^{-7})} = \frac{6}{6.6} = 0.91$$

$$p = 1 - q = 1 - 0.91 = 0.09$$

Thus, the frequencies are 0.0081 *AA*, 0.1638 *Aa*, and 0.8281 *aa*.

24.17

a. Calculate the effective population size (N_e) for a breeding population of 50 adult males and 50 adult females.

b. Calculate the effective population size (N_e) for a breeding population of 60 adult males and 40 adult females.

c. Calculate the effective population size (N_e) for a breeding population of 10 adult males and 90 adult females.

d. Calculate the effective population size (N_e) for a breeding population of 2 adult males and 98 adult females.

Answer: The effective breeding size of a population is given by the equation

$$N_e = \frac{4 \times N_f \times N_m}{N_f + N_m}$$

where N_f equals the number of breeding females and N_m equals the number of breeding males. Apply this equation to each of the situations described as follows:

a. $(4 \times 50 \times 50)/100 = 100$

b. $(4 \times 60 \times 40)/100 = 96$

c. $(4 \times 10 \times 90)/100 = 36$

d. $(4 \times 2 \times 98)/100 = 7.8$

24.18 In a population of 40 adult males and 40 adult females, the frequency of allele A is 0.6 and the frequency of allele a is 0.4.

a. Calculate the 95 percent confidence limits of the allelic frequency for A.

b. Another population with the same allelic frequencies consists of only 4 adult males and 4 adult females. Calculate the 95 percent confidence limits of the allelic frequency for A in this population.

c. What are the 95 percent confidence limits of A if the population consists of 76 females and 4 males?

Answer:

a.

$$Ne = \frac{4 \times N_f \times N_m}{N_f + N_m} = \frac{4 \times 40 \times 40}{40 + 40} = 80$$

$$2s_p = 2 \times \sqrt{\frac{pq}{2Ne}} = 2 \times \sqrt{\frac{0.6 \times 0.4}{2 \times 80}} = 2 \times 0.039 = 0.078$$

The 95 percent confidence limits of A are $p \pm 2s_p$, or 0.6 ± 0.078, or $0.522 \le p \le 0.678$.

b.

$$\frac{4 \times 4 \times 4}{4 + 4} = 8$$

$$2s_p = 2 \times \sqrt{\frac{0.6 \times 0.4}{2 \times 8}} = 0.245$$

The 95 percent confidence limits of A are 0.6 ± 0.245, or $0.355 \leq p \leq 0.845$.

c.

$$\frac{4 \times 76 \times 4}{76 + 4} = 15.2$$

$$2s_p = 2 \times \sqrt{\frac{0.6 \times 0.4}{2 \times 15.2}} = 0.178$$

The 95 percent confidence limits of A are 0.6 ± 0.178, or $0.422 \leq p \leq 0.778$.

24.19 The land snail *Cepaea nemoralis* is native to Europe but has been accidentally introduced into North America at several localities. These introductions occurred when a few snails were inadvertently transported on plants, building supplies, soil, or other cargo. The snails subsequently multiplied and established large, viable populations in North America.

Assume that today the average size of *Cepaea* populations found in North America is equal to the average size of *Cepaea* populations in Europe. What predictions can you make about the amounts of genetic variation present in European and North American populations of *Cepaea*? Explain your reasoning.

Answer: Since the gene pool in the present, large North American population is derived from a small number of individuals, a founder effect is likely to have occurred. Thus, genetic drift (random change in allelic frequency due to chance) will influence the North American populations to a greater degree than it will the European populations. One would expect to see less variation within and greater genetic differentiation among the North American populations. The magnitude of the difference between the native North American population and the introduced population will depend on how many times different European populations were introduced. For example, if relatively large numbers of European snails from multiple localities were introduced many times, the founder effect would not be as marked as if a small number of snails were introduced just once from one location.

24.20 A population of 80 adult squirrels resides on campus, and the frequency of the *Est*[1] allele among these squirrels is 0.70. Another population of squirrels is found in a nearby woods, and there, the frequency of the *Est*[1] allele is 0.5. During a severe winter, 20 of the squirrels from the woods population migrate to campus in search of food and join the campus population. What will be the allelic frequency of *Est*[1] in the campus population after migration?

Answer: Let p_I equal the frequency of A in population I, and p_{II} equal the frequency of A in population II. If individuals in population I migrate to population II and make up proportion m of population II', the new frequency of A in population II' (p'_{II}) is given by:

$$p'_{II} = mp_I + (1 - m)p_{II}$$

Here, $p'_{II} = [20/(20 + 80)](0.50) + \{1 - [20/(20 + 80)]\}(0.70) = 0.66$.

24.21 Upon sampling three populations and determining genotypes, you find the following three genotype distributions. What would each of these distributions imply with regard to selective advantages of population structure?

Population	AA	Aa	aa
1	0.04	0.32	0.64
2	0.12	0.87	0.01
3	0.45	0.10	0.45

Answer: For each population, determine the frequency of the *A* and *a* alleles, and assess how the population structure compares with that expected if no selection was occurring. To do this calculation, suppose there were 100 individuals in each population containing 200 alleles.

In population 1, there would be $(0.04)(100)(2) + (0.32)(100)(1) = 40$ *A* alleles and $(0.64)(100)(2) + (0.32)(100)(1) = 160$ *a* alleles. Thus, $f(A) = p = 40/200 = 0.20$, and $f(a) = q = 160/200 = 0.80$. If the population was in Hardy-Weinberg equilibrium, one would expect $p^2 = 0.04$ *AA*, $2pq = 0.32$ *Aa*, and 0.64 *aa* individuals, as is observed. If selection is acting at all, it may be acting to maintain the observed frequencies of the *A* and *a* alleles in equilibrium.

In population 2, there would be $(0.12)(100)(2) + (0.87)(100)(1) = 111$ *A* alleles, and $(0.01)(100)(2) + (0.87)(100)(1) = 89$ *a* alleles. Here, $p = 0.555$ and $q = 0.455$. If the population was in Hardy-Weinberg equilibrium, one would expect $p^2 = 0.308$ *AA*, $2pq = 0.505$ *Aa*, and 0.207 *aa* individuals. The population is not in equilibrium, as there are far more heterozygotes and far fewer homozygotes than would be expected. This suggests that the heterozygote is being selected for in this population.

In population 3, there would be $(0.45)(100)(2) + (0.10)(100)(1) = 100$ *A* alleles and $(0.10)(100)(1) + (0.45)(100)(2) = 100$ *a* alleles. Here, $p = q = 0.5$. If the population was in Hardy-Weinberg equilibrium, one would expect $f(AA) = f(aa) = p^2 = q^2 = 0.25$ and $f(Aa) = 2pq = 0.50$. The population is not in equilibrium, as there are far more homozygotes (of either type) and far fewer heterozygotes than expected. This suggests that the heterozygote is being selected against in this population.

24.22 The frequency of two adaptively neutral alleles in a large population is 70 percent *A* : 30 percent *a*. The population is wiped out by an epidemic, leaving only four individuals, who produce many offspring. What is the probability that the population several years later will be 100 percent *AA*? (Assume no new mutations occur.)

Answer: There are a variety of situations in which all individuals would become *AA*. One is that the four "founding" individuals are all *AA*. Since the probability of a single individual being *AA* is $(0.7)^2 = 0.49$, the probability of the four founding individuals being *AA* is $(0.49)^4 = 0.0576$, about $1/17$. Even if all four founding individuals are not *AA*, there will be some chance that the subsequent population may become all *AA*. For example, there is a low but distinct probability that only *AA* offspring may be had by *Aa* × *AA* parents. Given that "many" offspring are produced, the likelihood of this will be less than $1/17$ [e.g., if 10 offspring are produced, the likelihood of all being *AA* in such a cross would be $(1/2)^{10} = 1/1,024$]. Since the alleles are adaptively neutral, it would appear that $1/17$ is an upper bound for the likelihood of a population being 100 percent *AA*.

24.23 A completely recessive gene, through changed environmental circumstances, becomes lethal in a certain population. It was previously neutral, and its frequency was 0.5.

a. What was the genotype distribution when the recessive genotype was not selected against?

b. What will be the allelic frequency after one generation in the altered environment?

c. What will be the allelic frequency after two generations?

Answer:

a. Let $p = f(A)$, and $q = f(a)$. Initially, $p = q = 0.5$, so $p^2 = f(AA) = q^2 = f(aa) = (0.5)^2 = 0.25$. $f(Aa) = 2pq = 0.50$.

b. If aa individuals are now lethal, they will not contribute to the gene pool in the next generation. Only AA and Aa individuals will contribute to the gene pool in the next generation. The allele frequency will be

$$f(A) = \frac{(0.25 \times 2) + (0.50 \times 1)}{(0.25 \times 2) + (0.50 \times 2)} = \frac{1.0}{1.5} = 0.67$$

$$f(a) = \frac{(0.50 \times 1)}{(0.25 \times 2) + (0.50 \times 2)} = \frac{0.5}{1.5} = 0.33$$

c. Given the result of (b), the genotype of the progeny are $p^2 = (0.67)^2 = 0.449$ AA, $2pq = 2(0.67)(0.33) = 0.442$ Aa, and $q^2 = (0.33)^2 = 0.109$ aa. As in (b), only AA and Aa individuals will contribute to the gene pool in the next generation. The allele frequency will be

$$f(A) = \frac{(0.449 \times 2) + (0.442 \times 1)}{(0.449 \times 2) + (0.442 \times 2)} = 0.75$$

$$f(a) = \frac{(0.442 \times 1)}{(0.449 \times 2) + (0.442 \times 2)} = 0.25$$

Can you show that the following statement is true under the conditions of this problem? If the last generation in the unaltered environment is considered to be generation number 1, then after n generations, the frequency of A will be $n/(n + 1)$ and the frequency of a will be $1/(n + 1)$.

24.24 Human individuals homozygous for a certain recessive autosomal gene die before reaching reproductive age. Despite this removal of all affected individuals, there is no indication that homozygotes occur less frequently in succeeding generations. To what might you attribute the continued presence of appearance of these recessives?

Answer: There are several reasons why recessives may appear in a constant frequency. (1) Perhaps most important is that new mutations of A to a could occur at a low but constant rate. (2) There could also be a selective advantage to heterozygotes (overdominance). (3) There could be nonrandom mating within the population (e.g., positive assortative mating, inbreeding). (4) There could be a low but steady frequency of migration of heterozygotes into the population. Each of these possibilities violates assumptions made for a population in Hardy-Weinberg equilibrium, and allows maintenance of a recessive allele that is deleterious as a homozygote. (Problem 24.23 considers the consequences of such an allele if the assumptions were not violated.)

(5) If the homozygous recessive condition is a relatively rare occurrence, the frequency of a alleles in the population is very low. As a result, only a very small proportion of the a alleles would be found in homozygotes; almost all a alleles would be found in heterozygotes (protected polymorphism). The few generations for which data have been recorded may not be enough to see a decline in the number of homozygotes.

24.25 A completely recessive gene (Q^1) has a frequency of 0.7 in a large population, and the Q^1Q^1 homozygote has a relative fitness of 0.6.

    a. What will be the frequency of Q^1 after one generation of selection?

    b. If there is no dominance at this locus (the fitness of the heterozygote is intermediate to the fitnesses of the homozygotes), what will the allele frequency be after one generation of selection?

    c. If Q^1 is dominant, what will the allele frequency be after one generation of selection?

Answer: See Table 24.12, p. 693. For (a) and (b), $p = 0.3$, $q = 0.7$, and $s = 1 - W = 0.4$.

    a. If there is selection against the recessive homozygotes, after one generation,

$$\Delta q = \frac{-spq^2}{1-sq^2} = \frac{-(0.4)(0.3)(0.7)^2}{1-(0.4)(0.7)^2} = -0.073$$

$$q^1 = 0.7 - 0.073 = 0.627$$

    b. If there is selection with no dominance, so that the fitness of the heterozygote is intermediate between the two homozygotes,

$$\Delta q = \frac{-spq/2}{1-sq} = \frac{-(0.4)(0.3)(0.7)/2}{1-(0.4)(0.7)} = -0.0583$$

$$q^1 = 0.7 - 0.0583 = 0.642$$

    c. To select against Q^1 as a dominant allele, let p equal the frequency of the dominant allele being selected against: $p = 0.7$, $q = 0.3$, and $s = 0.4$.

$$\Delta p = \frac{-spq^2}{1-s+sq^2} = \frac{-(0.4)(0.7)(0.3)^3}{1-s+sq^2} = \frac{-(0.4)(0.7)(0.3)^2}{1-0.4+(0.4)(0.3)^2} = -0.04$$

$$p^1 = 0.70 - 0.04 = 0.66$$

24.26 As discussed earlier in this chapter, the gene for sickle-cell anemia exhibits heterozygote advantage. An individual who is an *Hb-A/Hb-S* heterozygote has increased resistance to malaria and therefore has greater fitness than the *Hb-A/Hb-A* homozygote, who is susceptible to malaria, and the *Hb-S/Hb-S* homozygote, who has sickle-cell anemia. Suppose that the fitness values of the genotypes in Africa are as presented here:

$$Hb\text{-}A/Hb\text{-}A = 0.88$$
$$Hb\text{-}A/Hb\text{-}S = 1.00$$
$$Hb\text{-}S/Hb\text{-}S = 0.14$$

Give the expected equilibrium frequencies of the sickle-cell gene (*Hb-S*).

Answer: From the text discussion, at equilibrium, $p = f(Hb\text{-}A) = t/(t + s)$, and $q = f(Hb\text{-}S) = s/(s + t)$, where t equals the selection coefficient of $Hb\text{-}S/Hb\text{-}S$, and s equals the selection coefficient of $Hb\text{-}A/Hb\text{-}A$. Since *fitness* $= 1 - selection\ coefficient$, $f(Hb\text{-}S) = 0.12/(0.12 + 0.86) = 0.122$.

24.27 Achondroplasia, a type of dwarfism in humans, is caused by an autosomal dominant gene. The mutation rate for achondroplasia is about 5×10^{-5}, and the fitness of achondroplastic dwarfs has been estimated to be about 0.2, compared with unaffected individuals. What is the equilibrium frequency of the achondroplasia gene based on this mutation rate and fitness value?

Answer: For a dominant allele, the frequency at equilibrium is $\hat{p} = u/s$, where u equals the mutation rate, and s is the selection coefficient ($= 1 - $ fitness). Here, $\hat{p} = 5 \times 10^{-5}/0.8 = 6.25 \times 10^{-5}$.

24.28 To answer the following questions, consider the spontaneous mutation frequencies tabulated in Table 24.6 (p. 678).
 a. In humans, why is the frequency of forward mutations to neurofibromatosis an order of magnitude larger than that for the other human diseases?
 b. In *E. coli*, why is the frequency of mutations to arabinose dependence two to four orders of magnitude larger than the frequency of mutations to leucine, arginine, or tryptophan independence?
 c. What factors influence the spontaneous mutation frequency for a specific trait?

Answer:
 a. There are many potential reasons. If all forms of the disease are caused by mutations in one gene, then that gene appears to be more highly mutable. This may be because the gene is very large, and so there is a larger amount of DNA that can be altered by randomly distributed point mutations, or because the gene may have some unusual sequence structure that results in mutation—for example, mutations due to DNA replication errors or transposon insertion. Alternatively, the rate could be large because mutations in any one of a number of genes can cause an identical disease phenotype.
 b. Arabinose dependence will result from mutations in any of the genes whose function is required to utilize arabinose, while mutations to leucine, arginine, or tryptophan independence require very specific synthetic functions to be "added" to the cell. The former is much more likely to occur by single new mutations than the latter.
 c. Factors include the number of genes that are required for the appearance of the trait, the size of those genes, the presence of genes that modify the mutation rate, and environmental factors.

24.29 The frequencies of the L^M and L^N blood group alleles are the same in each of the populations I, II, and III, but the genotypes' frequencies are not the same, as shown in the following table. Which of the populations is most likely to show each of the following characteristics: random mating, inbreeding, genetic drift? Explain your answers.

	$L^M L^M$	$L^M L^N$	$L^N L^N$
I	0.50	0.40	0.10
II	0.49	0.42	0.09
III	0.45	0.50	0.05

Answer: First calculate the frequencies of each allele. In population I, one has [(0.5 × 2) + (0.4 × 1)/ 2] = 0.7 L^M, and 0.3 L^N. Similarly, L^M = 0.7 and L^N = 0.3 in populations II and III also. A population in Hardy-Weinberg equilibrium would have p^2 = $(0.7)^2$ = 0.49 $L^M - L^M$, $2pq$ = 2(0.7)(0.3) = 0.42 $L^M - L^N$, and q^2 = $(0.3)^2$ = 0.09 $L^N - L^N$. As population II shows these features, it must be in Hardy-Weinberg equilibrium and would be expected to exhibit random mating. Inbreeding results in an increase in the frequency of homozygotes and would be associated with population I (but not III). Genetic drift is random change in allele frequency due to chance and can be explained by random effects, a small effective population size, founder or bottleneck effects, or sampling error. Genetic drift could be associated with either population I or population III. Because the frequencies in populations I and III are fairly close to those of an equilibrium population, sampling error in a population in Hardy-Weinberg equilibrium could also explain the results seen.

24.30   DNA was collected from 100 people randomly sampled from a given human population and was digested with the restriction enzyme *Bam*HI; the fragments were separated by electrophoresis and then transferred to a membrane filter using the Southern blot technique. The blots were probed with a particular cloned sequence. Three different patterns of hybridization were seen on the blots. Some DNA samples (56 of them) showed a single band of 6.3 kb; others (6) showed a single band at 4.1 kb; and others (38) showed both the 6.3- and the 4.1-kb bands.

a. Interpret these results in terms of *Bam*HI sites.

b. What are the frequencies of the restriction site alleles?

c. Does this population appear to be in Hardy-Weinberg equilibrium for the relevant restriction sites?

Answer:

a. One explanation is that a restriction fragment length polymorphism exists in the population. On some chromosomes, an RFLP having a size of 4.1 kb is found, while on others, the 6.3-kb fragment is found. Individuals having just one size band are homozygotes, while individuals with two different-sized bands are heterozygous. The difference in sizes of the fragments could result from a missing site in the 6.3-kb individuals or the insertion of a 2.2-kb piece of DNA between two sites that are normally 4.1 kb apart.

b. Homozygotes have two identical alleles, while heterozygotes have two different alleles. There are (56 × 2) + 38 = 150 6.3-kb alleles and (6 × 2) + 38 = 50 4.1-kb alleles. Let q equal the frequency of 6.3-kp alleles, and p equal the frequency of 4.1-kp alleles. Then q = 0.75 and p = 0.25.

c. For a population in Hardy-Weinberg equilibrium, one would expect q^2 = 0.5625 6.3-kb homozygotes (compared to the 0.56 seen), $2pq$ = 0.375 heterozygotes (compared to the 0.38 seen), and p^2 = 0.0625 4.1-kb homozygotes (compared to the 0.06 seen). The population appears to be in Hardy-Weinberg equilibrium.

24.31   Fifty tiger salamanders from one pond in west Texas were examined for genetic variation by using the technique of protein electrophoresis. The genotype of each salamander was determined for five loci (AmPep, ADH, PGM, MDH, and LDH-1). No variation was found at AmPep, ADH, and LDH-1; in other words, all individuals were homozygous for the same allele at these loci. The following numbers of genotypes were observed at the MDH and PGM loci:

MDH Genotypes	Number of Individuals		PGM Genotypes	Number of Individuals
AA	11		DD	35
AB	35		DE	10
BB	4		EE	5

Calculate the proportion of polymorphic loci and the heterozygosity for this population.

Answer: Since five loci were examined, and only two have more than one allele, (2/5)(100%) = 40% of the loci are polymorphic. Heterozygosity is calculated by averaging the frequency of heterozygotes for each locus. The frequency of heterozygotes for the AmPep, ADH, and LDH-1 loci is zero. At MDH, 35 out of 50 individuals were heterozygous (0.70). At PGM, 10 out of 50 individuals were heterozygous (0.20). Thus, the average heterozygosity is

$$\frac{0+0+0+0.7+2}{5} = 0.18$$

24.32 The success of a population depends in part on its reproductive rate, which may be affected by low genetic variability. Vyse and his colleagues have been interested in the conservation of small populations of grizzly bears, studying 304 members of 30 grizzly bear family groups in a population in northwestern Alaska. They have identified a set of polymorphic loci with these alleles, allele frequencies, and observed heterozygosities (obs. het.):

Locus *G1A*		Locus *G10X*		Locus *G10C*		Locus *G10L*	
Obs. het. : 0.776		Obs. het. : 0.783		Obs. het. : 0.770		Obs. het. : 0.651	
Allele	Freq.	Allele	Freq.	Allele	Freq.	Allele	Freq.
A194	0.398	X137	0.395	C105	0.355	L155	0.487
A184	0.240	X135	0.211	C103	0.257	L157	0.276
A192	0.211	X141	0.211	C111	0.240	L161	0.128
A180	0.086	X133	0.102	C113	0.092	L159	0.089
A190	0.036	X131	0.053	C107	0.043	L171	0.013
A200	0.016	X129	0.030	C101	0.010	L163	0.007
A186	0.007			C109	0.003		
A188	0.006						

The genotypes of a mother bear and her three cubs are shown here.

Mother	Cub #1	Cub #2	Cub #3
A184, A192	A184, A194	A184, A192	A184, A194
X135, X137	X135, X137	X133, X135	X137, X141
C105, C113	C105, C111	C105, C105	C111, C113
L155, L159	L155, L157	L159, L161	L155, L155

a. How do the observed heterozygosities compare with the expected heterozygosities? On the basis of this information, can you tell whether this grizzly bear population is in Hardy-Weinberg equilibrium?

b. What can you infer about the paternity of the mother's three cubs? How might paternity information affect the genetic variability and the effective population size in this population of grizzly bears?

Answer:

a. The expected heterozygosity is $1 -$ (frequency of expected homozygotes). If the frequency of alleles in the population are $p_1, p_2, p_3, \ldots p_n$, the expected frequency of homozygotes is $p_1^2 + p_2^2 + p_3^2 + \ldots p_n^2$. For locus *G1A*, the expected frequency of homozygotes is $(0.398)^2 + (0.240)^2 + (0.211)^2 + (0.086)^2 + (0.036)^2 + (0.016)^2 + (0.007)^2 + (0.006)^2 = 0.270$, and the expected heterozygosity is $1 - 0.270 = 0.730$. The expected heterozygosities for the other loci are *G10X*, 0.741; *G10C*, 0.740; *G10L*, 0.662. These are approximately the observed frequencies of heterozygosities. Since the numbers and types of different heterozygotes are not given, it is not possible to employ the chi-square test to directly evaluate if the population is in Hardy-Weinberg equilibrium. The population appears to be close to Hardy-Weinberg equilibrium.

b. The three cubs of the mother show evidence of multiple paternity. For each of the loci *G10X* and *G10L*, three alleles present in the cubs must have been contributed paternally (*G10X: X133, X135* or *X137, X141*; *G10L: L155, L157, L161*). This could only have happened if the cubs were sired by at least two different fathers. Multiple paternity within one set of cubs would tend to increase the genetic variability in the population, as it would allow a larger number of males to contribute gametes seen in the next generation. Since $N_e = (4 \times N_f \times N_m)/(N_f + N_m)$, a larger N_m will tend to increase the effective population size.

24.33   What factors cause genetic drift?

Answer: Genetic drift arises from random change in allele frequency due to chance. Random factors producing mortality in natural populations and sampling error can lead to genetic drift. Its causes include a small effective population size over many generations, a small number of founders (founder effect), and a reduction in population size (bottleneck effect).

24.34   What are the primary effects of the following evolutionary processes on the gene and genotypic frequencies of a population?

a. mutation

b. migration

c. genetic drift

d. inbreeding

Answer:

a. Mutation will lead to change in allele frequencies within a population if no other forces are acting. It will introduce genetic variation. If the effective population size is small, mutation may lead to genetic differentiation among populations.

b. Migration will increase the population size, and has the potential to disrupt a Hardy-Weinberg equilibrium. It can increase genetic variation and may influence the evolution of allele frequencies within populations. Over many generations migration will reduce divergence among populations, and equalize allele frequencies among populations.

c. Genetic drift produces changes in allele frequencies within a population. It can reduce genetic variation and increase the homozygosity within a population. Over time, it leads to genetic change. When several populations are compared, genetic drift can lead to increased genetic differences among populations.

d. Inbreeding will increase the homozygosity within a population and decrease its genetic variation.

24.35 Explain how overdominance leads to an increased frequency of sickle-cell anemia in areas where malaria is widespread.

Answer: Overdominance results when a heterozygote genotype has higher fitness than either of the homozygotes. The two alleles of the heterozygote are maintained in a population because both are favored in the heterozygote genotype. In the case of the sickle-cell allele, heterozygotes for *Hb-A/Hb-S* are at a selective advantage because the hemoglobin mixture in these individuals provides an unfavorable environment for the growth of malarial parasites. Thus, heterozygotes have higher fitness than do *Hb-A/Hb-A* homozygotes that are susceptible to malaria. Heterozygotes also have higher fitness than *Hb-S/Hb-S* homozygotes that suffer from sickle-cell anemia. The favoring of the sickle-cell allele *Hb-S* in the heterozygote results in its relatively high frequency in areas with malaria.

24.36 Since 1968, Pinter has studied the population dynamics of the montane vole, a small rodent in the Grand Teton mountains in Wyoming. For more than 25 years, severe periodic fluctuations in population density have been negatively correlated with precipitation levels: Vole density sharply declines every few years when spring precipitation is extremely high.

a. Propose several hypotheses concerning the genetic structure of the population of montane voles in two separate sampling sites if

   i. there is negligible migration of voles between sampling sites

   ii. there is substantial migration of voles between sampling sites

b. How would you gather data to evaluate these hypotheses?

Answer:

a. The cyclical decline in population density could cause a cyclically repeated bottleneck effect at each of the sampling sites. This would lead to the loss of some genes from the gene pool as a result of chance, and founder effects. If there were negligible migration between sampling sites, the populations would most likely diverge in their allele frequencies through genetic drift. There would be more variance in allele frequency among the small populations at each sampling site. If there were substantial migration between sampling sites, there would be reciprocal gene flow among the populations. This would increase the amount of genetic variation in the populations at each sampling site, and reduce the divergence between the populations. If there were cyclical, sharp declines in the vole population, it is unlikely that the population would be in Hardy-Weinberg equilibrium.

b. First, randomly trap a subset of the voles and sample tissue or blood from them. Then, identify a set of polymorphic loci (using protein electrophoresis, DNA sequencing, or RFLPs), and estimate the heterozygosity and allele frequencies at these loci. Use this information to compare populations sampled at each of the study sites, and estimate the degree of homogeneity or divergence of the two populations.

24.37   What are some of the advantages of using DNA sequences to infer the strength of evolutionary processes?

Answer: Using DNA sequence information to infer the strength of evolutionary processes has several advantages. DNA sequences provide highly accurate and reliable information, allow direct comparison of the genetic differences among organisms, are easily quantified, and can be used in all organisms.

24.38   Multiple, geographically isolated populations of tortoises exist on the Galapagos Islands off the coast of Ecuador. Several populations are endangered, partly because of hunting and partly because of illegal capture and trade. In principle, a significant number of tortoises in captivity (in zoos or private collections) could be used to repopulate some of the endangered populations.

a. Suppose you are interested in returning a particular captive tortoise to its native subpopulation, but you have no record of its capture. How could you determine its original subpopulation?

b. A researcher planned to characterize one subpopulation of tortoises. In her first field season, she tagged and collected blood samples from all animals in the subpopulation. When she returns a year later, she cannot locate two animals. She learns that two untagged, smuggled tortoises are being held by U.S. custom officials. How can she assess whether the smuggled animals are from her field site?

Answer: In both instances, protein electrophoresis, RFLP analyses, and microsatellite and DNA sequence analysis of specific genes could be used to gather information on the genotype of the captured individuals and members of each island population. In (a), the captured individual should be returned to the subpopulation from which it shows the least genetic variation. In (b), evaluate the genotype of the two missing tortoises using DNA from the previously collected blood samples, and compare these genotypes to the genotypes of the two captured tortoises. If a captured animal was taken from the field site, its genotype will exactly match a genotype obtained from one of the blood samples.

25

MOLECULAR EVOLUTION

CHAPTER OUTLINE

THINKING ANALYTICALLY

Perhaps the most striking element in the field of molecular evolution is the time scale in which the mechanisms of evolution operate. Almost throughout this text, we have considered genetic mechanisms that act over relatively short time frames. The replication of the genetic material, its transcription, and the production of gene products through translation occur in time scales that are on the order of seconds and minutes. Genetic crosses may take several generations, but even for organisms that have relatively long life spans, the time scales are on the order of years or tens of years. Molecular evolution occurs over many, many generations, and the time scales being considered are on the order of tens of thousands to millions of years. In this space of time, much can happen. Some say that if you can think of it, nature has had time to try it. With genomes serving as historical records of molecular evolution, it may be possible to find evidence of it as well. Having said this, consider that small alterations in events that occur quickly can, over a long period of time, result in major effects. For example, variation in the bonding energy associated with the instantaneous interaction between codons and the anticodons of isoacceptor tRNAs can effect the efficiency of translation. Over many, many generations, this may lead to the selection of a species' codon usage bias.

REVIEW OF KEY TERMS, SYMBOLS, AND CONCEPTS

In your own words, write a brief, precise definition of each term in the groups below. Check your definitions using the text. Then develop a concept map using the terms in each list.

1	2	3
molecular evolution	nucleotide substitution	molecular phylogeny
population, species	homologous proteins, genes	morphological phylogeny
gene	sequence alignment	evolutionary relationship
evolutionary time, history	computer algorithm	convergent evolution
population genetics	optimal alignment	phenotype
Hardy-Weinberg equilibrium	indel	phylogenetic tree
mutation vs. substitution	Jukes-Cantor model	branch, node
gene duplication	substitution rate	rooted, unrooted trees
gene conversion	evolutionary rate	taxa
transposition	divergence time	gene vs. species tree
unequal crossing over	transition, transversion	inferred tree
phylogenetic relationship	coding, noncoding sequence	distance matrix
natural selection	leader, trailer region	parsimony
multigene family	pseudogene	unweighted pair group
globin, myoglobin genes	synonymous site, codon	method with arithmetic
ancestral gene	nonsynonymous site, codon	averages (UPGMA)
domain shuffling	mutation vs. substitution	transformed distance
exon shuffing	comparative genome	method
internal duplication	analysis	neighbor-joining method
exon, intron	genome project	clustering
Bacteria, Archaea, Eukarya	codon usage bias	informative site
	stochastic factor	noninformative site
	natural selection	tree of maximum parsimony
	diversifying selection	bootstrap test
	McDonald-Kreitman test	confidence level
	functional constraint	tree of life
	major histocompatibility	evolutionary domain
	complex (MHC)	eubacteria
	mitochondrial DNA	archaebacteria
	error-prone replication	eukaryotes
	proofreading ability	"out of Africa" theory
	molecular clock	"mitochondrial Eve"
	molecular clock hypothesis	"Y chromosome Adam"
	relative rate test	
	outgroup	

It is therefore helpful to keep in mind the time scales associated with molecular evolution. The text presents data documenting that rates of evolutionary change are not constant between regions of a gene, between different positions of codons within the coding region of a gene, between genes, between species, and between mitochondrial and nuclear genomes. Where possible, relate differences in the rate of evolutionary change to the time frames associated with molecular evolution. Focus on why rates are different—what functional differences or constraints exist that preclude uniform rates. As you do this, draw on everything you have learned about the mechanisms of inheritance and gene expression, since these can not only influence molecular evolution, but have also been derived through molecular evolution and therefore represent the tuning of existing biological instruments.

QUESTIONS FOR PRACTICE

Multiple Choice Questions

1. Why does the molecular clock run at different rates in different proteins?
   a. Neutral substitution rates differ among proteins.
   b. Some proteins have resulted from gene duplication followed by sequence divergence.
   c. Some species have shorter generation times.
   d. Codon usage varies between proteins.

2. Two homologous proteins are found in two different species. These proteins
   a. perform identical functions in the different species
   b. share a common ancestor
   c. arose by gene duplication
   d. have only synonymous substitutions in their genes

3. An indel is
   a. a gap in a sequence alignment caused by either by an insertion or a deletion
   b. a gap in a sequence alignment caused by a deletion in just one of the sequences
   c. a gap in a sequence alignment caused by an insertion in just one of the sequences
   d. a deletion that removes an intron

4. What issues does the Jukes-Cantor model address?
   a. the rate of the molecular clock in different proteins
   b. whether nucleotide substitutions are adaptive or random
   c. the relationship between the observed and actual number of substitutions
   d. whether an inferred phylogenetic tree is robust

5. What is generally true about synonymous and nonsynonymous substitutions?
   a. In coding regions, synonymous substitutions are more frequent than nonsynonymous substitutions.
   b. In coding regions, synonymous and nonsynonymous substitutions are equally frequent.
   c. Synonymous substitutions in coding regions and substitutions in intronic regions are equally frequent.
   d. Both a and c are true.

6. Which one of the following sequences of a gene exhibits the highest rate of evolution?
   a. the 3' untranslated but transcribed region of the gene
   b. introns within the gene
   c. the 5' flanking region of the gene
   d. nonfunctional pseudogenes related to the gene

7. Why might codon usage bias have evolved?
   a. The tRNA for one codon is more abundant than an isoacceptor tRNA for another codon.
   b. The bonding energy between pairs of codons and anticodons differs, leading to increased translation efficiency of certain pairs.
   c. The required abundant expression of some genes favors selection of codons that provide efficient translation.
   d. all of the above

8. Which of the following statements is true?
   a. An outgroup is a species that a phylogenetic analysis using maximum parsimony shows is a common ancestor to two existing species.
   b. The molecular clock typically runs at the same rate in different proteins.
   c. The synonymous substitution rate in mammalian mitochondrial genes is about tenfold greater than that in nuclear genes.
   d. The length of a species' generation time is unlikely to affect its evolutionary rate.

9. What is the best description of convergent evolution?
   a. mutations that introduce the same polymorphism into two species after they diverged from a common ancestor
   b. the evolution of similar phenotypes in distantly related species
   c. the finding that there are parallel gene duplications in two unrelated species
   d. the slowing of the molecular clock in an existing species such as humans

10. In phylogenetic analysis, how do rooted and unrooted trees differ?
    a. When a given number of taxa are being analyzed, there are more possible unrooted than rooted trees.
    b. A rooted tree is based on relationships between genes, while an unrooted tree is based on relationships between species.
    c. Outgroups can be used only in unrooted trees.
    d. A rooted tree specifies one internal node as representing a common ancestor to all the other nodes on the tree; an unrooted tree specifies only the relationship between the nodes and does not indicate the evolutionary path that was taken.

11. Which approach to phylogenetic reconstruction first determines the distance between taxa, and then repeatedly clusters the two taxa with the smallest distance separating them into a single, composite taxa until all the taxa are grouped together?
    a. the maximum parsimony method
    b. the UPGMA method
    c. the bootstrap method
    d. the neighbor-joining method

Answers: 1a, 2b, 3a, 4c, 5a, 6d, 7d, 8c, 9b, 10d, 11b

Thought Questions

1. Gene duplication provides a mechanism that allows for the evolution of new protein functions. How does this mechanism affect our view of which proteins are homologous? How can one infer which proteins encoded by a multigene family present in each of two organisms are homologous?

2. What are the relationships between the fields of comparative genomics and molecular evolution?

3. How do the results of the *Arabidopsis* genome project add to our understanding of molecular evolution and how new gene functions evolve?

4. Suppose you have identified a sequence that can be translated and appears to encode an open reading frame (ORF). How could you use an analysis of codon usage bias to infer whether it was likely to be translated?

5. What problems in dating the times of existence of common ancestors are posed by variable molecular clocks in different taxonomic groups?

6. Why is the parsimony approach to phylogenetic tree reconstruction able to give some insight into the nature of long-dead organisms? Are other methods, such as distance matrix-based approaches, also able to do this?

7. What role does gene conversion play in the evolution of new gene functions?

8. Does the proposition that domain shuffling provides a mechanism for the generation of new protein function require that introns once existed in primitive life forms? If so, why might they no longer exist in the simpler Bacteria and Archaea?

9. Do phylogenetic reconstructions tell us anything about the mechanisms of speciation? If so, what?

10. What are the general types of functional constraints that restrict proteins from undergoing rapid evolution? Consider in your answer the histones, which evolve so slowly that the human histone H4 protein can functionally replace its homolog in yeast.

Thought Questions for Media

After reviewing the media on the *iGenetics* CD-ROM, try to answer these questions

1. In a phylogenetic tree, what is the difference between a terminal node and an internal node? Which of these types of nodes represent organisms for which we have data? What does the other type of node represent?

2. Under what circumstances would an unrooted tree be constructed?

3. What type of tree, rooted or unrooted, should be drawn if we want to depict the evolutionary history of a set of organisms?

4. How can an unrooted tree be rooted by using an outgroup? Why, in general terms, does using an outgroup lead to a greater number of rooted trees rather than unrooted trees?

5. The rate of evolution in two branches of a tree is often not identical. What implications does this have for drawing evolutionary trees?

1. The iActivity illustrates how the sequence of ancient DNA is obtained by using PCR. Fragments averaging 100 to 200 base pairs in length are cloned, the cloned segments are sequenced and aligned, and a *consensus nucleotide sequence* is determined from the aligned sequences.

   a. What is a *consensus nucleotide sequence,* and why is it necessary to use one in this type of analysis?

   b. In this analysis, sequences from the control region of mitochondrial DNA were amplified. Why is mitochondrial DNA used, and why might the control region be specifically chosen for analysis?

   c. What types of errors might be introduced when amplifying ancient DNA templates? How might you ensure that a sequence accurately reflects the sequence in the ancient DNA and is not an artifact?

2. Suppose a bone sample from a new, non-Neanderthal, extinct hominid species called *ML* (for *missing link*) was made available to you.

    a. Outline how you would test if *ML* was a direct ancestor of modern humans.

    b. Suppose you undertook an analysis similar to that of the iActivity and compared DNA sequences in the control region of mitochondrial DNA. You find 12 differences in mitochondrial DNA between *ML* and modern European sequences, 11 differences between *ML* and modern African sequences, and 7 differences between the modern European and modern African DNA sequences. What would you conclude? Be as specific as possible.

3. In the iActivity, you were able to estimate the time of divergence of modern Europeans and Africans and the time of divergence of Neanderthals and modern humans.

    a. What assumptions did you make in this analysis?

    b. How did you estimate the rate of nucleotide evolution? That is, what was the basis for relating the number of observed mutations to the times of divergence between the European, African, and Neanderthal populations?

    c. How would your estimates of the times of divergence between the European, African, and Neanderthal populations differ if the rate of nucleotide evolution in hominid species is not constant, but tripled, relative to a previously constant rate, in the last 300,000 years?

SOLUTIONS TO TEXT QUESTIONS

25.1   The following sequence is that of the first 45 codons from the human gene for preproinsulin. Using the genetic code (Figure 14.7, p. 362), determine what fraction of mutations at the first, second, and third positions of these 45 codons will be synonymous.

```
ATG GCC CTG TGG ATG CGC CTC CTG CCC CTG CTG GCG CTG CTG GCC CTC
TGG GGA CCT GAC CCA GCC GCA GCC TTT GTG AAC CAA CAC CTG TGC GGC
TCA CAC CTG GTG GAA GCT CTC TAC CTA GTG TGC GGG GAA
```

At which position is natural selection likely to have the greatest effect and nucleotides are most likely to be conserved?

Answer: A synonymous mutation or substitution is a change in nucleotide sequence that does not affect the amino acid sequence in a polypeptide chain. Due to the redundancy of the genetic code, more than one codon can code for one amino acid. Inspection of a table of the genetic code shows that at one extreme, some amino acids, such as methionine and tryptophan, are coded only by one codon (AUG and UGG, respectively), while at the other extreme, some amino acids, such as leucine and serine, are coded for by six codons. Other amino acids are coded for by two, three, or four codons. Say, the codon in question codes for leucine and is CUG. Then, since the codons for leucine are UAA, UGG, CUU, CUC, CUA, and CUG, any of three coding strand changes in the third nucleotide (G to T, G to A, or G to C) and a change of C to T in the first nucleotide would still result in leucine being inserted into the polypeptide chain. Hence, any of these substitutions would be synonymous.

To get some sense of which substitutions will be synonymous, inspect the genetic code and look for some general principles. For amino acids coded for by four codons, variation in the third position results in a synonymous substitution; for amino acids coded for by two codons, variation in the third position by one base only (a purine or a pyrimidine) results in a synonymous mutation. For amino acids coded for by either two or four codons, variation in the first or second positions results in a nonsynonymous substitution. Similar but not identical patterns are seen for amino acids coded for by either three or six codons. The sense that synonymous changes are mostly associated with substitutions in the third position is confirmed by a more systematic analysis. The following table lists the number of possible nonsynonymous (NS) and synonymous (S) mutations at each position of each codon for the amino acids in question.

Coding Strand	mRNA Codon	Amino Acid	Position 1 NS	Position 1 S	Position 2 NS	Position 2 S	Position 3 NS	Position 3 S
ATG	AUG	methionine	3	0	3	0	3	0
GCC	GCC	alanine	3	0	3	0	0	3
CTG	CUG	leucine	2	1	3	0	0	3
TGG	UGG	tryptophan	3	0	3	0	3	0
ATG	AUG	methionine	3	0	3	0	3	0
CGC	CGC	arginine	3	0	3	0	0	3
CTC	CUC	leucine	3	0	3	0	0	3
CTG	CUG	leucine	2	1	3	0	0	3
CCC	CCC	proline	3	0	3	0	0	3

(continued)

Coding Strand	mRNA Codon	Amino Acid	Position 1		Position 2		Position 3	
			NS	S	NS	S	NS	S
CTG	CUG	leucine	2	1	3	0	0	3
CTG	CUG	leucine	2	1	3	0	0	3
GCG	GCG	alanine	3	0	3	0	0	3
CTG	CUG	leucine	2	1	3	0	0	3
CTG	CUG	leucine	2	1	3	0	0	3
GCC	GCC	alanine	3	0	3	0	0	3
CTC	CUC	leucine	3	0	3	0	0	3
TGG	UGG	tryptophan	3	0	3	0	3	0
GGA	GGA	glycine	3	0	3	0	0	3
CCT	CCU	proline	3	0	3	0	0	3
GAC	GAC	aspartate	3	0	3	0	2	1
CCA	CCA	proline	3	0	3	0	0	3
GCC	GCC	alanine	3	0	3	0	0	3
GCA	GCA	alanine	3	0	3	0	0	3
GCC	GCC	alanine	3	0	3	0	0	3
TTT	UUU	phenylalanine	3	0	3	0	2	1
GTG	GUG	valine	3	0	3	0	0	3
AAC	AAC	asparagine	3	0	3	0	2	1
CAA	CAA	glutamine	3	0	3	0	2	1
CAC	CAC	histidine	3	0	3	0	2	1
CTG	CUG	leucine	2	1	3	0	0	3
TGC	UGC	cysteine	3	0	3	0	2	1
GGC	GGC	glycine	3	0	3	0	0	3
TCA	UCA	serine	3	0	3	0	0	3
CAC	CAC	histidine	3	0	3	0	2	1
CTG	CUG	leucine	2	1	3	0	0	3
GTG	GUG	valine	3	0	3	0	0	3
GAA	GAA	glutamate	3	0	3	0	2	1
GCT	GCU	alanine	3	0	3	0	0	3
CTC	CUC	leucine	3	0	3	0	0	3
TAC	UAC	tyrosine	3	0	3	0	2	1
CTA	CUA	leucine	2	1	3	0	0	3
GTG	GUG	valine	3	0	3	0	0	3
TGC	UGC	cysteine	3	0	3	0	2	1
GGG	GGG	glycine	3	0	3	0	0	3
GAA	GAA	glutamate	3	0	3	0	2	1
	TOTAL		124	11	135	0	34	101
	Percent		91.8	8.2	100	0	25.2	74.8

From this analysis, 8.2 percent of mutations in the first position, none of the mutations in the second position, and 74.8 percent of mutations in the third position will be synonymous. Since none of the mutations in the second position will be synonymous, all mutations in this position will affect the amino acid inserted into a polypeptide chain. Natural selection is therefore likely to have the greatest effect on mutations in the second position, and nucleotides in this position are the most likely to be conserved.

25.2   The following sequences represent an optimal alignment of the first 50 nucleotides from the human and sheep preproinsulin genes. Estimate the number of substitutions that have occurred in this region since humans and sheep last shared a common ancestor, using the Jukes-Cantor model.

```
Human: ATGGCCCTGT GGATGCGCCT CCTGCCCCTG CTGGCGCTGC TGGCCCTCTG
Sheep: ATGGCCCTGT GGACACGCCT GGTGCCCCTG CTGGCCCTGC TGGCACTCTG
```

Answer: Inspection of the sequences reveals that there are six differences (underlined):

```
Human: ATGGCCCTGT GGATGCGCCT CCTGCCCCTG CTGGCGCTGC TGGCCCTCTG
Sheep: ATGGCCCTGT GGACACGCCT GGTGCCCCTG CTGGCCCTGC TGGCACTCTG
```

The Jukes-Cantor model estimates the number of substitutions per site (K) based on the fraction of nucleotides that are different between two sequences (p), taking into account that a particular site can undergo multiple changes. According to this model, $K = -3/4 \ln[1 - (4/3)(p)]$. Here, $p = 6/50 = 0.12$, so $K = -3/4 \ln[1 - (4/3)(0.12)] = 0.1308$. Since $K = 0.1308$, the estimated number of substitutions is $50 \times 0.1308 = 6.54$.

25.3   Using the alignment in Question 25.2 and assuming that humans and sheep last shared a common ancestor 80 million years ago, estimate the rate at which the sequence of the first 50 nucleotides in their preproinsulin genes have been accumulating substitutions.

Answer: Since substitutions are assumed to accumulate simultaneously and independently in both sequences, the substitution rate (r) is obtained by dividing the number of substitutions between the homologous sequences by $2T$, where T = divergence time: $r = K/2T$. Here, $K = 0.1308$ and $T = 80 \times 10^6$, so $r = 0.1308/(2 \times 80 \times 10^6) = 8.175 \times 10^{-10}$ substitutions/year.

25.4   Would the mutation rate be greater or less than the observed substitution rate for a sequence of a gene such as the one shown in Question 25.2? Why?

Answer: The mutation rate reflects changes in DNA sequence due to errors in replication or repair processes, while the substitution rate reflects changes in DNA sequence (i.e., mutations) that have passed through the filter of natural selection. Since mutations that are not synonymous may be selected against, an estimate of the number of substitutions in a coding region, such as in this analysis, is likely to be lower than the mutation rate. Therefore, the mutation rate is likely to be greater than the observed substitution rate.

25.5   If the rate of nucleotide evolution along a lineage is 1.0 percent per million years, what is the rate of substitution per nucleotide per year? What would be the observed rate of divergence between two species evolving at that rate since they last shared a common ancestor?

Answer: Substitutions are mutations that have passed through the filter of selection, so the rate of nucleotide evolution corresponds to the rate of nucleotide substitution. A rate of nucleotide evolution of 1.0 percent per million years equals 0.01 substitutions/nucleotide/10^6 years, which equals 1×10^{-8} substitutions/nucleotide/year. Since substitutions accumulate simultaneously and independently in two diverging sequences, the observed rate of divergence will 2×10^{-8} substitutions/nucleotide/year.

25.6   The average synonymous substitution rate in mammalian mitochondrial genes is approximately 10 times the average value for synonymous substitutions in nuclear genes. Why would it be better to use comparisons of mitochondrial sequences to study human migration patterns and nuclear genes when studying the phylogenetic relationships of mammalian species that diverged 80 million years ago?

Answer: There are two broad issues to consider here. First, sequence differences must be able to be observed within the time period being assessed, and there must be a sufficient number of them so that meaningful conclusions can be drawn from their analysis. For the analysis of human migration patterns, the time scale is on the order of tens of thousands of years, while for the analysis of mammalian phylogenetic relationships, the time scale is on the order of millions of years. The average synonymous substitution rate in the mitochondrial genome is high enough for differences between different human populations to be observed and provides for a sufficient number of differences for a meaningful analysis to be performed. The average synonymous substitution rate in the nuclear genome may be too low to identify a sufficient number of differences for a meaningful analysis to be performed.

Second, over time, synonymous nucleotide substitutions will change repeatedly and will include changes back to their ancestral state. If the substitution rate is too high relative to the time period being examined, it may be difficult to clearly identify relationships due to noise in the data from this phenomenon. The average synonymous substitution rate in nuclear genes is high enough for differences to be observed among mammalian species over periods of millions of years, yet not so high that multiple substitutions cloud their phylogenetic interpretation. The rate of synonymous mutations in mitochondrial genes is so high that multiple substitutions cloud their phylogenetic interpretation. In summary, the rapid and regular rate of accumulation of nucleotide sequence differences in mitochondrial genomes allows for the phylogenetic analysis of closely related lineages, but for more distant lineages, it is better to use less rapidly changing sequences, such as those in the nuclear genome.

25.7   Why might mitochondrial DNA sequences accumulate substitutions at a faster rate than nuclear genes in the same organism?

Answer: The increased rate of substitutions in animal mtDNA may result from a higher mutation rate due to error-prone replication and DNA repair, since mitochondrial DNA polymerases lack proofreading ability; from increased concentrations of mutagens, such as free radicals resulting from mitochondrial metabolic processes; and/or from relaxed selection pressure, since most cells have at least several dozen mitochondria, each of which has multiple copies of the mitochondrial genome.

25.8   Why might substitution rates differ from one species to another, and how would such differences depart from Zuckerkandl and Pauling's assumptions for molecular clocks?

Answer: Substitution rates may differ between species because their generation times are different, and so the number of germ-line DNA replications (which occur once per generation) are different; because the species are different in their average repair efficiency; because the species have different average exposure to mutagens; and/or because the species are exposed to different opportunities to adapt to new ecological niches and environments.

Zuckerkandl and Pauling's molecular clock hypothesis was based on initial data suggesting that substitution rates were constant in homologous proteins, that is, that there is a steady rate of change between two sequences. If this were generally true, the number of differences between two homologous proteins could be correlated with the amount of time since speciation caused them to diverge independently. If the hypothesis were correct, it would facilitate the determination of the phylogenetic relationships between species and the times of their divergence. However, the molecular clock is not uniform for all species, and so molecular divergence cannot always be used for phylogenetic analysis and to date the times that recent common ancestors existed. Before making these types of inferences, it is necessary to demonstrate that the species being examined have a uniform molecular clock.

25.9   Suppose we examine the rates of nucleotide substitution in two nucleotide sequences isolated from humans. In the first sequence (sequence A) we find a nucleotide substitution rate of 4.88×10^{-9} substitutions per site per year. The substitution rate is the same for synonymous and nonsynonymous substitutions. In the second sequence (sequence B), we find a synonymous substitution rate of 4.66×10^{-9} substitutions per site per year and a nonsynonymous substitution rate of 0.70×10^{-9} substitutions per site per year. Referring to Table 25.1 (p. 708), what might you conclude about the possible roles of sequence A and sequence B?

Answer: Nonsynonymous substitutions are those that code for different amino acids, while synonymous substitutions code for the same amino acid. In sequence A, the finding of the same, relatively high rate for synonymous and nonsynonymous substitutions suggests that this sequence may not code for a functional protein and may derive from a pseudogene. The rate seen for synonymous and nonsynonymous substitutions in sequence A is similar to that seen for pseudogenes shown in Table 25.1.

The different rates seen for synonymous and nonsynonymous substitutions in sequence B (and in particular, the much lower rate of nonsynonymous substitutions) suggest that this sequence encodes a protein. A low nonsynonymous substitution rate would be expected if nonsynonymous changes resulted in changes in protein function that were detrimental to fitness. Since most nonsynonymous substitutions would not "improve" a protein's function, such mutations would be eliminated by the filter of natural selection. Synonymous substitutions would be tolerated and seen at a higher frequency, as they would not alter protein function.

25.10   What evolutionary process might explain a coding region in which the rate of amino acid replacement is greater than the rate of synonymous substitution?

Answer: If there is evolutionary pressure, or natural selection, for diversity, then the rate of amino acid replacement, driven by nonsynonymous substitution, may be greater than the rate of synonymous substitution. This is seen at the major histocompatibility complex (MHC) in mammals, which is involved in immune function where diversity favors fewer individuals vulnerable to an infection by any single virus, and in viruses, which utilize error-prone replication coupled with diversifying selection. In both cases, this serves as a response to pressure for rapid evolution.

25.11   Natural selection does not always act just at the level of amino acid sequences in proteins. Ribosomal RNAs, for instance, are functionally dependent on extensive and specific intramolecular secondary structures that form when complementary nucleotide sequences within a single rRNA interact. Would the regions involved in such pairing accumulate mutations at the same rate as unpaired regions? Why?

Answer: While the sequences of rRNA regions that interact and provide for ribosomal function by pairing will be subject to mutation at the same rates as sequences that do not pair, mutations that disrupt pairing will be selected against. Such mutations will alter ribosomal function, and this provides a basis for selection against them. Consequently, substitutions in paired regions will not accumulate unless there are two simultaneous, complementary mutations—one in each of the two sites that pair. A high degree of conservation of nucleotides at particular sites within the rRNA genes therefore indicates that these nucleotides are functionally important.

25.12   The following three-way alignment is of the nucleotide sequences from the beginning of the human, rabbit, and duck α-globin genes. Given that humans and rabbits are known to be more closely related to each other than they are to ducks (an outgroup), use the relative rate test to determine whether there has been a change in the rate of substitution in this region of the human and rabbit genomes since they last shared a common ancestor.

```
Human:  ATGGTGCTCT CTCCTGCCGA CAAGACCAAC GTCAAGGCCG CCTGGGACAA
Rabbit: ATGGTGCTGT CTCCCGCTGA CAAGACCAAC ATCAAGACTG CCTGGGAAAA
Duck:   ATGGTGCTGT CTGCGGCTGA CAAGACCAAC GTCAAGGGTG TCTTCTCCAA
```

Answer: First identify how many differences exist between pairs of the sequences. In the annotation below, H represents human; R represents rabbit; D represents duck; an underlined base indicates a base that is different between the mammalian and duck sequence; and an 'x' below the aligned sequences identifies the position of a base that is different between the indicated pair of species.

```
H:        ATGGTGCTCT CTCCTGCCGA CAAGACCAAC GTCAAGGCCG CCTGGGACAA
R:        ATGGTGCTGT CTCCCGCTGA CAAGACCAAC ATCAAGACTG CCTGGGAAAA
D:        ATGGTGCTGT CTGCGGCTGA CAAGACCAAC GTCAAGGGTG TCTTCTCCAA
H vs. D:            x x  x                        xx  x  xxxx
R vs. D:            x x                     x      xx     xxxxx
H vs. R:              x  x                   x      x x        x
```

Based on this analysis, there are 10 substitutions between humans and ducks, 11 substitutions between rabbits and ducks, and 6 substitutions between humans and rabbits. With ducks as an outgroup, we can draw a species tree as follows:

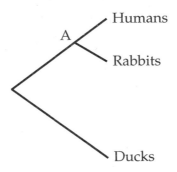

If the molecular clock hypothesis is correct, then $d_{\text{A-Humans}} = d_{\text{A-Rabbits}}$ [d in these equations represents distance; $d_{\text{A-Humans}}$ is the distance between the point of common ancestry (A) to humans]. The path by which the distances are calculated should not matter if the molecular clocks throughout the tree are uniform. Check this by inserting the numbers of substitutions determined from the analysis of the sequences into equations that trace the paths in the species tree. These equations are presented in the text, and reflect the relationships between distances in the species tree.

$$d_{\text{A-Humans}} = (d_{\text{Humans-Rabbits}} + d_{\text{Humans-Ducks}} - d_{\text{Rabbits-Ducks}})/2$$
$$= (6 + 10 - 11)/2$$
$$= 2.5$$

$$d_{\text{A-Rabbits}} = (d_{\text{Humans-Rabbits}} + d_{\text{Rabbits-Ducks}} - d_{\text{Humans-Ducks}})/2$$
$$= (6 + 11 - 10)/2$$
$$= 3.5$$

This analysis indicates that although $d_{\text{A-Humans}}$ and $d_{\text{A-Rabbits}}$ are similar, they are not identical. This suggests that there is a difference in the rate of substitution in this region of the rabbit and human genomes since they last shared a common ancestor, and that the human lineage appears to have accumulated substitutions at a slower rate.

25.13   What are some of the advantages of using DNA sequences to infer evolutionary relationships?

Answer: Using DNA sequence information to infer evolutionary relationships has several advantages. First, DNA sequences serve as historical records that can be unraveled to identify the dynamics of evolutionary history. Second, since they do not rely on phenotypes for comparison of genetic relatedness, they circumvent problems associated with phylogenetic classification in organisms based solely on phenotypes. These problems include the identification of usable phenotypes in organisms that lack phenotypes correlating with their degree of genetic relatedness or for which there is no fossil record evidence (e.g., eubacteria and archaebacteria); problems in the identification of usable phenotypes in groups of organisms where few characters are shared (e.g., eubacteria and mammals); and problems associated with convergent evolution, where distantly related organisms can share similar phenotypes. Third, they allow for an abundance of parameters that can be measured so that theories can be tested, as they provide highly accurate and reliable information, allow direct comparison of the genetic differences among organisms, are easily quantified, and can be used for all organisms.

25.14   As suggested by the popular movie *Jurassic Park,* organisms trapped in amber have proven to be a good source of DNA from tens and even hundreds of millions of years ago. However, when using such sequences in phylogenetic analyses it is usually not possible to distinguish between samples that come from evolutionary dead ends and those that are the ancestors of organisms still alive today. Why would the former be no more useful than simply including the DNA sequence of another living species in an analysis?

Answer: Phylogenetic analyses cannot distinguish between existing and extinct species—they only describe species relatedness. The appeal of analyzing ancient DNA samples is that they might allow ancestral sequences to be determined (and not just inferred). However, it is almost impossible to prove that an ancient organism is from the same lineage as an extant species. If the taxon is an evolutionary dead end, it is on a separate, no-longer-existing branch of the tree of life. It therefore serves as just another species for comparison, much as does a living species on a separate existing branch of the tree of life. Increasing the number of taxa in any analysis increases the robustness of any phylogenetic inferences, but extant taxa are almost invariably easier to obtain.

25.15   In the phylogenetic analysis of a group of closely related organisms, the conclusions drawn from one locus were found to be at odds with those from several others. What might account for the discordant locus?

Answer: Sequence polymorphisms that result in divergence within genes often predate the splitting of populations that occurs during speciation. If an ancestral polymorphism at a locus predates the split between two species, some members of a species may have retained the ancestral polymorphism, while others may have not retained it, having accumulated a newer polymorphism at the locus that occurred following speciation. Consequently, for certain loci, some members of a species will be more similar to other species than to their own or more closely related species. For these loci, phylogenetic trees made from a single gene will not always reflect the relationships between species. This is why it is important to use multiple loci when constructing species trees.

25.16   What is the chance of randomly picking the one rooted phylogenetic tree that describes the true relationship between a group of six organisms? Are the odds better or worse for randomly picking from among all the possible unrooted trees for those organisms?

Answer: Refer to Table 25.4 and Figure 25.5, p. 717, and the accompanying text. The number of possible rooted (N_R) and unrooted (N_U) trees for n taxa are given by the equations:

$$N_R = (2n - 3)!/[2^{n-2}(n-2)!] \quad \text{and} \quad N_U = (2n - 5)!/[2^{n-3}(n-3)!]$$

For $n = 6$, $N_R = 9!/(2^4 \, 4!) = 945$ and $N_U = 7!/(2^3 \, 3!) = 105$. The chance of picking the one rooted tree that describes the true relationship between a group of six organisms is $1/945$ (0.11%). The odds are better in picking a random unrooted tree: $1/105$ (0.95%).

25.17 Draw all the possible unrooted trees for four species: A, B, C and D. How many rooted trees are there for the same four species?

Answer: For four species, $N_U = (2n - 5)!/[2^{n-3}(n - 3)!] = 3!/(2^1 1!) = 3$. The three unrooted trees are drawn below and differ from each other in the common ancestors that are shared by each species. In the tree on the left, A and D share a different common ancestor than do B and C. In the tree in the middle, A and C share a different common ancestor than do B and D. In the tree on the right, A and B share a different common ancestor than do D and C.

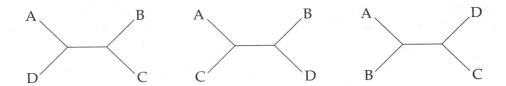

The number of rooted trees $N_R = (2n - 3)!/[2^{n-2}(n - 2)!] = 5!/(2^2 2!) = 15$.

25.18 Use the same sequence alignment provided for the Analytical Question 25.1 to generate a distance matrix, but do so by weighting transversions (As or Gs changing to Cs or Ts) twice as heavily as transitions (Cs changing to Ts, Ts changing to Cs, As changing to Gs or Gs changing to As).

Answer: Start by constructing a table of the distances between all pairwise combinations of sequences: Compare two sequences at a time, determine how many of the differences are transversions and how many are transitions, and determine the distance between the sequences by weighting transversions twice as much as transversions. For example, the differences between taxa A and B are underlined in the following sequence alignment:

```
             10         20         30         40         50
A:  GCCAACGTCC ATACCACGTT GTTTAGCACC GGTTCTCGTC CGATCACCGA
B:  GCCAACGTCC ATACCACGTT GTCAAACACC GGTTCTCGTC CGATCACCGA
C:  GGCAACGTCC ATACCACGTT GTTATACACC GGTTCTCGTC AGGTCACCGA
D:  GCTAACGTCC ATATCACGCT GTCATGTACC GGTCCTCGTC AGATCCCCAA
E:  GCTGGTGTCC ATATCACGTT ATCATGTACC GGTACTCGTC CGATCACCGA
```

Taxa A and B differ by two transitions (T changes to C, G changes to A) and one transversion (T changing to A), giving a distance score $d_{AB} = 2$ (1×2 transitions) + 2 (2×1 transversion) = 4. The complete distance matrix is shown next (R = transversion = 2, T = transition = 1):

Taxa	A	B	C	D
B	2T + 1R = 4	–	–	–
C	2T + 4R = 10	2T + 3R = 8	–	–
D	2T + 4R = 15	7T + 3R = 13	9T + 2R = 13	–
E	2T + 3R = 14	8T + 2R = 12	10T + 3R = 16	6T + 3R = 12

The smallest distance separating any of the two sequences is d_{AB}, so group together taxon A and B. Calculate a new distance matrix in which the composite group (AB) takes their place. Calculate the distances between (AB) and the remaining taxa by taking the average distance between A and B and the remaining taxa. For example, $d_{(AB)D} = \frac{1}{2}(d_{AD} + d_{BD}) = \frac{1}{2}(15 + 13) = 14$. A matrix giving the distances between (AB) and the remaining taxa is:

Taxa	AB	C	D
C	9	–	–
D	14	13	–
E	13	16	12

The smallest distance separating any two taxa in this new matrix is $d_{(AB)C} = 9$, so create a new combined taxon (AB)C, and calculate the distances between this taxon and the others to give a new distance matrix:

Taxa	(AB)C	D
D	13.5	–
E	14.5	12

In this last matrix, the smallest distance is between taxa D and E ($d_{DE} = 12$), so group these taxa together as (DE). This gives the final clustering of taxa as ((AB)C)(DE), which, for this example, is the same clustering as was obtained when base changes were equally weighted. See text p. 726.

25.19   Increasing the amount of sequence information available for analysis usually has little effect on the length of time computer programs use to generate phylogenetic trees with the parsimony approach. Why doesn't the amount of sequence information affect the total number of possible rooted and unrooted trees?

Answer: The equations that describe the total number of possible rooted and unrooted trees are based only on the number of taxa, not the amount of sequence information provided for each taxon. If n is the number of taxa considered, then the total number of rooted trees is $N_R = (2n - 3)!/[2^{n-2} (n - 2)!]$ and the total number of unrooted trees is $N_U = (2n - 5)!/[2^{n-3} (n - 3)!]$. The amount of time required for computer generation of phylogenetic trees with the parsimony approach is not affected much by the amount of sequence information because in parsimony-based approaches, not all sites are used when considering molecular data. Rather, only informative sites—sites that have at least two different nucleotides present at least twice—are used. Simply increasing the amount of sequence information does not necessarily dramatically increase the number of informative sites, since there is no guarantee that the sequences that are added harbor informative sites. It is more important to identify regions that have informative sites and add these to the analysis. Once informative sites within an alignment are determined, the parsimony approach is used; the unrooted tree that invokes the fewest number of mutations for each of the sites is determined. Restricting the analysis to informative-only sites and invoking parsimony quickly eliminates many trees from further consideration.

25.20   When bootstrapping is used to assess the robustness of branching patterns in a tree of maximum parsimony, why is it more important to use sequences that have as many informative sites as possible than simply to use longer sequences?

Answer: Bootstrap tests draw repeatedly on different subsets of the original data and infer trees from the newly selected incomplete data set. Repeating this process hundreds or thousands of times identifies which portions of the inferred, maximum parsimony tree have the same groupings as many of the repetitions. These are the regions of the inferred tree that are especially well supported by the entire data set. When maximum parsimony is used, only informative sites—sites that have at least two different nucleotides present at least twice—are used for the analysis. Other sequence sites are not used in the analysis, as they are not considered biologically informative characters. Consequently, simply using longer sequences does not necessarily add any additional information to the analysis. It is more important to identify subsets of sequences that harbor a high density of informative sites and add these to the analysis. This will increase the size of the data set that can be selected from during the bootstrap analysis and therefore be more useful in assessing the robustness of the inferred tree.

25.21 What are the advantages of gene duplication (in whole or in part) in generating genes with new functions? Suggest an alternative way in which genes with new functions could arise.

Answer: The chance of a sequence randomly accumulating mutations that give it an open reading frame and appropriate promoter elements all at the same time is extremely small. It is more plausible to generate genes with new functions following a gene duplication event when two functional copies of a gene are present. As long as one of them continues to provide a required function, the other is essentially redundant, so it may undergo changes in sequence. It is free from the constraint of providing a required function. Not all of the changes, of course, will be desirable or lead to new functions—many may result in a loss of function and result in a pseudogene. However, some may result in a modified function that serves the organism. Subsequent misalignments between the altered, duplicated gene or a pseudogene and the retained functional gene provide an opportunity for recombination and gene conversion events to "repair" inactivated copies of a gene. Thus, a mutated, duplicated gene/pseudogene can have its function restored.

When only part of a gene is duplicated, there is a potential for domain shuffling. Domain shuffling is the duplication and rearrangement of domains in proteins that provide specific functions. This can lead to the assemblage of proteins with more complex domain arrangements, which can result in proteins with novel functions.

Gene duplications are therefore advantageous as a mechanism to generate genes with new functions because they provide a shortcut for modifying existing proteins. Rather than starting to build a gene "from scratch," duplications provide a template that already serves a function. The template can be tinkered with, repeatedly if recombination and gene conversion occur, in a manner that is largely free from selection. Of course, if the new function is deleterious to the organism (e.g., a dominant mutation resulting from a gain-of-function or new function), it will be selected against. Partial gene duplications are advantageous, since they allow combinations of existing functional domains to be reassembled into new arrangements and so effectively provide a set of building blocks for the assembly of new functions.

There are at least four additional mechanisms for the generation of genes with new functions. First, new gene functions could be obtained following the rearrangement of chromosomal segments through transposition, inversion, or translocation. Chromosomal rearrangements could alter the transcriptional structure of an existing gene, create a novel fusion protein, or introduce new sites for mRNA processing. Chromosomal rearrangements could also place the gene under the control of another gene's regulatory elements and introduce new functions by altering where or when it is expressed during development. Second, new gene functions could be obtained by inserting copies of existing or foreign genes via viral retrotransposons. Much like duplications, these insertions could change over time to generate new functions. Third, new gene functions could be obtained by inserting and/or partially deleting mobile repetitive elements into exonic or intronic regions. These could introduce new regulatory and/or RNA processing signals. Fourth, new gene functions can be obtained following the introduction of point, small deletion, and insertion mutations. These could alter the function of an existing gene or modify the way that existing RNA processing sites are used.